A Visual Guide to Stata Graphics

A Visual Guide to Stata Graphics

MICHAEL N. MITCHELL
University of California, Los Angeles

A Stata Press Publication
StataCorp LP
College Station, Texas

Stata Press, 4905 Lakeway Drive, College Station, Texas 77845

Copyright © 2004 by StataCorp LP
All rights reserved
Typeset in LaTeX 2_ε
Printed in the United States of America
10 9 8 7 6 5 4 3 2 1

ISBN 1-881228-85-1

This book is protected by copyright. All rights are reserved. No part of this book may be reproduced, stored in a retrieval system, or transcribed, in any form or by any means—electronic, mechanical, photocopying, recording, or otherwise—without the prior written permission of StataCorp LP.

Stata is a registered trademark of StataCorp LP. LaTeX 2_ε is a trademark of the American Mathematical Society.

Dedication

I would like to dedicate this book to Paul Hoffman. Although he was my supervisor for the last nine years, it always felt much more like he was a trusted friend always there to help me do the best work that I could. I am so sorry he had so leave us so soon. In my own way, I hope that I can give to others the same kinds of things he gave to me. I am really going to miss you, Paul.

Acknowledgments

Although there is a single name on the cover of this book, many people have helped to make this book possible. Without them, this book would have remained a dream, and I could have never shared it with you. I want to thank those people who helped that dream become the book you are now holding.

I want to thank the warm people at Stata, who were very generous in their assistance and who always find a way to be friendly and helpful. In particular, I wish to thank Vince Wiggins for his generosity of time, insightful advice, boundless enthusiasm, and commitment to help make this book the best that it could be. I am very grateful to Jeff Pitblado, who created the LaTeX tools that made the layout of this book possible. Without the benefit of his time and talent, I would still be learning LaTeX instead of writing these acknowledgments. Also, I would like to thank the Stata technical support team, especially Derek Wagner, for patiently working with me on my numerous questions. I am also very grateful to John Williams for his thoroughness and alacrity in editing the book and to Chinh Nguyen for his creative and clever cover design.

I also want to thank, in alphabetical order, Xiao Chen, Phil Ender, Frauke Kreuter, and Christine Wells for their support and suggestions.

Last, and certainly not least, I would like to thank the teachers who have added to my life in very special ways. I have been very fortunate to have been touched by many special teachers, and I will always be grateful for what they kindly gave to me. I want to thank (in order of appearance) Larry Grossman, Fred Perske, Rosemary Sheridan, Donald Butler, Jim Torcivia, Richard O'Connell, Linda Fidell, and Jim Sidanius. These teachers all left me gifts of knowledge and life lessons that help me every day. Even if they do not all remember me, I will always remember them.

Contents

Preface

It is obvious to say that graphics are a visual medium for communication. This book takes a visual approach to help you learn about how to use Stata graphics. While you can read this book in a linear fashion or use the table of contents to find what you are seeking, it is designed to be "thumbed through" and visually scanned. For example, the right margin of each right page has what I call a *Visual Table of Contents* to guide you through the chapters and sections of the book. Generally, each page has three graphs on it, allowing you to see and compare as many as six graphs at a time on facing pages. For a given graph, you can see the command that produced it, and next to each graph is some commentary. But don't feel compelled to read the commentary; often, it may be sufficient just to see the graph and the command that made it.

This is an informal book and is written in an informal style. As I write this, I picture myself sitting at the computer with you, and I am showing you examples that illustrate how to use Stata graphics. The comments are written very much as if we were sitting down together and I had a couple of points to make about the graph that I thought you might find useful. Sometimes, the comments might seem obvious, but since I am not there to hear your questions, I hope it is comforting to have the obvious stated just in case there was a bit of doubt.

While this book does not spend much time discussing the syntax of the graph commands (since you will be able to infer the rules for yourself after seeing a number of examples), the Intro : Options (20) section discusses some of the unique ways that options are used in Stata graph commands and compares them to the way that options are used in other Stata commands.

I strived to find a balance to make this book comprehensive but not overwhelming. As a result, I have omitted some options I thought would be seldom used. So, just because a feature is not illustrated in this book, this does not mean that Stata cannot do that task, and I would refer to [G] **graph** for more details. I try to include frequent cross-references to [G] **graph**; for example, see also [G] ***axis_options***. I view this book as a complement to the *Stata Graphics Reference Manual*, and I hope that these cross-references will help you use these two books in a complementary manner. Note that, whenever you see references to [G] **xyz**, you can either find "xyz" in the *Stata Graphics Reference Manual* or type `whelp xyz` within Stata. The manual and the help have the same information, although the help may be more up to date and allows hyperlinking to related topics.

Each chapter is broken into a number of sections showing different features and options for the particular kind of graph being discussed in the chapter. The examples illustrate how these options or features can be used, focusing on examples that isolate these features so you are not distracted by irrelevant aspects of the Stata command or graph. While this approach improves the clarity of presentation, it does sacrifice some realism since graphs frequently have many options used together. To address this, there is a section addressing strategies for building up more complicated graphs, Intro : Building graphs (29), and a section giving tips on creating more complicated graphs, Appendix : More examples (366). These sections are geared to help you see how you can combine options to make more complex and feature-rich graphs.

While this book is printed in color, this does not mean that it ignores how to create monochrome (black & white) graphs. Some of the examples are shown using monochrome graphs illustrating how you can vary colors using multiple shades of gray and how you can vary other attributes, such as marker symbol and size, line width, and pattern, and so forth. I have tried to show options that would appeal to those creating color or monochrome graphs.

The graphs in this book were created using a set of schemes specifically created for this book. Despite differences in their appearance, all the schemes increase the size of textual and other elements in the graphs (e.g., titles) to make them more readable, given the small size of the graphs in this book. You can see more about the schemes in Intro : Schemes (14) and how to obtain them in Appendix : Online supplements (382). While one purpose of the different schemes is to aid in your visual enjoyment of the book, they are also used to illustrate the utility of schemes for setting up the look and default settings for your graphs. See Appendix : Online supplements (382) for information about how you can obtain these schemes.

Stata has a number of graph commands for producing special-purpose statistical graphs. Examples include graphs for examining the distributions of variables (e.g., `kdensity`, `pnorm`, or `gladder`), regression diagnostic plots (e.g., `rvfplot` or `lvr2plot`), survival plots (e.g., `sts` or `ltable`), time series plots (e.g., `ac` or `pac`), and ROC plots (e.g., `roctab` or `lsens`). To cover these graphs in enough detail to add something worthwhile would have expanded the scope and size of this book and detracted from its utility. Instead, I have included a section, Appendix : Stat graphs (345), that illustrates a number of these kinds of graphs to help you see the kinds of graphs these commands create. This is followed by Appendix : Stat graph options (352), which illustrates how you can customize these kinds of graphs using the options illustrated in this book.

If I may close on a more personal note, writing this book has been very rewarding and exciting. While writing, I kept thinking about the kind of book you would want to help you take full advantage of the powerful, but surprisingly easy to use, features of Stata graphics. I hope you like it!

Simi Valley, California
February 2004

1 Introduction

This chapter starts off by telling you a little bit about the organization of this book and giving you tips to help you use it most effectively. The next section gives a brief overview of the different kinds of Stata graphs we will be examining in this book, followed by an overview of the different kinds of schemes that will be used for showing the graphs in this book. The fourth section illustrates the structure of options in Stata graph commands. In a sense, the second to fourth sections of this chapter are a thumbnail preview of the entire book, showing the types of graphs covered, how you can control their overall look, and the general structure of options used within those graphs. By contrast, the final section is about the process of creating graphs.

1.1 Using this book

I hope that you are eager to start reading this book but will take just a couple of minutes to read this section to get some suggestions that will make the book more useful to you.

First of all, there are many ways you might read this book, but perhaps I can suggest some tips:

- Please consider reading this chapter before reading the other chapters, as it provides key information that will make the rest of the book more understandable.
- While you might read a traditional book cover to cover, this book has been written so that the chapters stand on their own. You should feel free to dive into any chapter or section of any chapter.
- Sometimes you might find it useful to visually scan the graphs rather than to read. I think this is a good way to familiarize yourself with the kinds of features available in Stata graphs. If a certain feature catches your eye, you can stop and see the command that made the graph and perhaps even read the text explaining the command.
- Likewise, you might scan a chapter just by looking at the graphs and the part of the command in red, which is the part of the command we are discussing for that graph. For example, scanning the chapter on bar charts in this way would quickly familiarize you with the kinds of features available for bar graphs and show you how to obtain those features.

As you have probably noticed, the right margin contains what I call the *Visual Table of Contents*. I hope you will find it a useful tool for quickly finding the information you seek. I frequently use the *Visual Table of Contents* to cross-reference information within the book. By design, Stata graphs share many features in common. For example, you use the same kinds of options to control legends across different types of graphs. It would be repetitive to go into detail about legends for bar charts, box plots, and so on. Within each kind of graph, legends are briefly described and illustrated, but the details are described in the *Options* chapter in the section titled *Legend*. This is cross-referenced in the book by saying something like "for more details, see Options : Legend (287)", which indicates that you should look to the *Visual Table of Contents* and thumb to the *Options* chapter and then to the *Legend* section, which begins on page 287.

Sometimes it may take an extra cross-reference to get the information you need. Say that you want to make the `ytitle()` large for a bar chart, so you first consult Bar : Y-axis (143). This gives you some information about using `ytitle()`, but then that section refers you to Options : Axis titles (254), where more details about axis titles are described. This section then refers you to Options : Textboxes (303) for more complete details about options you can use to control the display of text. That section shows more details but then refers to Styles : Textsize (344), where all of the possible text sizes are described. I know this sounds like a lot of jumping around, but I hope that it feels more like drilling down for additional detail, that you feel you are in control of the level of detail that you want, and that the *Visual Table of Contents* eases the process of getting the additional details.

Most pages of this book have three graphs per page, each graph being composed of the graph itself, the command that produced it, and some descriptive text. An example is shown below, followed by some points to note.

```
graph twoway scatter propval100 ownhome, msymbol(Sh)
```

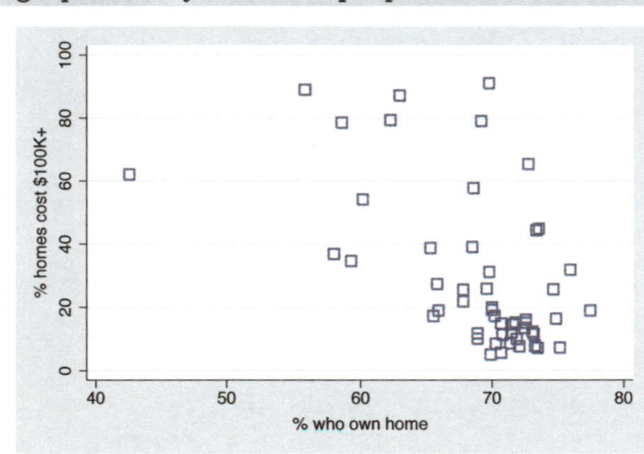

In this example, we use the `msymbol()` (marker symbol) option to make the symbols large hollow squares; see Options : Markers (235) for more details. Note that the `msymbol()` option is only useful for the types of graphs that have marker symbols, and Stata will ignore this option if you use it with a command like the **`graph twoway histogram`** command.
Uses allstates.dta & scheme vg_s2c

- Note that the command itself is displayed in a `typewriter font`, and the part of the command we are discussing (i.e., `msymbol(Sh)`) is in `this color`, both in the command and when referenced in the descriptive text.
- When commands or parts of commands are given in the descriptive text (e.g., **`graph twoway histogram`**), they are displayed in `typewriter font`.
- Many of the descriptions contain cross-references, for example, Options : Markers (235), which means to flip to the *Options* chapter and then to the section *Markers*. Equivalently, go to page 235.
- The names of some options are shorthand for two or more words that are sometimes explained; for instance, "we use the `msymbol()` (marker symbol) option to make . . . ".

- The descriptive text always concludes by telling you the name of the data file and scheme used for making the graph. In this case, the data file was *allstates.dta*, and the scheme was *vg_s2c.scheme*. You can read the data file over the Internet by using the **vguse** command, a command added to Stata when you install the online supplements; see Appendix: Online supplements (382). If you are connected to the Internet, and your Stata is fully up to date, you can simply type **vguse allstates** to use that file over the Internet, and you can run the graph command shown to create the graph.

If you want your graphs to look like the ones in the book, you can display them using the same schemes. See Appendix: Online supplements (382) for information about how to download the schemes used in this book. Once you have downloaded the schemes, you can then type the following in the Stata Command window:

```
. set scheme vg_s2c
. vguse allstates
. graph twoway scatter propval100 ownhome, msymbol(Sh)
```

After you issue the **set scheme vg_s2c** command, subsequent graph commands will show graphs using the **vg_s2c** scheme. If you prefer, you could add the **scheme(vg_sc2)** option to the graph command to specify the scheme used just for that graph; for example,

```
. graph twoway scatter propval100 ownhome, msymbol(Sh) scheme(vg_s2c)
```

In general, all commands and options are provided in their complete form. Commands and options are generally not abbreviated. However, for purposes of typing, you may wish to use abbreviations. The previous example could have been abbreviated to

```
. gr tw sc propval100 ownhome, m(Sh)
```

and even the **gr** could have been omitted, leaving

```
. tw sc propval100 ownhome, m(Sh)
```

The **tw** could also have been omitted, leaving

```
. sc propval100 ownhome, m(Sh)
```

For guidance on appropriate abbreviations, consult [G] **graph**.

I should note that, while this book is designed for creating graphs in Stata version 8 and beyond, many of the examples take advantage of numerous enhancements that have been released as online updates subsequent to the initial version 8 release. As a result, some features will either look different or may not work at all in Stata 8.0 or 8.1. Therefore, it is very important that your copy of Stata be fully up to date. Please verify that your copy of Stata is up to date and obtain any free updates; to do this, enter Stata, type

```
. update query
```

and follow the instructions. After the update is complete, you can use the **help whatsnew** command to learn about the updates you have just received, as well as prior updates documenting the evolution of Stata. Because Stata sometimes evolves beyond the printed manual, you might find that some commands or options are documented via the online help but not in your manual. For example, **graph twoway tsline** was released after the printed manual and, as of the first printing of this book, is only documented via the online help (**help tsline**).

What if you are using a newer version of Stata than version 8.2? It is possible that, in the future, Stata may evolve to make the behavior of some of these commands change. If this happens, you can use the `version` command to ask Stata to run the graph commands as though they were run under version 8.2. For example, if you were running Stata version 9 but wanted a graph command to run as though you were running Stata 8.2, you could type

```
. version 8.2 : graph twoway scatter propval100 ownhome
```

and the command would be executed as if you were running version 8.2.

This book has a number of associated online resources to complement the book. Appendix : Online supplements (382) has more information about these online resources and how to access them. I strongly suggest that you install the online supplements, which make it easier to run the examples from the book. To install the supplemental programs, schemes, and help files, just type from within Stata

```
. net from http://www.stata-press.com/data/vgsg
. net install vgsg
```

For an overview of what you have installed, type `whelp vgsg` within Stata. Then, with the `vguse` command, you can use any dataset from the book. Likewise, all the custom schemes used in the book will be installed into your copy of Stata and can be used to display the graphs, as described earlier in this section.

1.2 Types of Stata graphs

Stata has a wide variety of graph types. This section introduces the types of graphs Stata produces and covers twoway plots (including scatterplots, line plots, fit plots, fit plots with confidence intervals, area plots, bar plots, range plots, and distribution plots), scatterplot matrices, bar charts, box plots, dot plots, and pie charts. We will start off with a section showing the variety of twoway plots that can be created with `graph twoway`. For this introduction, we have combined them into six families of related plots: scatterplots and fit plots, line plots, area plots, bar plots, range plots, and distribution plots. We will start by illustrating scatterplots and fit plots.

```
graph twoway scatter propval100 popden
```

Here is a basic scatterplot. The variable `propval100` is placed on the y-axis, and `popden` is placed on the x-axis. See Twoway : Scatter (35) for more details about these kinds of plots.
Uses allstates.dta & scheme vg_s2c

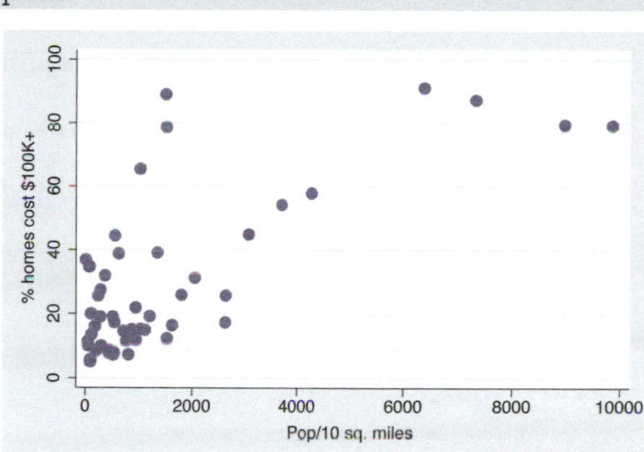


```
twoway scatter propval100 popden
```

We can start this command with just `twoway`, and Stata understands that this is shorthand for `graph twoway`.
Uses allstates.dta & scheme vg_s2c

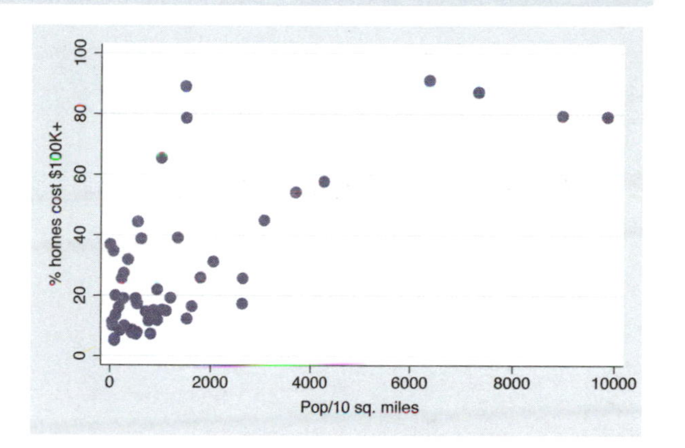


```
twoway lfit propval100 popden
```

We can make a linear fit line (`lfit`) predicting `propval100` from `popden`. See Twoway : Fit (49) for more information about these kinds of plots.
Uses allstates.dta & scheme vg_s2c

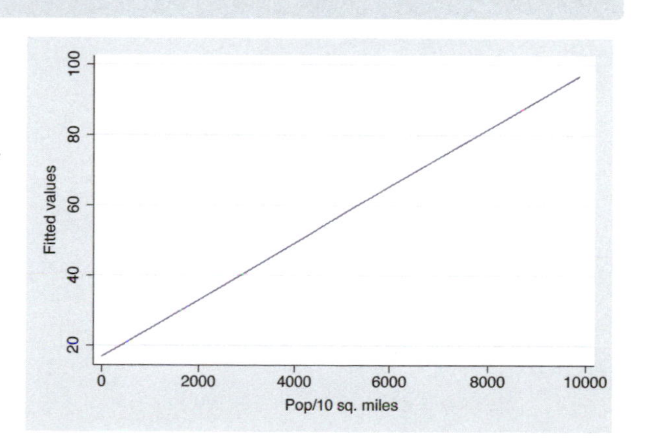

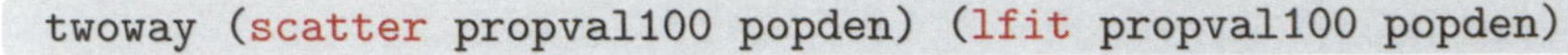

```
twoway (scatter propval100 popden) (lfit propval100 popden)
```

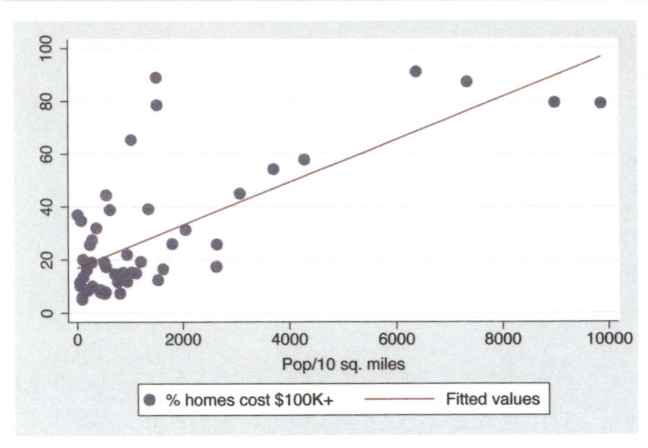

Stata allows us to overlay `twoway` graphs. In this case, we make a classic plot showing a scatterplot overlaid with a fit line using the `scatter` and `lfit` commands. For more details about overlaying graphs, see Twoway : Overlaying (87).
Uses allstates.dta & scheme vg_s2c

```
twoway (scatter propval100 popden) (lfit propval100 popden)
   (qfit propval100 popden)
```

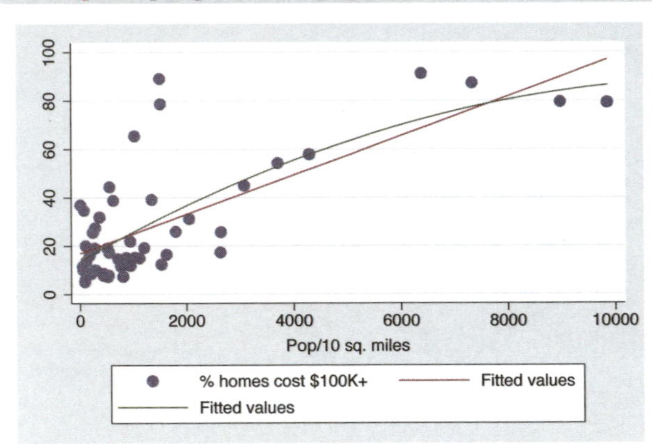

The ability to combine `twoway` plots is not limited to just overlaying two plots; we can overlay multiple plots. Here, we overlay a scatterplot with a linear fit line (`lfit`) and a quadratic fit line (`qfit`).
Uses allstates.dta & scheme vg_s2c

```
twoway (scatter propval100 popden) (mspline propval100 popden)
   (fpfit propval100 popden) (mband propval100 popden)
   (lowess propval100 popden)
```

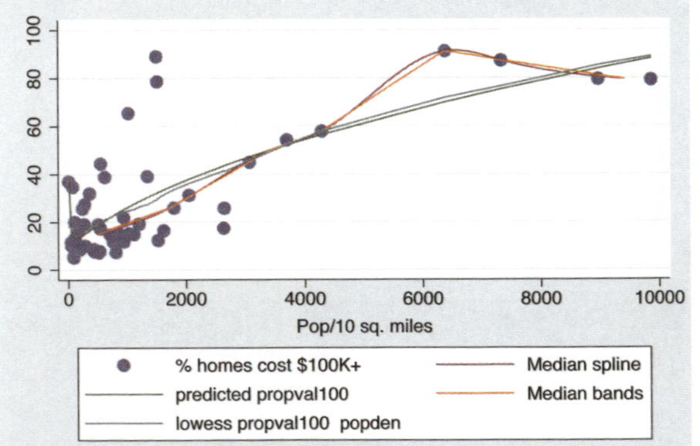

Stata has other kinds of fit methods in addition to linear and quadratic fits. This example includes a median spline (`mspline`), fractional polynomial fit (`fpfit`), median band (`mband`), and lowess (`lowess`). For more details, see Twoway : Fit (49).
Uses allstates.dta & scheme vg_s2c

```
twoway (lfitci propval100 popden) (scatter propval100 popden)
```

In addition to being able to plot a fit line, we can also plot a linear fit line with a confidence interval using the lfitci command. We also overlay the linear fit and confidence interval with a scatterplot. See Twoway : CI fit (50) for more information about fit lines with confidence intervals.
Uses allstates.dta & scheme vg_s2c

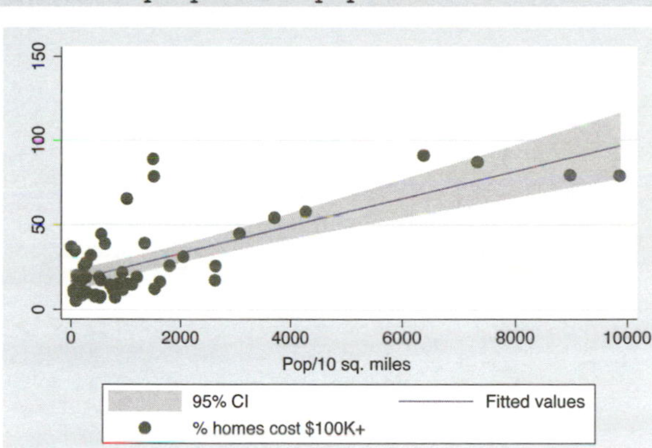


```
twoway dropline close tradeday
```

This dropline graph shows the closing prices of the S&P 500 by trading day for the first 40 days of 2001. A dropline graph is like a scatter plot since each data point is shown with a marker, but a dropline for each marker is shown as well. For more details, see Twoway : Scatter (35).
Uses spjanfeb2001.dta & scheme vg_s2c

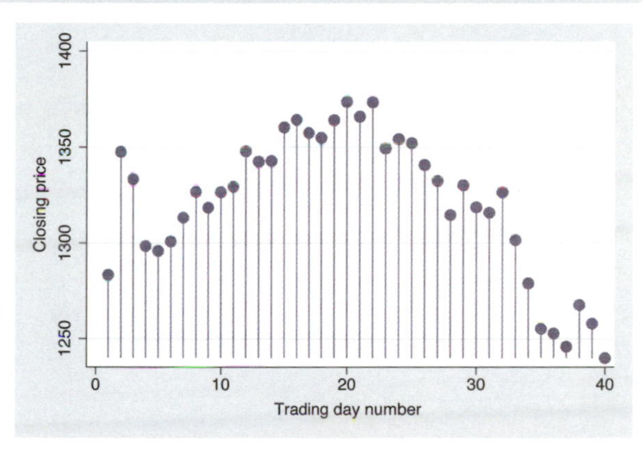


```
twoway spike close tradeday
```

Here, we use a spike graph to show the same graph as the previous graph. It is like the dropline plot, but no markers are put on the top. For more details, see Twoway : Scatter (35).
Uses spjanfeb2001.dta & scheme vg_s2c

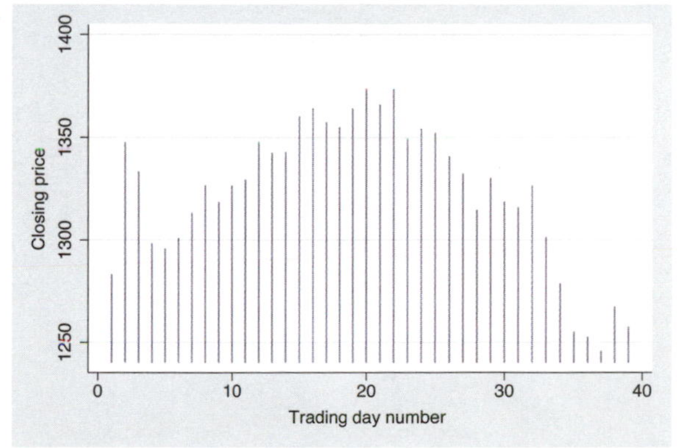

Introduction
Using this book
Twoway
Types of Stata graphs
Matrix
Bar
Schemes
Box
Options
Dot
Pie
Building graphs
Options
Standard options
Styles
Appendix

```
twoway dot close tradeday
```

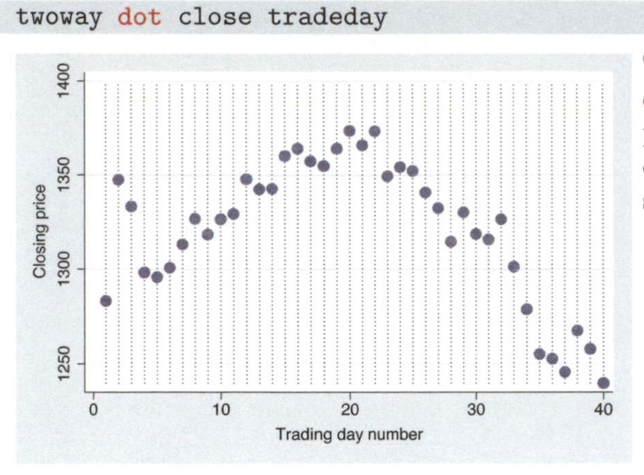

The dot plot, like the **scatter** command, shows markers for each data point but also adds a dotted line for each of the x-values. For more details, see Twoway : Scatter (35).
Uses spjanfeb2001.dta & scheme vg_s2c

```
twoway line close tradeday, sort
```

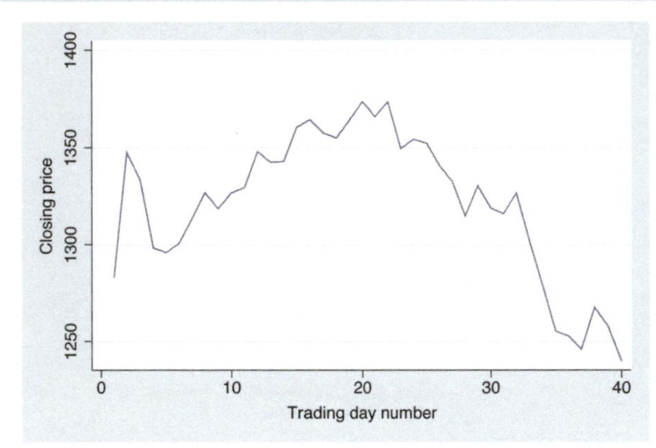

The line command is used in this example to make a simple line graph. See Twoway : Line (54) for more details about line graphs.
Uses spjanfeb2001.dta & scheme vg_s2c

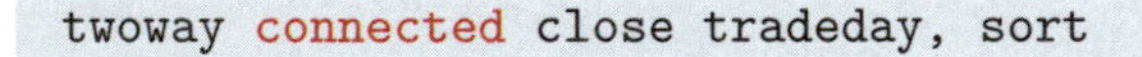

```
twoway connected close tradeday, sort
```

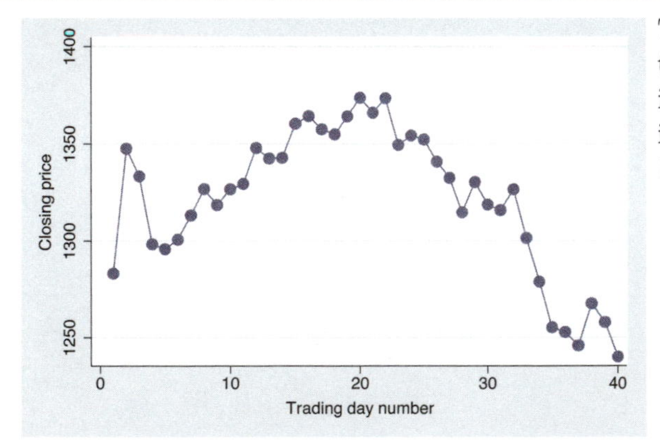

The twoway connected graph is similar to **twoway line**, except that a symbol is shown for each data point. For more information, see Twoway : Line (54).
Uses spjanfeb2001.dta & scheme vg_s2c

```
twoway tsline close, sort
```

The **tsline** (time-series line) command makes a line graph where the x-variable is a date variable that has previously been declared using `tsset`; see [TS] **tsset**. This example shows the closing price of the S&P 500 by trading date. For more information, see Twoway : Line (54).
Uses sp2001ts.dta & scheme vg_s2c

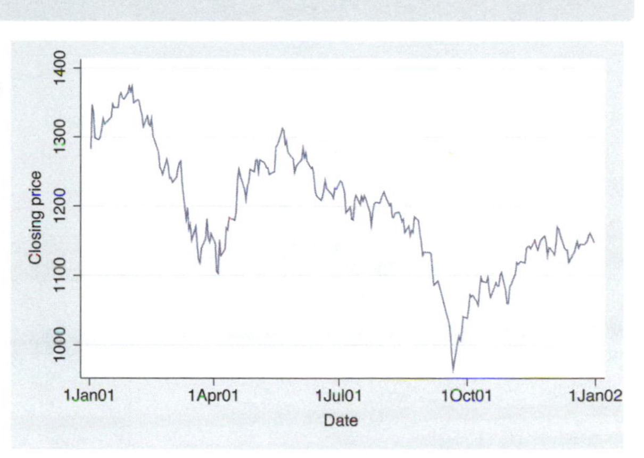


```
twoway tsrline high low, sort
```

This command uses **tsrline** (time series range line) to make a line graph showing the high and low prices of the S&P 500 by trading date. For more information, see Twoway : Line (54).
Uses sp2001ts.dta & scheme vg_s2c

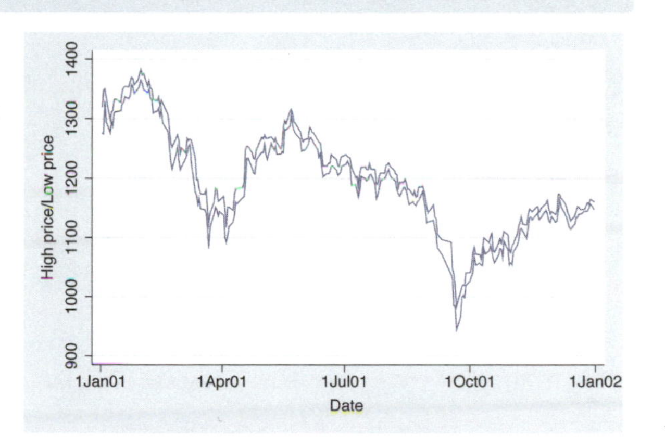


```
twoway area close tradeday, sort
```

An **area** plot is similar to a `line` plot, but the area under the line is shaded. See Twoway : Area (61) for more information about area plots.
Uses spjanfeb2001.dta & scheme vg_s2c

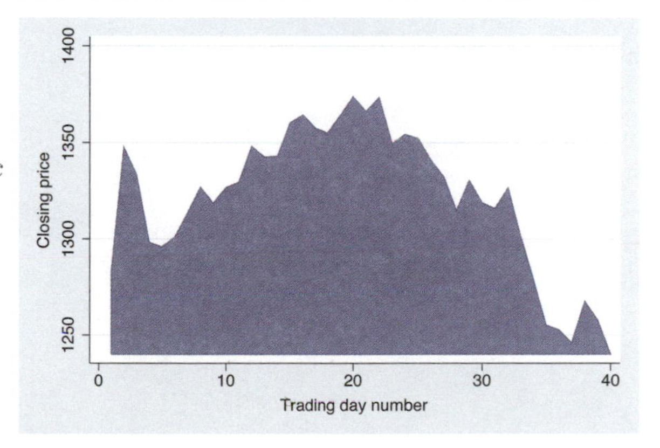

```
twoway bar close tradeday
```

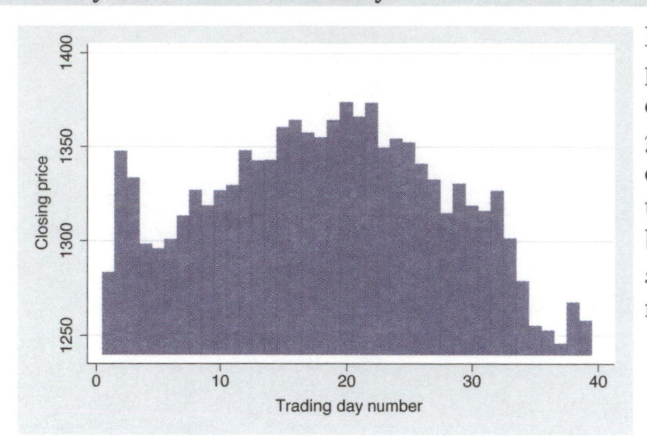

Here is an example of a **twoway bar** plot. For each x-value, a bar is shown corresponding to the height of the y-variable. Note that this shows a continuous x-variable as compared with the **graph bar** command, which would be useful when we have a categorical x-variable. See Twoway : Bar (62) for more details about bar plots.
Uses spjanfeb2001.dta & scheme vg_s2c

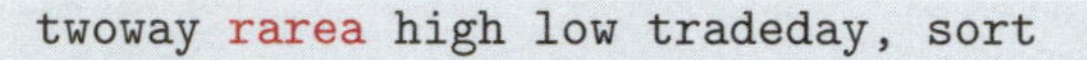

```
twoway rarea high low tradeday, sort
```

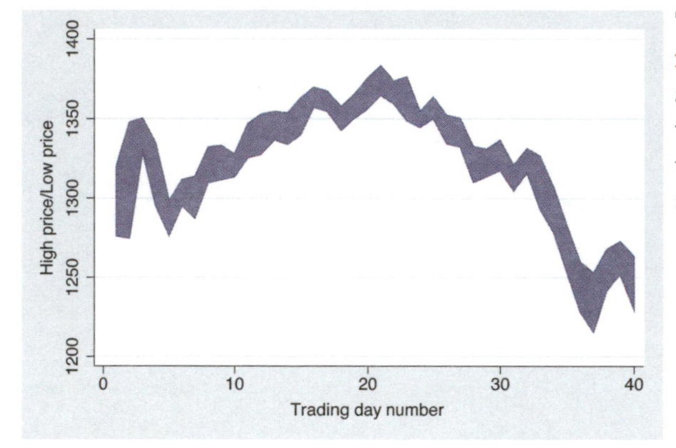

This example illustrates the use of **rarea** (range area) to graph the high and low prices with the area filled. If we used **rline** (range line), the area would not be filled. See Twoway : Range (64) for more details.
Uses spjanfeb2001.dta & scheme vg_s2c

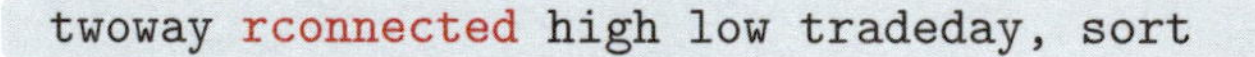

```
twoway rconnected high low tradeday, sort
```

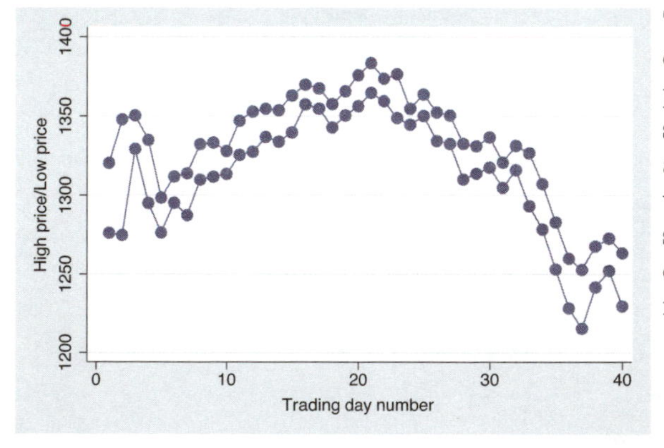

The **rconnected** (range connected) command makes a graph similar to the previous one, except that a marker is shown at each value of the x-variable and the area in between is not filled. If we instead used **rscatter** (range scatter), the points would not be connected. See Twoway : Range (64) for more details.
Uses spjanfeb2001.dta & scheme vg_s2c

```
twoway rcap high low tradeday, sort
```

Here, we use `rcap` (range cap) to graph the high and low prices with a spike and a cap at each value of the x-variable. If you used `rspike` instead, spikes would be displayed but not caps. If we used `rcapsym`, the caps would be symbols and you could modify the symbol. See Twoway : Range (64) for more details.
Uses spjanfeb2001.dta & scheme vg_s2c

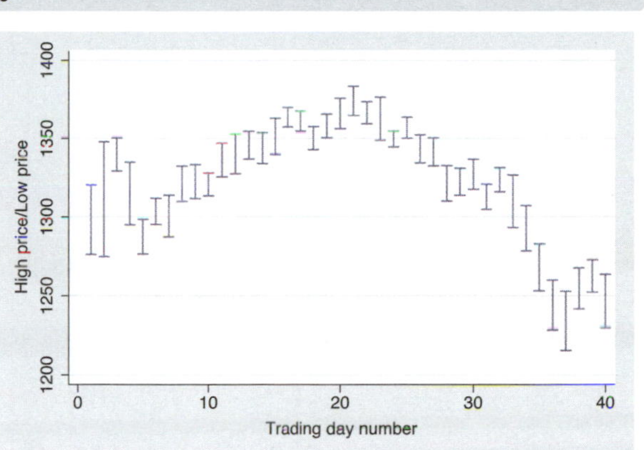


```
twoway rbar high low tradeday, sort
```

Here, we use the `rbar` to graph the high and low prices with bars at each value of the x-variable. See Twoway : Range (64) for more details.
Uses spjanfeb2001.dta & scheme vg_s2c

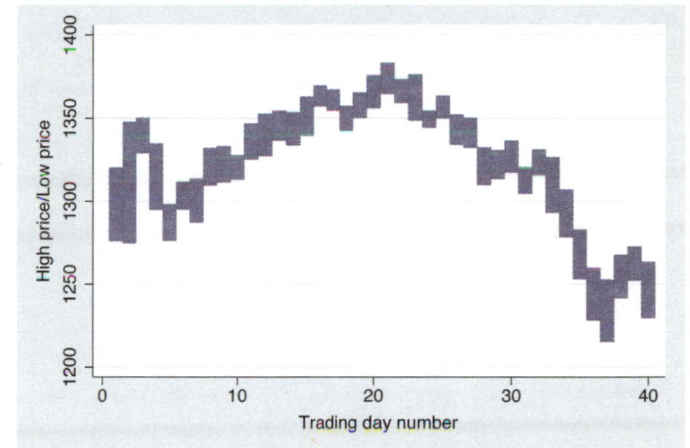


```
twoway histogram popk, freq
```

The `twoway histogram` command can be used to show the distribution of a single variable. It is often useful when overlaid with other twoway plots; otherwise, the `histogram` command would be preferable. See Twoway : Distribution (74) for more details.
Uses allstates.dta & scheme vg_s2c

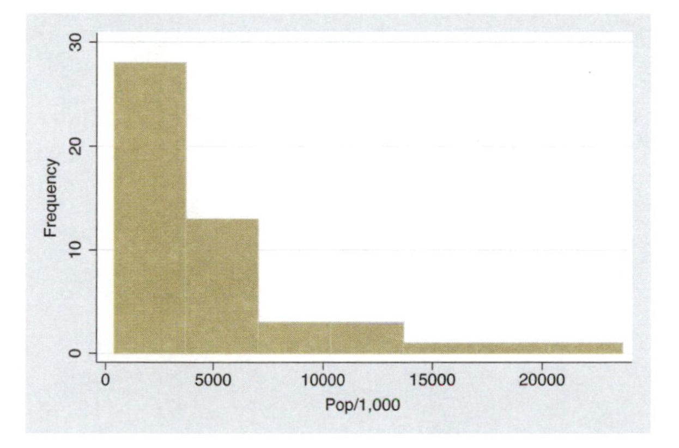

```
twoway kdensity popk
```

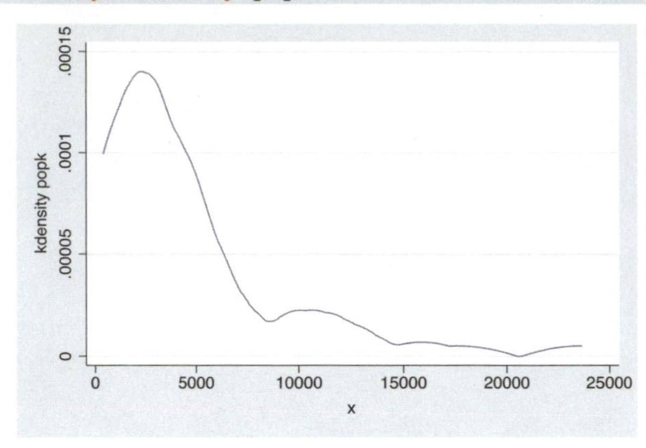

The twoway kdensity command shows a kernel-density plot and is useful for examining the distribution of a single variable. It can be overlaid with other twoway plots; otherwise, the kdensity command would be preferable. See Twoway : Distribution (74) for more details.

Uses allstates.dta & scheme vg_s2c

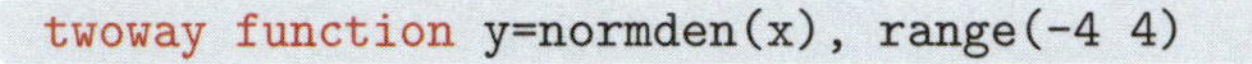

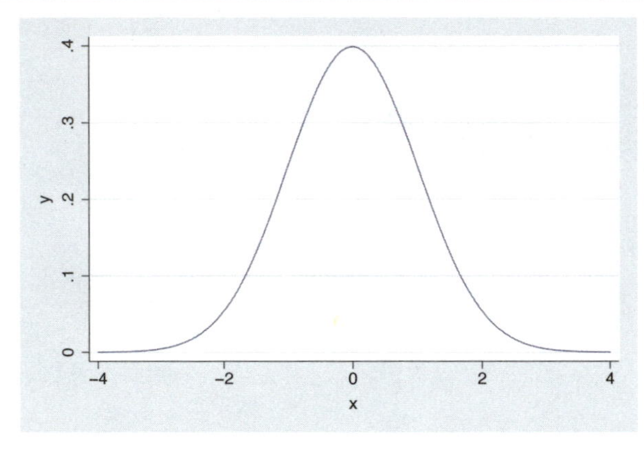

The twoway function command allows us to graph an arbitrary function over a range of values we specify. See Twoway : Distribution (74) for more details.

Uses allstates.dta & scheme vg_s2c

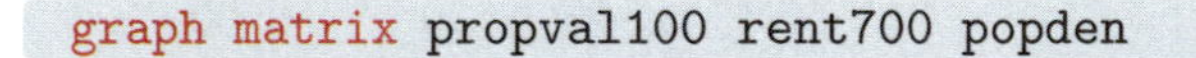

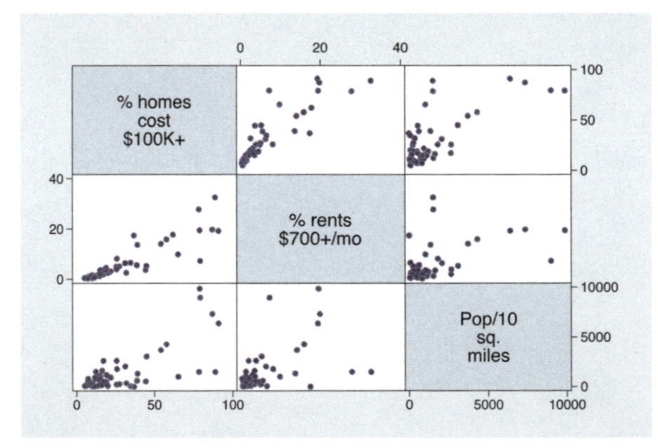

We can use the graph matrix command to show a scatterplot matrix. See Matrix (95) for more details.

Uses allstates.dta & scheme vg_s2c

```
graph hbar popk, over(division)
```

The `graph hbar` (horizontal bar) command is often used to show the values of a continuous variable broken down by one or more categorical variables. Note that `graph hbar` is merely a rotated version of `graph bar`. See Bar (107) for more details.
Uses allstates.dta & scheme vg_s2c

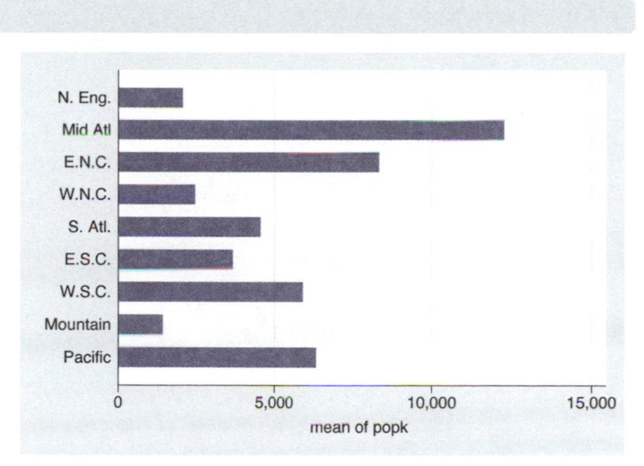


```
graph hbox popk, over(division)
```

We can show the previous graph as a box plot using the `graph hbox` (horizontal box) command. The `graph hbox` command is commonly used for showing the distribution of one or more continuous variables, broken down by one or more categorical variables. Note that `graph hbox` is merely a rotated version of `graph box`. See Box (157) for more details.
Uses allstates.dta & scheme vg_s2c

```
graph dot popk, over(division)
```

The previous plot could also be shown as a dot plot using `graph dot`. Dot plots are often used to show one or more summary statistics for one or more continuous variables, broken down by one or more categorical variables. See Dot (193) for more details.
Uses allstates.dta & scheme vg_s2c

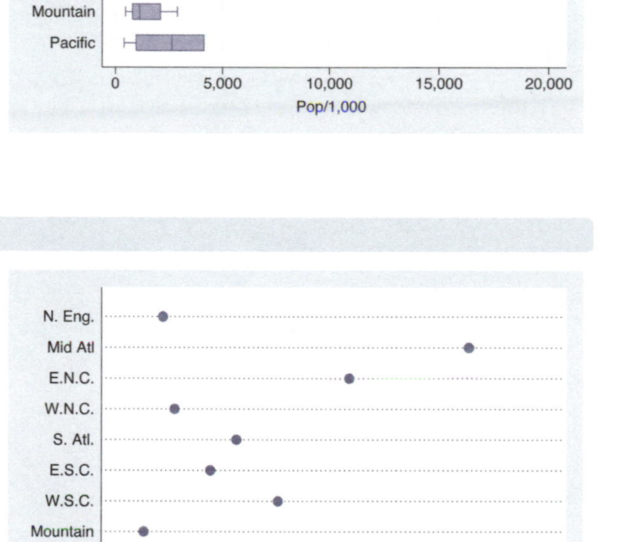

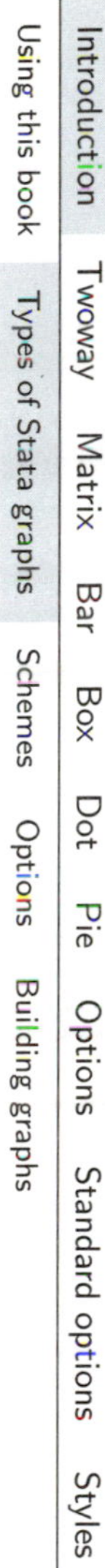

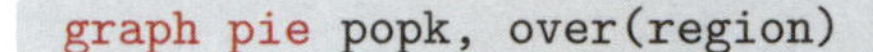

```
graph pie popk, over(region)
```

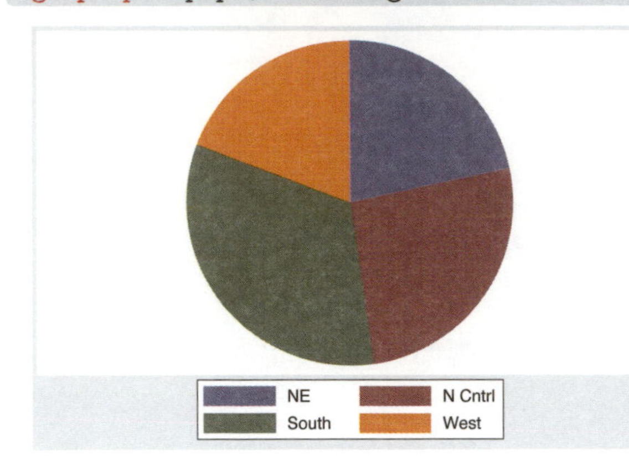

The graph pie command can be used to show pie charts. See Pie (217) for more details.
Uses allstates.dta & scheme vg_s2c

1.3 Schemes

While the previous section was about the different types of graphs Stata can make, this section is about the different kinds of looks that you can have for Stata graphs. The basic starting point for the look of a graph is a scheme, which controls just about every aspect of the look of the graph. A scheme sets the stage for the graph, but you can use options to override the settings in a scheme. As you might surmise, if you choose (or develop) a scheme that produces graphs similar to the final graph you wish to make, you can reduce the need to customize your graphs using options. Here, we give you a basic flavor of what schemes can do and introduce you to the schemes you will be seeing throughout the book. See Intro : Using this book (1) for more details about how to select and use schemes and Appendix : Online supplements (382) for more information about how to download them.

```
twoway scatter propval100 rent700 ownhome
```

This scatterplot illustrates the vg_s1c scheme. It is based on the s1color scheme but increases the sizes of elements in the graph to make them more readable. This scheme is in color and has a white background, both inside the plot region and in the surrounding area.
Uses allstates.dta & scheme vg_s1c

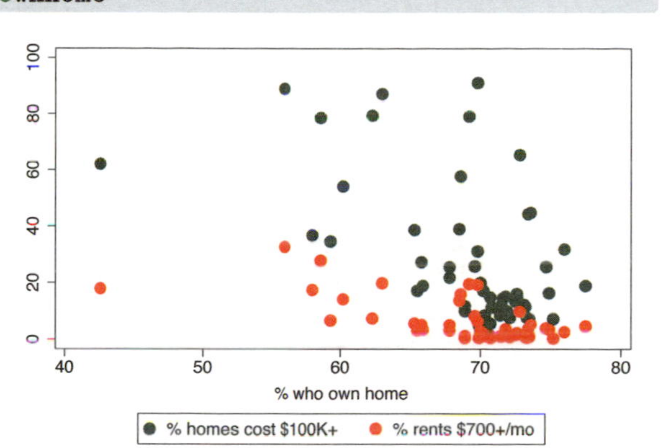


```
twoway scatter propval100 rent700 ownhome
```

This scatterplot is similar to the last one but uses the vg_s1m scheme, the monochrome equivalent of the vg_s1c scheme. It is based on the s1mono scheme but increases the sizes of elements in the graph to make them more readable. This scheme is in black and white and has a white background, both inside the plot region and in the surrounding area.
Uses allstates.dta & scheme vg_s1m

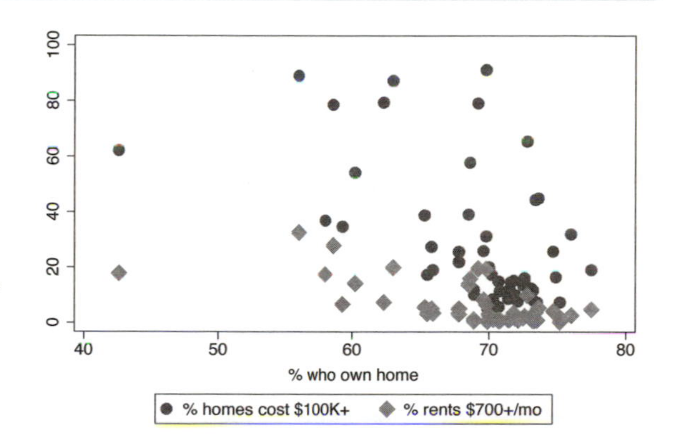


```
graph hbox wage, over(grade) asyvar nooutsides legend(rows(2))
```

This box plot shows an example of the vg_s2c scheme. It is based on the s2color scheme but increases the sizes of elements in the graph to make them more readable. When we use this scheme, the plot region has a white background, but the surrounding area (the graph region) is light blue.
Uses nlsw.dta & scheme vg_s2c

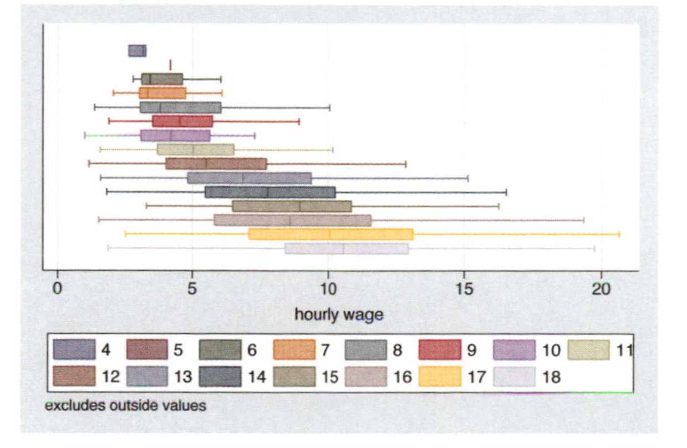

```
graph hbox wage, over(grade) asyvar nooutsides legend(rows(2))
```

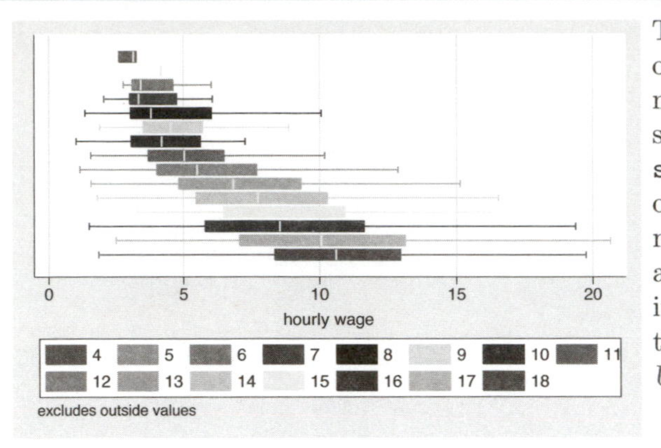

This box plot is similar to the previous one but uses the vg_s2m scheme, the monochrome equivalent of the vg_s2c scheme. This scheme is based on the s2mono scheme but increases the sizes of elements in the graph to make them more readable. This scheme is in black and white and has a white background in the plot region but is light gray in the surrounding graph region.
Uses nlsw.dta & scheme vg_s2m

```
graph hbar wage, over(occ7, label(nolabels)) blabel(group, position(base))
```

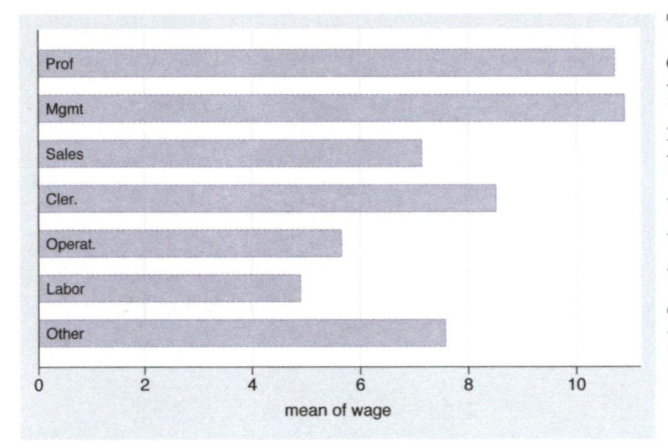

This horizontal bar chart shows an example of the vg_palec scheme. It is based on the s2color scheme but makes the colors of the bars/boxes/markers paler by decreasing the intensity of the colors. As shown in this example, one use of this scheme is to make the colors of the bars pale enough to include text labels inside of bars.
Uses nlsw.dta & scheme vg_palec

```
graph hbar wage, over(occ7, label(nolabels)) blabel(group, position(base))
```

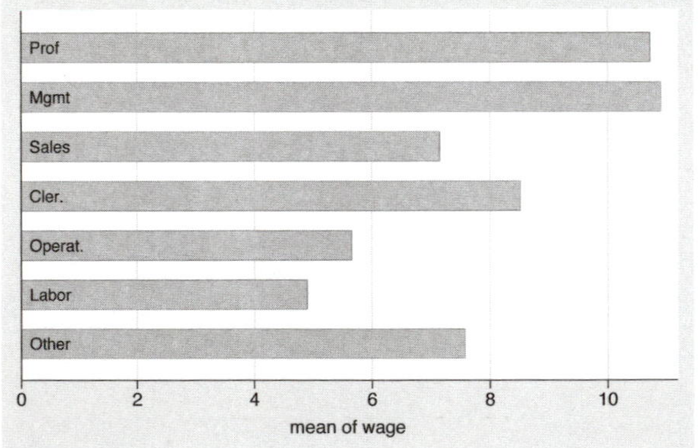

This example is the same as the last example but uses the vg_palem scheme, the monochrome equivalent of the vg_palec scheme. This scheme is based on the s2mono scheme but makes the colors of the bars/boxes/markers paler by decreasing the intensity of the colors.
Uses nlsw.dta & scheme vg_palem

```
scatter propval100 rent700 ownhome
```

This scatterplot illustrates the vg_outc scheme. It is based on the s2color scheme but makes the fill color of the bars/boxes/markers white, so they appear hollow. The plot region is a light blue to contrast with the white fill color. In this case, this scheme is useful to help us see number of markers present where numerous markers are close or partially overlapping.
Uses allstates.dta & scheme vg_outc

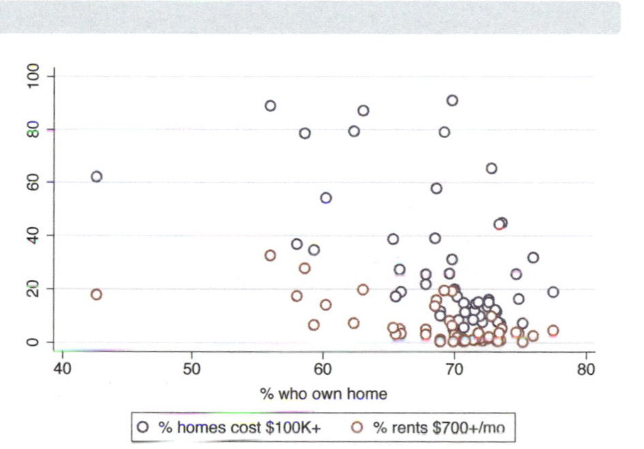


```
scatter propval100 rent700 ownhome
```

This example is similar to the previous one but illustrates the vg_outm scheme, the monochrome equivalent of the vg_outc scheme. It is based on the s2mono scheme but makes the fill color of the bars/boxes/markers white, so they appear hollow.
Uses allstates.dta & scheme vg_outm

```
twoway (scatter ownhome borninstate if stateab=="DC", mlabel(stateab))
    (scatter ownhome borninstate), legend(off)
```

This is an example of the vg_samec scheme, based on s2color, and makes all of the markers, lines, bars, etc., the same color, shape, and pattern. Here, the second scatter command labels Washington, DC, which normally would be shown in a different color, but with this scheme, the marker is the same. This scheme has a monochrome equivalent called vg_samem that is not illustrated.
Uses allstates.dta & scheme vg_samec

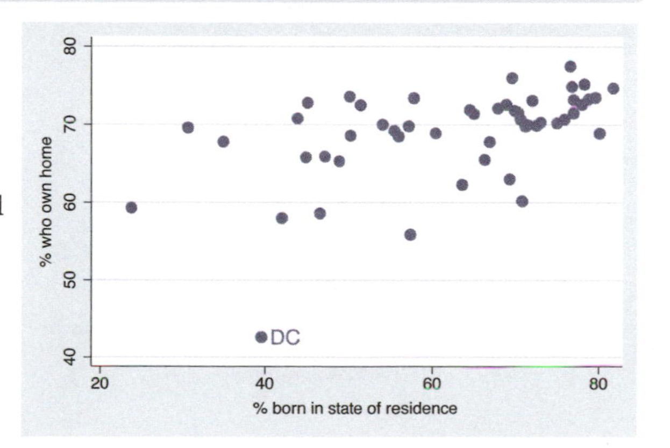

```
graph hbar commute, over(division) asyvar
```

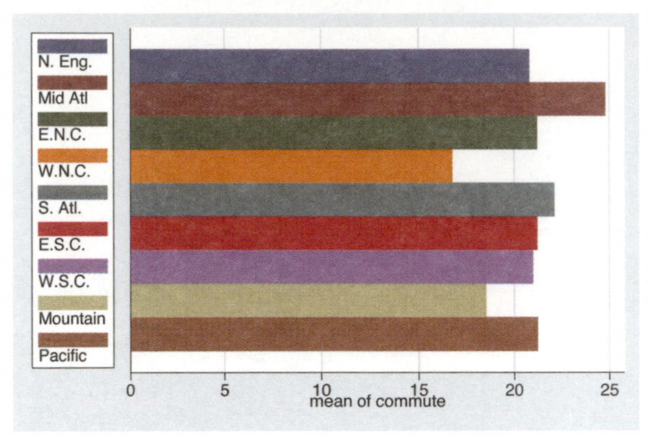

This horizontal bar chart shows an example of the vg_lgndc scheme. It is based on the s2color scheme but changes the default attributes of the legend, namely, showing the legend in one column to the left of the plot region, with the key and symbols placed atop each other. This can be an efficient way to place the legend to the left of the graph. There is also a vg_lgndm scheme, which is monochrome and is not illustrated here.
Uses allstates.dta & scheme vg_lgndc

```
graph bar commute, over(division) asyvar legend(rows(3))
```

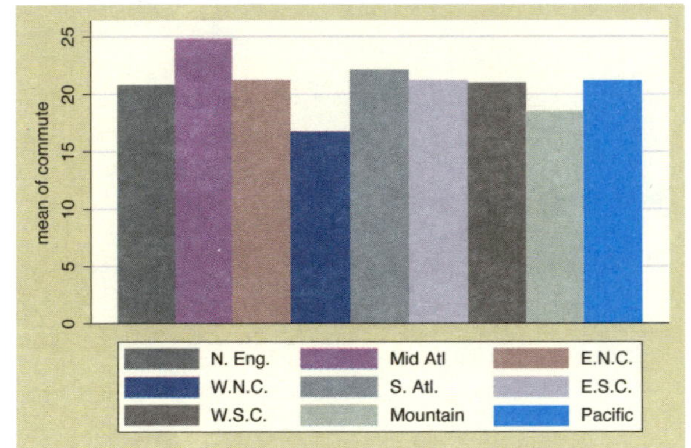

This bar chart shows an example of the vg_past scheme. It is based on the s2color scheme but selects subdued pastel colors and provides a sand background for the surrounding graph region and an eggshell color for the inner plot region and legend area.
Uses allstates.dta & scheme vg_past

```
twoway scatter rent700 propval100
```

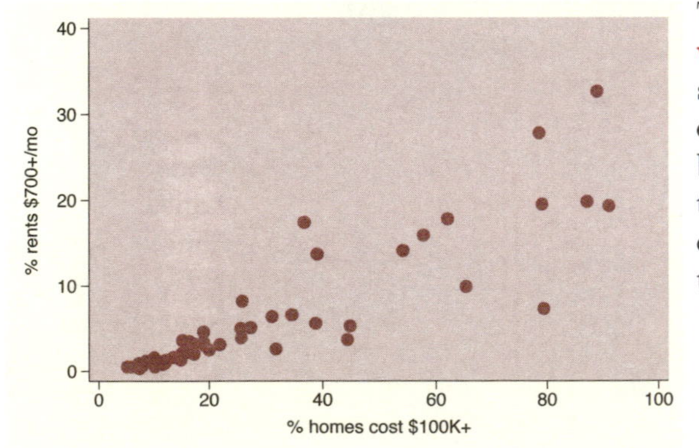

This bar chart shows an example of the vg_rose scheme. It is based on the s2color scheme but uses a different set of colors, having an eggshell background and a light rose color for the plot area. The grid lines are omitted by default, and the labels for the y-axis are horizontal by default.
Uses allstates.dta & scheme vg_rose

```
graph bar commute, over(division) asyvar legend(rows(3))
```

This bar chart shows an example of the **vg_blue** scheme. It is based on the **s2color** scheme but uses a set of blue colors, with a light blue background and a light blue-gray color for the plot area. The grid lines are omitted by default, and the labels for the y-axis are horizontal by default.
Uses allstates.dta & scheme vg_blue

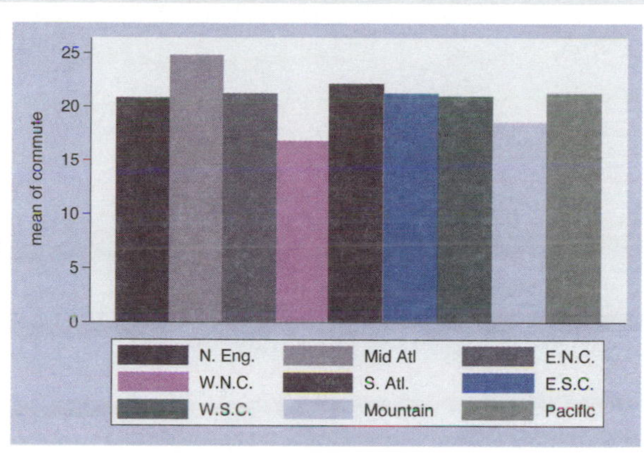


```
graph bar commute, over(division) asyvar legend(rows(3))
```

This is an example using the **vg_teal** scheme. This scheme is also based on the **s2color** scheme but uses an olive–teal background. It also suppresses the display of grid lines and makes the labels for the y-axis display horizontally by default.
Uses allstates.dta & scheme vg_teal

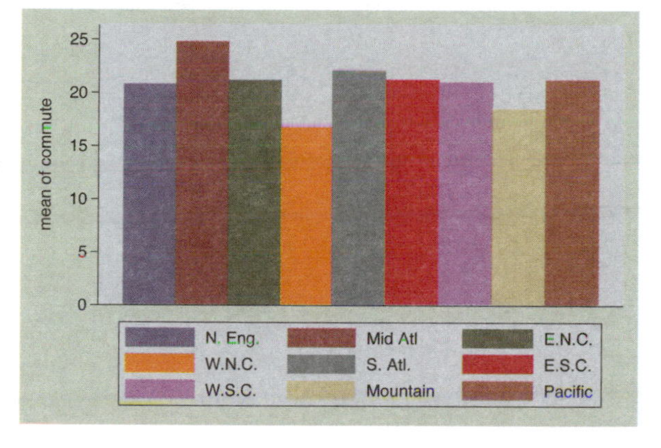


```
graph bar commute, over(division) asyvar legend(rows(3))
```

This bar chart shows an example of the **vg_brite** scheme. It is based on the **s2color** scheme but selects a bright set of colors and changes the background to light khaki.
Uses allstates.dta & scheme vg_brite

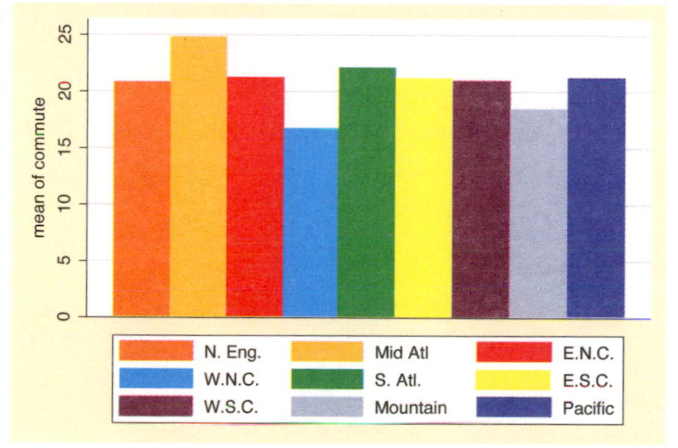

This section has just scratched the surface of all there is to know about schemes in Stata, but I hope that it helps you see how schemes create a starting point for your graph and that, by choosing a scheme that is most similar to the look you want, you can save time and effort in customizing your graphs.

1.4 Options

Learning to create effective Stata graphs is ultimately about using options to customize the look of a graph until you are pleased with it. This section illustrates the general rules and syntax for Stata graph commands, starting with their general structure, followed by illustrations showing how options work in the same way across different kinds of commands. Stata graph options work much like other options in Stata; however, there are additional features that extend their power and functionality. While we will use the `twoway scatter` command for illustration, most of the principles illustrated extend to all kinds of Stata graph commands.

```
twoway scatter propval100 rent700
```

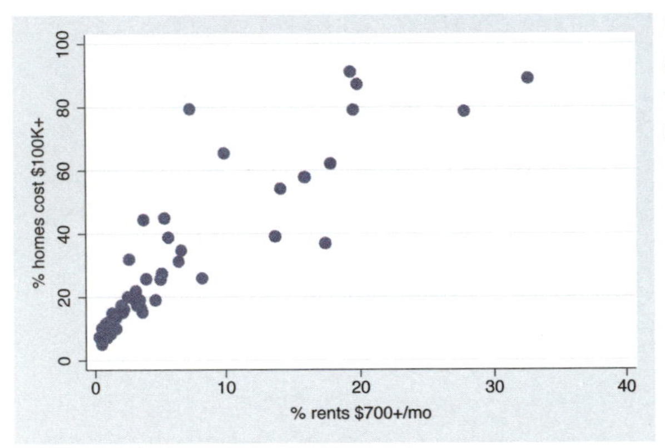

Consider this basic scatterplot. To add a title to this graph, we can use the `title()` option as illustrated in the next example.
Uses allstates.dta & scheme vg_s2c

```
twoway scatter propval100 rent700,
   title("This is a title for the graph")
```

Just as with any Stata command, the `title()` option comes after a comma, and in this case, it contains a quoted string that becomes the title of the graph.
Uses allstates.dta & scheme vg_s2c

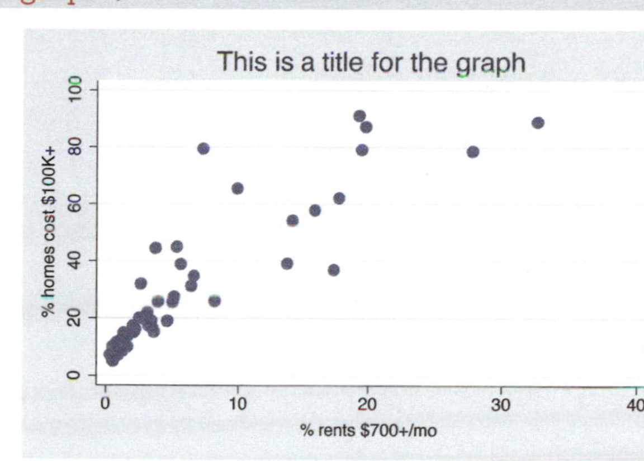


```
twoway scatter propval100 rent700,
   title("This is a title for the graph", box)
```

Starting with Stata 8, options can have options of their own. Let's put a box around the title of the graph. We can use `title(, box)`, placing `box` as an option within `title()`. If the default for the current scheme had included a box, then we could have used the `nobox` option to suppress it.
Uses allstates.dta & scheme vg_s2c

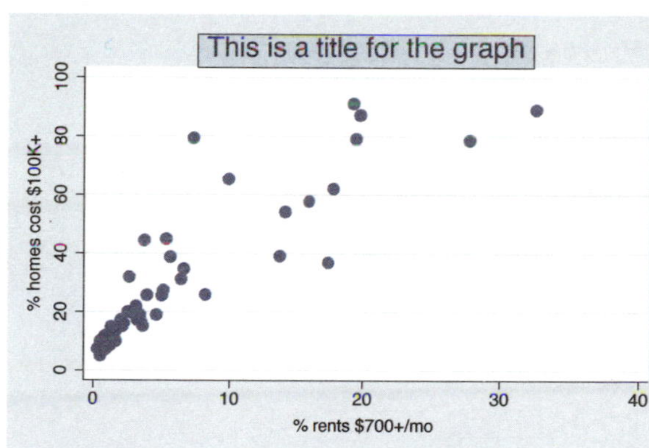


```
twoway scatter propval100 rent700,
   title("This is a title for the graph", box size(small))
```

Let's take the last graph and modify the title to make it small. We can add another option to the `title()` option by adding the `size(small)` option. Here, we see that one of the options is a keyword (`box`) and that another option allows us to supply a value (`size(small)`).
Uses allstates.dta & scheme vg_s2c

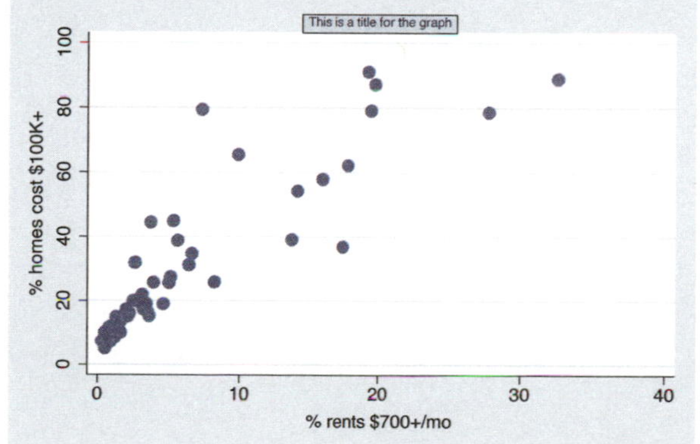

```
twoway scatter propval100 rent700,
  title("This is a title for the graph", box size(small))
  msymbol(S)
```

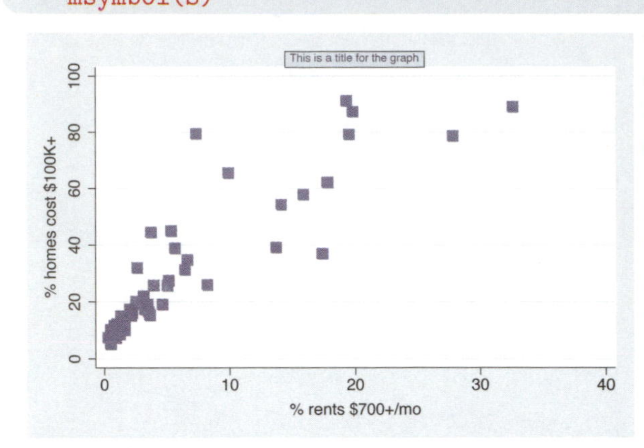

Say that we want the symbols to be displayed as squares. We can add another option called `msymbol(S)` to indicate that we want the marker symbol to be displayed as a square (`S` for square). Adding one option at a time is a common way to build a Stata graph. In the next graph, we will change gears and start building a new graph to show other aspects of options.
Uses allstates.dta & scheme vg_s2c

```
twoway scatter propval100 rent700
```

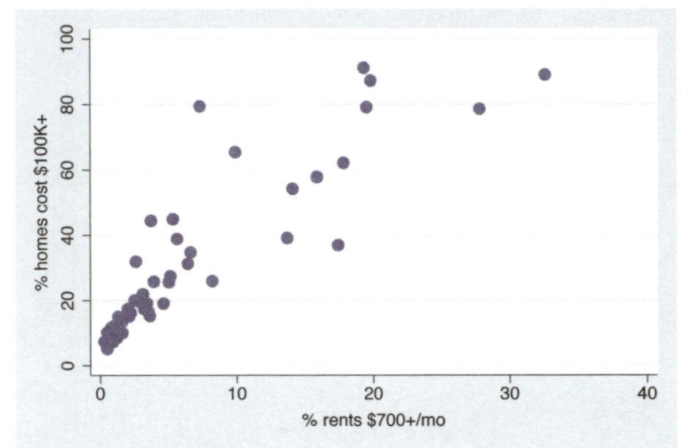

Let's return to this simple scatterplot. Say that we want the labels for the x-axis to change from 0 10 20 30 40 to 0 5 10 15 20 25 30 35 40.
Uses allstates.dta & scheme vg_s2c

```
twoway scatter propval100 rent700, xlabel(0(5)40)
```

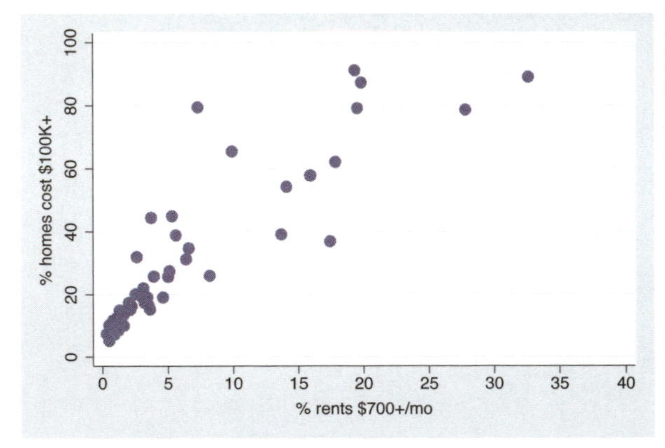

Here, we add the `xlabel()` option to label the x-axis from 0 to 40, incrementing by 5. But say that we want the labels to be displayed larger.
Uses allstates.dta & scheme vg_s2c

```
twoway scatter propval100 rent700, xlabel(0(5)40, labsize(huge))
```

Here, we add the `labsize()` (label size) option to increase the size of the labels for the x-axis. Say that we were happy with the original numbering (0 10 20 30 40) but wanted the labels to be huge. How would we do that?
Uses allstates.dta & scheme vg_s2c

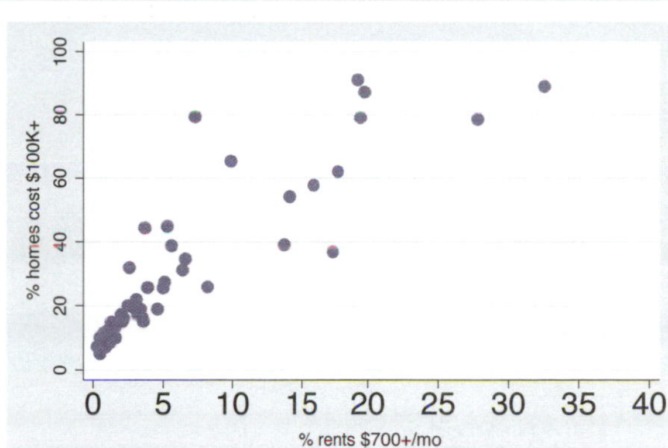

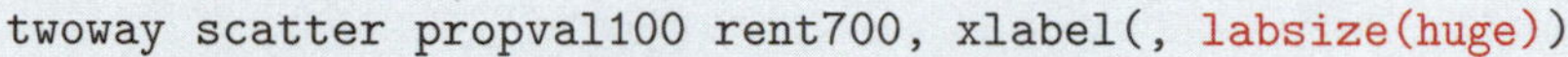

```
twoway scatter propval100 rent700, xlabel(, labsize(huge))
```

The `xlabel()` option we use here indicates that we are content with the numbers chosen for the label of the x-axis because we have nothing before the comma. After the comma, we add the `labsize()` option to increase the size of the labels for the x-axis.
Uses allstates.dta & scheme vg_s2c

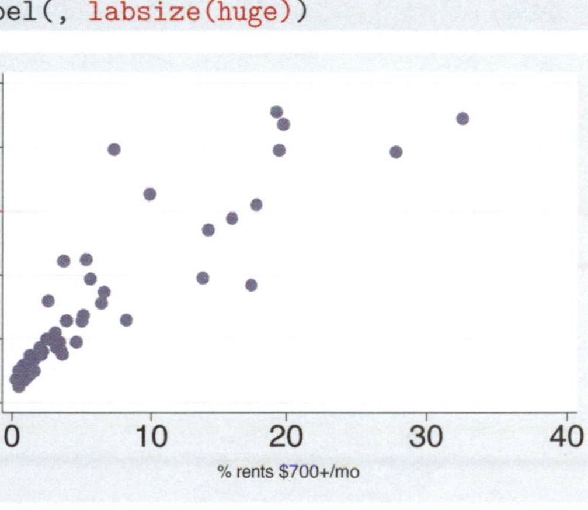

Let's consider some examples using the `legend()` option to show that some options do not require or permit the use of commas within them. Also, this allows us to show a case where you might properly specify an option over and over again.

```
twoway scatter propval100 rent700 popden
```

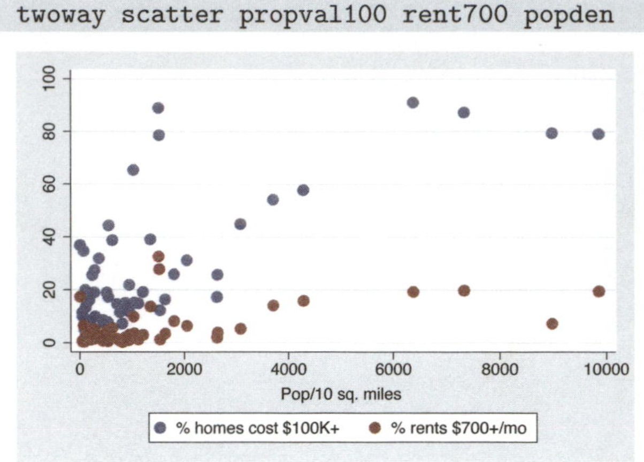

Here, we show two y-variables, **propval100** and **rent700**, graphed against population density, **popden**. Note that Stata has created a legend, helping us see which symbols correspond to which variables. We can use the **legend()** option to customize it.
Uses allstates.dta & scheme vg_s2c

```
twoway scatter propval100 rent700 popden, legend(cols(1))
```

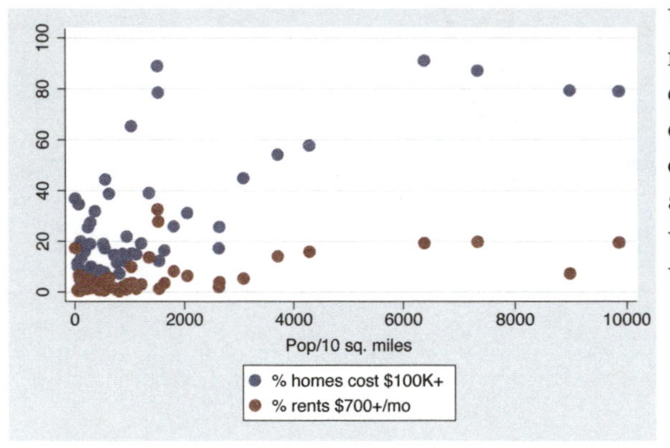

Using the legend(cols(1)) option, we make the legend display in a single column. Note that we did not use a comma because, with the **legend()** option, there is no natural default argument. If we had included a comma within the **legend()** option, Stata would have reported this as an error.
Uses allstates.dta & scheme vg_s2c

```
twoway scatter propval100 rent700 popden,
   legend(cols(1) label(1 "Property Value"))
```

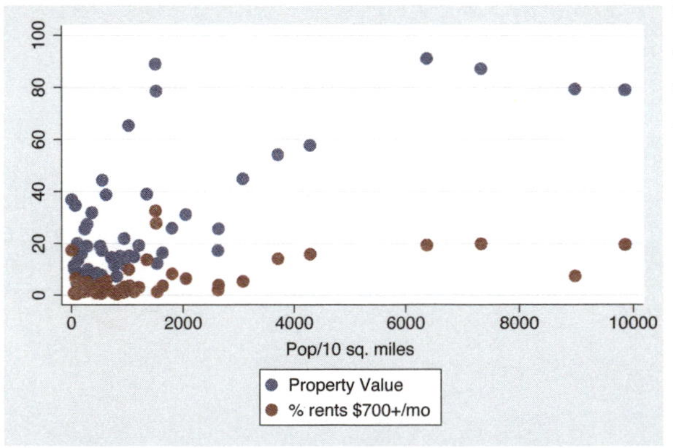

This example adds another option within the **legend()** option, label(), which changes the label for the first variable.
Uses allstates.dta & scheme vg_s2c

```
twoway scatter propval100 rent700 popden,
  legend(cols(1) label(1 "Property Value") label(2 "Rent"))
```

Here, we add another `label()` option for the `legend()` option, but in this case, we change the label for the second variable. Note that we can use the `label()` option repeatedly to change the label for the different variables.
Uses allstates.dta & scheme vg_s2c

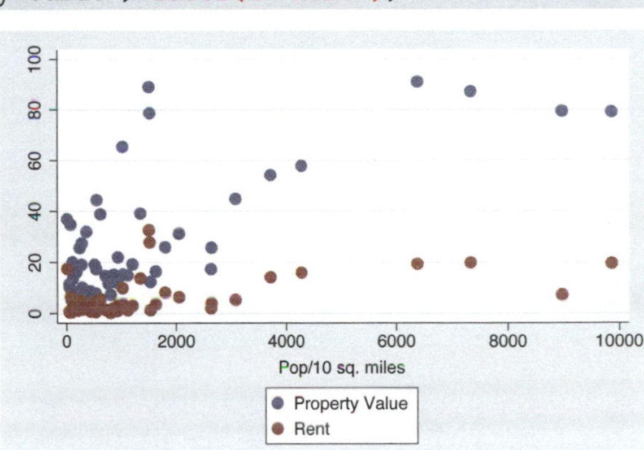

Finally, let's consider an example that shows how to use the **twoway** command to overlay two plots, how each graph can have its own options, and how options can apply to the overall graph.

```
twoway (scatter propval100 popden)
  (lfit propval100 popden)
```

Consider this graph, which shows a scatterplot predicting property value from population density and shows a linear fit between these two variables. Say that we wanted to change the symbol displayed in the scatterplot and the thickness of the line for the linear fit.
Uses allstates.dta & scheme vg_s2c

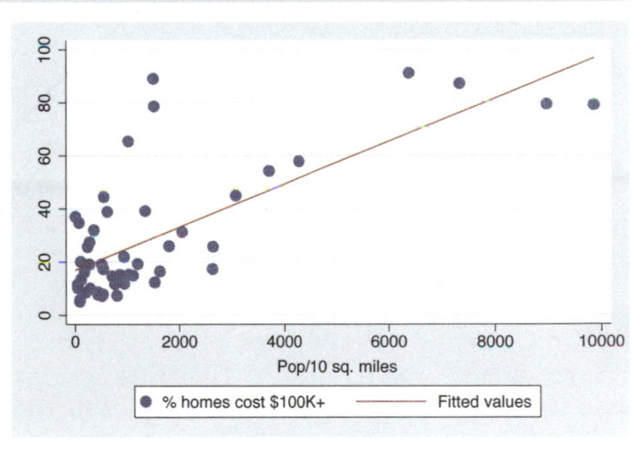

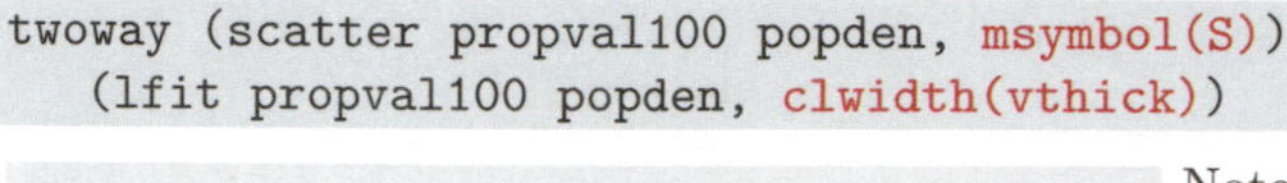

```
twoway (scatter propval100 popden, msymbol(S))
   (lfit propval100 popden, clwidth(vthick))
```

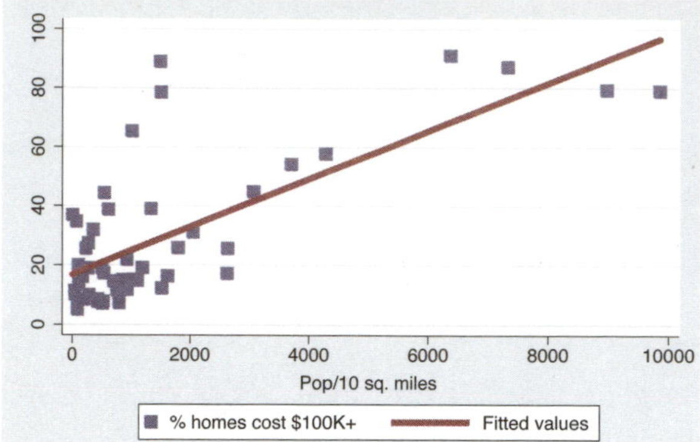

Note that we add the `msymbol()` option to the `scatter` command to change the symbol to a square, and we add the `clwidth()` (connect line width) option to the `lfit` command to make the line very thick. When we overlay two plots, each plot can have its own options that operate on its respective parts of the graph. However, some parts of the graph are shared, for example, the title.
Uses allstates.dta & scheme vg_s2c

```
twoway (scatter propval100 popden, msymbol(S))
   (lfit propval100 popden, clwidth(vthick)),
   title("This is the title of the graph")
```

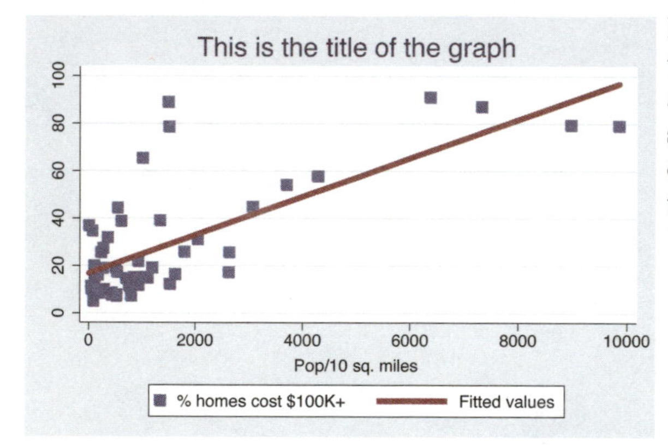

Note that we add the `title()` option to the very end of the command placed after a comma. That final comma signals that options concerning the overall graph are to follow, in this case, the `title()` option.
Uses allstates.dta & scheme vg_s2c

One of the beauties of Stata graph commands is the way that different graph commands share common options. If we want to customize the display of a legend, we do it using the same options, whether we are using a bar graph, a box plot, a scatterplot, or any other kind of Stata graph. Once we learn how to control legends with one type of graph, we have learned how to control legends for all types of graphs. Let's look at a couple of examples.

```
twoway scatter propval100 rent700 popden, legend(position(1))
```

Consider this scatterplot. We have added a `legend()` option to make the legend display in the one o'clock position on the graph, putting the legend in the top right corner.
Uses allstates.dta & scheme vg_s2c

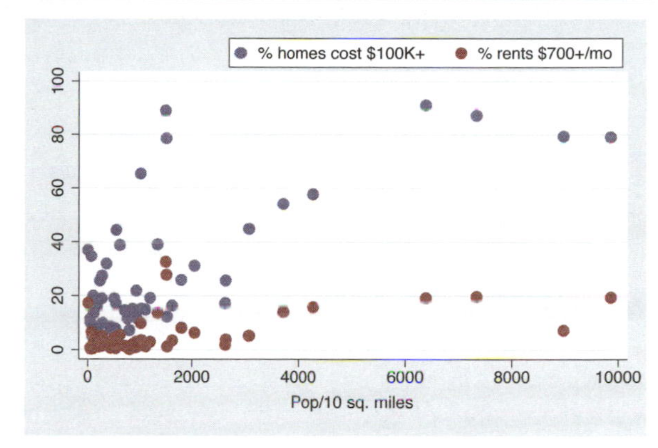


```
graph bar propval100 rent700, over(nsw) legend(position(1))
```

Here, we use the **graph bar** command, which is a completely different command from the previous one. Even though the graphs are different, the `legend()` option we supply is the same and has the same effect. Many (but not all) options function in this way, sharing a common syntax and having common effects.
Uses allstates.dta & scheme vg_s2c

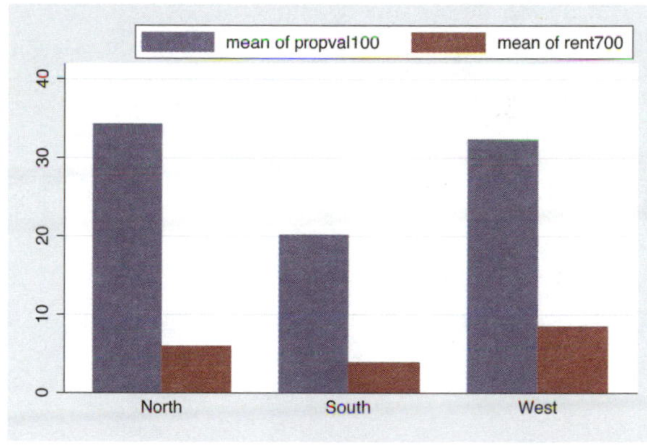


```
graph matrix propval100 rent700 popden, legend(position(1))
```

Contrast this example with the previous two. The **graph matrix** command does not support the `legend()` option because this graph does not need or produce legends. In the Matrix (95) chapter, for example, there are no references to legends, an indication that this is not a relevant option for this kind of graph. Note that, even though we included this additional irrelevant option, Stata ignored it and produced an appropriate graph anyway.
Uses allstates.dta & scheme vg_s2c

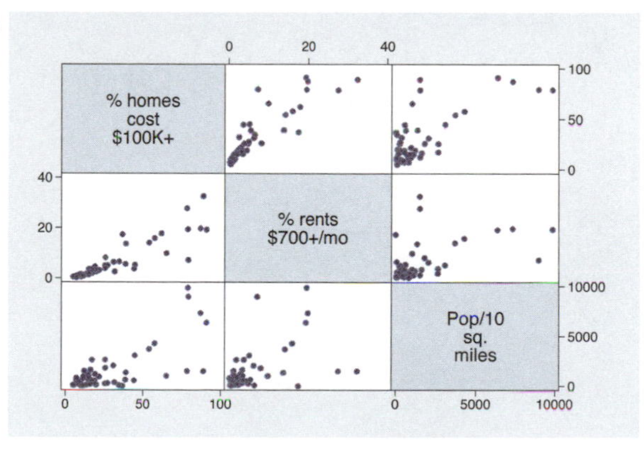

Because legends work the same way with different types of Stata graph commands, we can save pages by describing legends in detail in one place: Options : Legend (287). However, it is useful to see examples of legends for each type of graph that uses them. Each chapter, therefore, provides a brief section describing legends for each type of graph discussed in that chapter. Likewise, most options are described in detail in Options (235) with a brief section in every chapter discussing how each option works in specific types of graphs. As we saw in the case of legends, some options are not appropriate for some types of graphs, so those options will not be discussed with the commands that do not support them.

While an option like `legend()` can be used with many, but not all, kinds of Stata graph commands, other kinds of options can be used with almost every kind of Stata graph. These are called *Standard Options*. To help you differentiate these kinds of options, they are discussed in their own chapter, Standard options (313). Since these options can be used with most types of graph commands, they are generally not discussed in the chapters about the different types of graphs, except when their usage interacts with the options illustrated. For example, `subtitle()` is a *Standard Option*, but its behavior takes on a special meaning when used with the `legend()` option, so the `subtitle()` option is discussed in the context of legends. Consistent with what we have seen before, the syntax of *Standard Options* follows the same kinds of rules we have illustrated, and their usage and behavior are uniform across the many types of Stata graph commands.

To recap, this section was not about any particular options, but about some of the rules for using these options and how they behave. Some options permit options. In some cases, you may want to specify only options. Some options allow you to include one or more options, but no comma is required. When you overlay multiple graphs using `twoway`, you may have options that go along with each graph, as well as overall options that appear at the end of the command. Finally, the syntax of a certain option is the same across the different graph commands that use the options, but not all options are useful for all kinds of graph commands.

1.5 Building graphs

I have three agendas in writing this section. First, I will show the process of building complex graphs a little bit at a time. At the same time, I illustrate how to use the resources of this book to get the bits of information needed to build these graphs. Finally, I show that, even though a complete Stata graph command might look complicated and overwhelming, the process of building it slowly is actually very straightforward and logical. Let's first build a bar chart that looks at property values broken down by region of the country. Then, we will modify the legend and bar characteristics, add titles, and so forth.

```
graph display
```

Say that we want to create this graph. For now, the syntax is concealed, just showing the `graph display` command to show the previously drawn graph. It might be overwhelming at first to determine all of the options needed to make this graph. To ease our task, we will build it one bit at a time, refining the graph and fixing any problems we find.

Uses allstates.dta & scheme vg_past

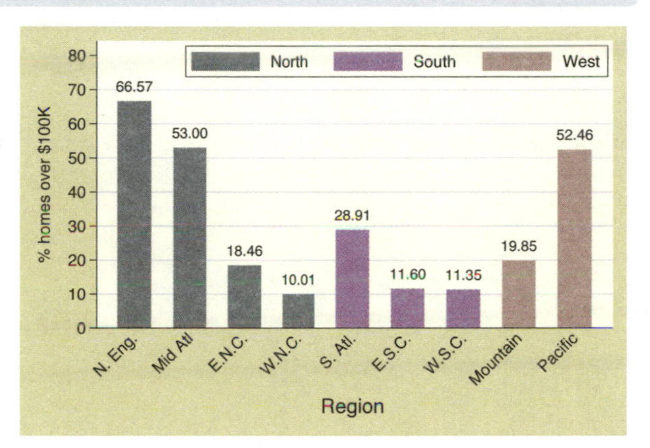


```
graph bar propval100, over(nsw) over(division)
```

We begin by seeing that this is a bar chart and look at Bar : Y-variables (107) and Bar : Over (111). We take our first step towards making this graph by making a bar chart showing `propval100` and adding `over(nsw)` and `over(division)` to break the means down by `nsw` and `division`.

Uses allstates.dta & scheme vg_past

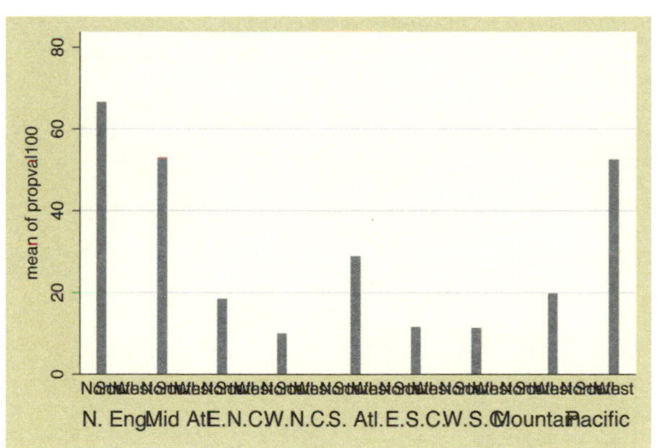

Introduction
Using this book
Twoway
Types of Stata graphs
Matrix
Bar
Schemes
Box
Dot
Options
Pie
Options
Building graphs
Standard options
Styles
Appendix

```
graph bar propval100, over(nsw) over(division) nofill
```

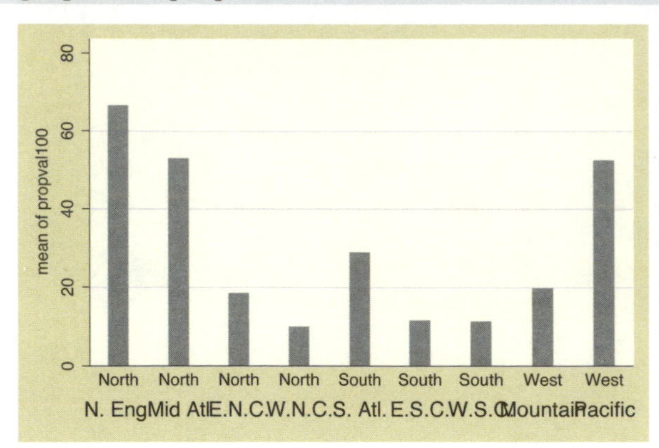

The previous graph is not quite what we want because we see every `division` shown with every `nsw`, but for example, the Pacific region only appears in the West. In Bar : Over (111), we see that we can add the `nofill` option to show only the combinations of `nsw` and `division` that exist in the data file. Next, we will look at the colors of the bars.

Uses allstates.dta & scheme vg_past

```
graph bar propval100, over(nsw) over(division) nofill asyvars
```

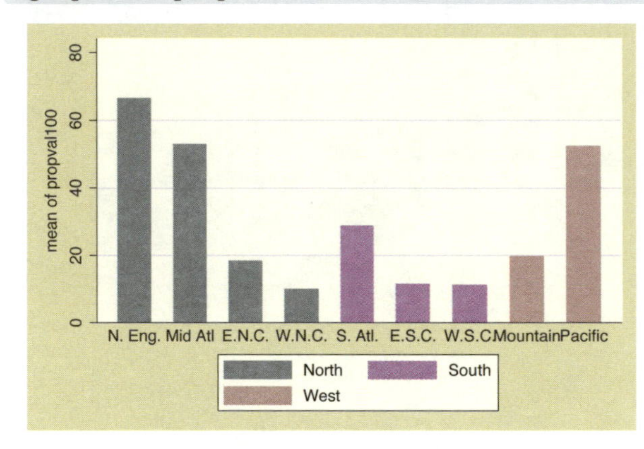

The last graph is getting closer, but we want the bars for North, South, and West to be displayed in different colors and labeled with a legend. In Bar : Y-variables (107), we see that the `asyvars` option will accomplish this. Next, we will change the title for the y-axis.

Uses allstates.dta & scheme vg_past

```
graph bar propval100, over(nsw) over(division) nofill asyvars
   ytitle("% homes over $100K")
```

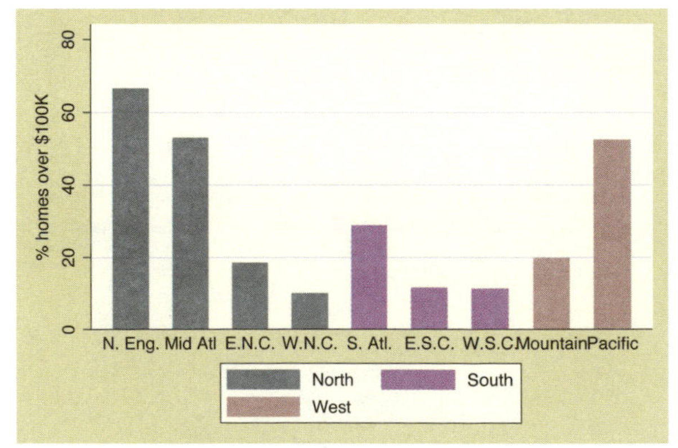

Now, we want to put a title on the y-axis. In Bar : Y-axis (143), we see examples illustrating the use of `ytitle()` for putting a title on the y-axis. Here, we put a title on the y-axis, but now we want to change the labels for the y-axis to go from 0 to 80, incrementing by 10.

Uses allstates.dta & scheme vg_past

```
graph bar propval100, over(nsw) over(division) nofill asyvars
   ytitle("% homes over $100K") ylabel(0(10)80, angle(0))
```

The Bar : Y-axis (143) section also tells us about the ylabel() option. In addition to changing the labels, we also want to change the angle of the labels, and in that section, we see that we can use the angle() option to change the angle of the labels. Now that we have the y-axis labeled as we wish, let's next look at the title for the x-axis.
Uses allstates.dta & scheme vg_past

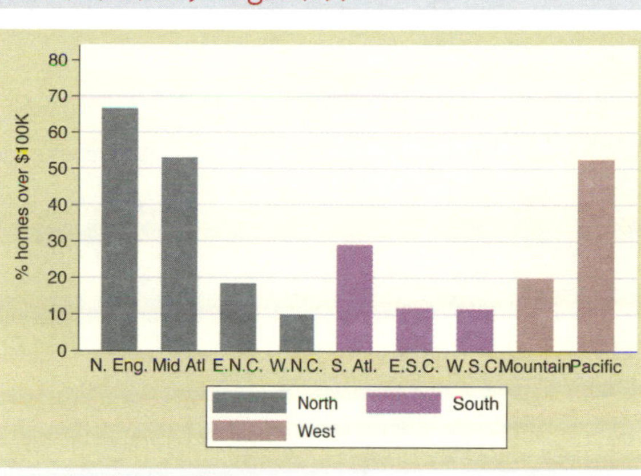


```
graph bar propval100, over(nsw) over(division) nofill asyvars
   ytitle("% homes over $100K") ylabel(0(10)80, angle(0)) b1title(Region)
```

After having used the ytitle() option to label the y-axis, we might be tempted to use the xtitle() option to label the x-axis, but this axis is a categorical variable. In Bar : Cat axis (123), we see that this axis is treated quite differently because of that. To put a title below the graph, we use the b1title() option. Now, let's turn our attention to formatting the legend.
Uses allstates.dta & scheme vg_past

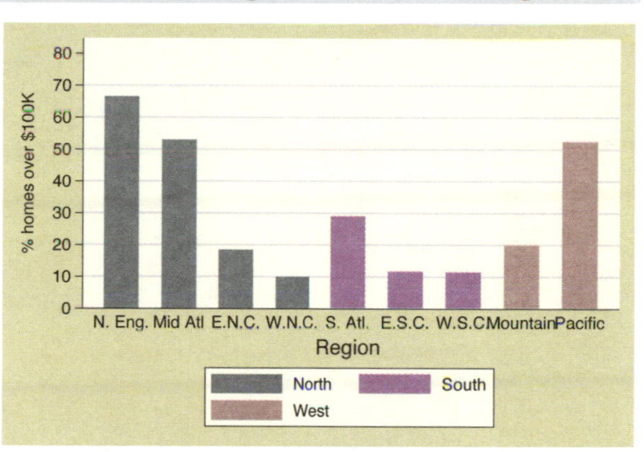


```
graph bar propval100, over(nsw) over(division) nofill asyvars
   ytitle("% homes over $100K") ylabel(0(10)80, angle(0)) b1title(Region)
   legend(rows(1) position(1) ring(0))
```

Here, we want to use the legend() option to make the legend have one row in the top right corner within the plot area. In Bar : Legend (130), we see that the rows(1) option makes the legend appear in one row and that the position(1) option puts the legend in the 1 o'clock position. The ring(0) option puts the legend inside the plot region. Next, let's label the bars.
Uses allstates.dta & scheme vg_past

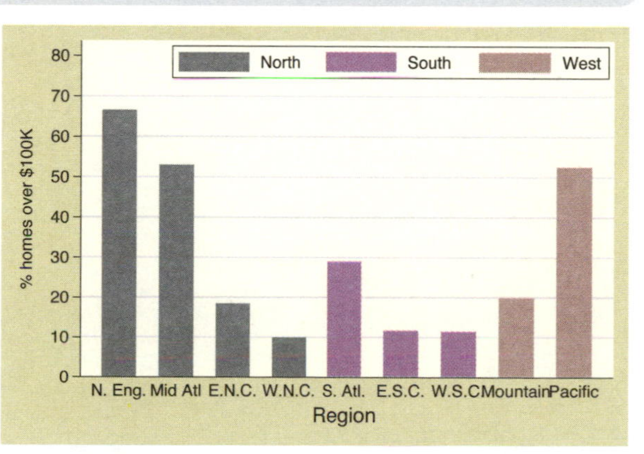

```
graph bar propval100, over(nsw) over(division) nofill asyvars
   ytitle("% homes over $100K") ylabel(0(10)80, angle(0)) b1title(Region)
   legend(rows(1) position(1) ring(0)) blabel(bar)
```

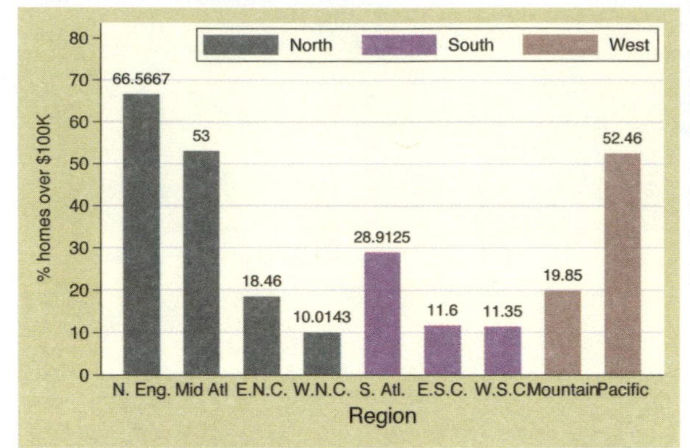

We want each bar to be labeled with the height of the bar, and Bar : Legend (130) shows how we can do this. This section shows how to use the `blabel()` (bar label) option to label the bars in lieu of legends. `blabel()` also can label the bars with their height, using `blabel(bar)`.
Uses allstates.dta & scheme vg_past

```
graph bar propval100, over(nsw) over(division) nofill asyvars
   ytitle("% homes over $100K") ylabel(0(10)80, angle(0)) b1title(Region)
   legend(rows(1) position(1) ring(0)) blabel(bar, format(%4.2f))
```

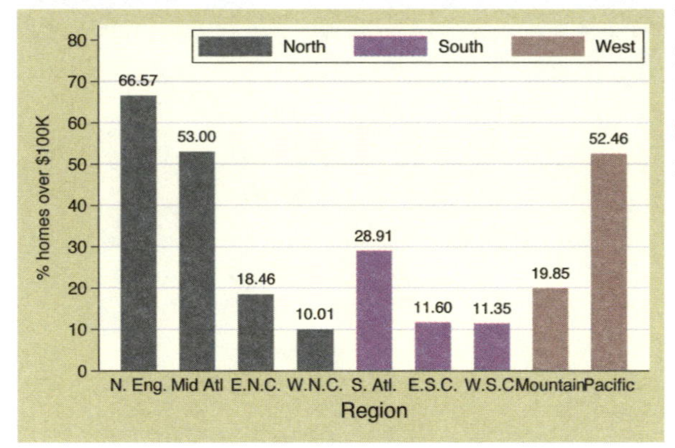

We want the labels for each bar to end in two decimal places, and we see in Bar : Legend (130) that we can use the `format()` option to format these numbers as we wish.
Uses allstates.dta & scheme vg_past

```
graph bar propval100, over(nsw) over(division, label(angle(45))) nofill
   ytitle("% homes over $100K") ylabel(0(10)80, angle(0)) b1title(Region)
   legend(rows(1) position(1) ring(0)) blabel(bar, format(%4.2f)) asyvars
```

Finally, in Bar : Cat axis (123), we see that we can add the `label(angle(45))` option to the `over()` option to specify that labels for that variable be shown at a 45-degree angle so they do not overlap each other.
Uses allstates.dta & scheme vg_past

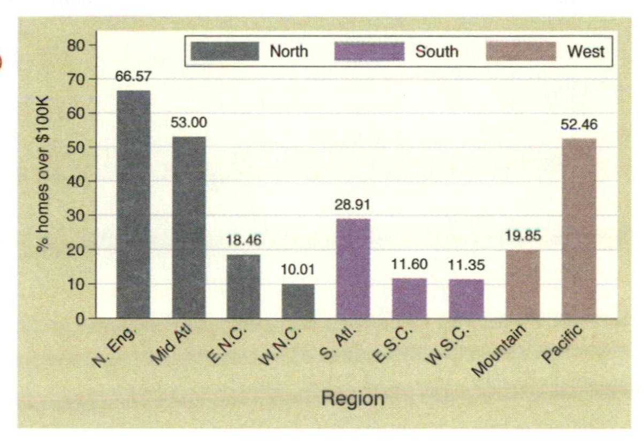

This section has shown that it is not that difficult to create a complex graph by building it one step at a time. You can use the resources in this book to seek out each piece of information you need and then put those pieces together the way you want to create your own graphs. For more information about how to integrate options to create complex Stata graphs, see Appendix : More examples (366).

2 Twoway graphs

The `graph twoway` command represents not just a single kind of graph but actually over thirty different kinds of graphs. Many of these graphs are similar in appearance and function, so I have grouped them into eight families, which form the first eight sections of this chapter. These first eight sections, which cover scatterplots to distribution plots, cover the general features of these plots and briefly mention some important options. These are followed by a section giving an overview of the options that can be used with `twoway` graphs. (For further details about the options that can be used with `twoway` graphs, see Options (235) and the sections within that chapter.) The chapter concludes with a section illustrating how you can overlay `twoway` graphs. For more details about `graph twoway`, see [G] **graph twoway**.

2.1 Scatterplots

This section covers the use of scatterplots. Because scatterplots are so commonly used, this section will cover more details about the use of these graphs than subsequent sections. Also, this section will introduce some of the kinds of options that can be used with many kinds of `twoway` plots, with cross-references to Options (235).

```
graph twoway scatter ownhome propval100
```

Here is a basic scatterplot. Note that this command starts with `graph twoway`, which indicates that this is a `twoway` graph. `scatter` indicates that we are creating a twoway scatterplot. These are followed by the variable to be placed on the y-axis and then the variable for the x-axis.
Uses allstates.dta & scheme vg_s2c

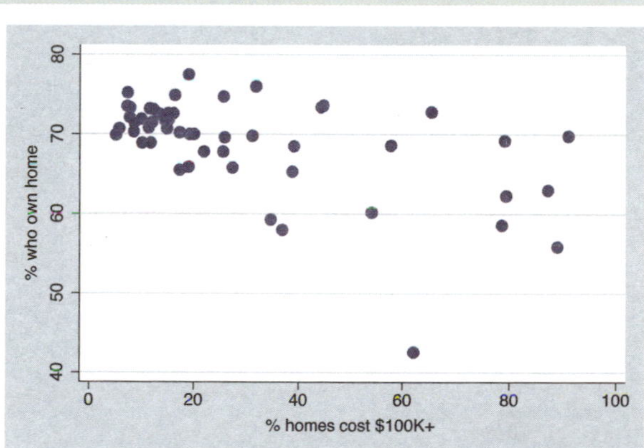

```
twoway scatter ownhome propval100
```

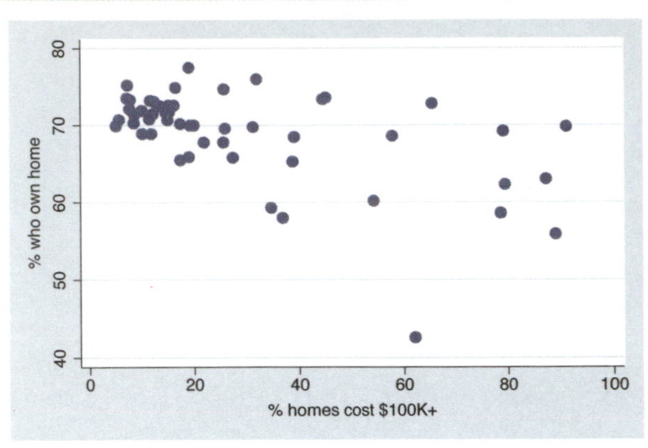

Since it can be cumbersome to type **graph twoway scatter**, Stata allows you to shorten this to **twoway scatter**. *Uses allstates.dta & scheme vg_s2c*

```
scatter ownhome propval100
```

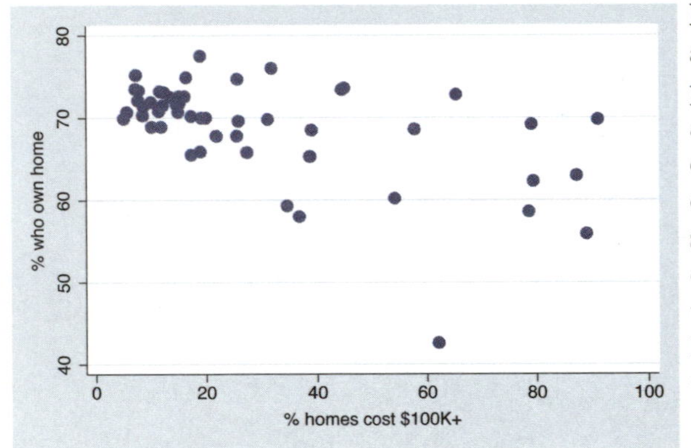

In fact, some **graph twoway** commands are so frequently used that Stata permits you to omit the **graph twoway**, as we have done here, and just start the command with **scatter**. While this can save some typing, this can sometimes conceal the fact that the command is really a **twoway** graph and that these are a special class of graphs. For clarity, I will generally present these graphs starting with **twoway**. *Uses allstates.dta & scheme vg_s2c*

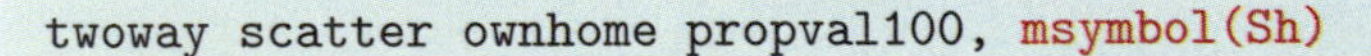

```
twoway scatter ownhome propval100, msymbol(Sh)
```

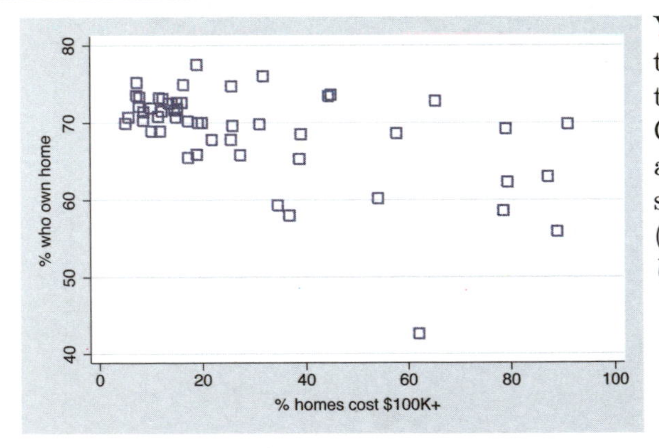

You can control the marker symbol with the **msymbol()** option. Here, we make the symbols large, hollow squares. See Options : Markers (235) for more details about controlling the marker symbol, size, and color, and see Styles : Symbols (342) for the symbols you can select. *Uses allstates.dta & scheme vg_s2c*

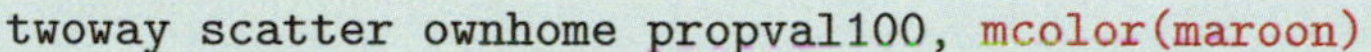

```
twoway scatter ownhome propval100, mcolor(maroon)
```

You can control the marker color with the `mcolor()` option. Here, we make the markers maroon. See Styles : Colors (328) for other colors you could choose and also Options : Markers (235) for more details.
Uses allstates.dta & scheme vg_s2c

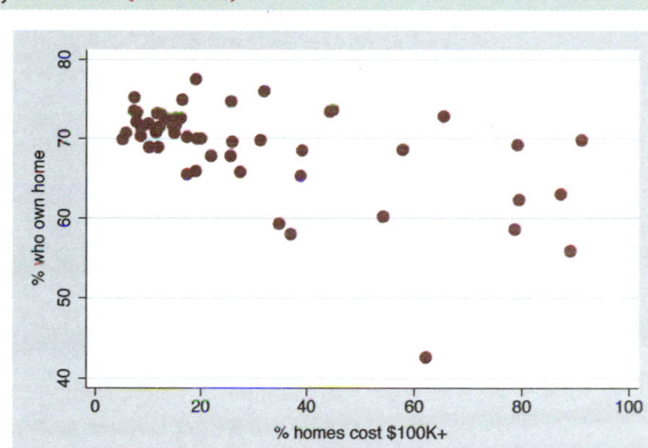


```
twoway scatter ownhome propval100, msize(vlarge)
```

You can control the marker size with the `msize()` option. Using `msize(vlarge)`, we make the markers very large. Note that we switched to the `vg_outc` scheme, showing white-filled markers, which can be useful when the markers are large. See Styles : Markersize (340) for other sizes you could choose and also Options : Markers (235) for more details.
Uses allstates.dta & scheme vg_outc

```
twoway scatter ownhome propval100 [aweight=rent700], msize(small)
```

You can also use a weight variable to determine the size of the symbols. Using `[aweight=rent700]`, we size the symbols according to the proportion of rents that exceed 700 dollars per month, allowing us to graph three variables at once. We add the `msize(small)` option to shrink the size of all the markers so they do not get too large. See Options : Markers (235) for more details.
Uses allstates.dta & scheme vg_outc

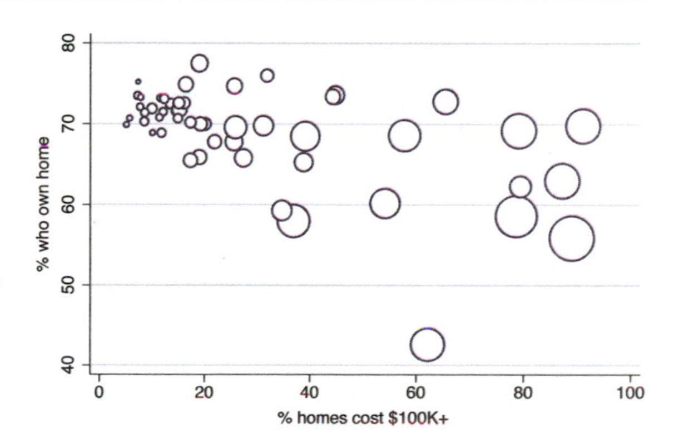

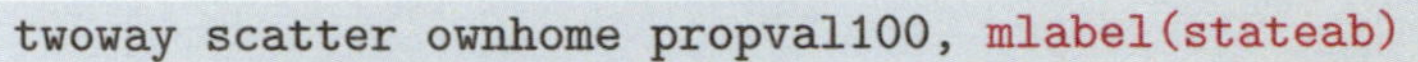

```
twoway scatter ownhome propval100, mlabel(stateab)
```

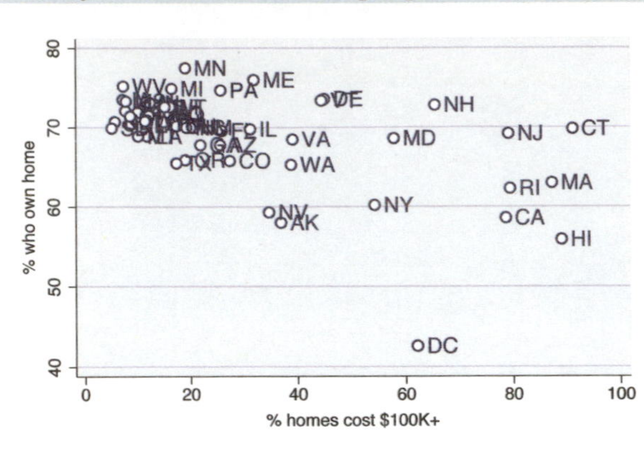

The `mlabel(stateab)` option can be used to add a marker label with the state abbreviation. See Options : Marker labels (247) for more details about how you can control the size, position, color, and angle of marker labels.
Uses allstates.dta & scheme vg_outc

```
twoway scatter ownhome propval100, mlabel(stateab) mlabsize(vlarge)
```

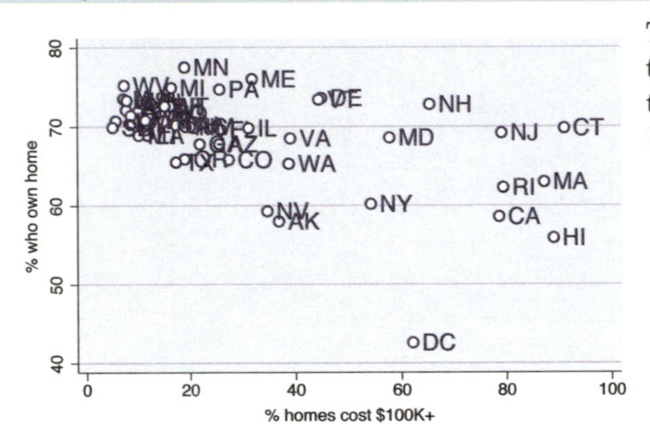

The `mlabsize(vlarge)` option controls the marker label size. Here, we make the marker label very large.
Uses allstates.dta & scheme vg_outc

```
twoway scatter ownhome propval100, mlabel(stateab) mlabposition(12)
```

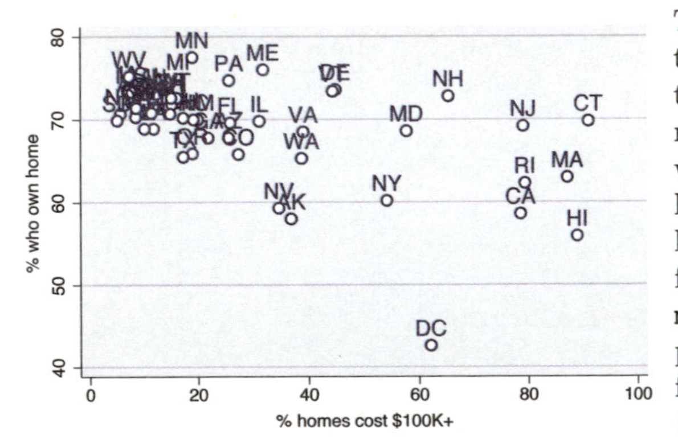

The `mlabposition()` option controls the marker label position with respect to the marker. Here, we place the marker labels at the 12 o'clock position with respect to the markers, placing the labels directly above the points they label. See Options : Marker labels (247) for examples illustrating the `mlabvposition()` option, which permits different marker label positions for different observations.
Uses allstates.dta & scheme vg_outc

```
twoway scatter ownhome propval100, mlabel(stateab)
   mlabposition(0) msymbol(i)
```

The `mlabposition(0)` option places the marker label in the center position. To keep it from being obscured by the marker symbol, we also add the `msymbol(i)` option to make the marker symbol invisible. In effect, the marker symbols have been replaced by the marker labels.
Uses allstates.dta & scheme vg_outc

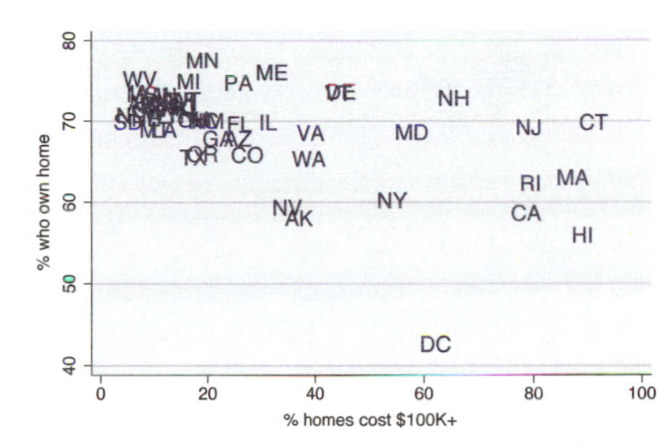


```
twoway scatter fv propval100
```

Say that we ran the following commands:

```
. regress ownhome propval100
. predict fv
```

The variable `fv` represents the fit values, and here we graph `fv` against `propval100`. As we expect, all of the points fall along a line, but they are not connected. The next few examples will consider options you can use to connect points; see Options : Connecting (250) for more details. For variety, we have switched to the `vg_past` scheme.
Uses allstates.dta & scheme vg_past

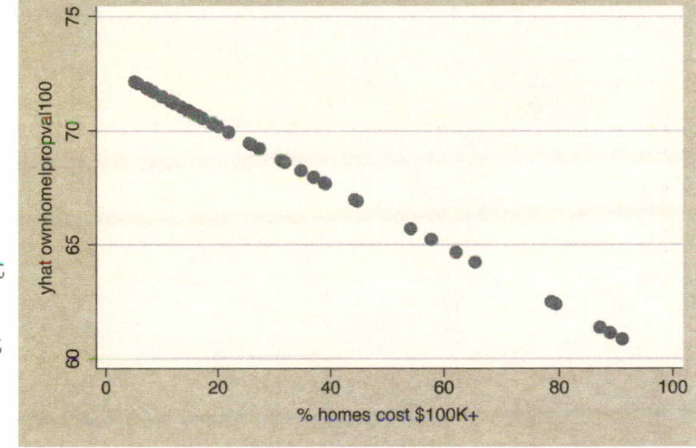


```
twoway scatter fv propval100, connect(l) sort
```

We add the `connect(l)` option to indicate that the points should be connected with a line. We also add the `sort` option, which is generally recommended when you connect observations and the data are not already sorted on the x-variable.
Uses allstates.dta & scheme vg_past

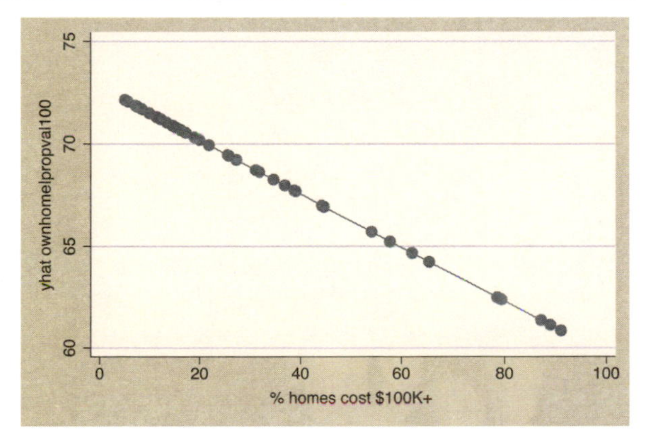

Introduction Twoway Matrix Bar Box Dot Pie Options Standard options Styles Appendix
Scatter Fit CI fit Line Area Bar Range Distribution Options Overlaying

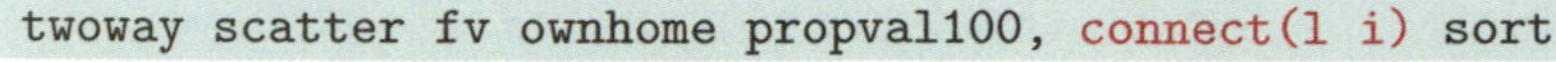

```
twoway scatter fv ownhome propval100, connect(l i) sort
```

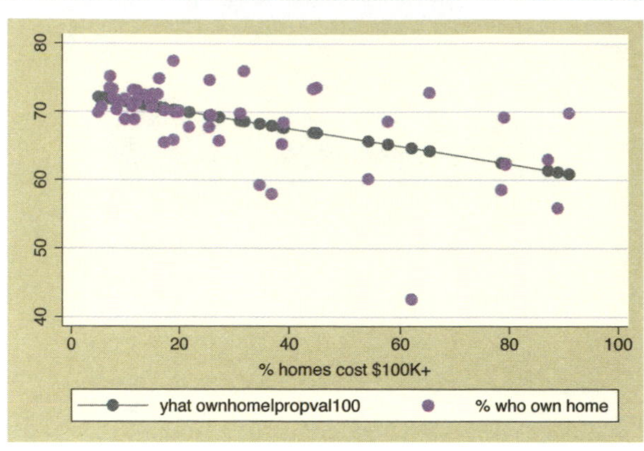

We can show both the observations and the fit values in one graph. The `connect(l i)` option specifies that the first y-variable should be connected with straight lines (`l` for line) and the second y-variable should not be connected (`i` for invisible connection).
Uses allstates.dta & scheme vg_past

```
twoway scatter fv ownhome propval100, msymbol(i .)  connect(l i)
   sort
```

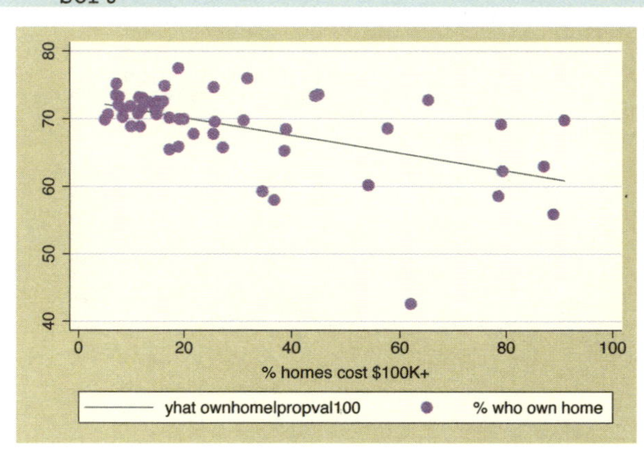

The `msymbol(i .)` option specifies that the first y-variable should not have symbols displayed (`i` for invisible symbol) and that the second y-variable should have the default symbols displayed.
Uses allstates.dta & scheme vg_past

```
twoway scatter fv ownhome propval100, msymbol(i .)  connect(l i)
   sort legend(label(1 Pred.  Perc.  Own))
```

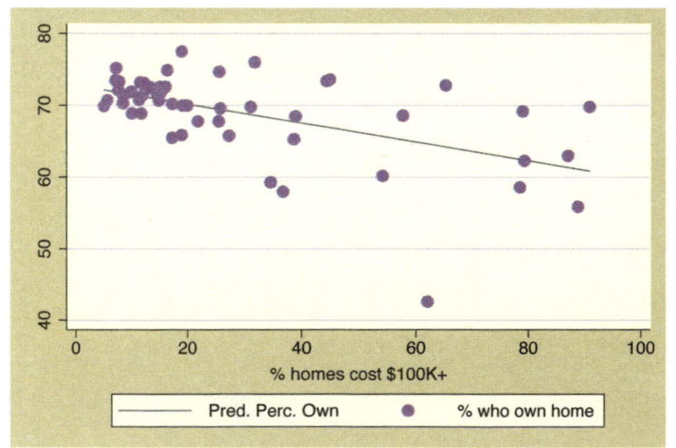

The `legend()` option can be used to control the legend. We use the `label()` option to specify the contents of the first item in the legend. See Options : Legend (287) for more details on legends.
Uses allstates.dta & scheme vg_past

```
twoway scatter fv ownhome propval100, msymbol(i .)  connect(l i)
   sort legend(label(1 Pred.  Perc.  Own) order(2 1))
```

The order() option can be used to specify the order in which the items in the legend are displayed.
Uses allstates.dta & scheme vg_past

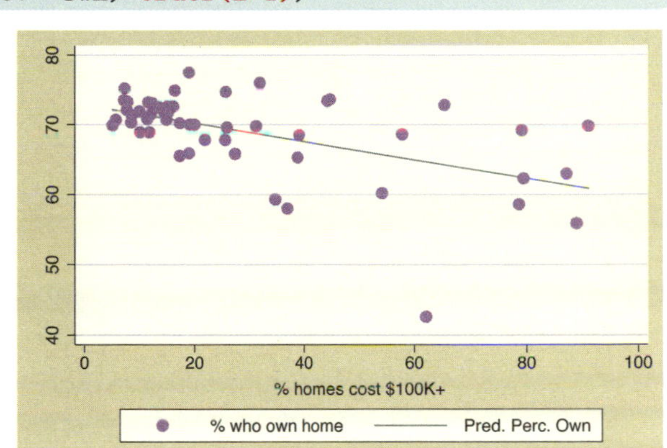


```
twoway scatter fv ownhome propval100, msymbol(i .)  connect(l i)
   sort legend(label(1 Pred.  Perc.  Own) order(2 1) cols(1))
```

The cols(1) option makes the items in the legend display in a single column.
Uses allstates.dta & scheme vg_past

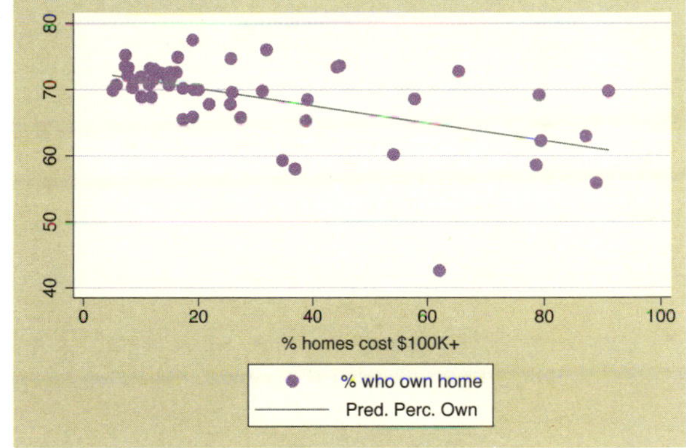


```
twoway scatter ownhome propval100,
   xtitle("Percent homes over $100K") ytitle("Percent who own home")
```

The xtitle() and ytitle() option can be used to specify the titles for the x- and y-axes. See Options : Axis titles (254) for more details about how to control the display of axes. Note that we are now using the vg_s2m scheme, one you might favor for graphs that will be printed in black and white.
Uses allstates.dta & scheme vg_s2m

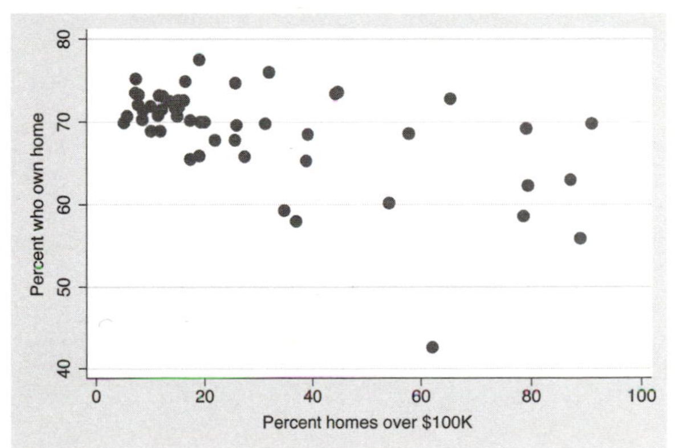

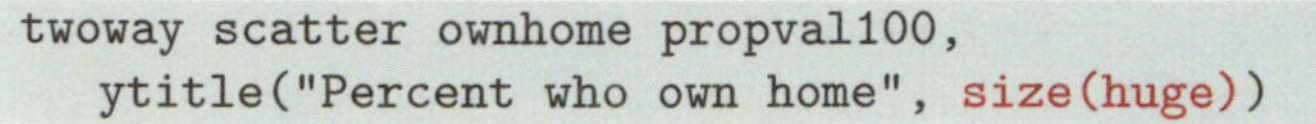

```
twoway scatter ownhome propval100,
   ytitle("Percent who own home", size(huge))
```

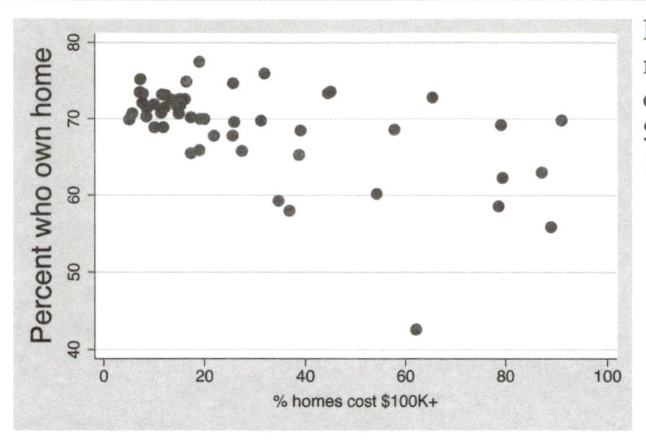

Here, we use the `size(huge)` option to make the title on the y-axis huge. For other text sizes you could choose, see Styles : Textsize (344).
Uses allstates.dta & scheme vg_s2m

```
twoway scatter ownhome propval100, xlabel(0(10)100) ylabel(40(5)80)
```

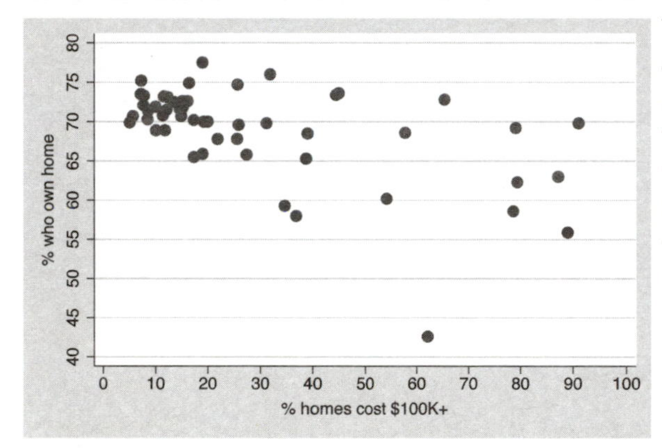

We use the `ylabel()` and `xlabel()` options to control the labeling of the x- and y-axes. We label the x-axis from 0 to 100, incrementing by 10, and the y-axis from 40 to 80, incrementing by 5. See Options : Axis labels (256) for more details on labeling axes.
Uses allstates.dta & scheme vg_s2m

```
twoway scatter ownhome propval100, xlabel(#10) ylabel(#5)
```

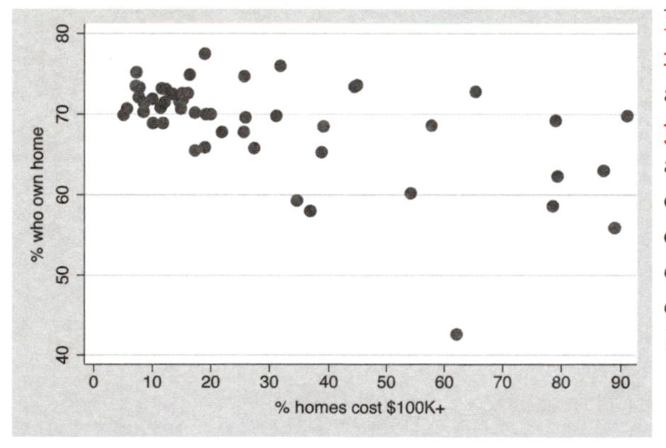

In this example, we use the `xlabel(#10)` option to ask Stata to use approximately 10 nice labels and the `ylabel(#5)` option to use approximately 5 nice labels. In this case, our gentle request was observed exactly, but in some cases, Stata will choose somewhat different values to create axis labels it believes are logical.
Uses allstates.dta & scheme vg_s2m

```
twoway scatter ownhome propval100, xlabel(#10) ylabel(#5, nogrid)
```

Using the **nogrid** option, we can suppress the display of the grid. Note that this option is placed within the **ylabel()** option, thus suppressing the grid for the y-axis. If the grid were absent, and we wished to include it, we could add the **grid** option. (You can also specify **grid** or **nogrid** within the **xlabel()** option to control grids for the x-axis.) For more details, see Options : Axis labels (256).
Uses allstates.dta & scheme vg_s2m

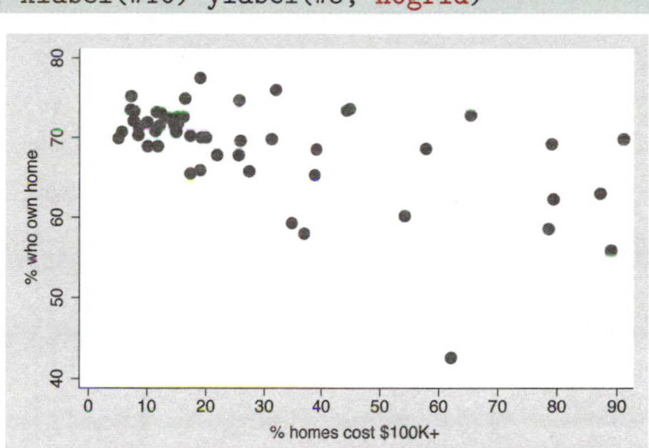


```
twoway scatter ownhome propval100, xlabel(#10) ylabel(#5, nogrid)
   yline(55 75, lwidth(thin) lcolor(black) lpattern(dash))
```

The **yline()** option is used to add a thin, black, dashed line to the graph where y equals 55 and 75.
Uses allstates.dta & scheme vg_s2m

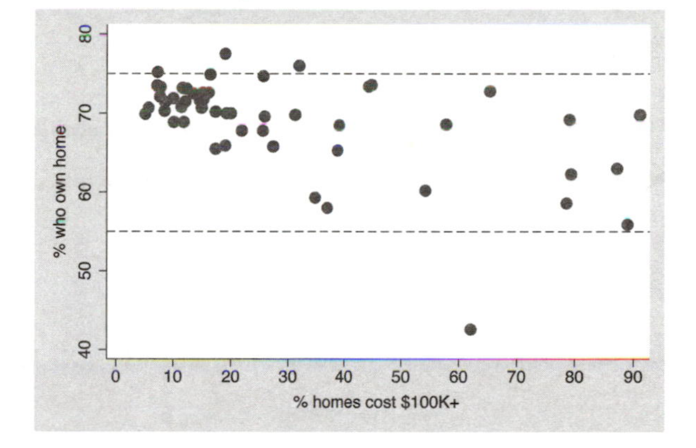


```
twoway scatter ownhome propval100, xscale(alt)
```

Here, we use the **xscale()** option to request that the x-axis be placed in its alternate position, in this case at the top instead of at the bottom. To learn more about axis scales, including suppressing, extending, or relocating them, see Options : Axis scales (265).
Uses allstates.dta & scheme vg_s2m

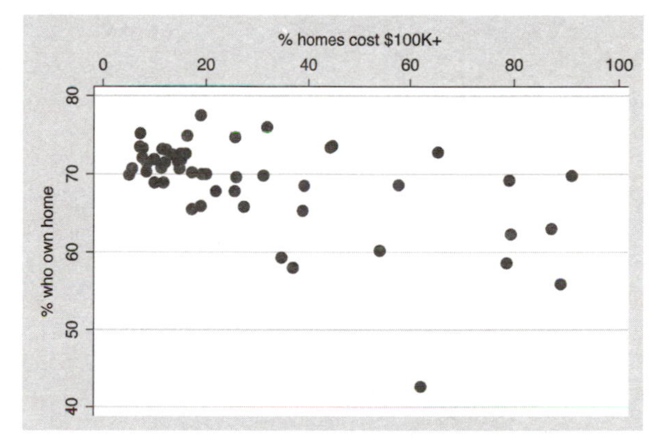

Introduction
Twoway
Matrix
Bar
Box
Dot
Pie
Options
Standard options
Styles
Appendix
Scatter
Fit
CI fit
Line
Area
Bar
Range
Distribution
Options
Overlaying

```
twoway scatter ownhome propval100, by(nsw)
```

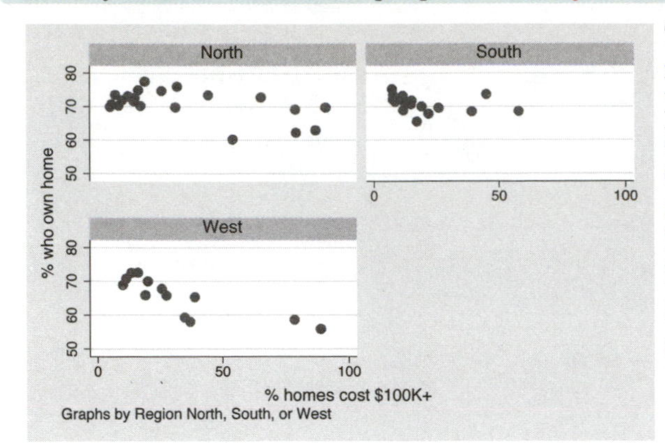

The `by(nsw)` option is used here to make separate graphs for states in the North, South, and West. At the bottom left corner, you can see a note that describes how the separate graphs arose. This is based on the variable label for `nsw`; if this variable had not been labeled, it would have read *Graphs by nsw*. See **Options : By** (272) for more details about using the `by()` option. *Uses allstates.dta & scheme vg_s2m*

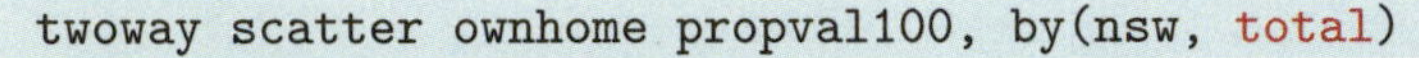

```
twoway scatter ownhome propval100, by(nsw, total)
```

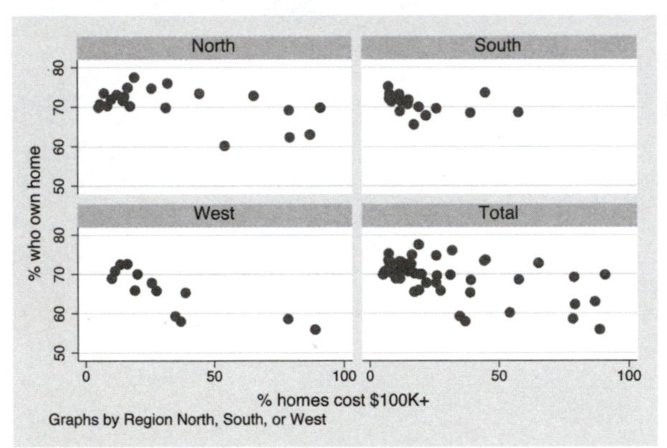

The `total` option can be used within the `by()` option to add an additional graph showing all the observations. *Uses allstates.dta & scheme vg_s2m*

```
twoway scatter ownhome propval100, by(nsw, total compact)
```

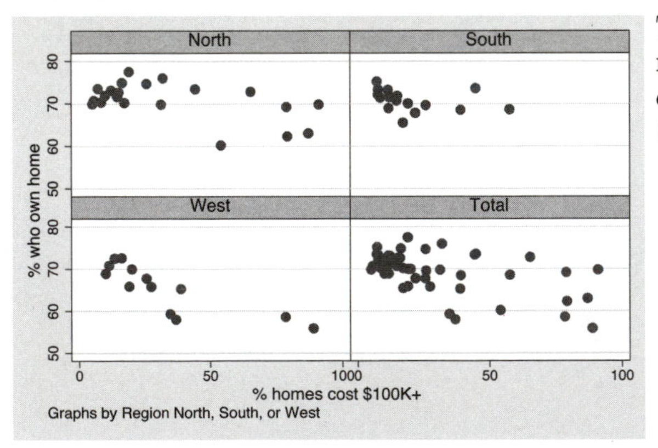

The `compact` option can be used to make the graphs display more compactly. *Uses allstates.dta & scheme vg_s2m*

```
twoway scatter ownhome propval100, text(47 62 "Washington, DC")
```

We can use the `text()` option to add text to the graph. We add text to label the observation belonging to Washington, DC. See Options : Adding text (299) for more information about adding text in the section.
Uses allstates.dta & scheme vg_s2m

```
twoway scatter ownhome propval100, text(47 62 "Washington, DC", size(large)
   margin(medsmall) blwidth(vthick) box)
```

Stata gives you considerable control over the display of text you add to the graph, as well as the ability to enclose the text in a box and control the characteristics of the box. See Options : Textboxes (303) for more details.
Uses allstates.dta & scheme vg_s2m

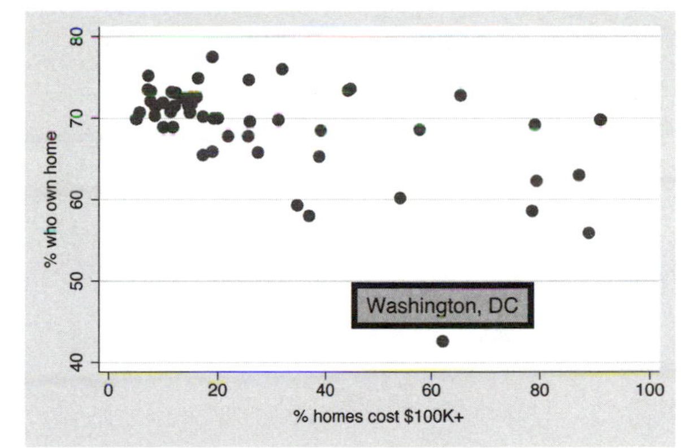


```
twoway (scatter ownhome propval100) (scatteri 42.6 62.1 "DC")
```

This graph uses the `scatteri` (scatter immediate) command to plot and label a point for Washington, DC. The values `42.6` and `62.1` are the values for `ownhome` and `propval100` for Washington, DC, and are followed by `"DC"`, which acts as a marker label for that point. If we had instead specified `(9) "DC"`, then "DC" would have been plotted at the 9 o'clock position.
Uses allstates.dta & scheme vg_s2m

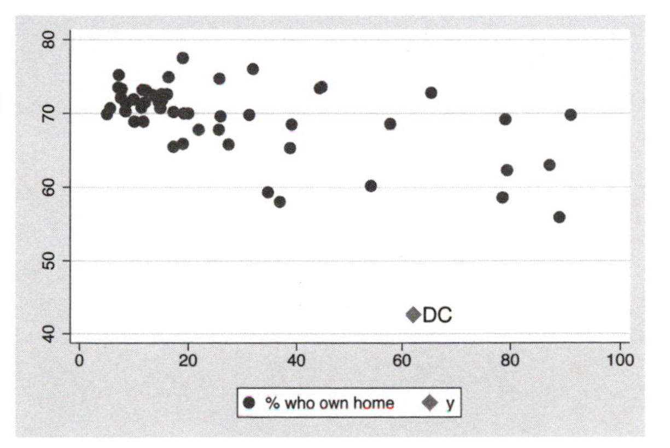

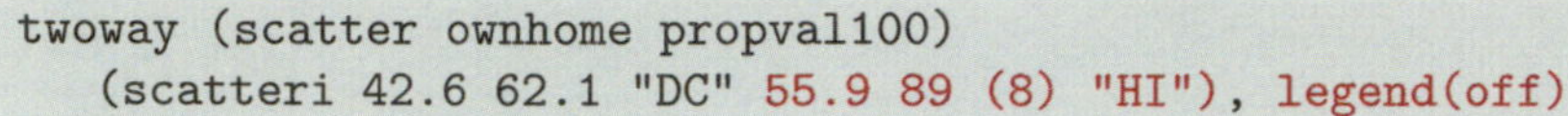

```
twoway (scatter ownhome propval100)
   (scatteri 42.6 62.1 "DC" 55.9 89 (8) "HI"), legend(off)
```

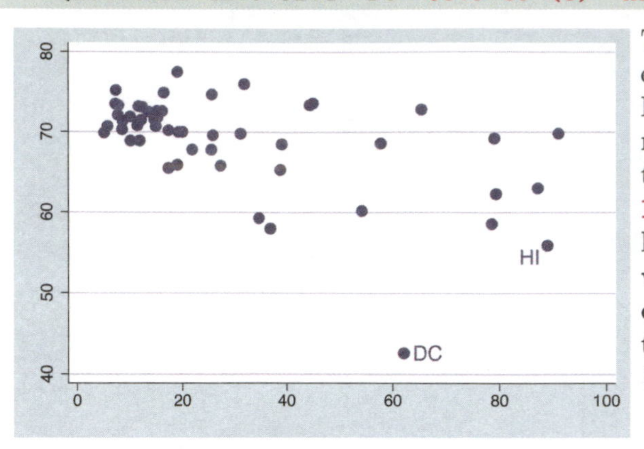

This graph extends the previous example by adding a second point for Hawaii and providing a position for the marker label for Hawaii, placing it at the 8 o'clock position. In addition, the `legend(off)` option suppresses the legend. Finally, this graph uses the `vg_samec` scheme, so the markers created via `scatteri` look identical to the other markers.
Uses allstates.dta & scheme vg_samec

This section concludes by looking at some additional graph commands that make graphs similar to `twoway scatter`, namely, `twoway spike`, `twoway dropline`, and `twoway dot`. Most of the options we have illustrated before apply to these graphs as well, so they will not be repeated here. We will switch to using the `vg_blue` scheme for the rest of the graphs in this section.

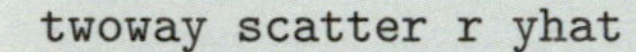

```
twoway scatter r yhat
```

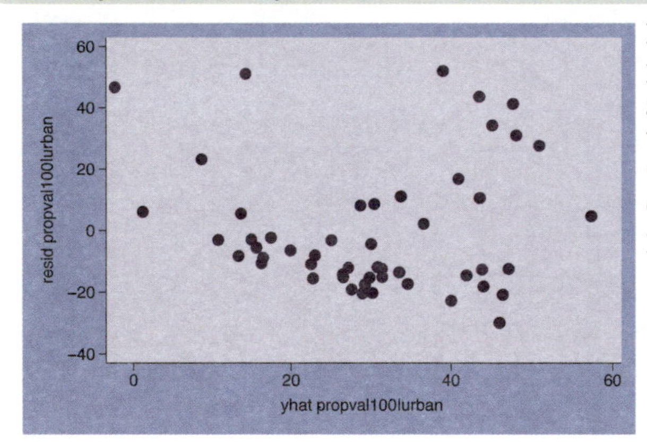

Imagine that we ran a regression predicting `propval100` from `urban` and generated the residual, calling it `r`, and the predicted value, calling it `yhat`. Consider this graph using the `scatter` command to display the residual by the predicted value.
Uses allstates.dta & scheme vg_blue

```
twoway spike r yhat
```

This same graph could be shown using the **spike** command. This produces a spike plot, and each spike, by default, originates from 0.
Uses allstates.dta & scheme vg_blue

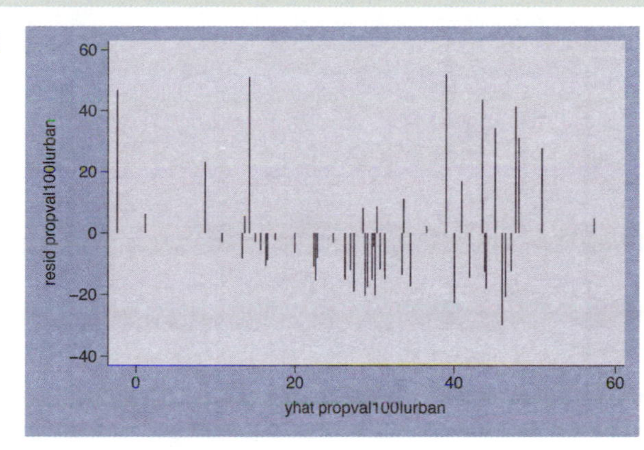


```
twoway spike r yhat, blcolor(navy) blwidth(thick)
```

You can use the **blcolor()** (bar line color) option to set the color of the spikes and the **blwidth()** (bar line width) option to set the width of the spikes. Here, we make the spikes thick and navy. See Styles : Colors (328) for more details about specifying colors and see Styles : Linewidth (337) for more details about specifying line widths.
Uses allstates.dta & scheme vg_blue

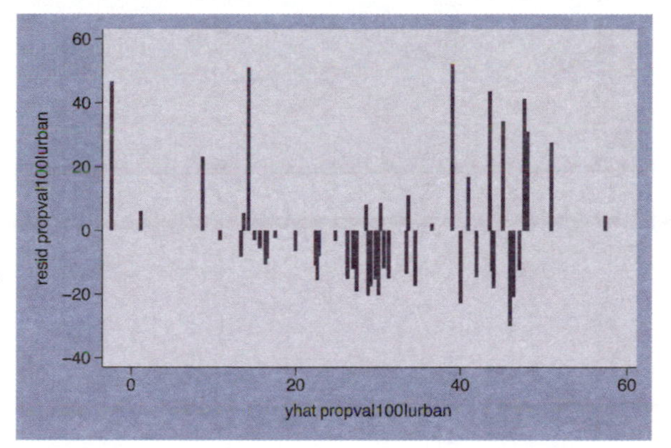


```
twoway spike r yhat, base(10)
```

By default, the base is placed at 0, which is a very logical choice when displaying residuals since our interest is in deviations from 0. For illustration, we use the **base(10)** option to set the base of the y-axis to be 10, and the spikes are displayed with respect to 10.
Uses allstates.dta & scheme vg_blue

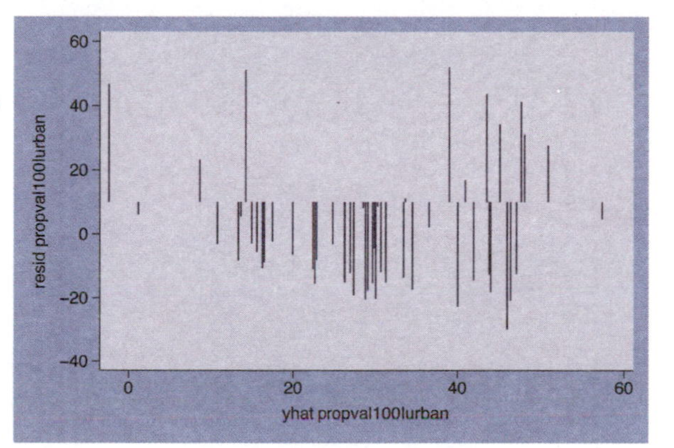

```
twoway spike r yhat, horizontal xtitle(Title for x-axis)
   ytitle(Title for y-axis)
```

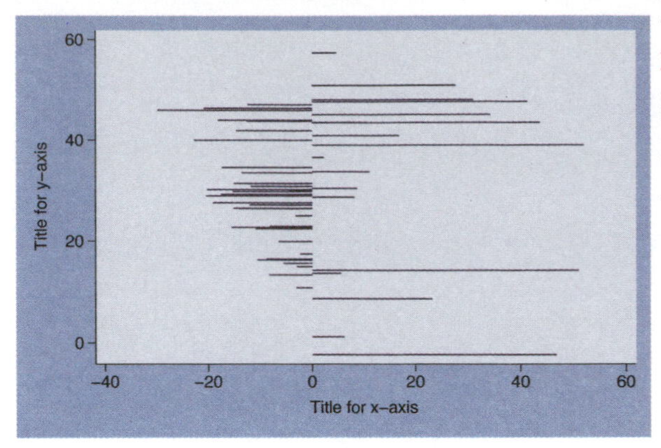

The **spike** command supports the **horizontal** option, which swaps the position of the **r** and **yhat** variables. Note that the x-axis still remains at the bottom and the y-axis still remains at the left.
Uses allstates.dta & scheme vg_blue

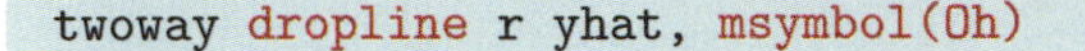

```
twoway dropline r yhat, msymbol(Oh)
```

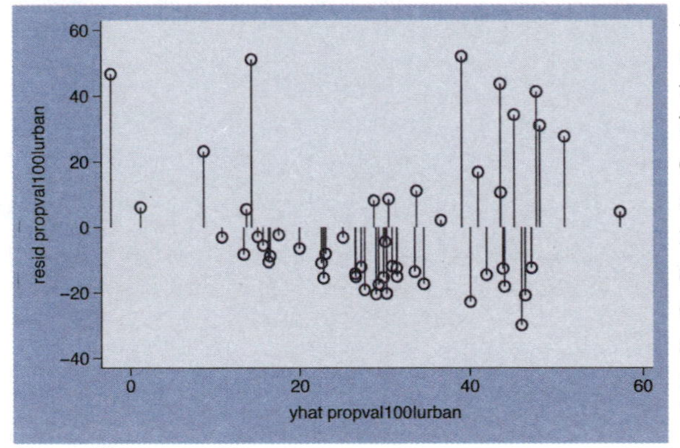

A **twoway dropline** plot is much like a **spike** plot but permits a symbol, as well. It supports the **horizontal**, **base()**, **blcolor()**, and **blwidth()** options just like **twoway spike**, so these are not illustrated. But you can use marker symbol options to control the symbol. Here, we add the **msymbol(Oh)** option to obtain hollow circles as the symbols; see Options : Markers (235) for more details.
Uses allstates.dta & scheme vg_blue

```
twoway dropline r yhat, msymbol(O) msize(vlarge)
   mfcolor(gold) mlcolor(olive) mlwidth(thick)
```

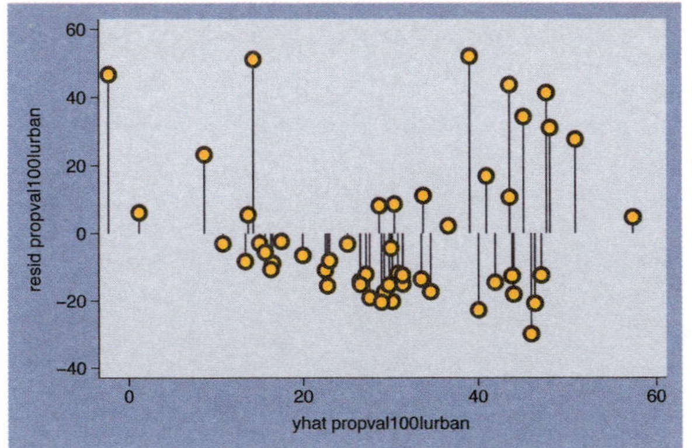

Here, we make the symbols very large circles and use **mfcolor()** to make the marker fill color gold, **mlcolor()** to make the marker line color olive, and **mlwidth()** to make the marker line width thick. For more information, see Options : Markers (235).
Uses allstates.dta & scheme vg_blue

```
twoway dot close tradeday, msize(large) msymbol(O)
   mfcolor(eltgreen) mlcolor(emerald) mlwidth(thick)
```

The dot command is similar to a scatterplot but shows dotted lines for each value of the x-variable, making it more useful when the x-values are equally spaced. In this example, we look at the closing price of the S&P 500 by trading day and make the markers filled with eltgreen with thick emerald outlines.
Uses spjanfeb2001.dta & scheme vg_blue

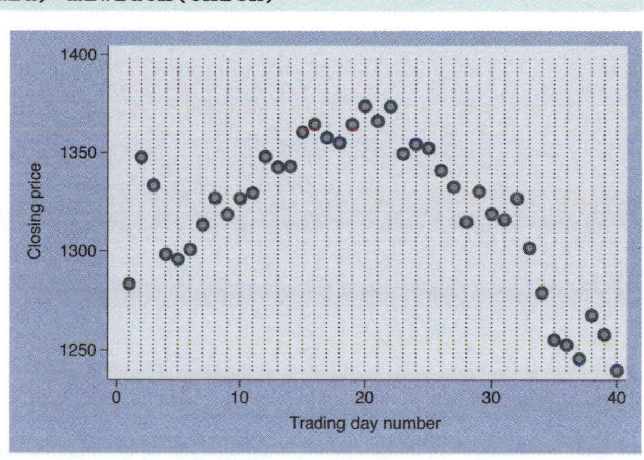

2.2 Regression fits and splines

This section focuses on the twoway commands that are used for displaying fit values: `lfit`, `qfit`, `fpfit`, `mband`, `mspline`, and `lowess`. For more information, see [G] **graph twoway lfit**, [G] **graph twoway qfit**, [G] **graph twoway fpfit**, [G] **graph twoway mband**, [G] **graph twoway mspline**, and [G] **graph twoway lowess**. We use the `allstates` data file, omitting Washington, DC, and show the graphs using the `vg_s2c` scheme.

```
twoway (scatter ownhome pcturban80) (lfit ownhome pcturban80)
```

Here, we show a scatterplot of `ownhome` by `pcturban80`. In addition, we overlay a linear fit `lfit` predicting `ownhome` from `pcturban80`. See Twoway : Overlaying (87) if you would like more information about overlaying twoway graphs.
Uses allstatesdc.dta & scheme vg_s2c

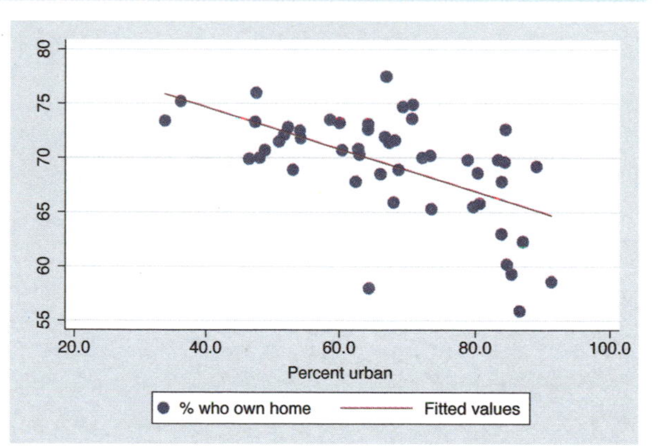

```
twoway (scatter ownhome pcturban80) (lfit ownhome pcturban80)
   (qfit ownhome pcturban80)
```

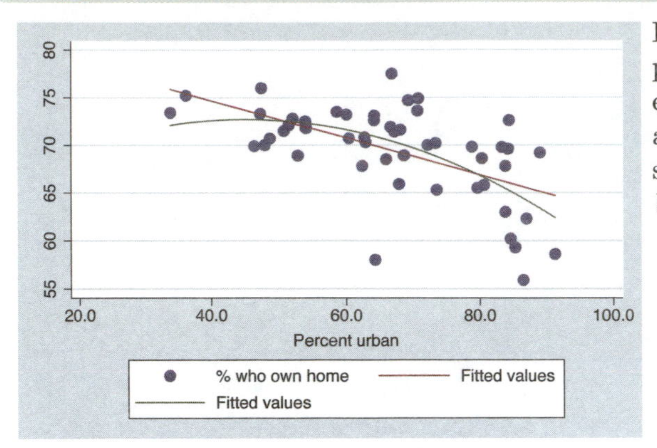

It is sometimes useful to overlay fit plots to compare the fit values. In this example, we overlay a linear fit `lfit` and quadratic fit `qfit` and can see some discrepancies between them.
Uses allstatesdc.dta & scheme vg_s2c

```
twoway (scatter ownhome pcturban80) (mspline ownhome pcturban80)
   (fpfit ownhome pcturban80) (lowess ownhome pcturban80)
```

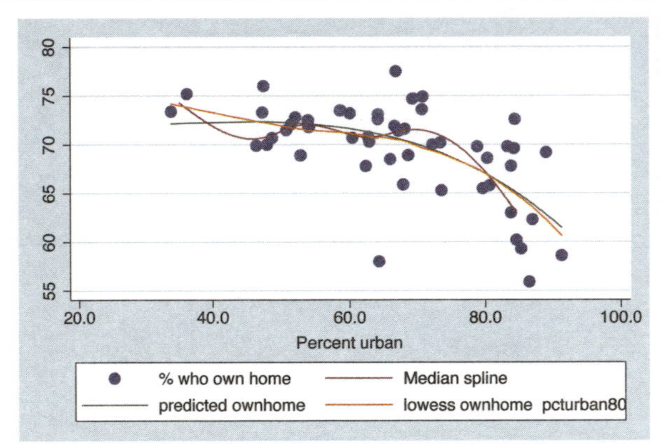

Stata supports a number of other fit methods. Here, we show an `mspline` (median spline) overlaid with `fpfit` (fractional polynomial fit) and `lowess`. Another similar command, not shown, is `mband` (median band).
Uses allstatesdc.dta & scheme vg_s2c

2.3 Regression confidence interval (CI) fits

This section focuses on the twoway commands that are used for displaying confidence intervals around fit values: `lfitci`, `qfitci`, and `fpfitci`. The options permitted by these three commands are virtually identical so we will use `lfitci` to illustrate these options. (Note, however, that `fpfitci` does not permit the options `stdp`, `stdf`, and `stdr`.) For more information, see [G] **graph twoway lfitci**, [G] **graph twoway qfitci**, and [G] **graph twoway fpfitci**.

```
twoway (lfitci ownhome pcturban80) (scatter ownhome pcturban80)
```

This graph uses the `lfitci` command to produce a linear fit with confidence interval. The confidence interval, by default, is computed using the standard error of prediction. We overlay this with a scatterplot.
Uses allstatesdc.dta & scheme vg_rose

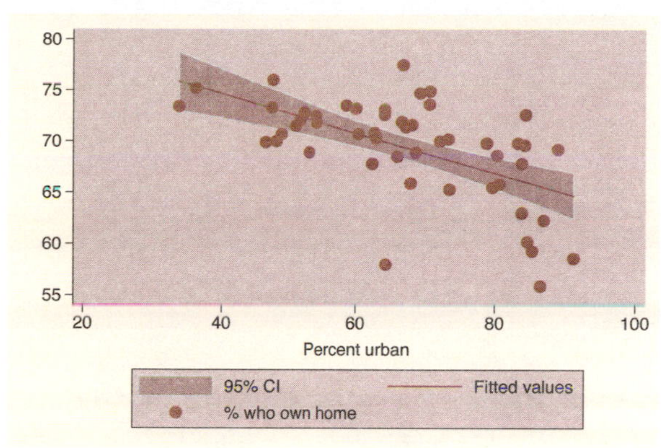


```
twoway (scatter ownhome pcturban80) (lfitci ownhome pcturban80)
```

This example is the same as the previous example; however, the order of the `scatter` and `lfitci` commands is reversed. Note that the order matters since the points that fell within the confidence interval are not displayed because they are masked by the shading of the confidence interval.
Uses allstatesdc.dta & scheme vg_rose

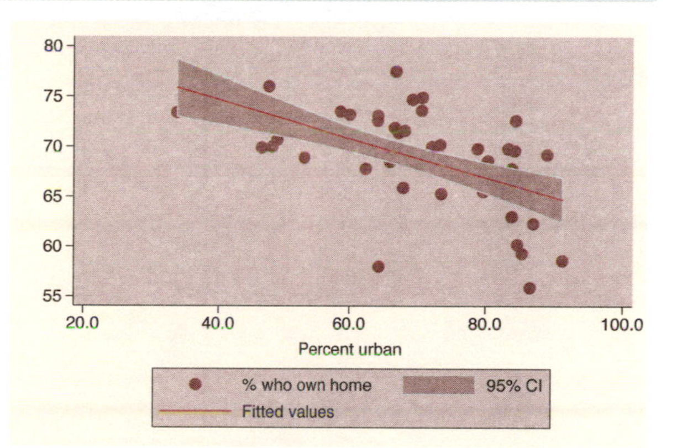


```
twoway (lfitci ownhome pcturban80, stdf)
   (scatter ownhome pcturban80)
```

Here, we add the `stdf` option, which computes the confidence intervals using the standard error of forecast. If samples were drawn repeatedly, this confidence interval would capture 95% of the observations. With 50 observations, we would expect 2 or 3 observations to fall outside of the confidence interval, and this corresponds to the data shown here.
Uses allstatesdc.dta & scheme vg_rose

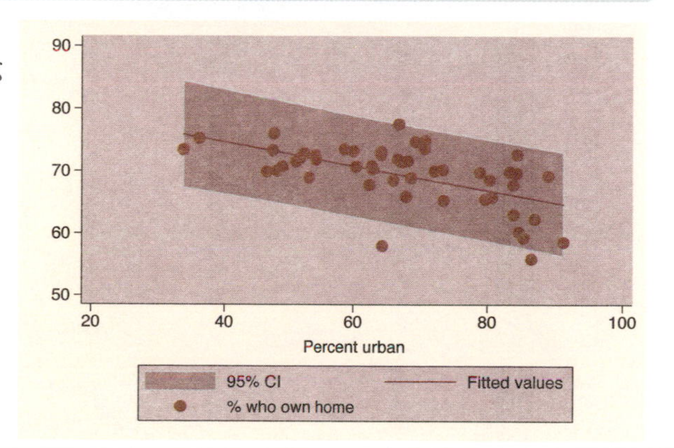

```
twoway (lfitci ownhome pcturban80, stdf level(90))
   (scatter ownhome pcturban80)
```

We can use the `level()` option to set the confidence level for the confidence interval. Here, we make the confidence level 90%.
Uses allstatesdc.dta & scheme vg_rose

```
twoway (lfitci ownhome pcturban80, nofit)
   (scatter ownhome pcturban80)
```

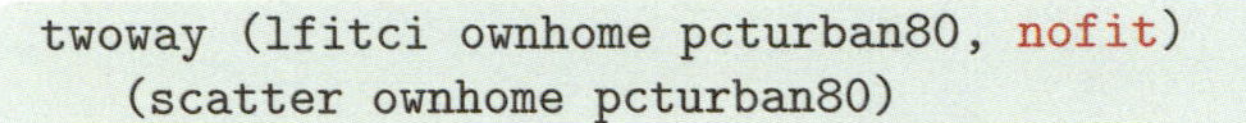

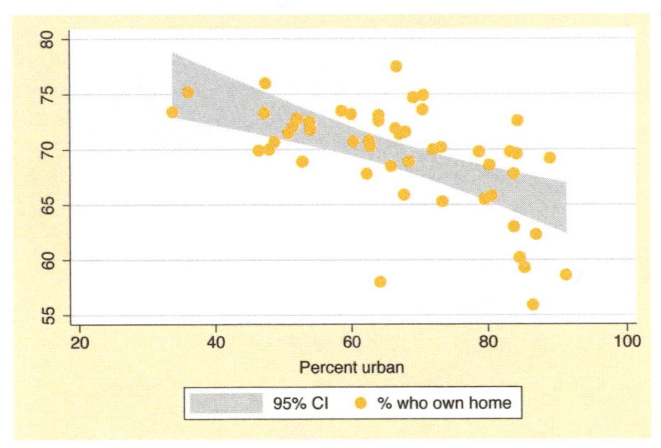

We now look at how you can control the display of the fit line. We can use the `nofit` option to suppress the display of the fit line. Note that we have switched to the **vg_brite** scheme for a different look for the graphs.
Uses allstatesdc.dta & scheme vg_brite

```
twoway (lfitci ownhome pcturban80, clpattern(dash) clwidth(thick))
   (scatter ownhome pcturban80)
```

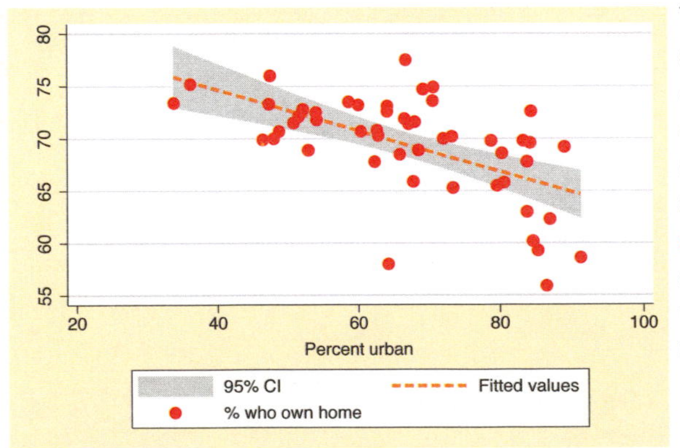

You can supply options like `connect()`, `clpattern()` (connect line pattern), `clwidth()` (connect line width), and `clcolor()` (connect line color) to control how the fit line will be displayed. Here, we use the `clpattern(dash)` and `clwidth(thick)` options to make the fit line dashed and thick. See Options : Connecting (250) for more details.
Uses allstatesdc.dta & scheme vg_brite

```
twoway (lfitci ownhome pcturban80, bcolor(stone))
   (scatter ownhome pcturban80)
```

We use the bcolor(stone) option to change the color of the area and outline of the confidence interval. You can use the options illustrated with `twoway rarea` to control the display of the area encompassing the confidence interval, namely, `bcolor()`, `bfcolor()`, `blcolor()`, `blwidth()`, and `blpattern()`. See Twoway : Range (64) and [G] **graph twoway rarea** for more details.
Uses allstatesdc.dta & scheme vg_brite

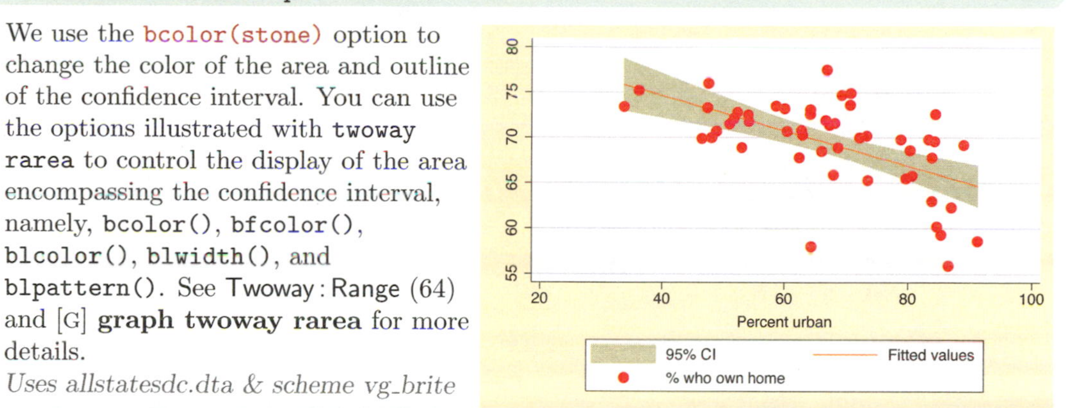


```
twoway (lfitci ownhome pcturban80, ciplot(rline))
   (scatter ownhome pcturban80)
```

The `ciplot()` option can be used to select a different command for displaying the confidence interval. The default command is `twoway rarea` and can be selected via the `ciplot(rarea)` option. Here, we use the `ciplot(rline)` option, which displays the confidence interval as two lines without any filled area. The valid options include `rarea`, `rbar`, `rspike`, `rcap`, `rcapsym`, `rscatter`, `rline`, and `rconnected`.

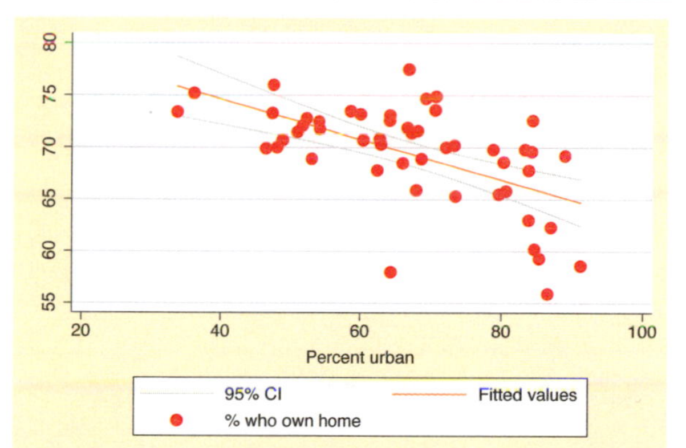

Uses allstatesdc.dta & scheme vg_brite

```
twoway
   (lfitci ownhome pcturban80, ciplot(rline) blcolor(green) blpattern(dash)
   blwidth(thick)) (scatter ownhome pcturban80)
```

By choosing the `rline` command for displaying the confidence interval, we can then use options appropriate for the `twoway rline` command. Here, we make the line green, dashed, and thick. See Styles : Colors (328), Styles : Linepatterns (336), and Styles : Linewidth (337) for more details about colors, line patterns, and line widths.
Uses allstatesdc.dta & scheme vg_brite

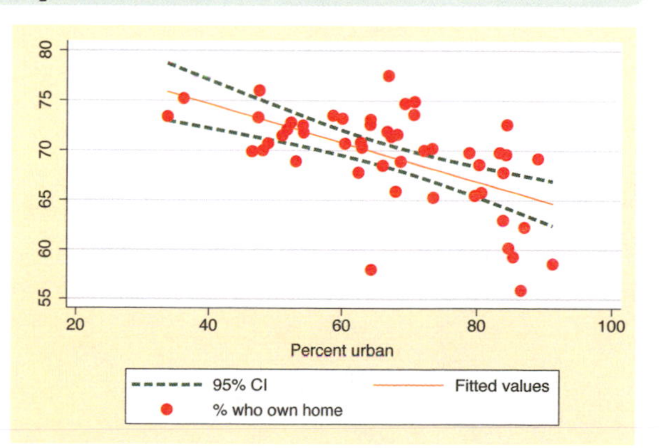

2.4 Line plots

This section focuses on the twoway commands for creating line plots, including the **twoway line** and **twoway connected** commands. The **line** command is the same as **scatter**, except that the points are connected by default and marker symbols are not permitted, whereas the **twoway connected** command permits marker symbols. This section also illustrates **twoway tsline** and **twoway tsrline**, which are useful for drawing line plots when the x-variable is a date variable. Since all these commands are related to the **twoway scatter** command, they support most of the options you would use with **twoway scatter**. For more information, see [G] **graph twoway line**, [G] **graph twoway connected**, and **help graph_tsline**.

```
twoway line close tradeday, sort
```

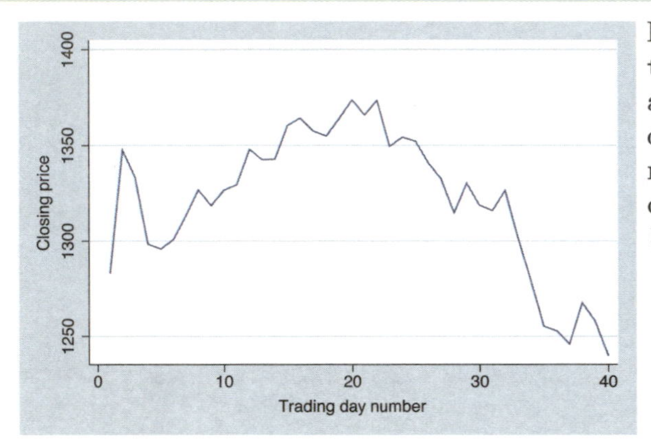

Here, we show an example using **twoway line** showing the closing price across trading days. Note the inclusion of the sort option, which is recommended when you have points connected in a Stata graph.
Uses spjanfeb2001.dta & scheme vg_s2c

```
twoway line close tradeday, sort clwidth(vthick) clcolor(maroon)
```

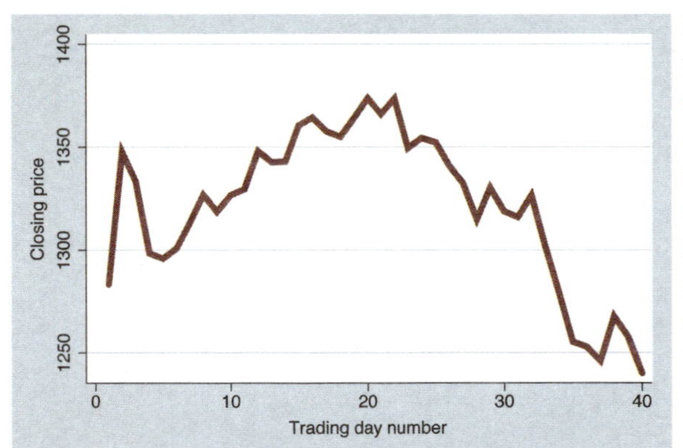

Here, we show options controlling the width and color of the lines. Using clwidth(vthick) (connect line width) and clcolor(maroon) (connect line color), we make the line very thick and maroon. See Options : Connecting (250) for more examples. Note that you cannot use options that control marker symbols with **graph twoway line**.
Uses spjanfeb2001.dta & scheme vg_s2c

```
twoway connected close tradeday, sort
```

This `twoway connected` graph is similar to the `twoway line` graphs we saw before, except that when you use `connected`, a marker is shown for each data point.
Uses spjanfeb2001.dta & scheme vg_s2c

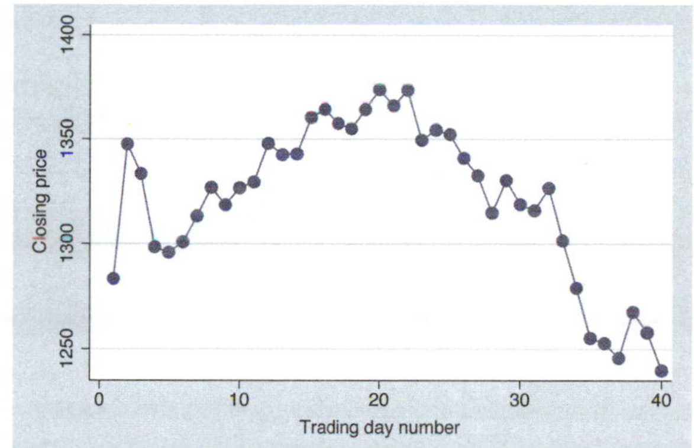


```
twoway scatter close tradeday, connect(l) sort
```

This graph is identical to the previous graph, except this graph is made with the `scatter` command using the `connect(l)` option. This illustrates the convenience of using `connected` since you do not need to manually specify the `connect()` option.
Uses spjanfeb2001.dta & scheme vg_s2c

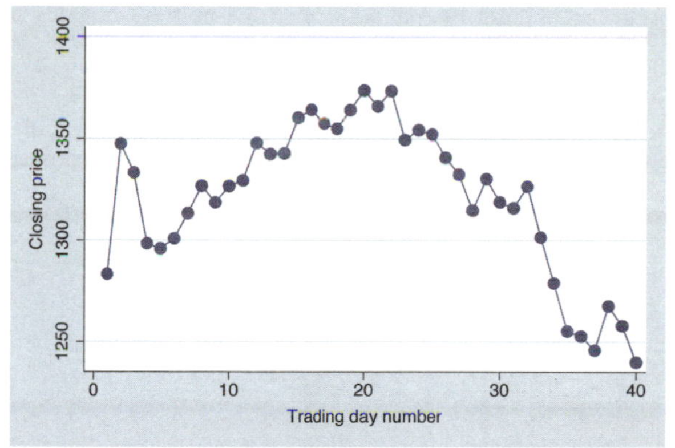


```
twoway connected close tradeday, sort
   msymbol(Dh) mcolor(blue) msize(large)
```

We can use marker symbol options, such as `msymbol()`, `mcolor()`, and `msize()` to control the marker symbols. Here, we make the symbols large, blue, hollow diamonds. See Options : Markers (235) for more examples.
Uses spjanfeb2001.dta & scheme vg_s2c

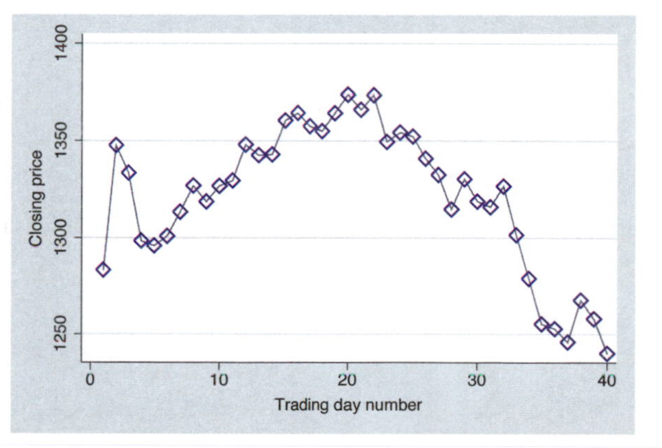

```
twoway connected close tradeday, sort
   clcolor(cranberry) clpattern(dash) clwidth(thick)
```

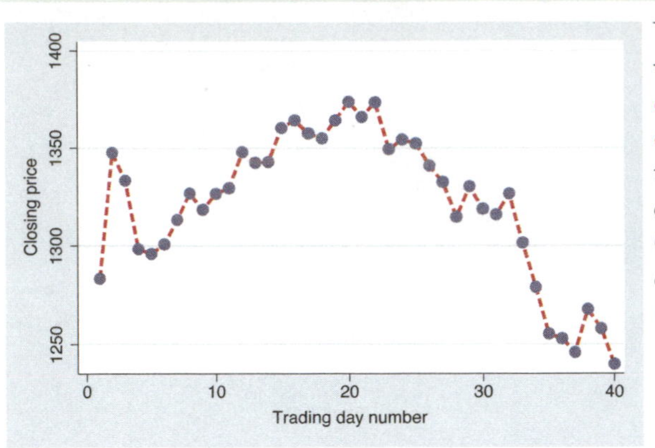

You can control the look of the lines with connect options such as `clwidth()`, `clcolor()`, and `clpattern()` (connect line pattern). In this example, we make the line cranberry, dashed, and thick. See Options : Connecting (250) for more details on connecting points.
Uses spjanfeb2001.dta & scheme vg_s2c

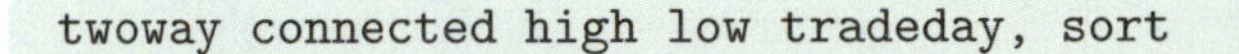

```
twoway connected high low tradeday, sort
```

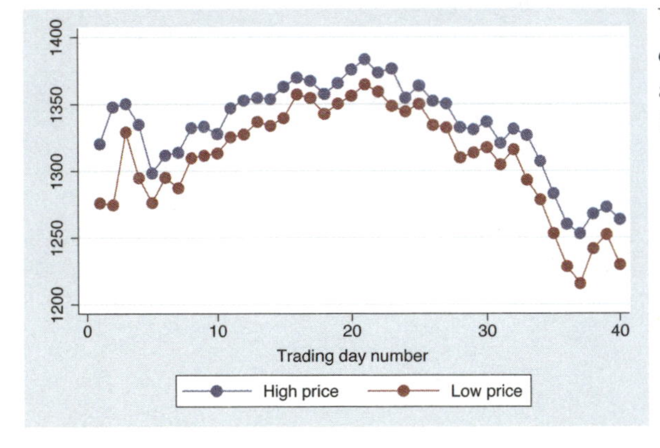

You can graph multiple variables at once. In this case, we graph the high and low prices across trading days.
Uses spjanfeb2001.dta & scheme vg_s2c

```
twoway connected high low tradeday, sort
   clwidth(thin thick) msymbol(Oh S)
```

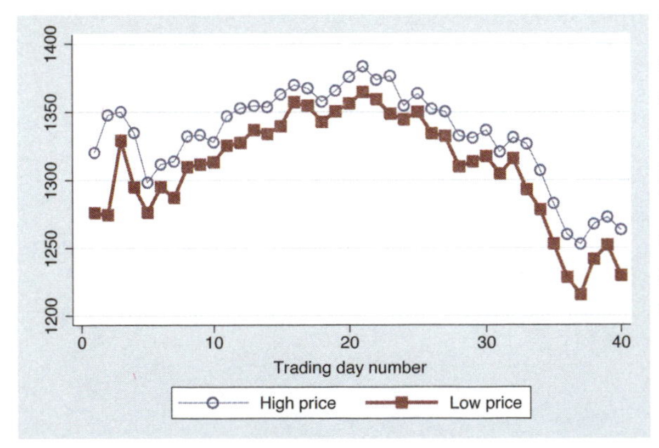

When graphing multiple variables, you can specify connect and marker symbol options to control each line. In this case, we use a thin line for the high price and a thick line for the low price. We also differentiate the two lines by using different marker symbols, hollow circles for the high price and squares for the low price.
Uses spjanfeb2001.dta & scheme vg_s2c

Stata has additional commands for creating line plots where the x-variable is a date variable, namely, **twoway tsline** and **twoway tsrline**. The **tsline** command is similar to the **line** command, and the **tsrline** is similar to the **rline** command, but both of these **ts** commands offer extra features, making it easier to reference the x-variable in terms of dates. Note that these commands are not currently documented in [G] **graph** but are documented via **help tsline**. We will use the **sp2001ts** data file, which has the prices for the S&P 500 index for 2001 with the trading date stored as a date variable named **date**. Before saving the file **sp2001ts**, the **tsset date, daily** command was used to tell Stata that the variable **date** represents the time variable and that it represents daily data.

```
twoway tsline close
```

The `tsline` (time-series line) graph shows the closing price on the y-axis and the date on the x-axis. Note that we did not specify the x-variable in the graph command. Stata knew the variable representing time because we previously issued the **tsset date, daily** command before saving the **sp2001ts** file. Note that if you save the data file, Stata remembers the time variable, and you do not need to set it again.
Uses sp2001ts.dta & scheme vg_s1c

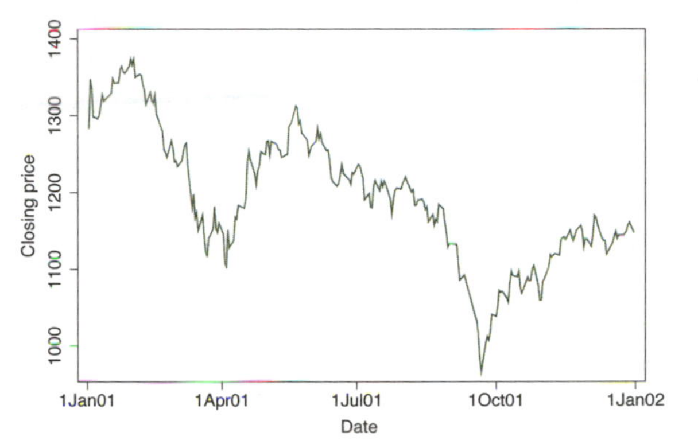


```
twoway tsrline low high
```

We can also use the `tsrline` (time-series range) graph to show the low price and high price for each day.
Uses sp2001ts.dta & scheme vg_s1c

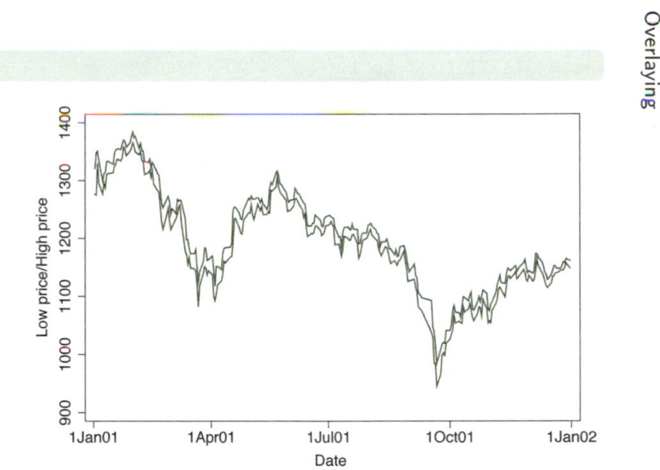

```
twoway tsline close, clwidth(thick) clcolor(navy)
```

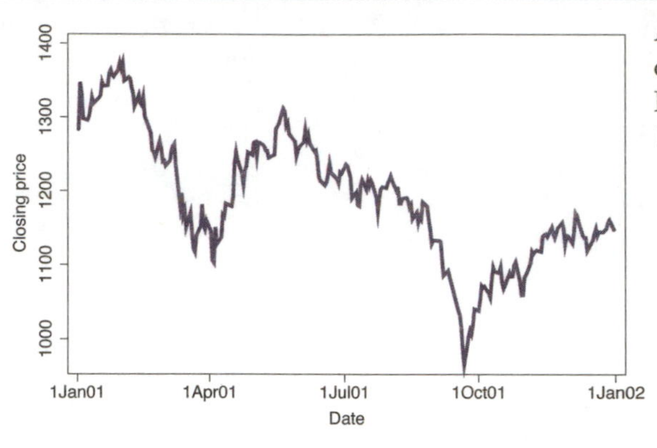

As with `twoway line`, you can use connect options to control the line. Here, we make the line thick and navy.
Uses sp2001ts.dta & scheme vg_s1c

```
twoway tsline close
   if (date >= mdy(1,1,2001)) & (date <= mdy(3,31,2001))
```

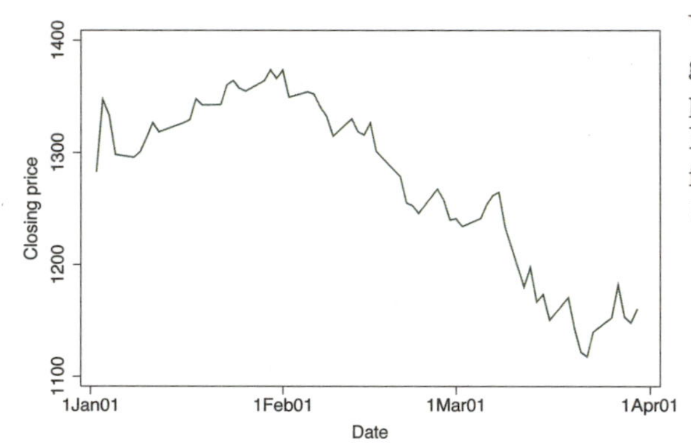

You can use `if` to subset cases to graph. Here, we graph the closing prices between January 1, 2001, and March 31, 2001. See the next example for an easier way of doing this.
Uses sp2001ts.dta & scheme vg_s1c

```
twoway tsline close if tin(01jan2001,31mar2001)
```

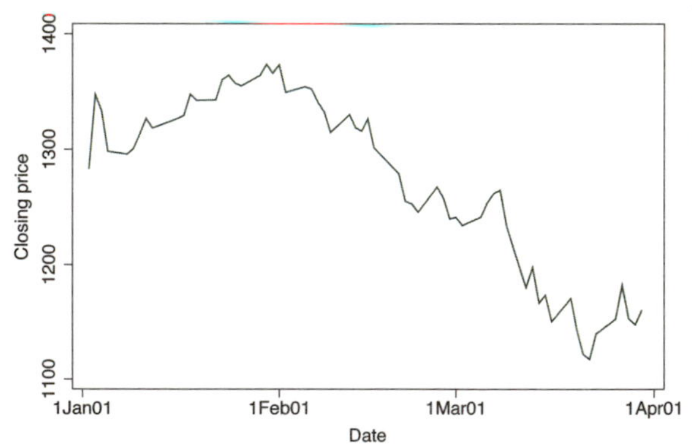

When using the `tsline` command, you can use `tin()` (time in between) to specify that you want to graph just the cases between January 1, 2001, and March 31, 2001, inclusively.
Uses sp2001ts.dta & scheme vg_s1c

```
twoway tsline close, ttitle(Day of Year)
```

We can use the `ttitle()` (time title) option to give a title to the time variable. We specify this as a `ttitle()` instead of `xtitle()` since this refers to the axis with the time variable.
Uses sp2001ts.dta & scheme vg_s1c

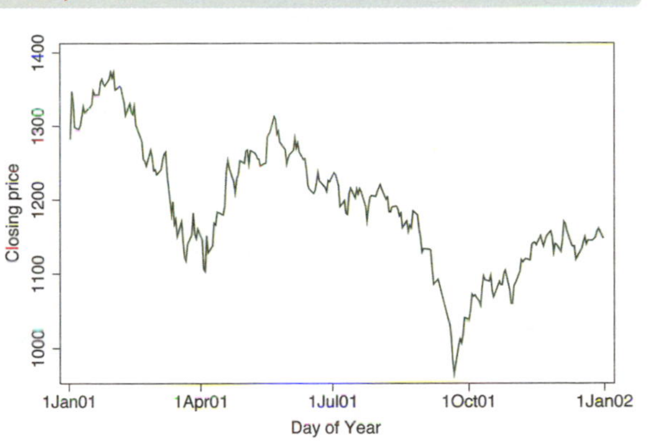


```
twoway tsline close,
   tlabel(01jan2001 31mar2001 30jun2001 30sep2001 01jan2002)
```

We can use the `tlabel()` option to label the time points on the time axis. Note that we specified these dates using date literals, and Stata knew how to interpret these and appropriately label the graph with these values.
Uses sp2001ts.dta & scheme vg_s1c

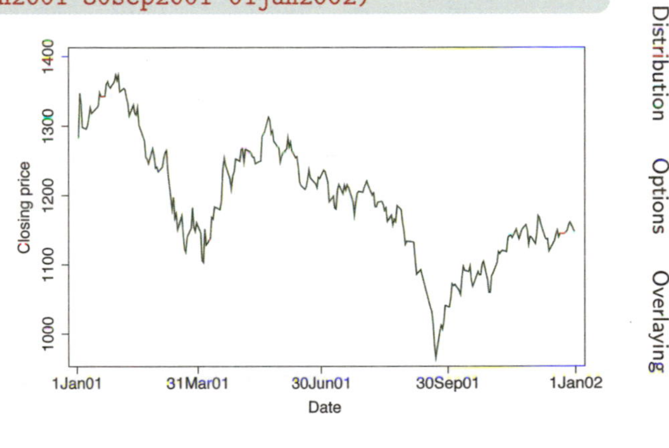


```
twoway tsline close,
   tlabel(01jan2001 30jun2001 01jan2002 ) tmlabel(31mar2001 30sep2001)
```

We can use the `tmlabel()` option to include minor labels.
Uses sp2001ts.dta & scheme vg_s1c

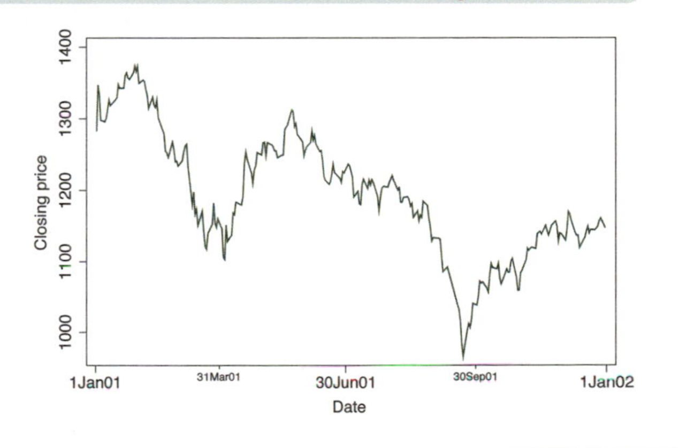

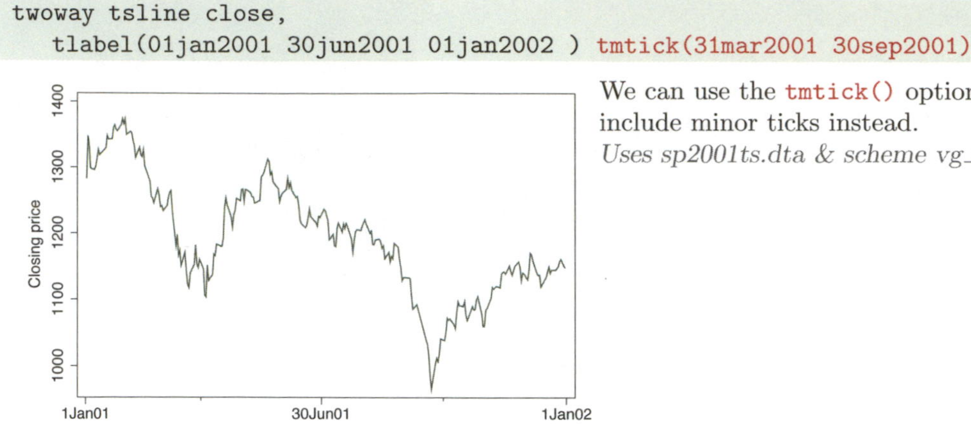

```
twoway tsline close,
  tlabel(01jan2001 30jun2001 01jan2002 ) tmtick(31mar2001 30sep2001)
```


We can use the `tmtick()` option to include minor ticks instead.
Uses sp2001ts.dta & scheme vg_s1c

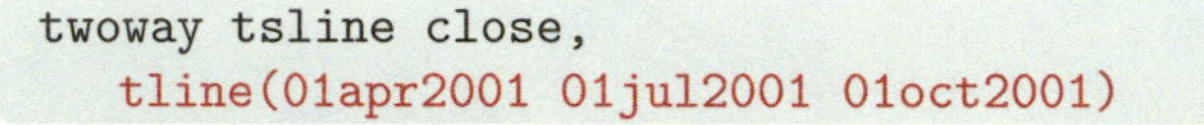

```
twoway tsline close,
  tline(01apr2001 01jul2001 01oct2001)
```

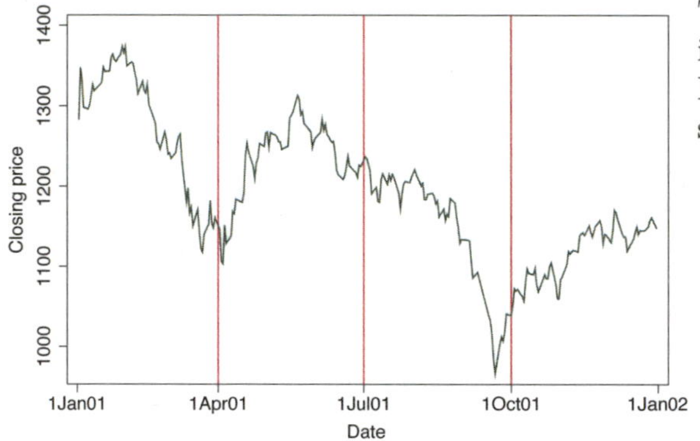

The `tline()` option can be used to include lines at certain time points. Here, we place lines at the start of the second, third, and fourth quarters.
Uses sp2001ts.dta & scheme vg_s1c

```
twoway tsline close,
  ttext(1035 01apr2001 "Start of Q2", orientation(vertical))
```

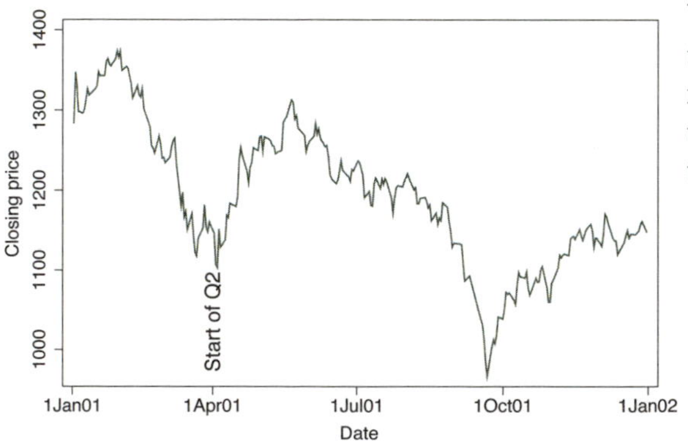

We can use the `ttext()` option to add text to the graph. The first coordinate refers to the position on the y-axis, and the second coordinate is the position on the time axis in terms of the date.
Uses sp2001ts.dta & scheme vg_s1c

2.5 Area plots

This section illustrates the use of area graphs using **twoway area**. These graphs are similar to **twoway line** graphs, except that the area under the line is shaded. As a result, many of the options that you would use with **twoway line** are applicable; see Twoway : Line (54) for more details. For even more details, see [G] **graph twoway area**. We will use the **spjanfeb2001** data file, which has the prices for the S&P 500 index for January and February 2001.

```
twoway area close tradeday, sort
```

This is an example of a **twoway area** graph. Because this graph is composed of connected points, the **sort** option is recommended in case the data are not already sorted by **tradeday**. If the data are not sorted, and the **sort** option is not specified, then the points are connected in the order they appear in the data file and will generally not be the graph you desire.
Uses spjanfeb2001.dta & scheme vg_palec

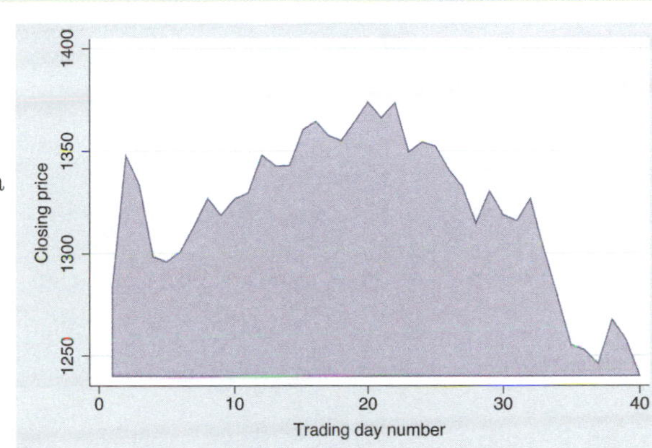


```
twoway area close tradeday, horizontal sort
   xtitle(Title for x-axis) ytitle(Title for y-axis)
```

The **horizontal** option swaps the position of the **close** and **tradeday** variables. Note that the x-axis remains at the bottom and the y-axis remains at the left.
Uses spjanfeb2001.dta & scheme vg_palec

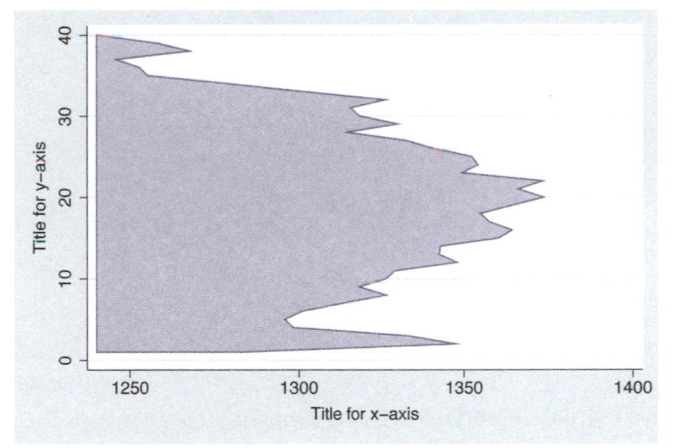

```
twoway area close tradeday, sort base(1320.28)
```

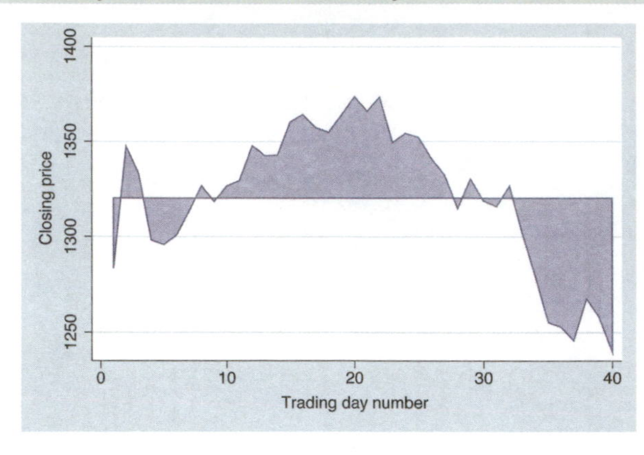

You can use the `base()` option to indicate a base from which the area is to be shaded. In this example, the base is the closing price on the first trading day, and thus all the subsequent points are a kind of deviation from the first day's closing price.
Uses spjanfeb2001.dta & scheme vg_palec

```
twoway area close tradeday, sort bcolor(emerald)
```

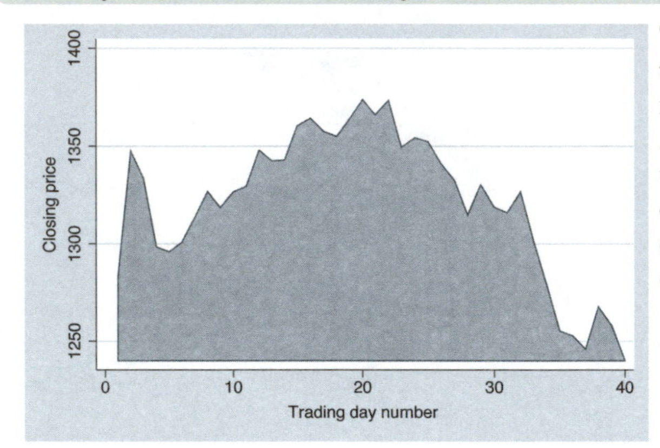

The `bcolor()` option sets the color of the shaded area and the line. Here, we make the shaded area and line emerald. Although it is not shown, you can also use the `bfcolor()` and `blcolor()` options to control the fill color and line color and the `blwidth()` option to control the thickness of the outline.
Uses spjanfeb2001.dta & scheme vg_palec

2.6 Bar plots

This section illustrates the use of twoway bar graphs using `twoway bar`. These graphs show a bar for each x-value where the height of the bar corresponds to the value of the y-variable. For more details, see [G] **graph twoway bar**. We will continue to use the `spjanfeb2001` data file, which has the prices for the S&P 500 index for January and February, 2001, but show the graphs using the `vg_s1m` scheme. `twoway bar` is useful for creating bar graphs with overlays of lines, points, or other plot types and can be useful with evenly spaced x-variable data. `graph bar` is more useful for creating bar graphs with categorical data.

```
twoway bar close tradeday
```

Consider this bar chart, which shows the closing prices of the S&P 500 broken down by the trading day of the year.
Uses spjanfeb2001.dta & scheme vg_s1m

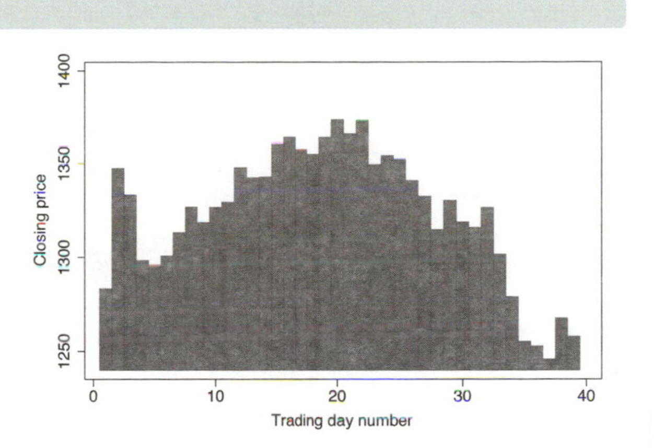


```
twoway bar close tradeday, horizontal
   xtitle(Title for x-axis) ytitle(Title for y-axis)
```

We can make the `close` and `tradeday` variables trade places with the `horizontal` option. Note that the x-axis still remains at the bottom and the y-axis still remains at the left.
Uses spjanfeb2001.dta & scheme vg_s1m

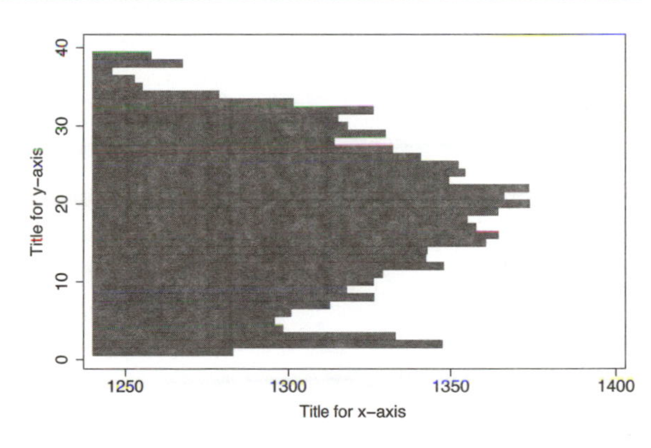


```
twoway bar close tradeday, base(1200)
```

Unless we specify otherwise, the base for the bar charts is the trading day with the lowest price. In this example, the closing price on day 40 was 1239.94, so unless we specify the `base()` option, the base would be 1239.94. As a result, the bar for day 40 would have a zero height. Here, we change the base to 1200 to give this bar a height.
Uses spjanfeb2001.dta & scheme vg_s1m

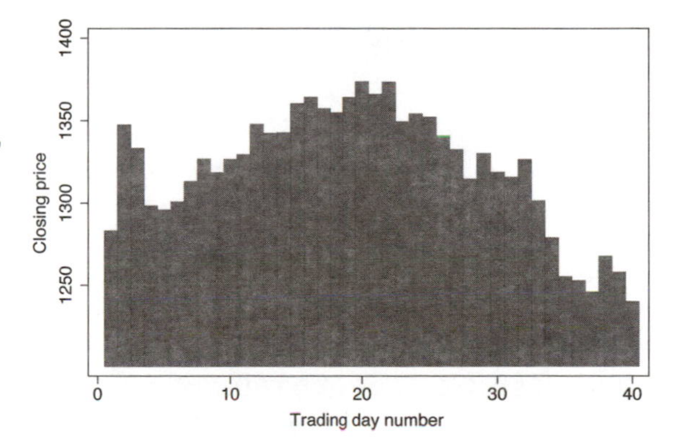

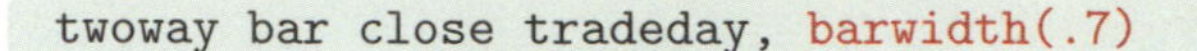

```
twoway bar close tradeday, barwidth(.7)
```

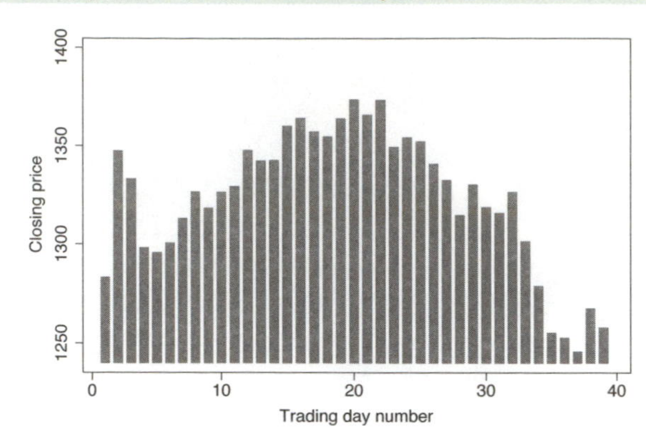

Unless otherwise specified, the width of each bar is one x-unit (in this case, one day). By making the width of the bars .7, we can obtain a small gap between the bars.
Uses spjanfeb2001.dta & scheme vg_s1m

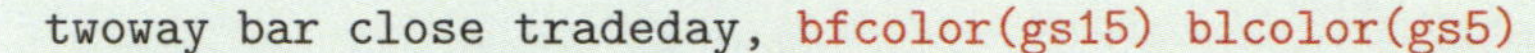

```
twoway bar close tradeday, bfcolor(gs15) blcolor(gs5)
```

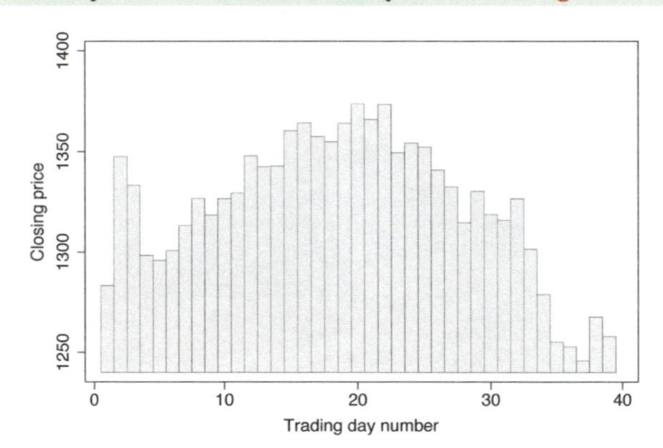

We can use the `bfcolor()` (bar fill color) option to set the color of the inside of the bars and the `blcolor` (bar line color) option to set the color of the bar outlines. Here, we make the bars light gray on the inside and dark gray on the outside. See Styles : Colors (328) for more colors you can choose.
Uses spjanfeb2001.dta & scheme vg_s1m

2.7 Range plots

This section focuses on twoway commands that display range plots. The major characteristic these graphs share is that, for each x-value, there are two corresponding y-values. A common example is a confidence interval where, for each x-value, there are upper and lower confidence limits. We first show examples of all of these types of graphs and then consider the options that can be used to customize them. For more information, see [G] **graph twoway rarea**, [G] **graph twoway rbar**, [G] **graph twoway rspike**, [G] **graph twoway rcap**, [G] **graph twoway rcapsym**, [G] **graph twoway rscatter**, [G] **graph twoway rline**, and [G] **graph twoway rconnected**. We will start by looking at the `rconnected`, `rscatter`, `rline`, and `rarea` graphs, which use combinations of lines, symbols, and shading to display range plots. These examples use the `spjanfeb2001` data file.

```
twoway rconnected high low tradeday, sort
```

The rconnected (range connected) graph shows the high and low prices by tradeday, the number of days stocks have been traded in the year. The rconnected plot shows a separate line for the high and low prices, and a marker appears for each x-value. The sort option is recommended because the points are connected by lines and is needed if the data were not already sorted on tradeday.
Uses spjanfeb2001.dta & scheme vg_rose

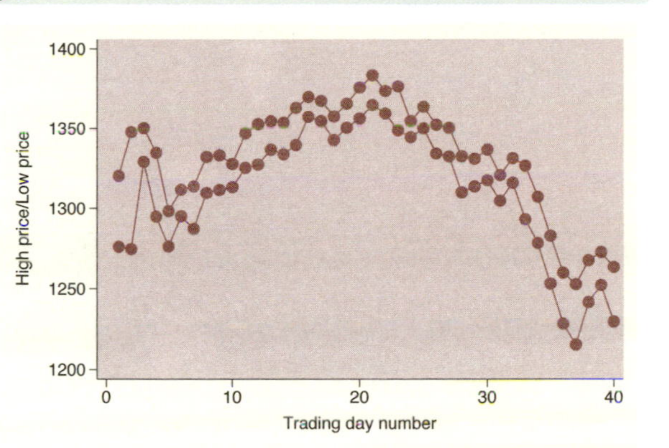


```
twoway rscatter high low tradeday
```

The rscatter graph is similar to the rconnected graph, except that lines connecting the symbols are not plotted.
Uses spjanfeb2001.dta & scheme vg_rose

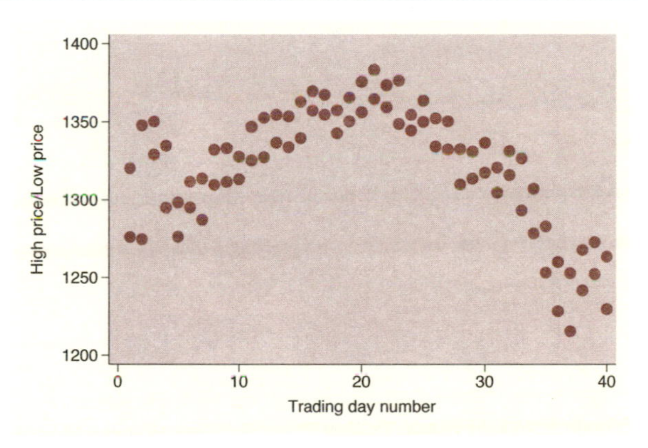


```
twoway rline high low tradeday, sort
```

The rline graph is similar to the rconnected graph, except that symbols are not plotted at each level of x. Note the inclusion of the sort option. This option is recommended because the points are connected by lines and is needed if the data were not already sorted on tradeday.
Uses spjanfeb2001.dta & scheme vg_rose

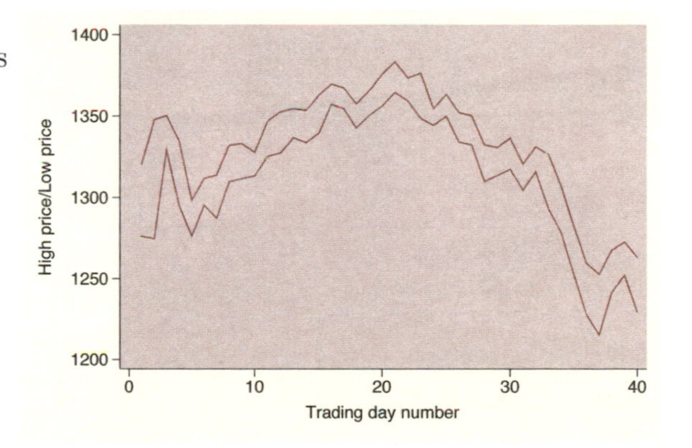

```
twoway rarea high low tradeday, sort
```

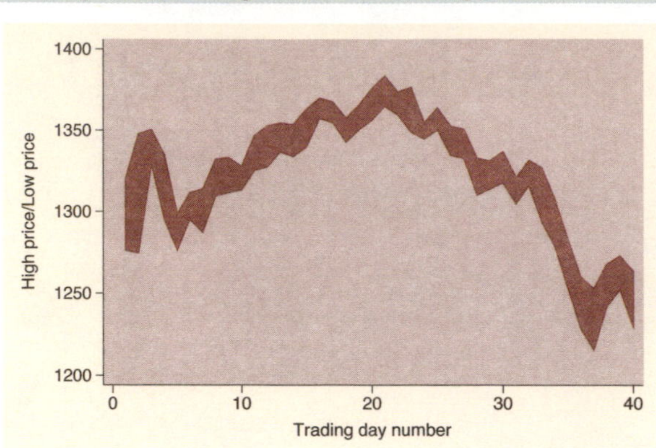

The **rarea** graph is similar to the **rline** graph, except that you can control the fill color of the area between the high and low values.
Uses spjanfeb2001.dta & scheme vg_rose

Next, we discuss the **rcap**, **rspike**, and **rcapsym** graphs, which use combinations of spikes, caps, and symbols to display range plots. These plots are followed by **rbar**, which uses bars to display range plots. These next examples are shown using the **vg_s2m** scheme.

```
twoway rcap high low tradeday
```

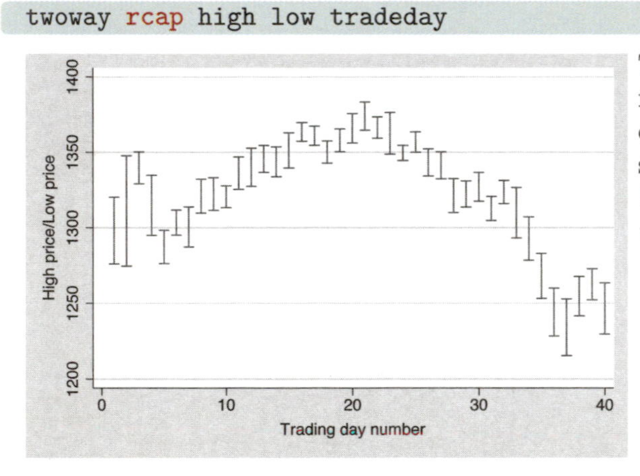

The **rcap** graph shows a spike ranging from the low to high values and puts a cap at the top and bottom of each spike.
Uses spjanfeb2001.dta & scheme vg_s2m

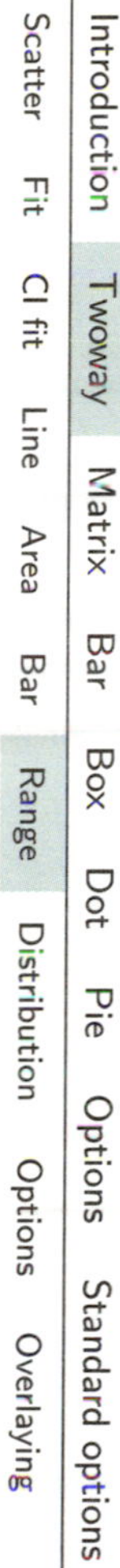

`twoway rspike high low tradeday`

The **rspike** graph is similar to the **rcap** graph, except that no caps are placed on the spikes.
Uses spjanfeb2001.dta & scheme vg_s2m

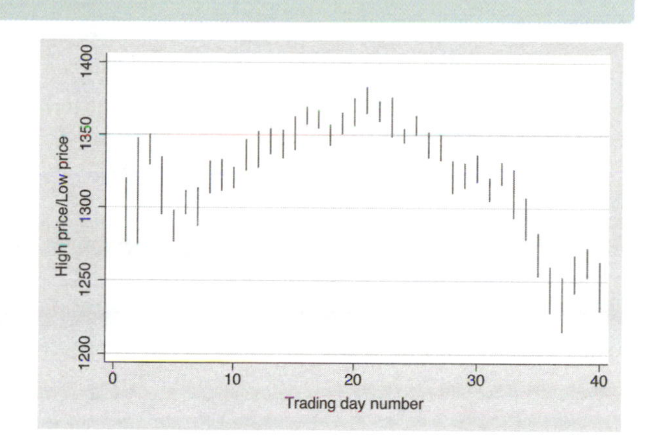

`twoway rcapsym high low tradeday`

The **rcapsym** graph is similar to the **rcap** graph, except that instead of caps, symbols are placed at the end of the spikes. You can choose among the symbols to use for a scatterplot.
Uses spjanfeb2001.dta & scheme vg_s2m

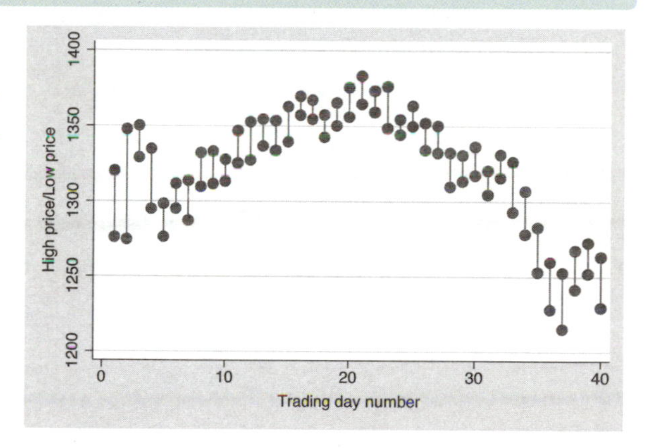

`twoway rbar high low tradeday`

The **rbar** graph uses bars for each value of x to show the high and low values of y.
Uses spjanfeb2001.dta & scheme vg_s2m

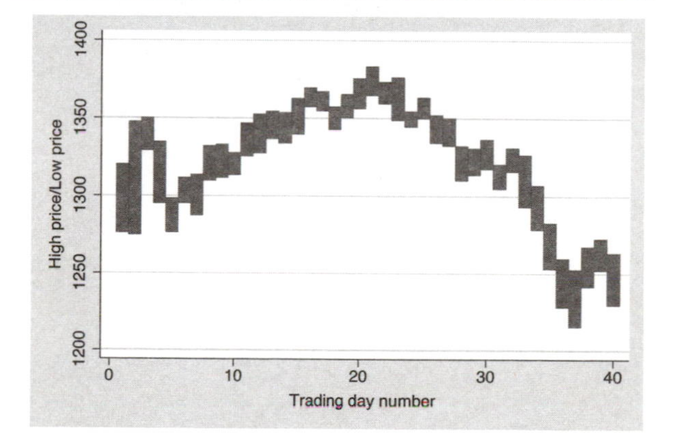

Let's now consider options you can use with the **rconnected**, **rscatter**, **rline**, and **rarea** graphs. We will start by looking at **rconnected** plots since many of the options used in that kind of graph also apply to **rscatter**, **rline**, and **rarea** graphs. These graphs will be shown using the **vg_s1c** scheme.

```
twoway rconnected high low tradeday, sort
```

Here is a general **rconnected** graph.
Uses spjanfeb2001.dta & scheme vg_s1c

```
twoway rconnected high low tradeday, sort horizontal
   xtitle(Title for x-axis) ytitle(Title for y-axis)
```

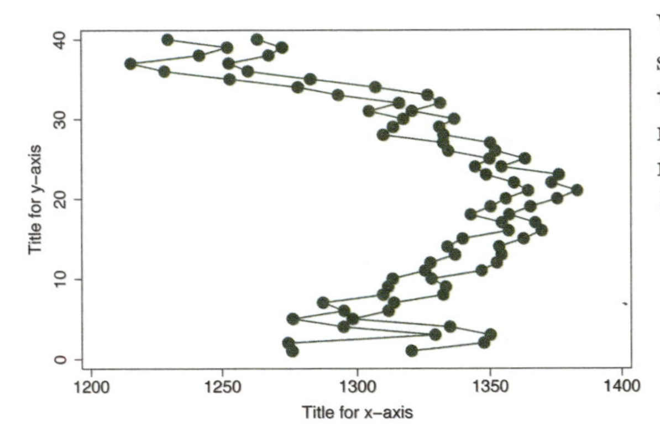

With the **horizontal** option, you can swap the axes where **high**/**low** and **tradeday** appear. Note that the x-axis remains at the bottom and the y-axis remains at the left.
Uses spjanfeb2001.dta & scheme vg_s1c

```
twoway rconnected high low tradeday, sort
   blwidth(thick) blcolor(dkgreen) blpattern(dash)
```

You can control the look of the lines with connect options such as `connect()`, `blwidth()`, `blcolor()`, and `blpattern()`. Here, we make the lines thick, dark green, and dashed. See Options : Connecting (250) for more examples, and see more details in Styles : Connect (332), Styles : Linewidth (337), Styles : Colors (328), and Styles : Linepatterns (336).
Uses spjanfeb2001.dta & scheme vg_s1c

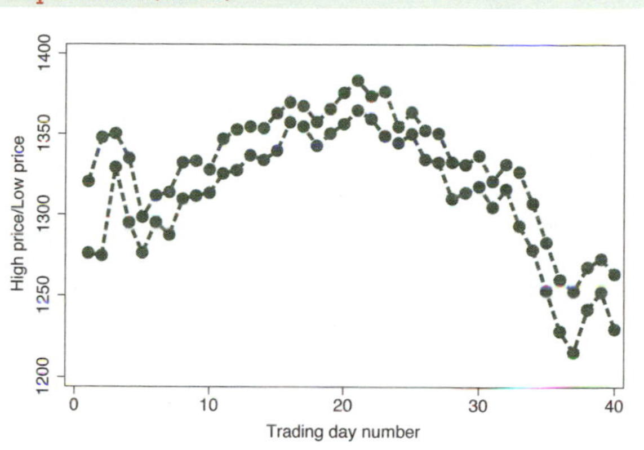


```
twoway rconnected high low tradeday, sort
   msymbol(Oh) msize(large) mcolor(lavender)
```

You can control the look of the marker symbols with options such as `msymbol()`, `msize()`, and `mcolor()`. Here, we make the marker symbols large, hollow, lavender circles. For more details about options related to symbols, see Options : Markers (235).
Uses spjanfeb2001.dta & scheme vg_s1c

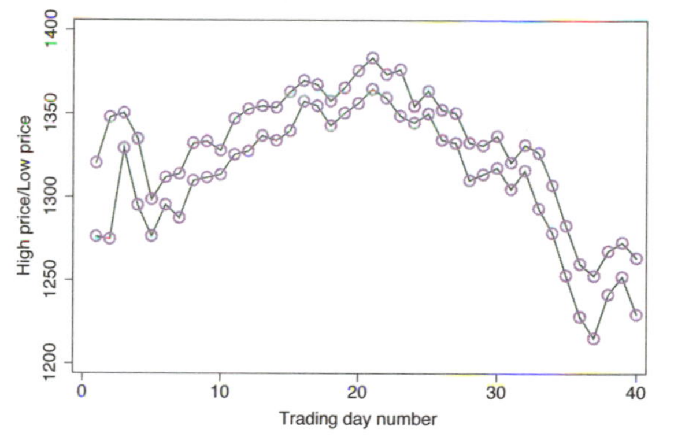


```
twoway rscatter high low tradeday, sort msymbol(Sh) msize(medium)
   mlwidth(thick)
```

The options you can use with `rscatter` are just a subset of those you would use with `rconnected`, where the connecting options would not be relevant. Here, we use the marker options to make the symbols medium, hollow squares with thick outlines. For more details about options related to marker symbols, see Options : Markers (235).
Uses spjanfeb2001.dta & scheme vg_s1c

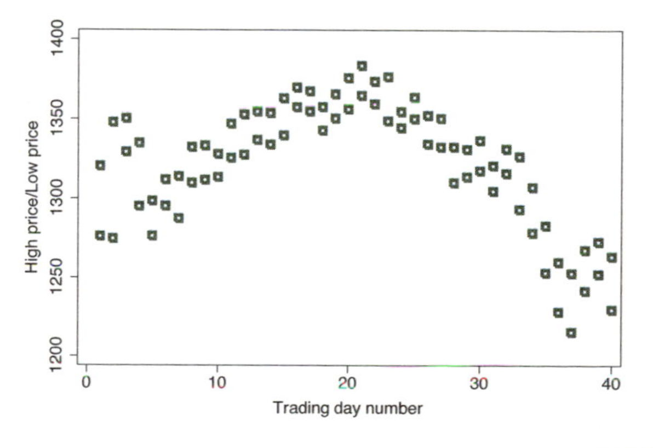

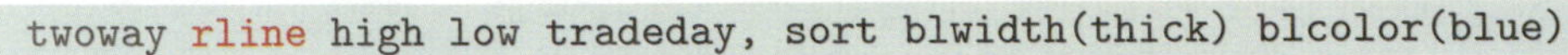

```
twoway rline high low tradeday, sort blwidth(thick) blcolor(blue)
```

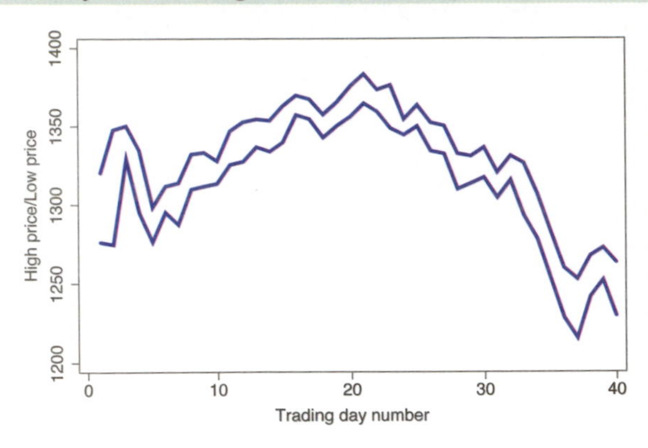

The options you can use with `rline` are a subset of those you would use with **`rconnected`**; namely, the marker symbol and marker label options are not relevant. Here, we show the use of connect options to make the lines thick and blue. For more details about connect options, see Options : Connecting (250).
Uses spjanfeb2001.dta & scheme vg_s1c

```
twoway rarea high low tradeday, sort bcolor(teal)
```

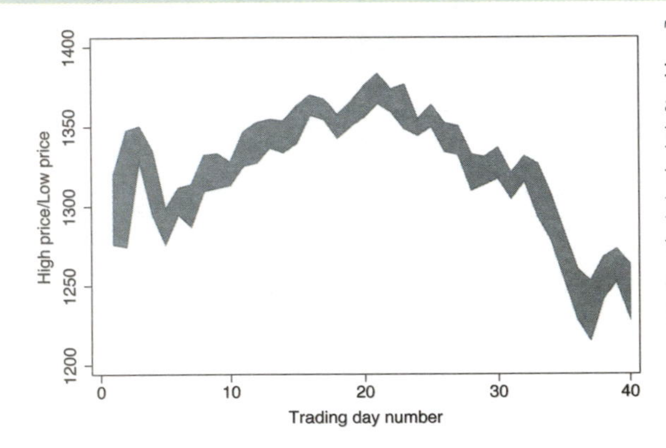

The `rarea` graph is similar to the **`rline`** graph, but in addition to being able to control the characteristics of the line, you can also control the color of the area between the low and high lines. Here, we use the `bcolor()` option to make the color of the line and the area teal.
Uses spjanfeb2001.dta & scheme vg_s1c

```
twoway rarea high low tradeday, sort
   blcolor(emerald) bfcolor(teal) blwidth(thick)
```

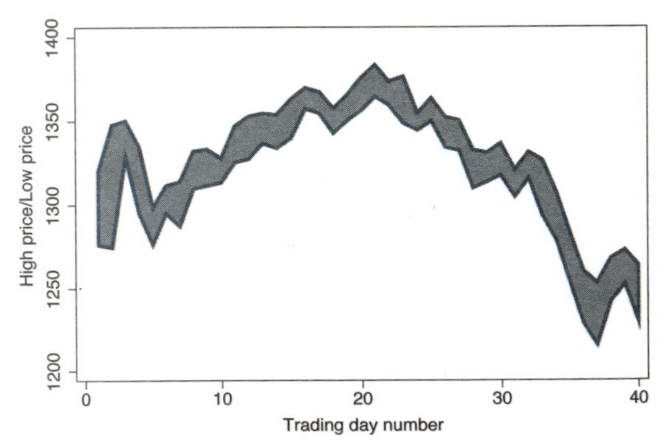

Here, we make the color of the line emerald with the `blcolor()` option, the fill color teal with the `bfcolor()` option, and the line thick with the `blwidth()` option.
Uses spjanfeb2001.dta & scheme vg_s1c

Now, let's look at options that can be used with the **rcap**, **rspike**, and **rcapsym** graphs. The options permitted by the **rcap** option are similar to the options used with the **rspike** and **rcapsym** graphs. For these examples, we will use the **vg_s2c** scheme.

```
twoway rcap high low tradeday
```

Here is an **rcap** graph with the default options. The **rcap** command supports the **horizontal** option, which would make the variables **high**/**low** and **tradeday** swap positions.
Uses spjanfeb2001.dta & scheme vg_s2c

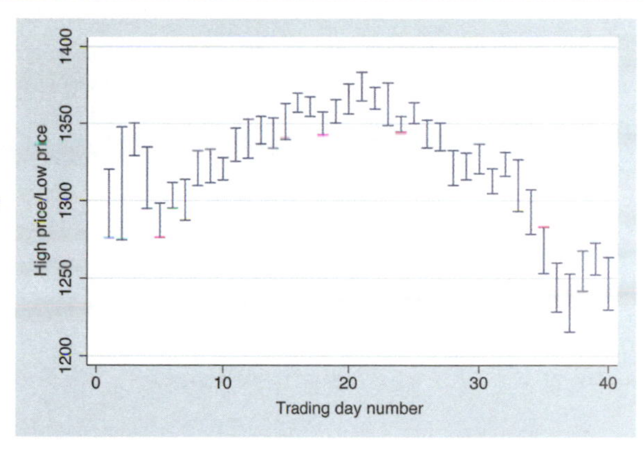


```
twoway rcap high low tradeday, msize(small)
```

The **msize()** option usually is used to control the size of a marker and is adapted for this kind of graph to control the size of the cap. In this case, the cap is made small.
Uses spjanfeb2001.dta & scheme vg_s2c

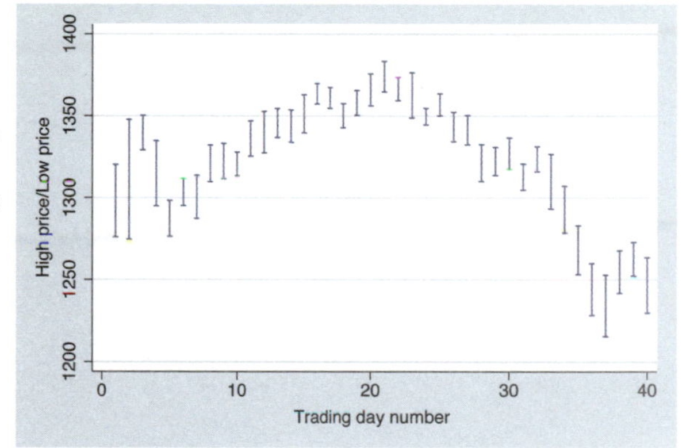

```
twoway rcap high low tradeday, blcolor(cranberry) blwidth(thick)
```

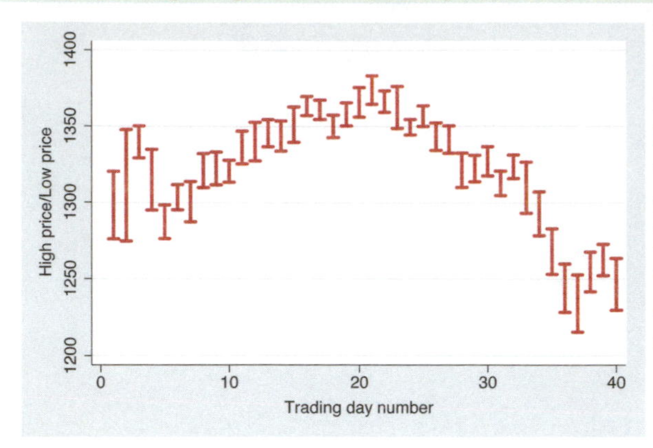

The `blcolor()` option is used to control the color of the line, in this case making the line cranberry. The `blwidth()` option is used to set the width of the line; in this case, the line is made thick. Although it is not shown here, you could also control the pattern of the line with the `blpattern()` option. See Options : Connecting (250) for more details.
Uses spjanfeb2001.dta & scheme vg_s2c

```
twoway rspike high low tradeday, blcolor(red) blwidth(thin)
```

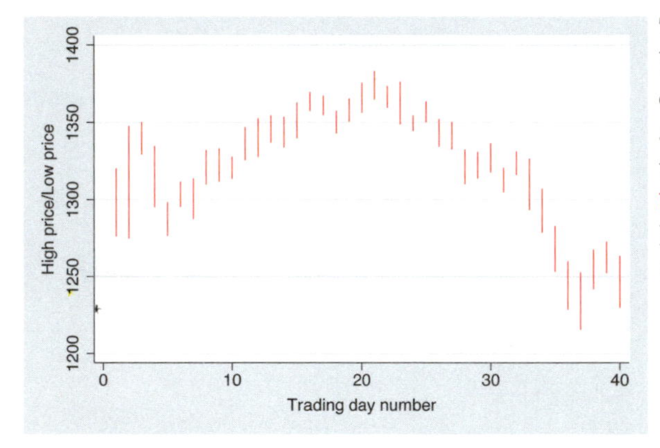

The options used for `rspike` are basically the same as those for `rcap`, except that the `msize()` option is not appropriate since there are no markers to size. Here, for example, we use `blcolor()` and `blwidth()` to make the lines red and thin.
Uses spjanfeb2001.dta & scheme vg_s2c

```
twoway rcapsym high low tradeday, msymbol(Oh) msize(large)
```

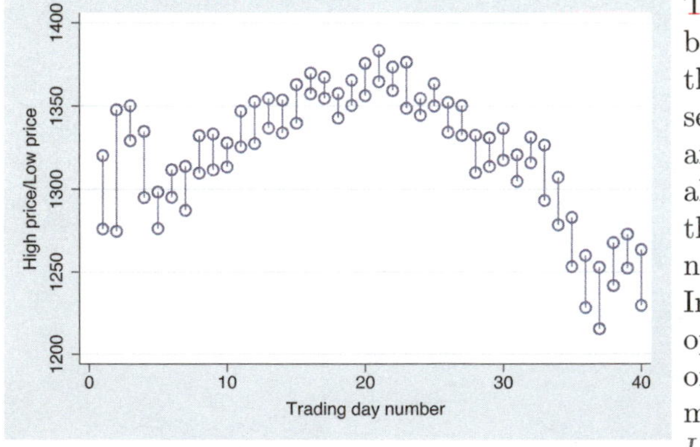

The options used for `rcapsym` are basically the same as for `rcap`, except that you can use marker options to select the marker that goes at the top and bottom of each spike, and you can also use marker label options to label the markers (however, this is probably not very useful and is not illustrated). In this case, we use the `msymbol()` option to place hollow circles at the end of the spikes and the `msize()` option to make the symbols large.
Uses spjanfeb2001.dta & scheme vg_s2c

We will now explore options that can be used with **twoway rbar**, and we will switch to using the **vg_brite** scheme.

```
twoway rbar high low tradeday
```

Here is a basic **rbar** graph with the default options. As with the other graphs in this family, we could have added the **horizontal** option to switch the position of the **high**/**low** and **tradeday** variables, but this is not shown.
Uses spjanfeb2001.dta & scheme vg_brite

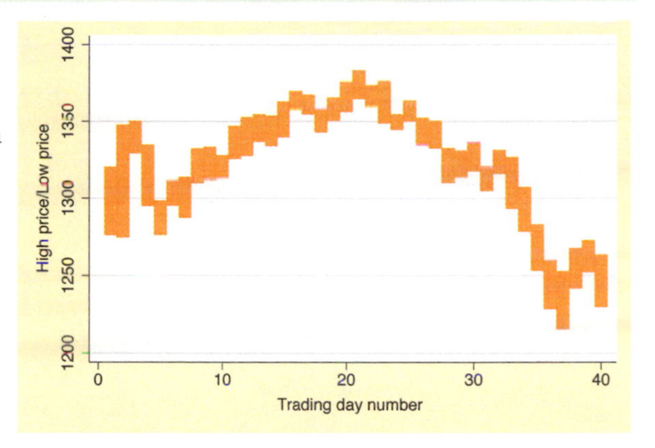


```
twoway rbar high low tradeday, barwidth(.7)
```

The **barwidth()** option can be used to set the width of the bar. This width is in units of the x-variable. We set the bars to be .7 units wide, so they no longer touch each other.
Uses spjanfeb2001.dta & scheme vg_brite

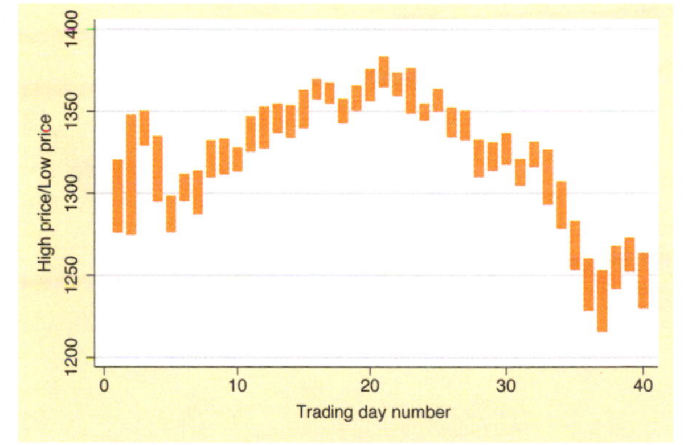

```
twoway rbar high low tradeday, bcolor(sienna)
```

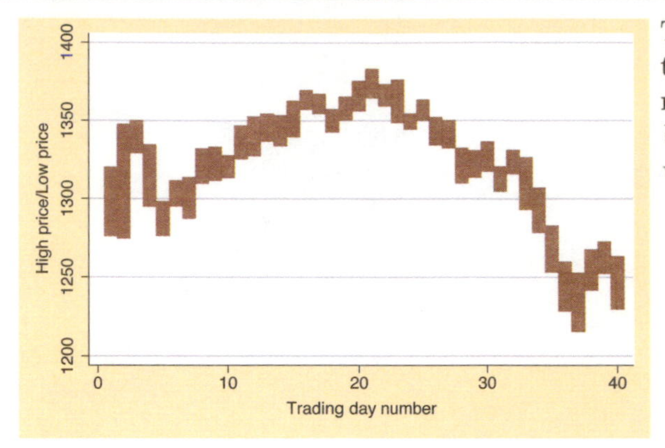

The `bcolor()` (bar color) option sets the color of the bar and the outline, making the color sienna.
Uses spjanfeb2001.dta & scheme vg_brite

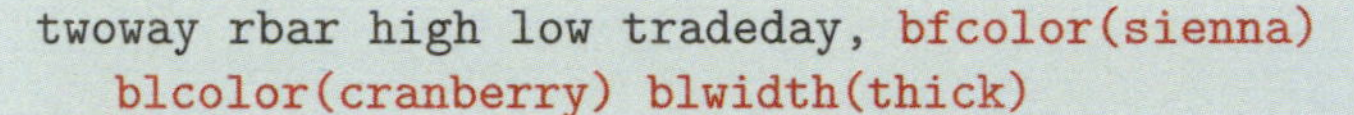

```
twoway rbar high low tradeday, bfcolor(sienna)
   blcolor(cranberry) blwidth(thick)
```

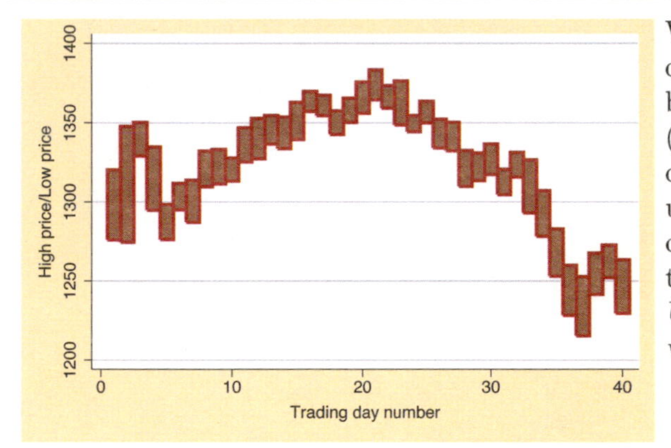

With the `bfcolor()` (bar fill color) option, we set the fill color of the bar to be sienna and then use the `blcolor()` (bar line color) option to set the color of the outline to be cranberry. We also use the `blwidth()` (bar line width) option to make the lines surrounding the bars thick.
Uses spjanfeb2001.dta & scheme vg_brite

2.8 Distribution plots

This section describes the use of `twoway histogram` and `twoway kdensity` for showing the distribution of a single variable. In addition, this section also shows the use of `twoway function` for showing the relationship between x and y using a function that you specify. See [G] **graph twoway histogram**, [G] **graph twoway kdensity**, and [G] **graph twoway function** for more information. We will start by showing the `twoway histogram` command and consider options that allow you to control such things as the number of bins, the width of the bins, and the starting point for the bins. Then, we will show options that control the scaling of the y-axis. The next few graphs use the `vg_past` scheme.

```
twoway histogram ttl_exp
```

We begin by showing a histogram of the variable total work experience. Note that, unlike many other twoway plots, this command takes only one variable that is graphed on the x-axis. The y-axis represents the density, such that the sum of the areas of the bars equals 1. If you are not going to combine this graph with other twoway graphs, the **histogram** command may be preferable to **twoway histogram**.
Uses nlsw.dta & scheme vg_past

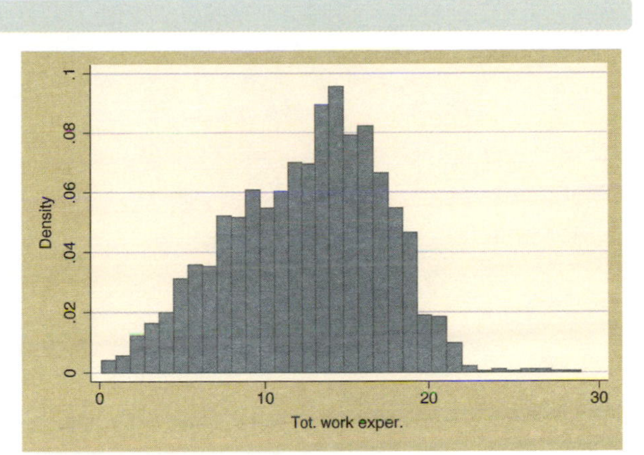


```
twoway histogram ttl_exp, bin(10)
```

We can control the number of bins that are used to display the histogram using the `bin()` option. Here, we request that 10 bins be used.
Uses nlsw.dta & scheme vg_past

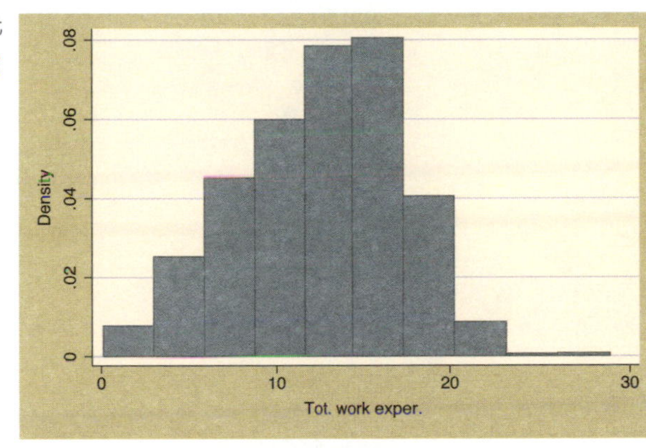


```
twoway histogram ttl_exp, width(5)
```

We can control the width of each bar using the `width()` option. Here, we make each bar 5 units wide. As you might imagine, you can use either the `bin()` option or the `width()` option but not both.
Uses nlsw.dta & scheme vg_past

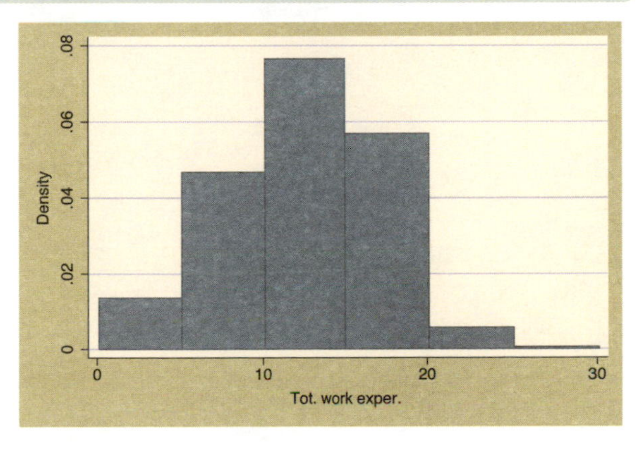

```
twoway histogram ttl_exp, start(-2.5) width(5)
```

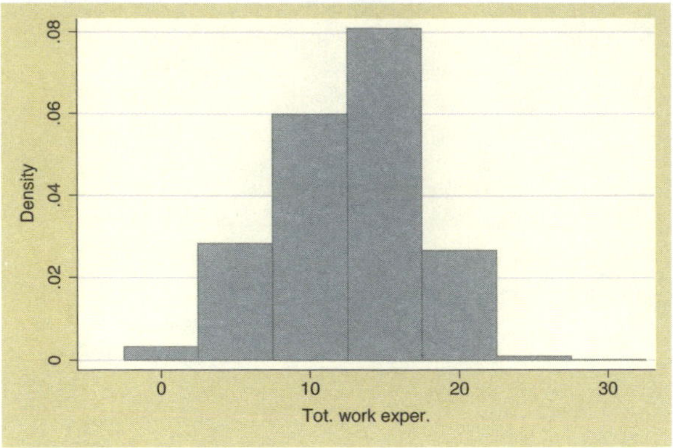

We add the `start()` option to indicate that we want the lower limit of the first bin to start at −2.5.
Uses nlsw.dta & scheme vg_past

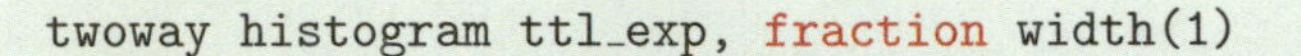

```
twoway histogram ttl_exp, fraction width(1)
```

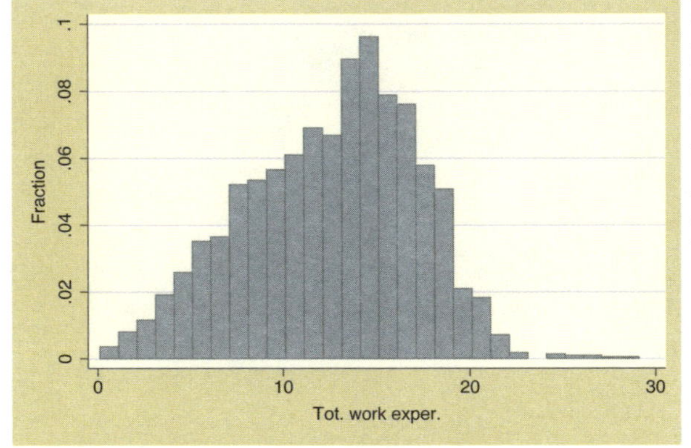

If we use the `fraction` option, the y-axis is scaled such that the height of each bar is the probability of falling within the range of x-values represented by the bar. Thus, if we specify the width of bars to be 1, the sum of the heights of the bars is 1.
Uses nlsw.dta & scheme vg_past

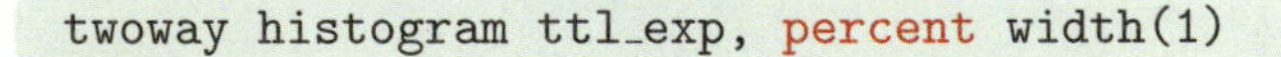

```
twoway histogram ttl_exp, percent width(1)
```

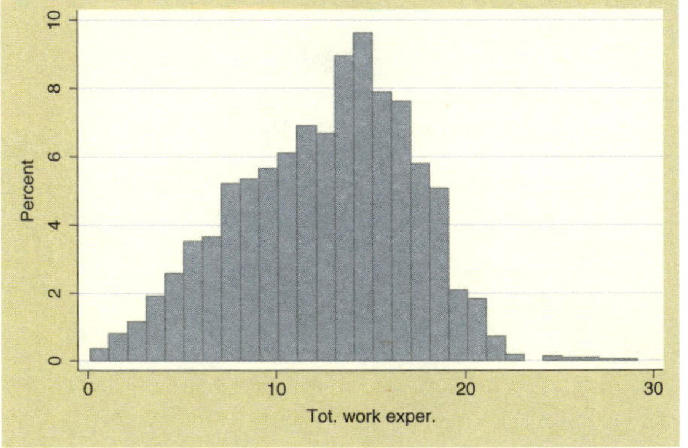

The `percent` option is similar to the `fraction` option, except that the y-axis is represented as a percentage instead of a proportion. If we also specify a bar width of 1, the sum of the heights of the bars is 100%.
Uses nlsw.dta & scheme vg_past

```
twoway histogram ttl_exp, frequency width(1)
```

The frequency option changes the scaling of the y-axis to represent the number of cases that fall within the range of x-values represented by the bar. If we specify a bar width of 1, the sum of the heights of the bars equals the number of nonmissing values for `ttl_exp`.
Uses nlsw.dta & scheme vg_past

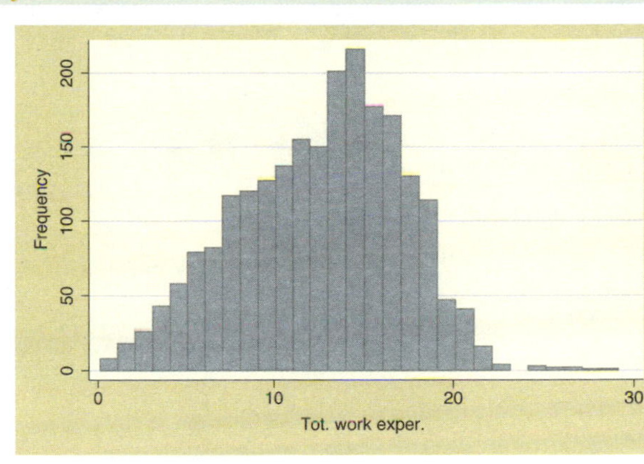

Let's now consider options that control the width of the bars and other characteristics of the bars, such as color. Then, we will show you how to display the graph as a horizontal histogram and demonstrate options that allow you to treat *varname* as a discrete variable. We will use the `vg_blue` scheme for these graphs.

```
twoway histogram ttl_exp, gap(20)
```

The `gap()` option specifies the gap between each of the bars. The gap is created by reducing the width of the bars. By default, the gap is 0, meaning that the bars touch exactly and the bars are reduced by 0%. Here, we reduce the size of the bars by 20%, making a small gap between the bars.
Uses nlsw.dta & scheme vg_blue

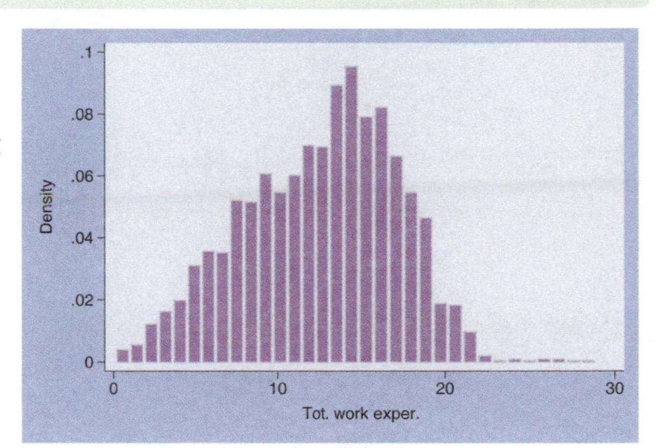

```
twoway histogram ttl_exp, gap(99.99)
```

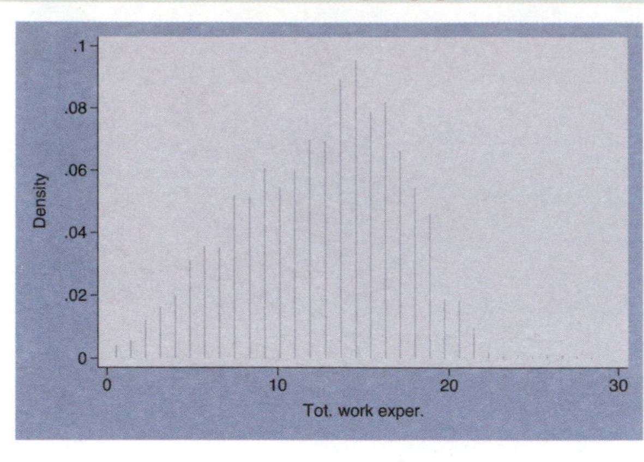

Here, we reduce the size of the bars 99.99%, making the bars 0.01% of their normal size.
Uses nlsw.dta & scheme vg_blue

```
twoway histogram ttl_exp, barwidth(.5)
```

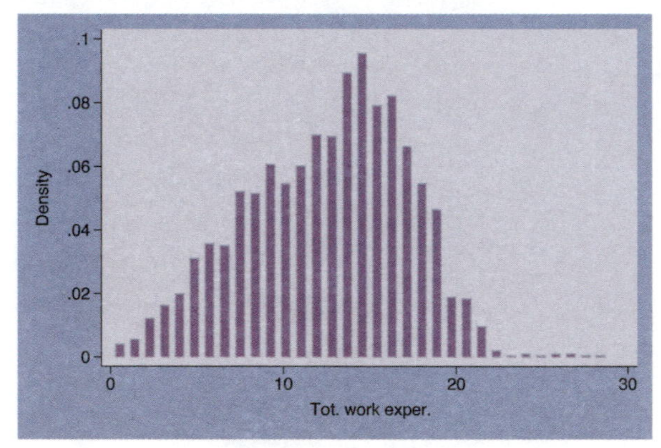

Another way you can control the width of the bars is though the `barwidth()` option. Here, we indicate that we wish each bar to be .5 x-units wide.
Uses nlsw.dta & scheme vg_blue

```
twoway histogram ttl_exp, bfcolor(olive_teal) blcolor(teal)
  blwidth(thick)
```

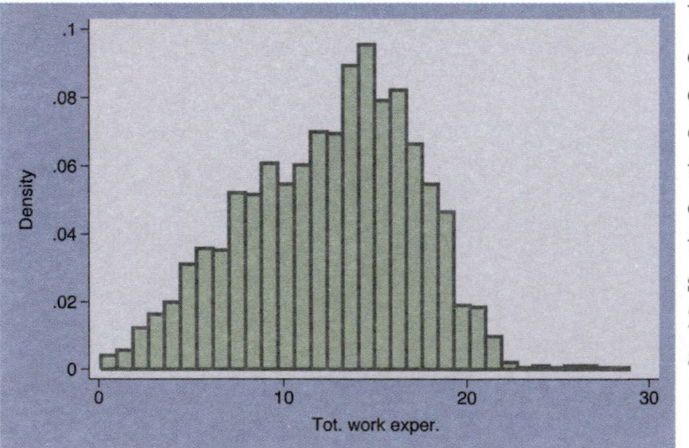

We use the `bfcolor()` (bar fill color) option to make the fill color of the bar olive–teal and the `blcolor()` (bar line color) option to make the bar line color teal. The `blwidth()` (bar line width) option makes the line around the bar thick. The section Styles : Colors (328) shows more about colors, and Styles : Linewidth (337) shows more about line widths.
Uses nlsw.dta & scheme vg_blue

We will now briefly consider some other options that can be used with **`twoway histogram`**, showing how you can swap the position of the x- and y-axes, and the **`discrete`** option for use with discrete variables. We will use the **`vg_s1m`** scheme for the next set of graphs.

```
twoway histogram ttl_exp, horizontal
```

We can use the `horizontal` option to swap the position of `ttl_exp` and its density, making a horizontal display of the histogram.
Uses nlsw.dta & scheme vg_s1m

```
twoway histogram grade, discrete
```

Here, we use the `discrete` option to tell Stata that the variable **`grade`** is a discrete variable and can take on only integer values. In this example, each bin has a width of 1, and the bars are too narrow to be useful.
Uses nlsw.dta & scheme vg_s1m

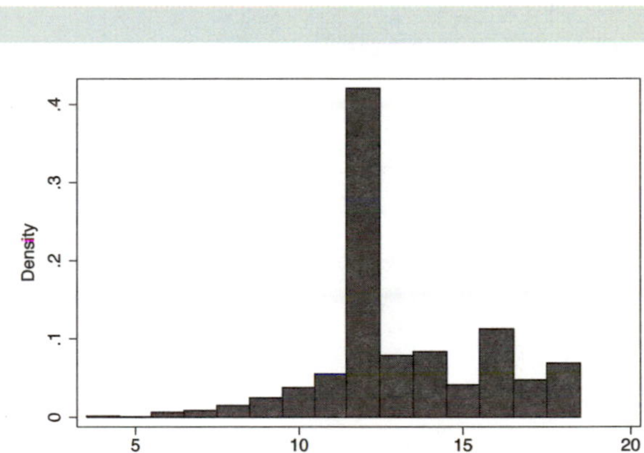

Introduction Twoway Matrix Bar Box Dot Pie Options Standard options Styles Appendix
Scatter Fit CI fit Line Area Bar Range Distribution Options Overlaying

```
twoway histogram grade, discrete width(2)
```

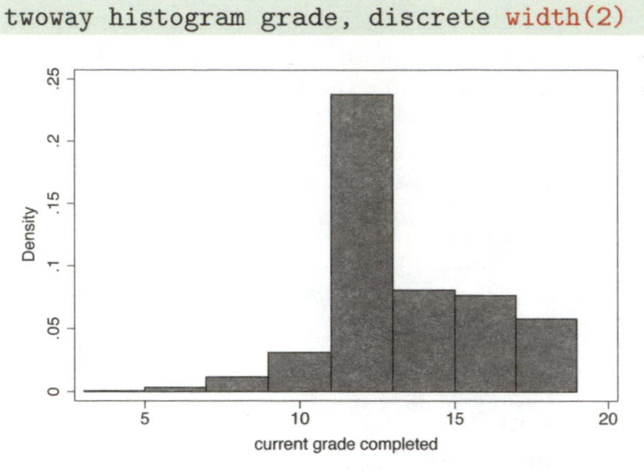

We add the `width()` option, and the bars have a width of 2.
Uses nlsw.dta & scheme vg_s1m

We will now consider kernel-density plots that can be created using **`twoway kdensity`**. For more details, see [G] **graph twoway kdensity**. As with histograms, if you are not going to combine the kernel-density plot with other twoway plots, and sometimes even when you are, the **`kdensity`** command is preferable to **`twoway kdensity`**. We will explore a handful of options that are useful for controlling the display of these graphs. These graphs will use the **`vg_s2c`** scheme.

```
twoway kdensity ttl_exp
```

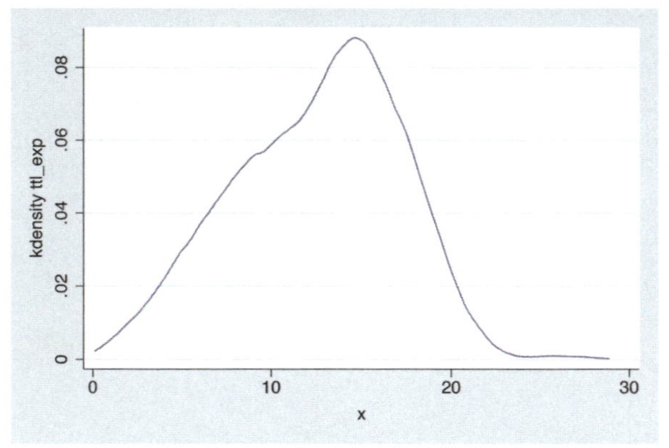

Here is a kernel-density plot of total work experience. We could have added the **`horizontal`** option to display the graph as a horizontal plot, but this option is not shown.
Uses nlsw.dta & scheme vg_s2c

```
twoway kdensity ttl_exp, biweight
```

By default, Stata uses a Epanechnikov kernel for computing the density estimates. Here, we use the `biweight` option to use the biweight kernel for computing the densities. Other methods include `cosine`, `gauss`, `parzen`, `rectangle`, and `triangle`.
Uses nlsw.dta & scheme vg_s2c

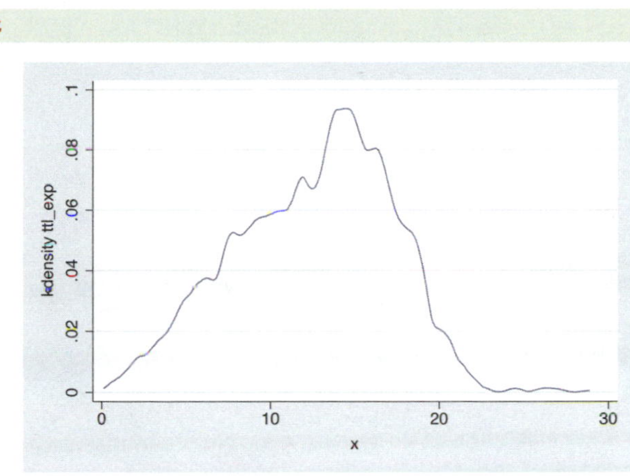


```
twoway kdensity ttl_exp, range(0 40)
```

You can use the `range()` option to specify the range of the x-values at which the kernel density is computed and displayed. Here, we expand the range to span from 0 to 40.
Uses nlsw.dta & scheme vg_s2c

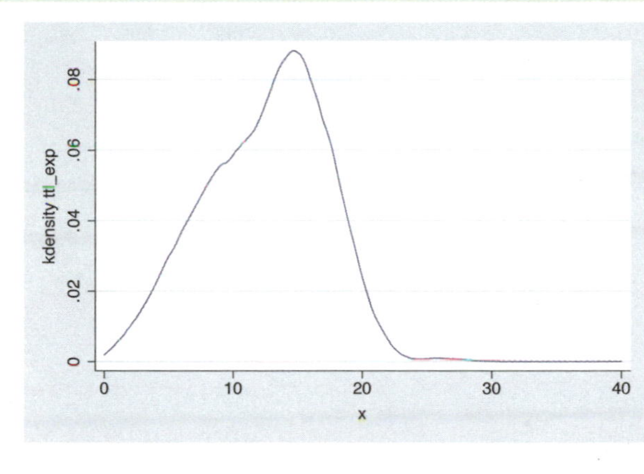


```
twoway (histogram ttl_exp, width(1) frequency)
   (kdensity ttl_exp, area(2246))
```

In this example, we overlay a histogram of `ttl_exp`, scaling the y-axis as the frequency of values in each bin. We overlay this with a `kdensity` plot but want to scale the y-axis in a commensurate manner. By using the `area()` option, we can specify that the sum of the area of the kernel density should sum to 2246, the sample size.
Uses nlsw.dta & scheme vg_s2c

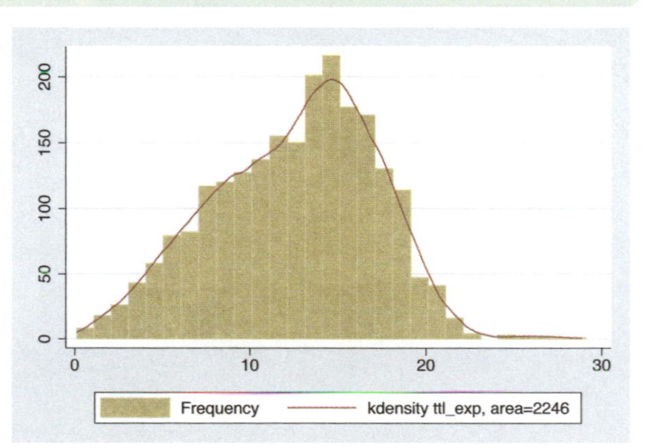

```
twoway kdensity ttl_exp, clwidth(thick) clpattern(dash)
```

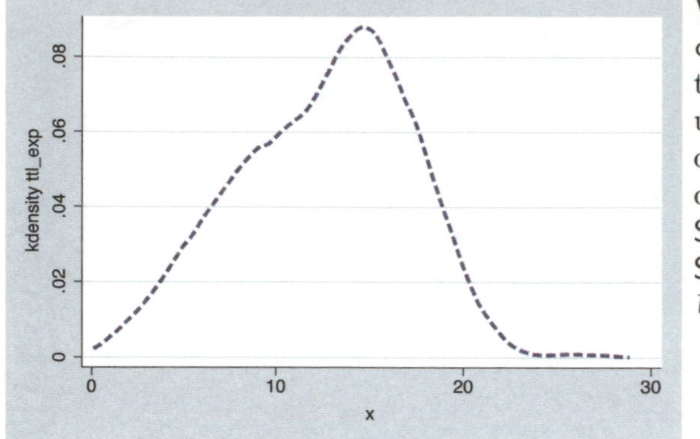

We can use options such as `clcolor()`, `clwidth()`, and `clpattern()` to alter the characteristics of the line. Here, we use the `clwidth()` and `clpattern()` options to make the line thick and dashed. See Styles : Linewidth (337), Styles : Linepatterns (336), and Styles : Colors (328) for more details.
Uses nlsw.dta & scheme vg_s2c

```
twoway function y=normden(x), range(-4 4)
```

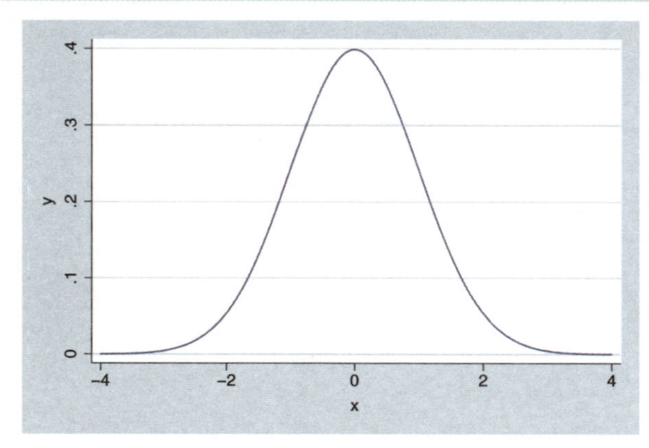

We conclude by showing how you can use `twoway function` to graph an arbitrary function. We graph the function `y=normden(x)` to show a normal curve. We add the `range(-4 4)` to specify that we want the x-values to range from -4 to 4. Otherwise, the graph would show the x-values ranging from 0 to 1.
Uses nlsw.dta & scheme vg_s2c

2.9 Options

This section discusses the use of options with `twoway`, showing the types of options you can use. For more details, see Options (235). This section uses the `vg_outm` scheme for displaying the graphs.

```
twoway scatter ownhome propval100
```

Consider this basic scatterplot. We will use this for illustrating options.
Uses allstates.dta & scheme vg_outm

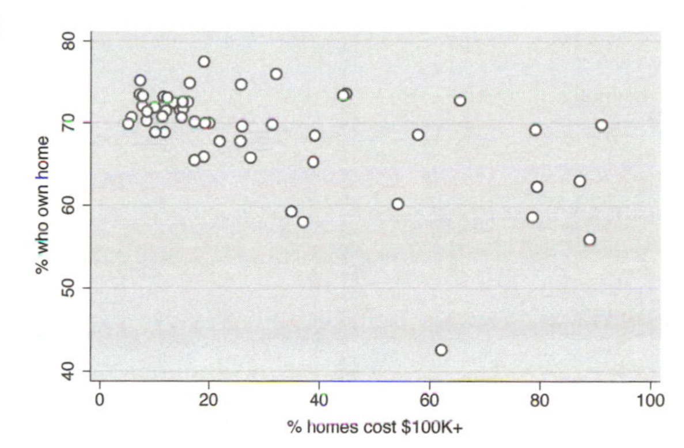


```
twoway scatter ownhome propval100, msymbol(S)
```

We can use the `msymbol()` option to control the marker symbols. Here, we use squares as symbols. See Options : Markers (235) for more details.
Uses allstates.dta & scheme vg_outm

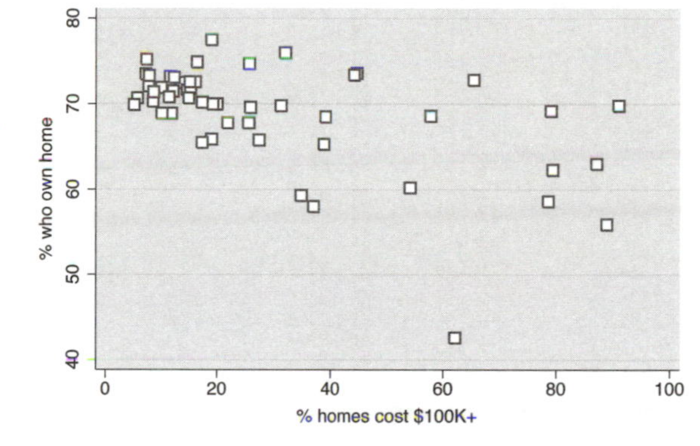


```
twoway scatter ownhome propval100, msymbol(S) mlabel(stateab)
```

We can use the `mlabel()` option to control the marker labels. Here, we label each of the markers with the variable **stateab** showing the two-letter abbreviation for each state next to each marker. See Options : Marker labels (247) for more information about marker labels.
Uses allstates.dta & scheme vg_outm

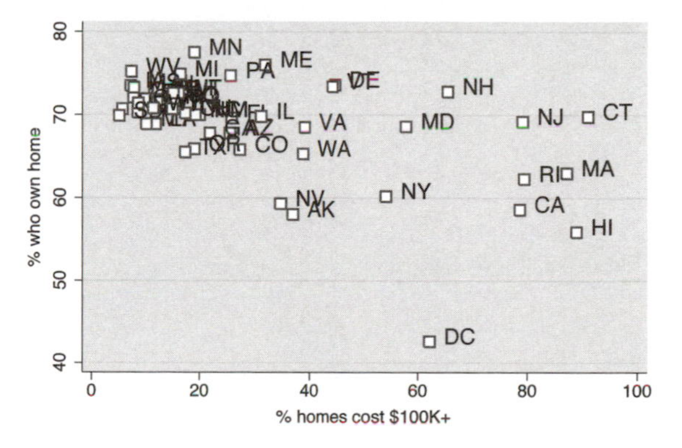

```
twoway scatter fv ownhome propval100, connect(l .) sort
```

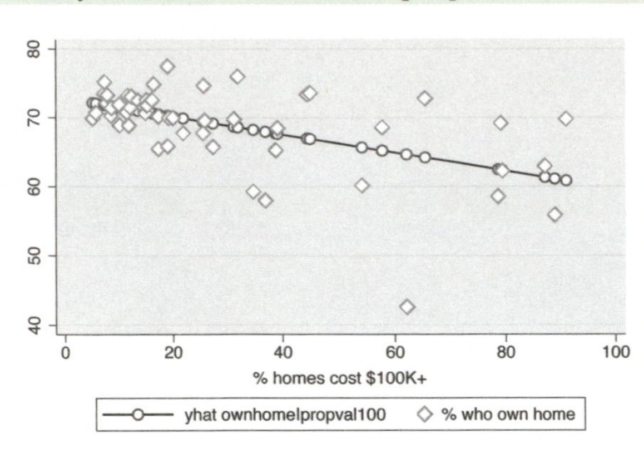

Say that we regressed **ownhome** on **propval100** and generated predicted values named **fv**. Here, we make a scatterplot and fit line in the same graph using the `connect(l .)` option to connect the values of **fv** but not the values of **ownhome**. We also add the **sort** option, which is generally recommended when using the `connect()` option. See Options : Connecting (250) for more details.
Uses allstates.dta & scheme vg_outm

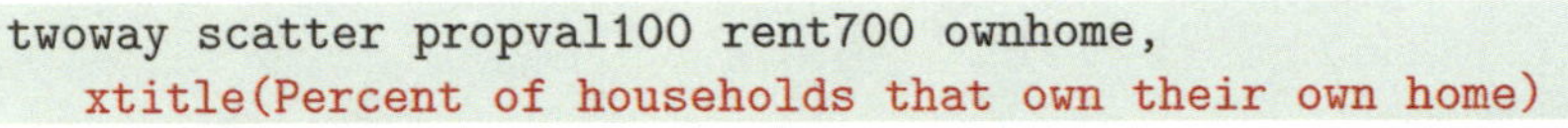

```
twoway scatter propval100 rent700 ownhome,
   xtitle(Percent of households that own their own home)
```

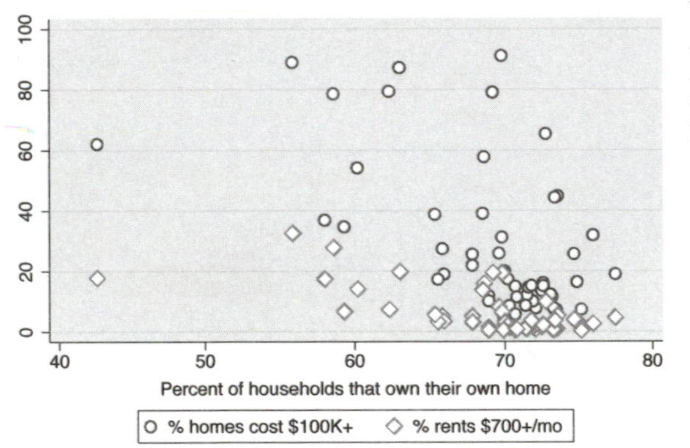

We can add a title to the x-axis using the `xtitle()` option, as illustrated here. See Options : Axis titles (254) for more details about titles.
Uses allstates.dta & scheme vg_outm

```
twoway scatter propval100 rent700 ownhome, ylabel(0(10)100)
```

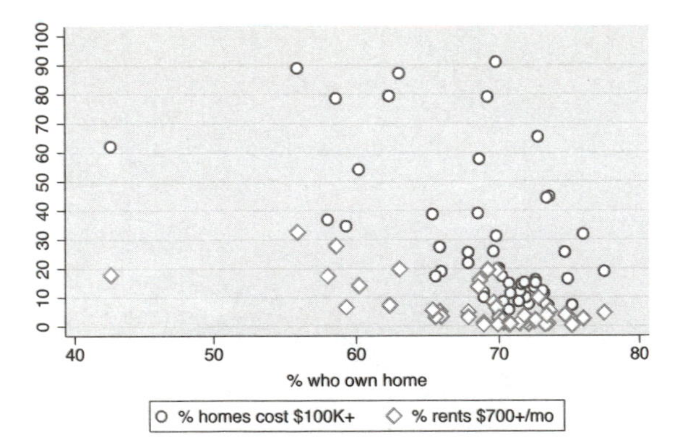

We can label the y-axis from 0 to 100, incrementing by 10, using the `ylabel(0(10)100)` option as shown here. See Options : Axis labels (256) for more information about labeling axes.
Uses allstates.dta & scheme vg_outm

```
twoway scatter propval100 rent700 ownhome,
   ylabel(0(10)100) yscale(alt)
```

Stata gives you a number of options that you can use to control the axis scale for both the x- and y-axes. For example, here we use `yscale(alt)` to move the y-axis to its alternate position, moving it from the left to the right. See Options : Axis scales (265) for more details about the options for controlling the axis scales.
Uses allstates.dta & scheme vg_outm

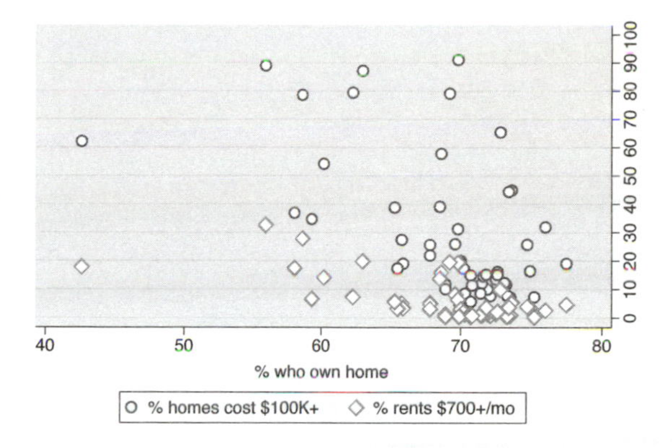


```
twoway (scatter propval100 ownhome)
   (scatter rent700 ownhome, yaxis(2))
```

In this example, we show `propval100` by `ownhome` and also `rent700` by `ownhome`, but for this second plot, we put the y-axis on the second y-axis with the `yaxis(2)` option. See Options : Axis selection (269) for more information about using and controlling additional axes.
Uses allstates.dta & scheme vg_outm

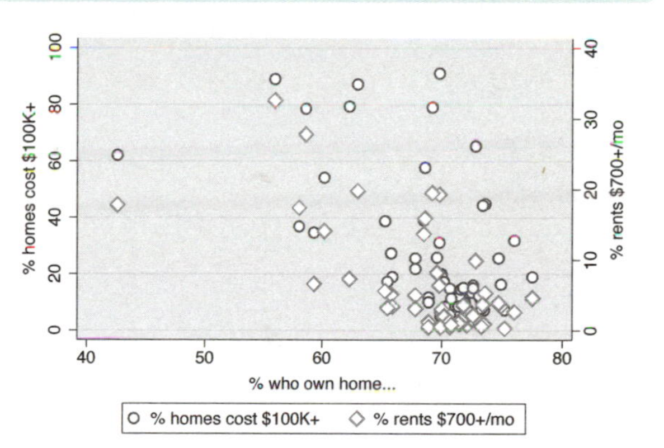


```
twoway scatter propval100 rent700 ownhome,
   ylabel(0(10)100) yscale(alt) by(north)
```

The `by()` option allows you to see a graph broken down by one or more `by()` variables. Here, we show the graph from above further broken down by whether the state was part of the North, making two graphs that are combined together into a single graph. The section Options : By (272) shows more details and examples about the use of the `by()` option.
Uses allstates.dta & scheme vg_outm

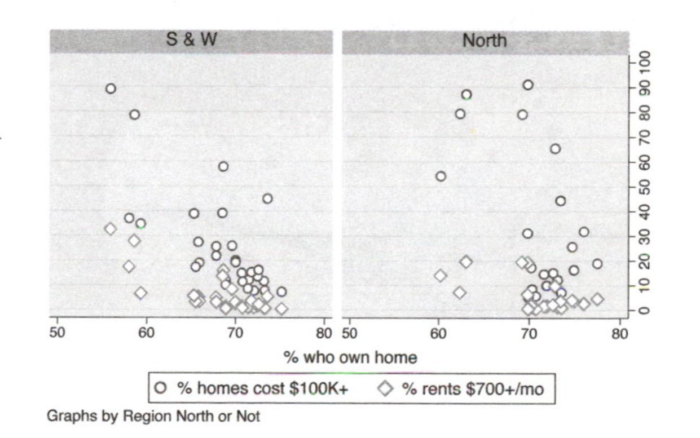

Introduction
Twoway
Matrix
Bar
Box
Dot
Pie
Options
Standard options
Styles
Appendix
Scatter
Fit
CI fit
Line
Area
Bar
Range
Distribution
Options
Overlaying

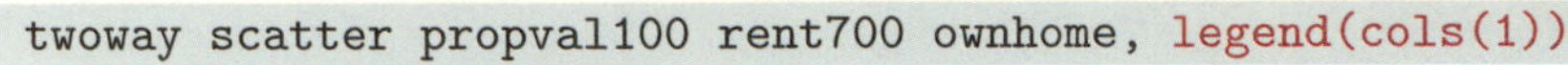

```
twoway scatter propval100 rent700 ownhome, legend(cols(1))
```

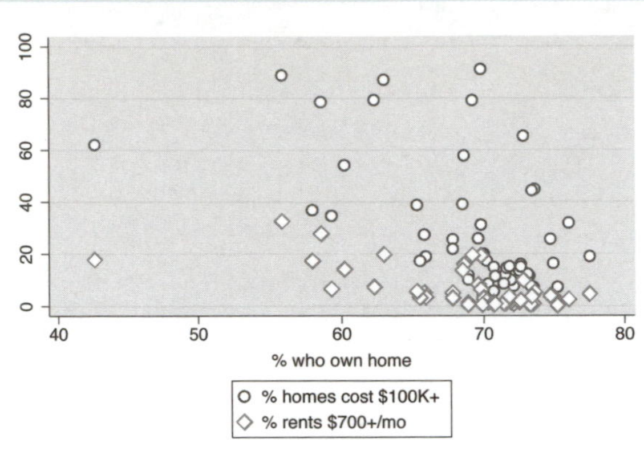

The `legend()` option allows you to control the contents and display of the legend. Here, we use the `legend(cols(1))` option to indicate that we want the legend to display as a single column. See Options : Legend (287) for more details about the `legend()` option.
Uses allstates.dta & scheme vg_outm

```
twoway scatter propval100 ownhome, text(62 45 "DC")
```

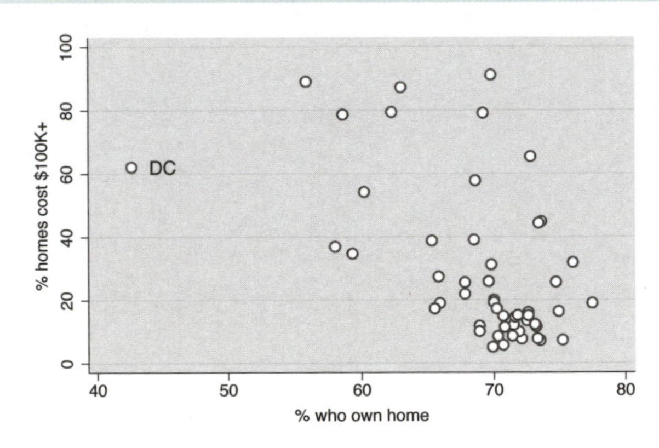

In this graph, there is a single observation that stands out from the rest. Rather than use the `mlabel()` option to label all of the markers, we may want to label just the outlying point. Here, we use the `text()` option to add the text DC at the (y,x) coordinates of (62,45), in effect labeling that point; see Options : Adding text (299) for more details.
Uses allstates.dta & scheme vg_outm

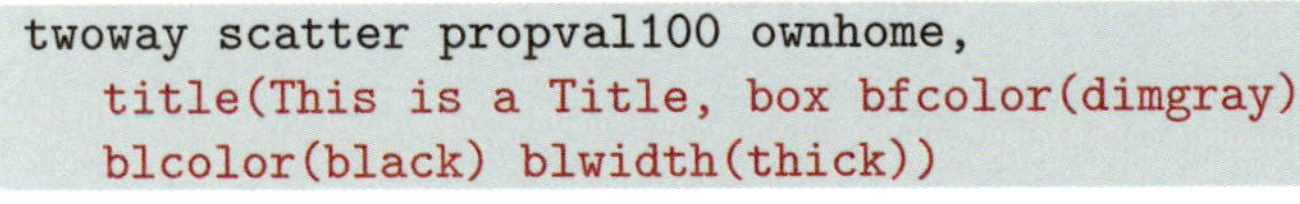

```
twoway scatter propval100 ownhome,
   title(This is a Title, box bfcolor(dimgray)
   blcolor(black) blwidth(thick))
```

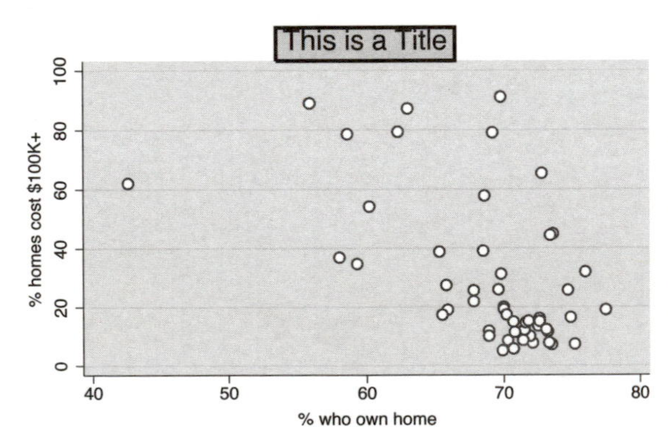

Most items of text on a Stata graph actually display within a box. We illustrate this with the `title()` option showing how we can place a box around this text. We make the background color of the box light gray and the outline thick and black. These options are described in more detail in Options : Textboxes (303).
Uses allstates.dta & scheme vg_outm

2.10 Overlaying plots

One of the terrific features of **twoway** graphs is the ability to overlay them, giving you the flexibility to create more complex graphs. This section shows two strategies you can use. The first strategy is graphing multiple y-variables against a single x-variable in a single **twoway** command. The second strategy is specifying multiple commands within a single **twoway** command, thus overlaying these graphs atop each other. It is also possible to create separate graphs and glue them together using the **graph combine** command, which is discussed in Appendix: Save/Redisplay/Combine (358). We first start by illustrating how you can specify multiple y-variables against a single x-variable using a single **twoway** command.

```
twoway scatter propval100 rent700 urban
```

We can use **twoway scatter** to graph multiple y-variables against a single x-variable in a single plot. Here, we show **propval100** and **rent700** against **urban**. Note that we are now using the **vg_teal** scheme.
Uses allstates.dta & scheme vg_teal

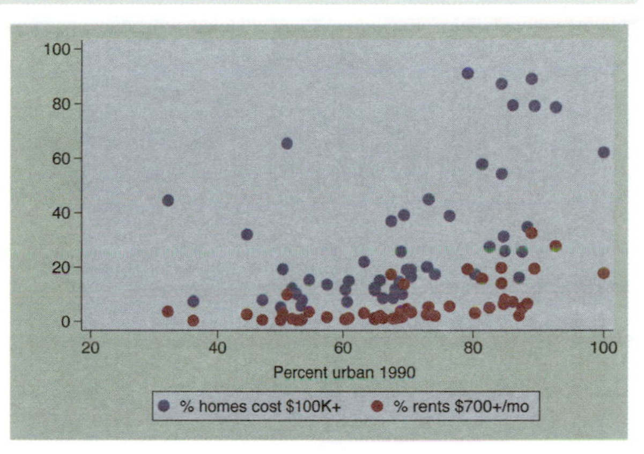


```
twoway scatter propval100 rent700 urban, msymbol(Oh t)
```

The **msymbol()** option can be used to select the marker symbols for the multiple y-variables. Here, we plot the variable **propval100** with hollow circles, and **rent700** is plotted with triangles.
Uses allstates.dta & scheme vg_teal

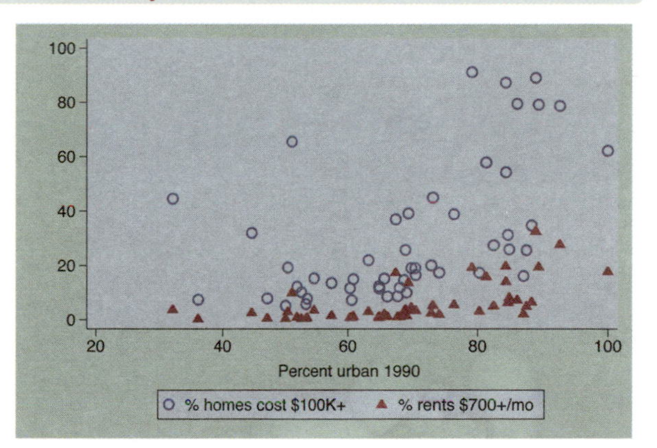

Introduction
Twoway
Matrix
Bar
Box
Dot
Pie
Options
Standard options
Styles
Appendix
Scatter
Fit
CI fit
Line
Area
Bar
Range
Distribution
Options
Overlaying

```
twoway scatter propval100 rent700 urban, mstyle(p2 p8)
```

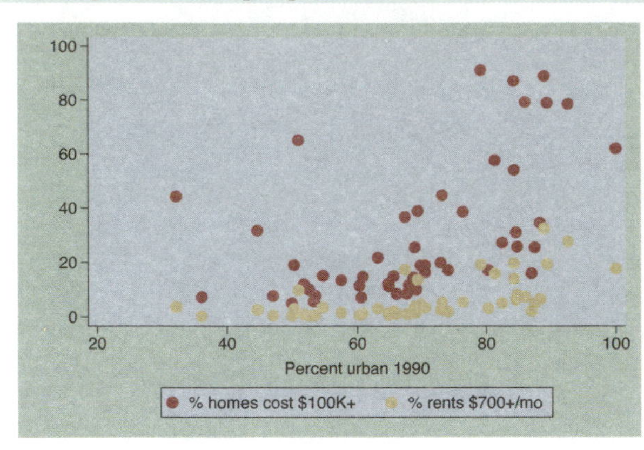

The `mstyle()` (marker style) option can be used to choose among marker styles. These composite styles set the symbol, size, fill, color, outline color, and outline width for the markers.
Uses allstates.dta & scheme vg_teal

```
twoway line high low close tradeday, sort
```

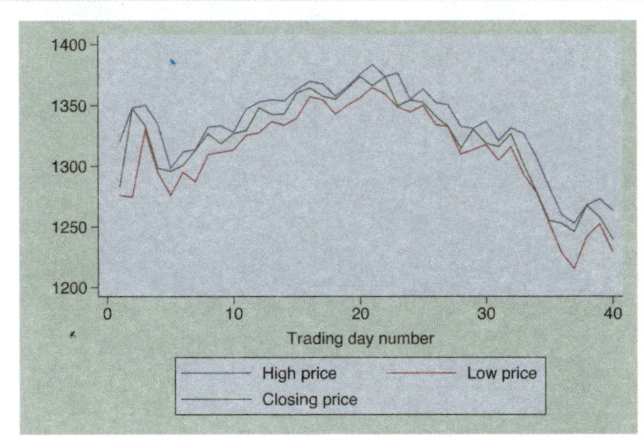

We will briefly switch to using the **spjanfeb2001** data file. You can also graph multiple y-variables against a single x-variable with a line graph. This works with **twoway line**, as illustrated here, as well as with **twoway connected** and **twoway tsline**.
Uses spjanfeb2001.dta & scheme vg_teal

```
twoway line high low close tradeday, sort clwidth(thick thick .)
```

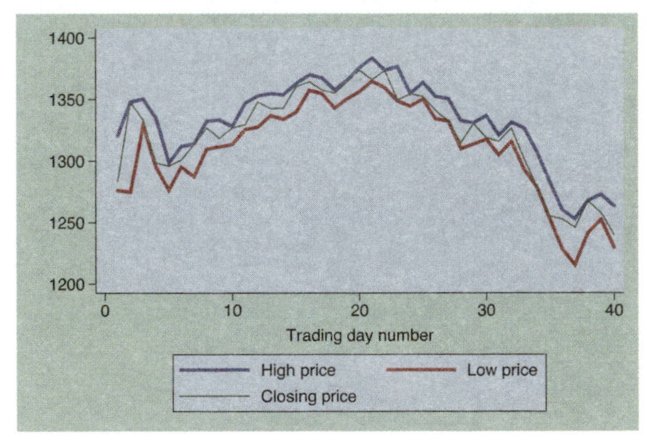

Here, we use the `clwidth()` option to change the width of the lines, making the lines for the high and low prices thick and leaving the line for the closing price at the default width.
Uses spjanfeb2001.dta & scheme vg_teal

```
twoway line high low close tradeday, sort clstyle(p1 p1 p2)
```

When we graph multiple y-variables, we can use `clstyle()` (connect line style) to control many characteristics of the lines at once. Here, we plot the high and low prices with the same style, `p1`, and the closing price printed with a second style, `p2`.
Uses spjanfeb2001.dta & scheme vg_teal

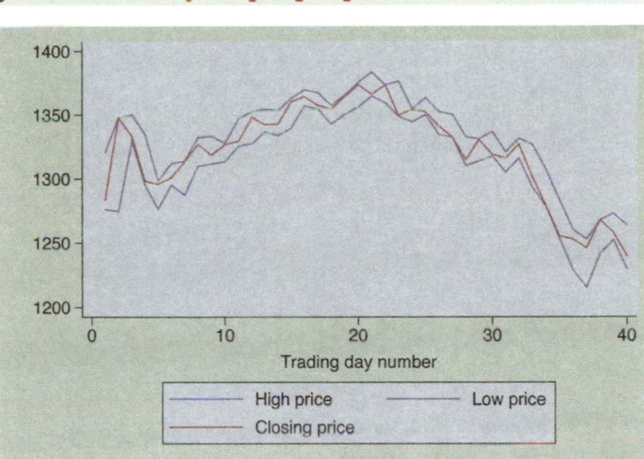


```
twoway line high low close tradeday, sort clstyle(p1 p1 p2)
   clwidth(thick thick .)
```

Here, we combine `clstyle()` and `clwidth()` to make the lines for the high and low prices the same style and make them both thick. The third line is drawn with the `p2` style, and the thickness is left at its default value.
Uses spjanfeb2001.dta & scheme vg_teal

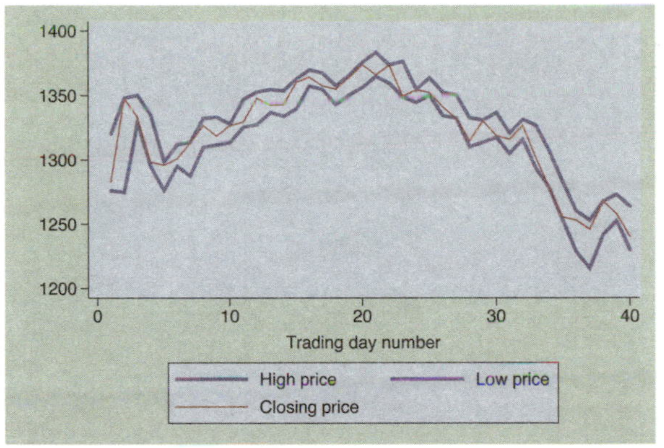


```
twoway (scatter propval100 urban) (lfit propval100 urban)
```

We return to the `allstates` data file. We can overlay multiple twoway graphs. Here, we show a common kind of overlay: scatterplot overlaid with a linear fit between the two variables. Note that both the `scatter` command and the `lfit` command are surrounded by parentheses.
Uses allstates.dta & scheme vg_teal

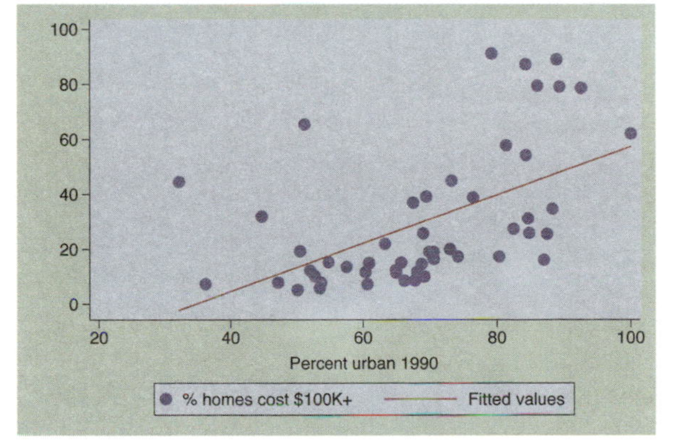

```
twoway (scatter propval100 urban) (lfit propval100 urban)
   (qfit propval100 urban)
```

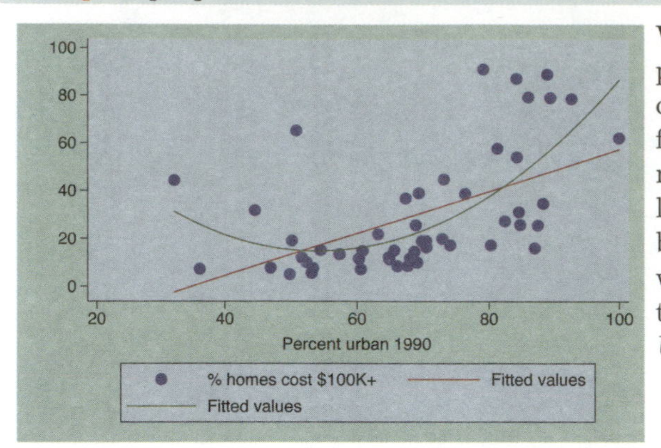

We can add a quadratic fit to the previous graph by adding a `qfit` command, so we can compare a linear fit and quadratic fit to see if there are nonlinearities in the fit. Note that the legend does not clearly differentiate between the linear and quadratic fit; we will show you how to modify the legend to label this more clearly below.
Uses allstates.dta & scheme vg_teal

```
twoway (scatter propval100 urban, msymbol(Oh))
   (lfit propval100 urban, clpattern(dash))
   (qfit propval100 urban, clwidth(thick))
```

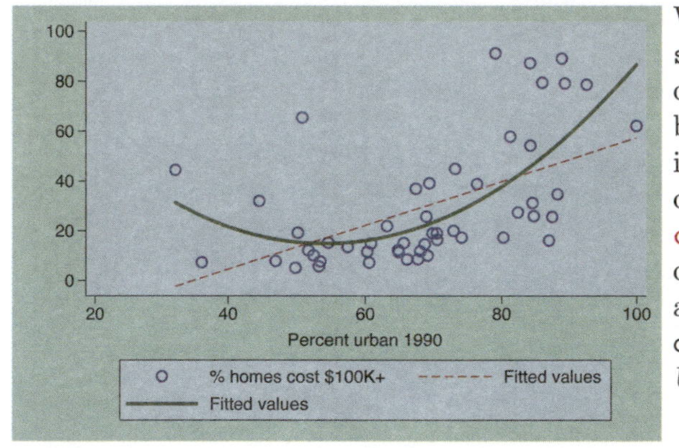

We add the `msymbol(Oh)` option to the `scatter` command, placing it after the comma, as it normally would be placed, but before the closing parenthesis that indicates the end of the `scatter` command. We also add the `clpattern(dash)` option to the `lfit` command to make the line dashed and add the `clwidth(thick)` option to the `qfit` command to make the line thick.
Uses allstates.dta & scheme vg_teal

```
twoway (scatter propval100 urban) (lfit propval100 urban)
   (qfit propval100 urban), legend(label(2 Linear Fit) label(3 Quad Fit))
```

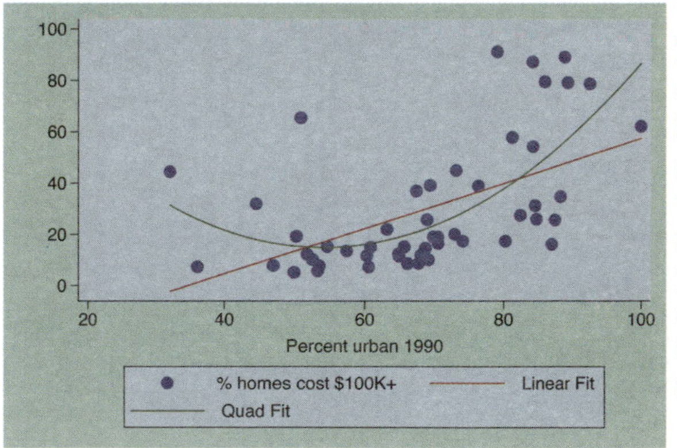

While each graph subcommand can have its own options, some options can apply to the entire graph. As illustrated here, we add a legend to the graph to clarify the difference in the fit values, and this option appears following a comma after the closing parenthesis following the `qfit` command. The `legend()` option appears at the end of the command since it applies to the entire graph.
Uses allstates.dta & scheme vg_teal

```
twoway (scatter propval100 urban) (lfit propval100 urban)
   (qfit propval100 urban, legend(label(2 Linear Fit) label(3 Quad Fit)))
```

We can make the previous graph in a different, but less appropriate, way. The `legend()` option is given as an option of the `qfit()` command, not at the very end as in the previous graph command. But Stata is forgiving of this, and even when such options are inappropriately given within a particular command, it treats them as though they were given at the end of the command.
Uses allstates.dta & scheme vg_teal

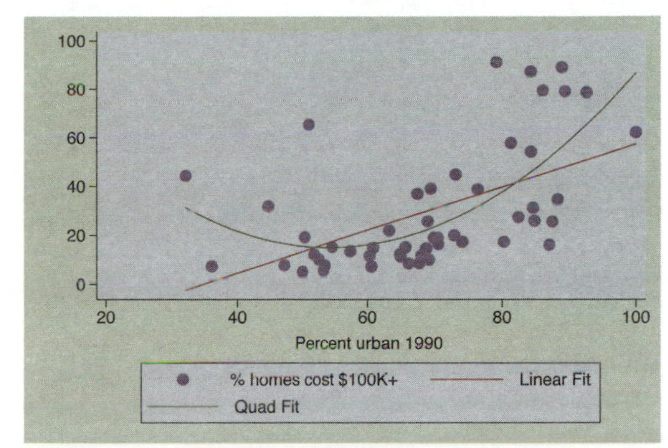


```
twoway (qfitci propval100 urban) (scatter propval100 urban)
```

Another common example of overlaying graphs is to overlay a fit line with confidence interval and a scatterplot.
Uses allstates.dta & scheme vg_teal

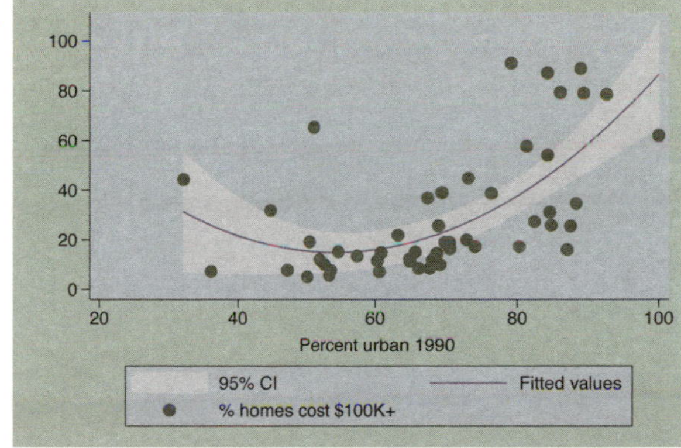


```
twoway (scatter propval100 urban) (qfitci propval100 urban)
```

However, note the order in which you overlay these two kinds of graphs. In this example, the `qfitci` was drawn after the `scatter`, and as a result, the points are obscured by the confidence interval.
Uses allstates.dta & scheme vg_teal

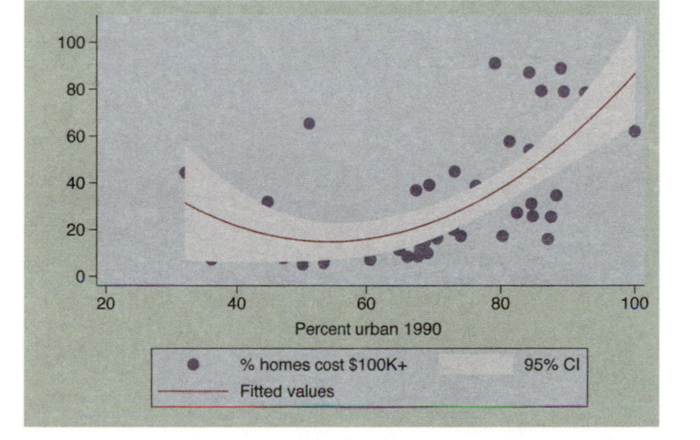

```
twoway (rarea high low date) (spike volmil date)
```

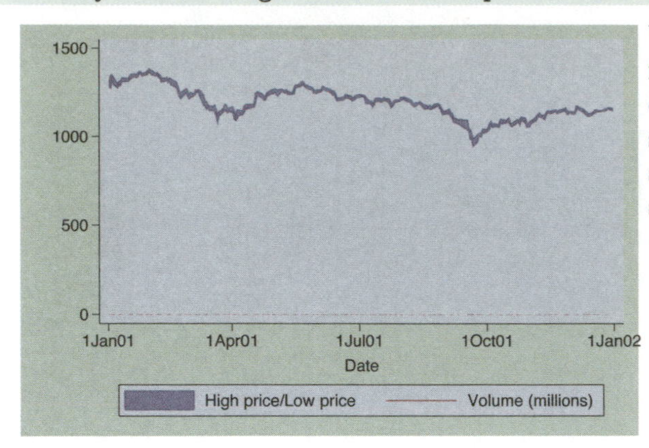

We now switch to the **sp2001ts** data file. Here, we overlay the high and low closing prices with the volume of shares sold. But, since both are placed on the same y-axis, it is difficult to see the spikes of **volmil**, volume in millions.
Uses sp2001ts.dta & scheme vg_teal

```
twoway (rarea high low date) (spike volmil date, yaxis(2)),
   legend(span)
```

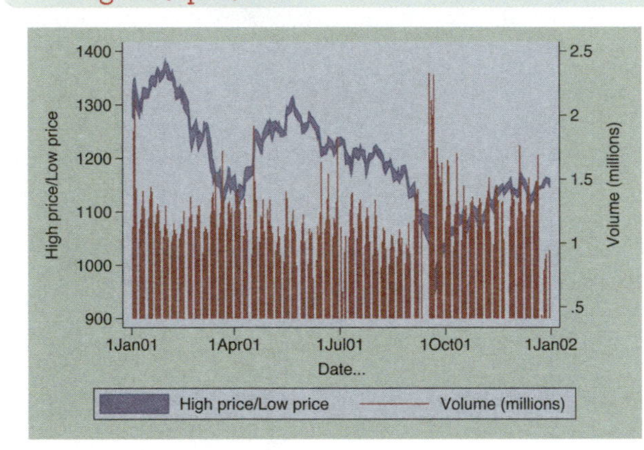

By placing **volmil** on the second y-axis using the **yaxis(2)** option, we can now see the volume, but it obstructs the stock prices. Note that we added the option **legend(span)** to allow the legend to be wider than the plot region of the graph.
Uses sp2001ts.dta & scheme vg_teal

```
twoway (rarea high low date) (spike volmil date, yaxis(2)),
   legend(span) yscale(range(500 1400) axis(1)) yscale(range(0 5) axis(2))
```

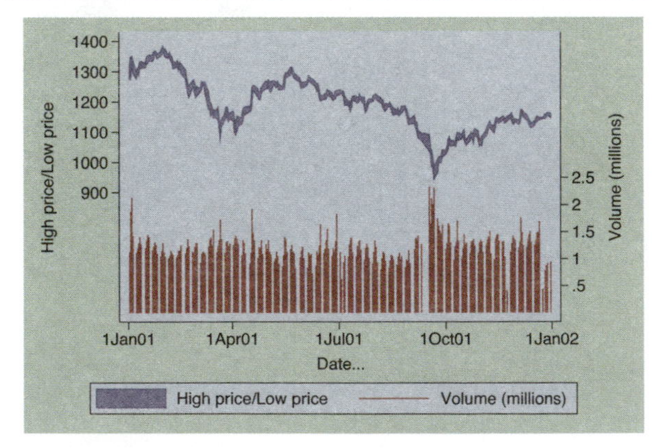

We use the **yscale()** option to modify the range for the first y-axis to lift its range into the top third of the graph, and another **yscale()** option to modify the range for the second y-axis, pushing the stock market volume down to the bottom third.
Uses sp2001ts.dta & scheme vg_teal

While the previous examples (and other examples in this book) have used the parenthetical notation for overlaid graphs, Stata also permits double vertical bars (||) for separating graphs. To illustrate this, some of the graphs from above will be repeated using this notation. These examples will be shown using the vg_s2m scheme.

```
twoway scatter propval100 urban || lfit propval100 urban
```

We switch back to the allstates data file. Here, the || notation is used to separate the scatter command from the lfit command.
Uses allstates.dta & scheme vg_s2m

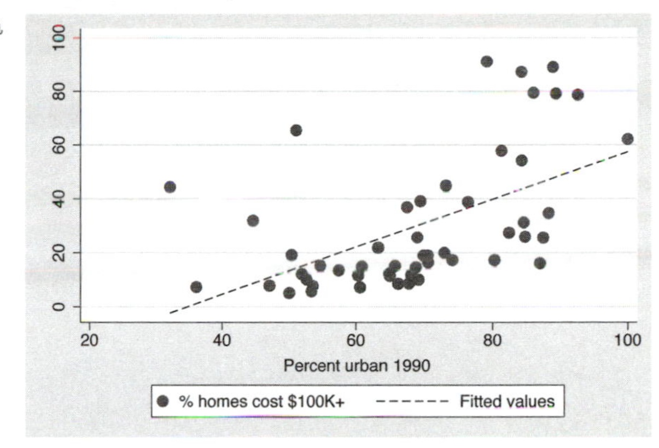


```
twoway scatter propval100 urban || lfit propval100 urban ||
   qfit propval100 urban
```

Here, we create three overlaid graphs using the || notation.
Uses allstates.dta & scheme vg_s2m

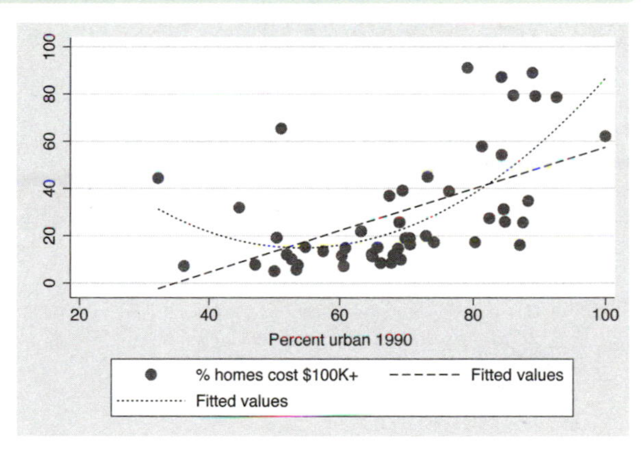

Introduction Twoway Matrix Bar Box Dot Pie Options Standard options Styles Appendix
Scatter Fit CI fit Line Area Bar Range Distribution Options Overlaying

```
twoway scatter propval100 urban, msymbol(Oh) ||
   lfit propval100 urban, clwidth(thick) ||
   qfit propval100 urban, clwidth(medium)
```

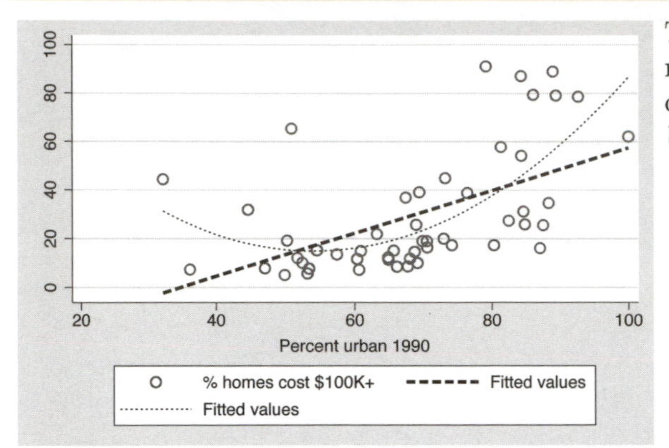

This example shows how to use the || notation with options for each of the commands.
Uses allstates.dta & scheme vg_s2m

```
twoway scatter propval100 urban, msymbol(Oh) ||
   lfit propval100 urban, clwidth(thick) ||
   qfit propval100 urban, clwidth(medium) ||,
   legend(label(2 Linear Fit) label(3 Quad Fit))
```

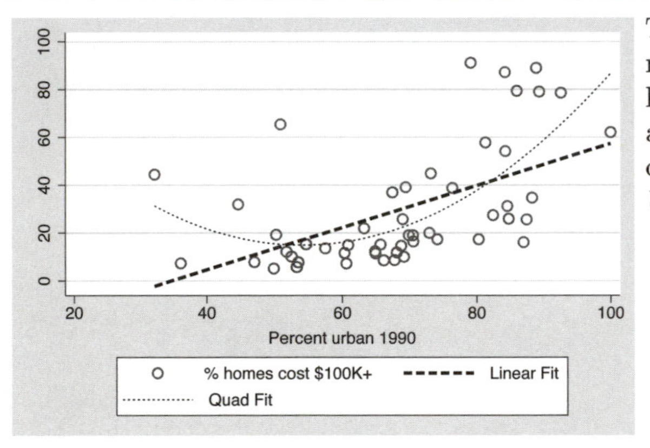

This is another example using the || notation, in this case illustrating how to have options on each of the commands, along with the `legend()` as an overall option.
Uses allstates.dta & scheme vg_s2m

3 Scatterplot matrix graphs

This chapter will explore the use of the **graph matrix** command for creating scatterplot matrices among two or more variables. Many of the options that you can use with **graph twoway scatter** apply to these kinds of graphs, as well; see Twoway : Scatter (35) and Options (235) for related information. This chapter illustrates the use of marker options and marker labels, as well as options for controlling the display of axes. It also includes options specific to the **graph matrix** command, as well as the use of the by() option. For more details about scatterplot matrices, see [G] **graph matrix**.

3.1 Marker options

This section looks at controlling and labeling the markers in scatterplot matrices. This section will show how to change the marker symbol, size, and color (both fill and outline color) and how to label the markers. You can label markers using the **graph matrix** command just as you could when using the **graph twoway scatter** command. See also Options : Markers (235) and Options : Marker labels (247) for more details. These examples will use the vg_s1m scheme.

```
graph matrix propval100 ownhome borninstate, msymbol(Oh)
```

You can control the marker symbol with the msymbol() (marker symbol) option. Here, we make the symbols hollow circles. Other values that we could specify include D (diamond), T (triangle), S (square), and X (x). Using a lowercase letter (d instead of D) makes the symbol smaller. For circles, diamonds, triangles, and squares, you can append an h (e.g., Oh) to indicate that the symbol should be hollow; see Styles : Symbols (342) for more examples.

Uses allstates.dta & scheme vg_s1m

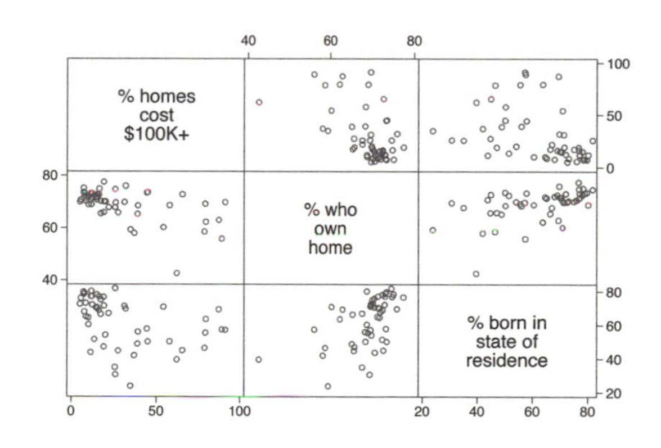

```
graph matrix heatdd cooldd tempjan tempjuly, msymbol(p)
```

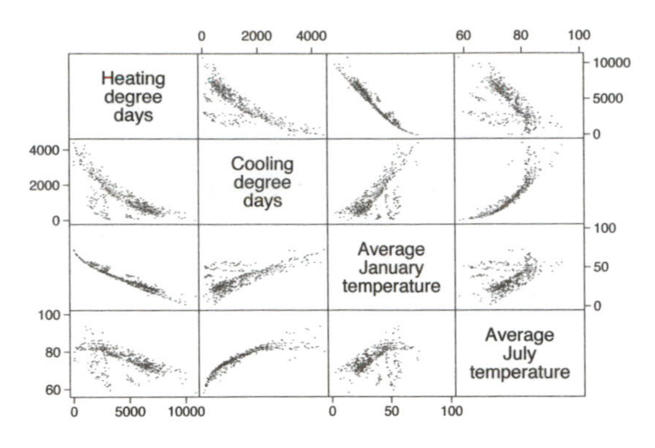

When you have a large number of observations, the `msymbol(p)` option can be very useful since it displays a very small point for each observation and can help you to see the overall relationships among the variables. Here, we switch to the `citytemp` data file to illustrate this.
Uses citytemp.dta & scheme vg_s1m

```
graph matrix propval100 ownhome borninstate, msize(vlarge)
```

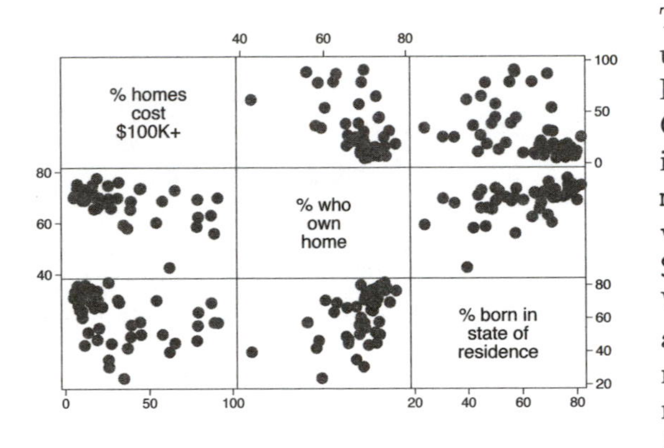

The size of the markers can be changed using the `msize()` (marker size) option. Here, we make the markers very large. Other values we could have chosen include `vtiny`, `tiny`, `vsmall`, `small`, `medsmall`, `medium`, `medlarge`, `large`, `vlarge`, `huge`, `vhuge`, and `ehuge`; see Styles : Markersize (340) for more details. We also could have specified the size as a multiple of the original size of the marker; e.g., `msize(*2)` makes the marker twice as big.
Uses allstates.dta & scheme vg_s1m

```
graph matrix propval100 ownhome borninstate, mcolor(gs8)
```

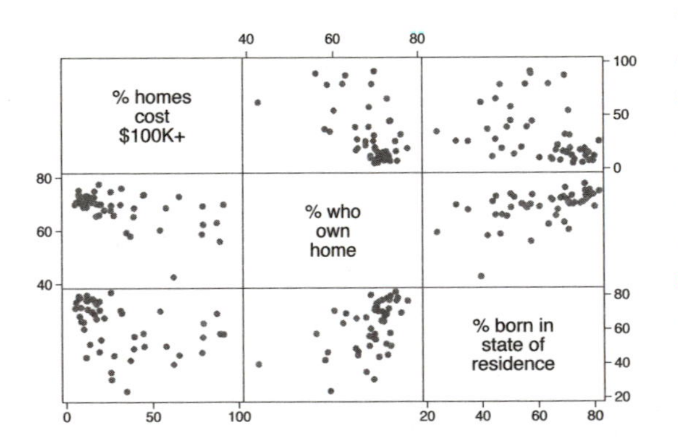

The `mcolor()` (marker color) option can be used to control the color of the symbols. Among the colors you can choose are 16 gray-scale colors named `gs0` (black) to `gs16` (white). We show a graph using symbols that are in the middle of this scale using the `mcolor(gs8)` option; see Styles : Colors (328) for more information about specifying colors.
Uses allstates.dta & scheme vg_s1m

```
graph matrix propval100 ownhome borninstate, msize(vlarge)
    mfcolor(gs13) mlcolor(gs0)
```

The `mfcolor()` (marker fill color) and `mlcolor()` (marker line color) options allow you to control the fill color (inside color) and outline color (periphery color) of the markers. Below, we make the fill color light gray by specifying `mfcolor(gs13)` and the line color black by specifying `mlcolor(gs0)`. We use the `msize()` option to make the markers very large to help see the effect of these options.
Uses allstates.dta & scheme vg_s1m

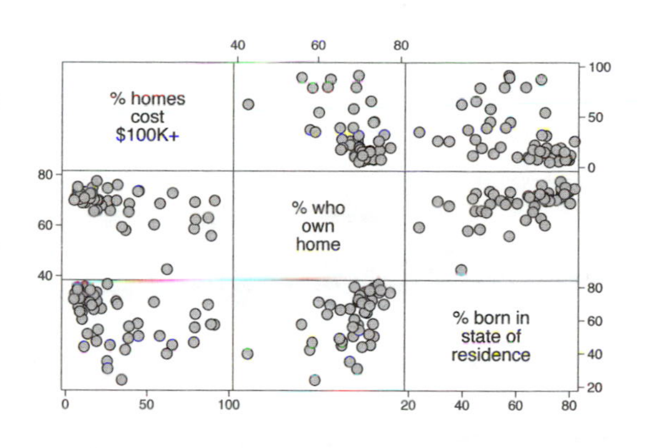


```
graph matrix propval100 ownhome borninstate, mlabel(stateab)
```

We can label the markers using the `mlabel()` (marker label) option. In this example, we label the markers with the two-letter postal abbreviation by supplying the option `mlabel(stateab)`. Even though many of the labels overlap, the most interesting observations are those that stand out and have readable labels, such as DC and NV. For additional details, see Options : Marker labels (247).
Uses allstates.dta & scheme vg_s1m

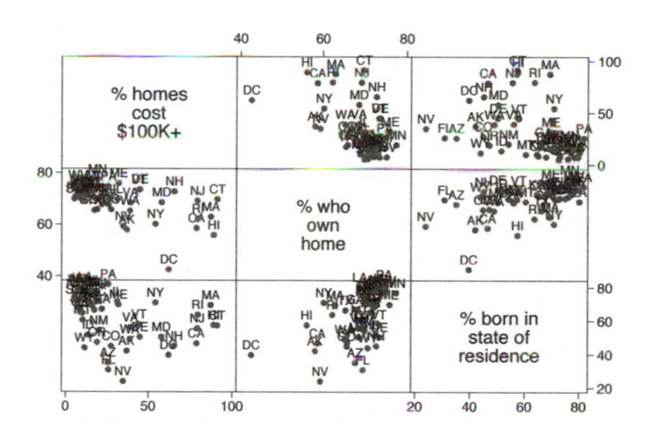


```
graph matrix propval100 ownhome borninstate, mlabel(stateab)
    mlabsize(large)
```

You can use the `mlabsize()` (marker label size) option to control the size of the marker label. Here, we indicate that the marker labels should be large. You can also specify the size of the marker label as a multiple of the original size of the marker label; e.g., specifying `mlabsize(*1.5)` would make the labels 1.5 times their normal size.
Uses allstates.dta & scheme vg_s1m

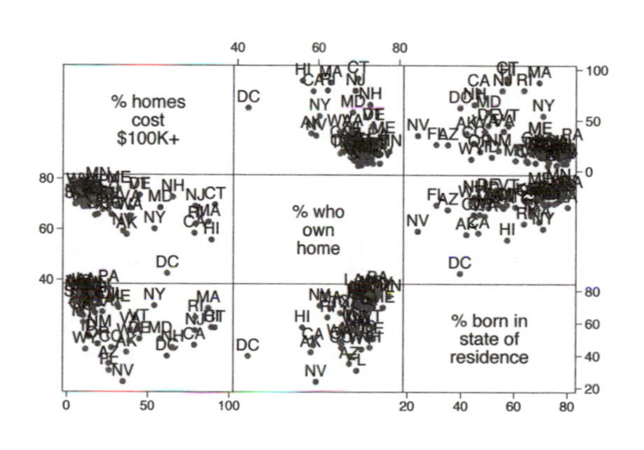

3.2 Controlling axes

This section looks at labeling axes in scatterplot matrices. It shows how to label axes of scatterplots, control the scale of axes, and insert titles along the diagonal. For more details, see Options : Axis labels (256), Options : Axis scales (265) and [G] ***axis_options***. This section uses the vg_s2c scheme.

```
graph matrix urban propval100 borninstate
```

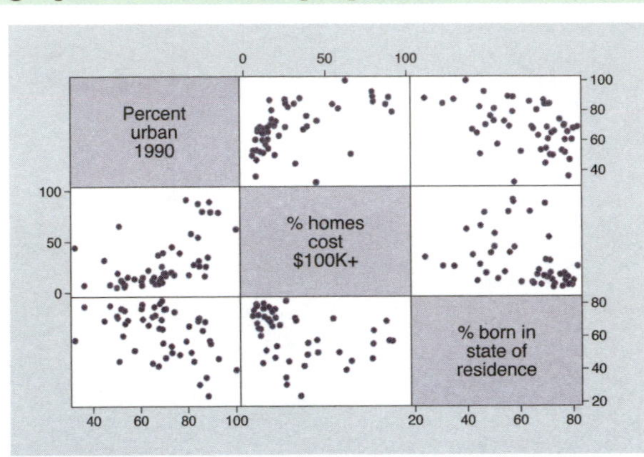

Let's look at a scatterplot matrix of three variables: `urban`, `propval100`, and `borninstate`.
Uses allstates.dta & scheme vg_s2c

```
graph matrix urban propval100 borninstate,
    xlabel(30(10)100, axis(1)) ylabel(30(10)100, axis(1))
```

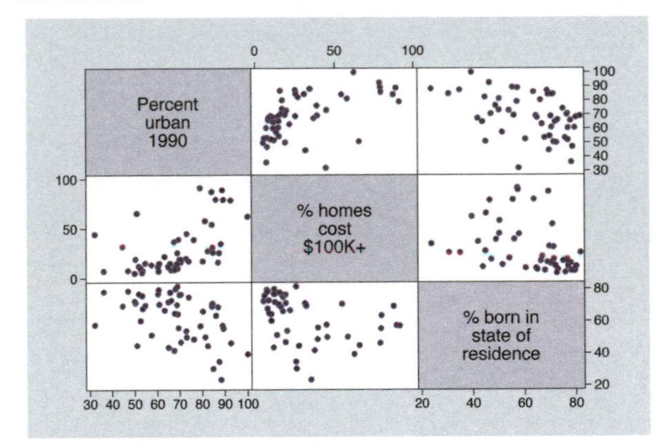

The way you control the axis labels with a scatterplot matrix is somewhat different than with other kinds of graphs. Here, we use the `xlabel()` and `ylabel()` options to control the x- and y-labels for the first variable, `urban`, to be scaled 30 to 100 in increments of 10. This applies to the first variable because we specified the `axis(1)` option.
Uses allstates.dta & scheme vg_s2c

```
graph matrix urban propval100 borninstate,
   xlabel(0(20)100, axis(2)) ylabel(0(20)100, axis(2))
```

We can change the label for the second variable, `propval100`, in a similar manner, but we need to specify `axis(2)`. In this example, we label the second variable ranging from 0 to 100 in increments of 20.
Uses allstates.dta & scheme vg_s2c

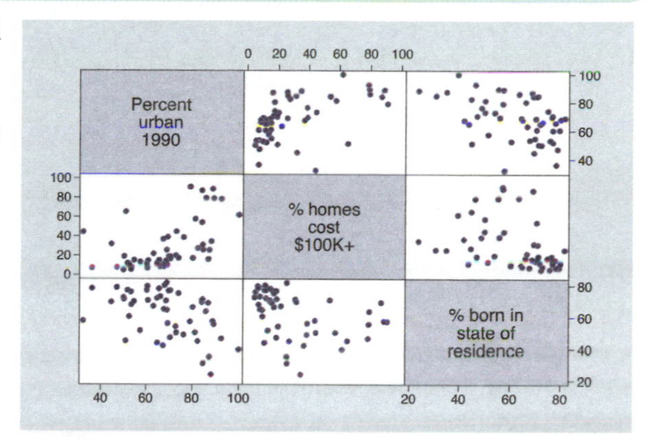


```
graph matrix urban propval100 borninstate,
   xlabel(0(20)100, axis(1)) ylabel(0(20)100, axis(1))
   xlabel(0(20)100, axis(2)) ylabel(0(20)100, axis(2))
   xlabel(0(20)100, axis(3)) ylabel(0(20) 100, axis(3))
```

Let's label all these variables using the same scale, from 0 to 100 in increments of 20. As you can see, this involves quite a bit of typing, applying the `xlabel()` and `ylabel()` for `axis(1)`, `axis(2)`, and `axis(3)`, which applies this to the first, second, and third variables. However, the next example shows a more efficient way to do this.
Uses allstates.dta & scheme vg_s2c

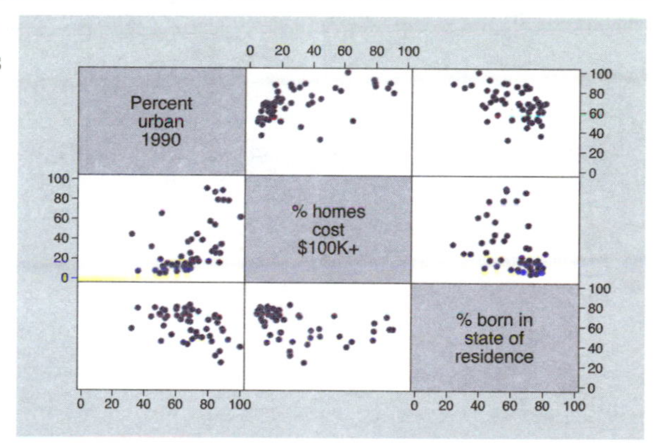

Introduction Twoway Matrix Bar Box Dot Pie Options Standard options Styles Appendix
Marker options Axes Matrix options By

```
graph matrix urban propval100 borninstate,
   maxes(xlabel(0(20)100) ylabel(0(20)100))
```

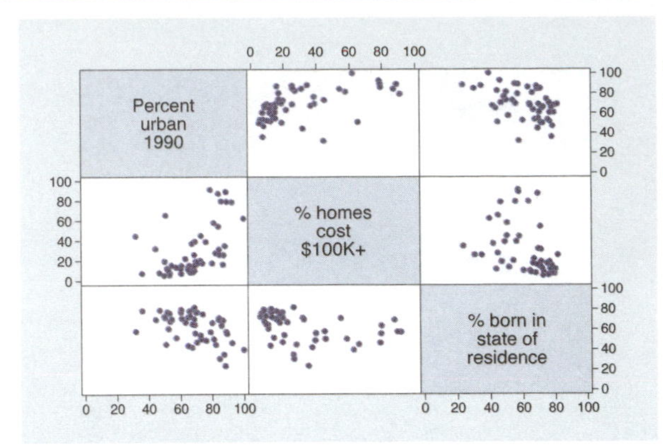

Stata has a simpler way of applying the same labels to all the variables in the scatterplot matrix by using the maxes() (multiple axes) option. This example labels the x- and y-axes from 0 to 100 with increments of 20 for all variables.
Uses allstates.dta & scheme vg_s2c

```
graph matrix urban propval100 borninstate,
   maxes(xlabel(0(20)100) ylabel(0(20)100))
   xlabel(20(20)100, axis(1)) ylabel(20(20)100, axis(1))
```

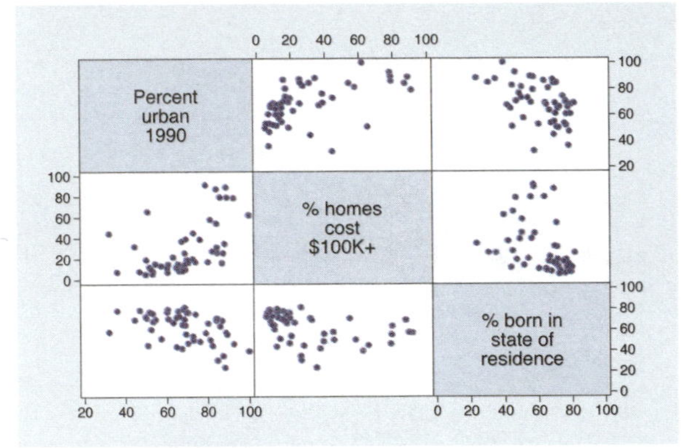

You might want to label most of the variables in the scatterplot matrix the same way but with one or more exceptions in a different way. In this example, we label all the variables from 0 to 100, incrementing by 20, but then override the labeling for urban to make it 20 to 100, incrementing by 20. We do this by adding additional xlabel() and ylabel() options that apply just for axis(1).
Uses allstates.dta & scheme vg_s2c

```
graph matrix urban propval100 borninstate,
   maxes(xlabel(0(20)100) ylabel(0(20)100) xtick(0(10)100) ytick(0(10)100))
```

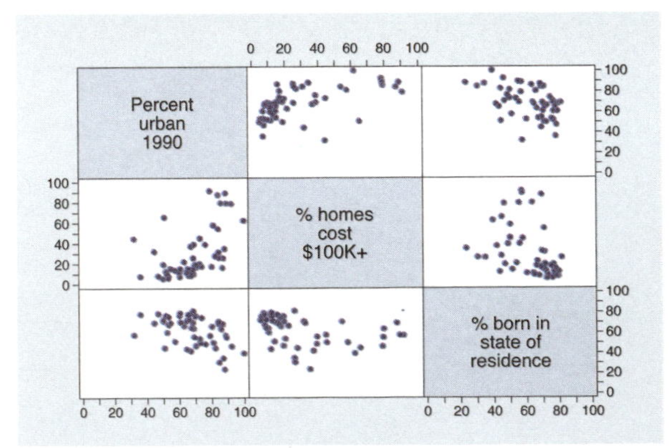

Here, we label all of the variables from 0 to 100, in increments of 20, and also add ticks from 0 to 100, in increments of 10. Note that the xtick() and ytick() options work the same way as the xlabel() and ylabel() options. We place these options within the maxes() option, and they apply to all of the axes. See Options : Axis labels (256) and Options : Axis scales (265) for more details.
Uses allstates.dta & scheme vg_s2c

```
graph matrix urban propval100 borninstate,
   diagonal("% Urban" "% Homes Over $100K" "% Born in State")
```

When you use `twoway scatter`, you can use `xtitle()` and `ytitle()` to control the titles for the axes. By contrast, when using `graph matrix`, you can control the titles that are displayed along the diagonal with the `diagonal()` option. We use the `diagonal()` option to change the titles for all variables.
Uses allstates.dta & scheme vg_s2c

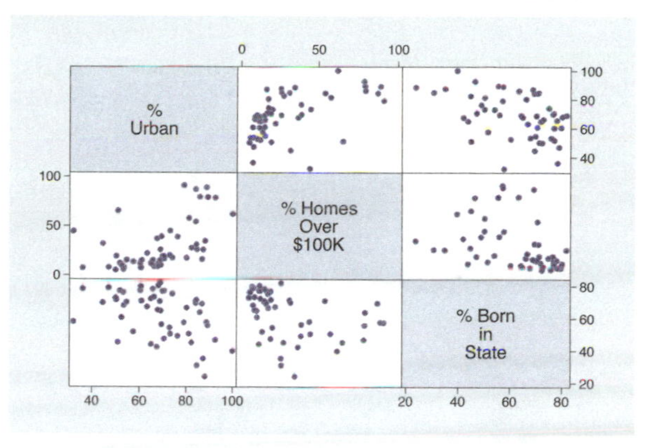


```
graph matrix urban propval100 borninstate,
   diagonal("% Urban" .  "% Born in State")
```

We do not have to change all the titles. If we want to change just some of the titles, we can place a period (.) for the labels where we want the label to stay the same. In this example, we change the titles for the first and third variables but leave the second as is.
Uses allstates.dta & scheme vg_s2c

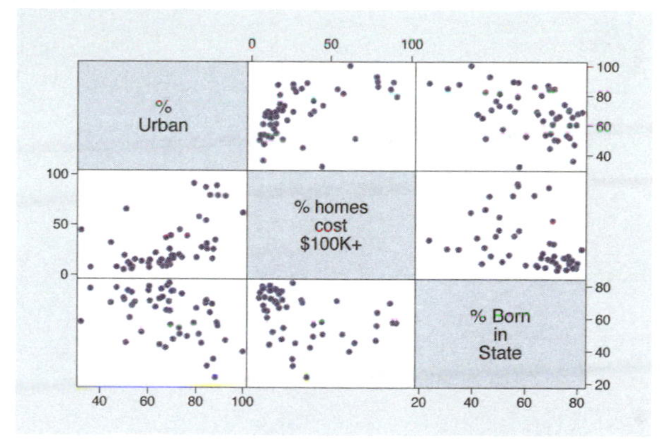


```
graph matrix urban propval100 borninstate,
   diagonal("% Urban" .  "% Born in State", bfcolor(eggshell))
```

We can control the display of the text on the diagonal using textbox options. For example, we make the background color of the text area eggshell using the `bfcolor(eggshell)` option. See Options : Textboxes (303) for more examples of textbox options.
Uses allstates.dta & scheme vg_s2c

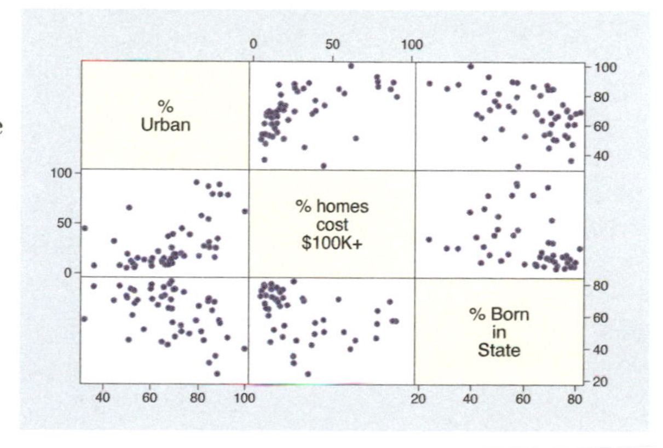

3.3 Matrix options

This section shows options that you can use to control the look of the scatterplot matrix, including showing just the lower half of the matrix, jittering markers, and scaling the size of marker text. For more details, see [G] **graph matrix**. These graphs use the **vg_s2m** scheme.

```
graph matrix propval100 ownhome region, half
```

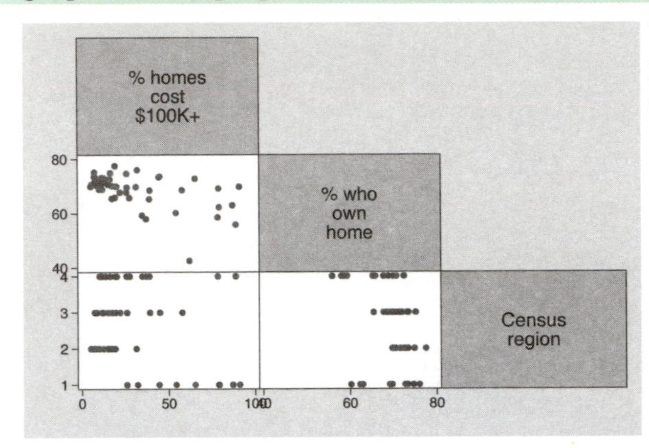

You can use the `half` option to display just the lower diagonal of the scatterplot matrix.
Uses allstates.dta & scheme vg_s2m

```
graph matrix propval100 ownhome region, jitter(3)
```

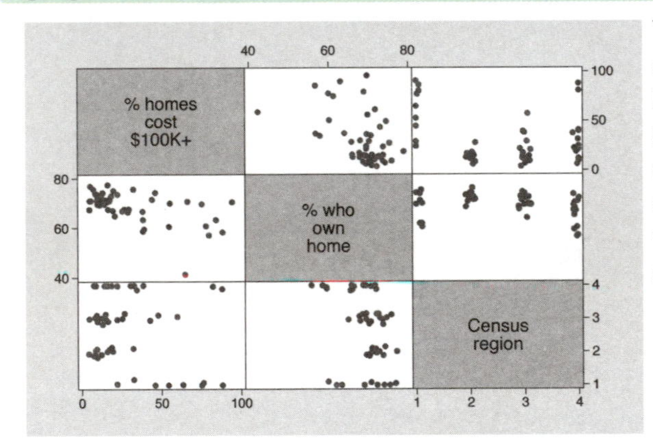

You can use the `jitter()` option to add random noise to the points; the higher the value given, the more random noise is added. This is especially useful when numerous observations have the same (x,y) values, so a number of observations can appear as a single point.
Uses allstates.dta & scheme vg_s2m

```
graph matrix propval100 ownhome region, scale(1.5)
```

The scale() option can be used to magnify the contents of the graph, including the markers, labels, and lines, but not the overall size of the graph. Here, we increase the size of these items, making them 1.5 times their normal size. Note that, unlike other similar options, this option does not take an asterisk preceding the multiplier; i.e., we specify 1.5 but not *1.5.
Uses allstates.dta & scheme vg_s2m

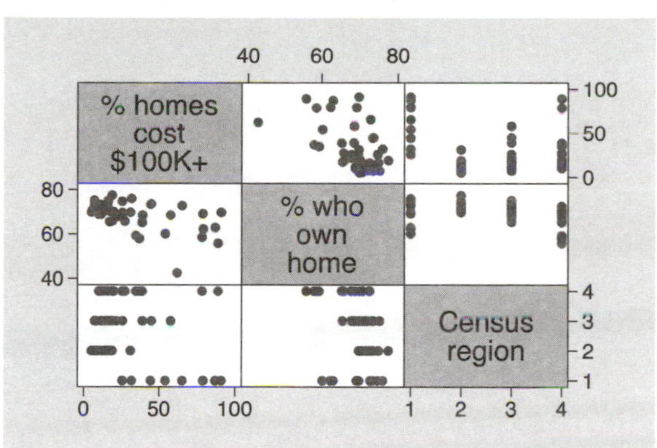

3.4 Graphing by groups

This section looks at the use of the by() option for showing separate graphs based on the levels of a by() variable. For more information, see Options: By (272) and [G] ***by_option***. This section uses the vg_brite scheme.

```
graph matrix propval100 ownhome borninstate, by(north)
```

The by() option can be used with **graph matrix** to show separate scatterplot matrices by a particular variable. Here, we show separate scatterplot matrices for households in northern states and non-northern states.
Uses allstates.dta & scheme vg_brite

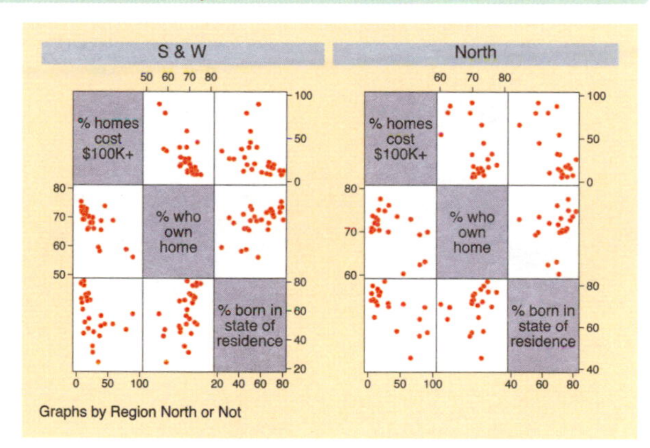

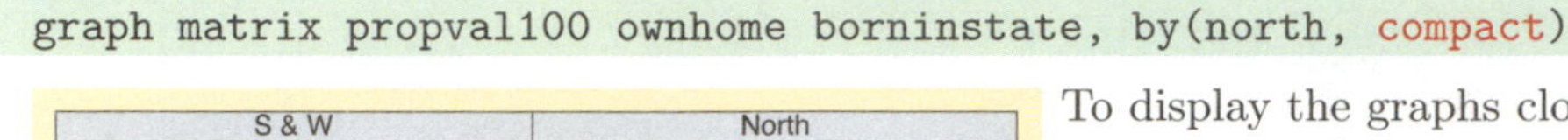

```
graph matrix propval100 ownhome borninstate, by(north, compact)
```

To display the graphs closer together, you can use the compact option.
Uses allstates.dta & scheme vg_brite

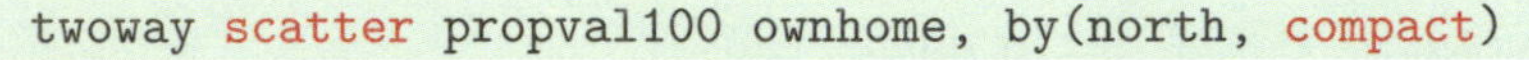

```
twoway scatter propval100 ownhome, by(north, compact)
```

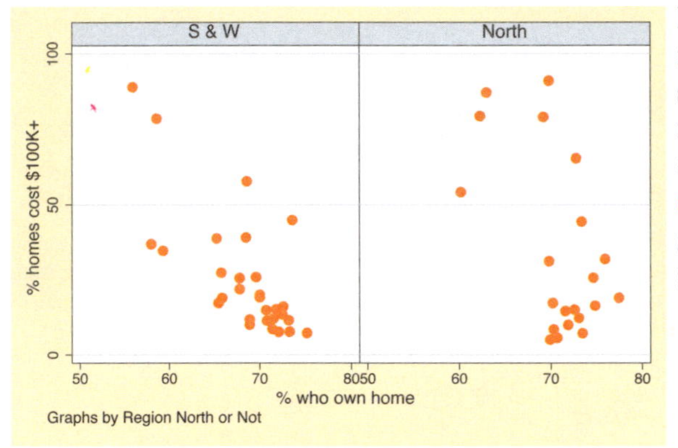

If we compare the previous scatterplot matrix to this twoway scatterplot, we see that the `compact` option does not make the scatterplot matrix as compact as it does with a regular `twoway scatter` command, which joins the two graphs on their edges by omitting the y-labels between the two graphs.
Uses allstates.dta & scheme vg_brite

```
graph matrix propval100 ownhome borninstate, by(north, compact)
    maxes(ylabel(, nolabels))
```

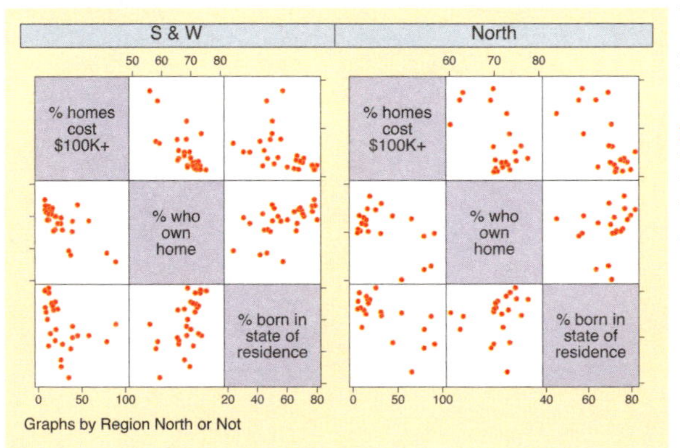

We can make the `graph matrix` display more compactly with the `by()` option by using the `maxes(ylabel(, nolabels))` option to suppress the labels on all of the y-axes. Then, when we use the `compact` option, the edges of the plots are pushed closer together.
Uses allstates.dta & scheme vg_brite

```
graph matrix propval100 ownhome borninstate, by(north, compact scale(*1.3))
   maxes(ylabel(, nolabels))
```

We can use the scale() option to increase the size of the markers, labels, and text to make them more readable. This is especially useful when graphs get small.

Uses allstates.dta & scheme vg_brite

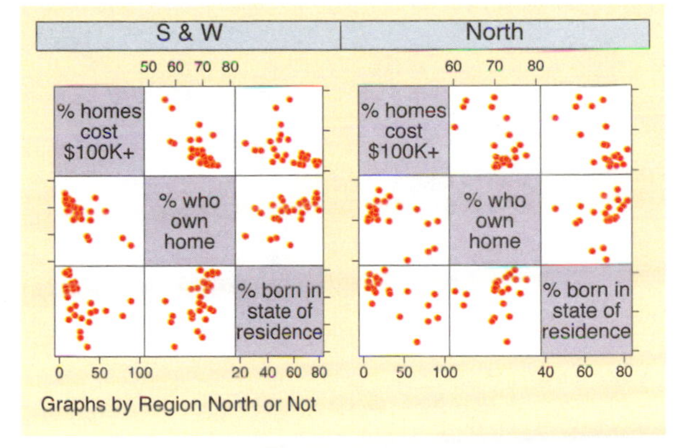

4 Bar graphs

This chapter will explore how to create bar charts using the **graph bar** command. It will show how you can use **graph bar** to graph one or more continuous y-variables and how you can break them down by one or more categorical variables. In addition, this chapter will illustrate how you can control the display of each of the axes, the legend, and the look of the bars, and how to use the `by()` option. We will start this chapter by looking at features related to graphing one or more y-variables. For this entire chapter, we will use the `nlsw` data file.

4.1 Y-variables

A bar chart graphs one or more continuous variables broken down by one or more categorical variables. The continuous variables are graphed on the y-axis and are referred to as y-variables. This section shows you how to specify the y-variables using the **graph bar** command, how to include one or more y-variables, and how to obtain different summary statistics for the y-variables. For more information, see [G] **graph bar**. This section begins using the `vg_past` scheme.

```
graph bar ttl_exp
```

This is probably the most basic bar chart that you can make (and perhaps the most boring, as well). It shows the average total work experience for all observations in the file. It graphs a single y-variable using the default summary statistic, the mean.
Uses nlsw.dta & scheme vg_past

```
graph bar prev_exp tenure ttl_exp
```

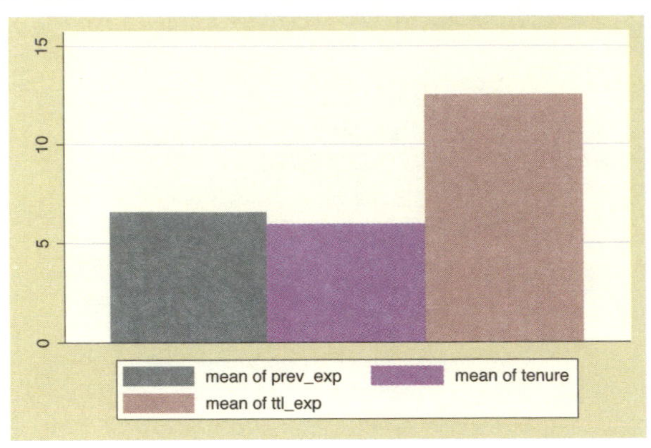

You can specify multiple y-variables to be plotted at one time. Here, we graph the mean of previous, current, and total work experience in the same plot. The bars are plotted touching each other, and a legend indicates which bar corresponds to which variable.
Uses nlsw.dta & scheme vg_past

```
graph bar (median) prev_exp tenure ttl_exp
```

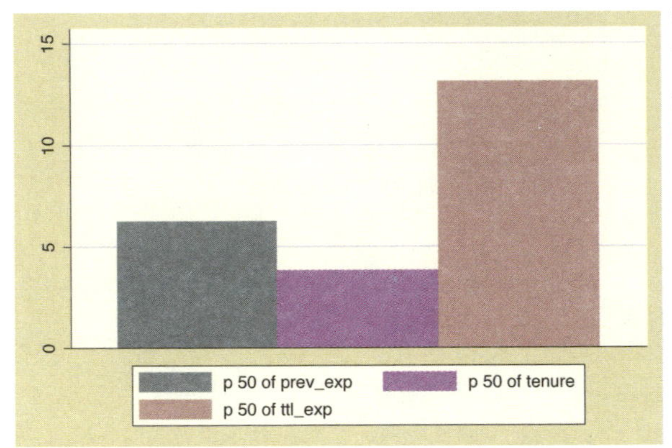

This graph is much like the last one, but it shows the median of these y-variables. Note that we only specified `(median)` before `prev_exp` but it applied to all the y-variables that follow. You can summarize the y-variables using any of the summary statistics permitted by the `collapse` command (e.g., `mean`, `sd`, `sum`, `median`, and `p10`); see [R] **collapse**.
Uses nlsw.dta & scheme vg_past

```
graph bar (median) prev_exp tenure (mean) ttl_exp
```

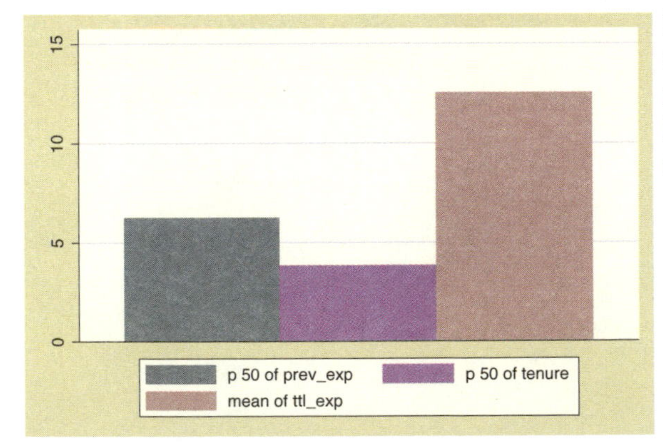

In this example, we get the median of the first two y-variables and then the mean of the last y-variable. I don't know, however, how often you would do this.
Uses nlsw.dta & scheme vg_past

```
graph bar (mean) meanwage=wage (median) medwage=wage
```

You can plot different summary statistics for the same y-variable, but you must specify a target name for the statistic being created. Here, we create `meanwage` for the mean of `wage` and `medwage` for the median of `wage`. If we omitted the `meanwage=` and `medwage=` from this command, Stata would return an error indicating that the name for the mean of `wage` conflicts with the median of `wage`.
Uses nlsw.dta & scheme vg_past

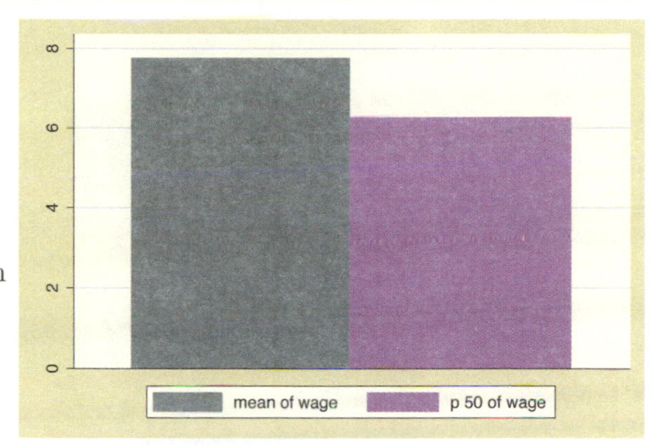

We now consider a handful of options that are useful when you have multiple y-variables. These options allow you to display the y-variables as though they were categories of the same variable, to create stacked bar charts, and to display the y-variables as percentages of the total y-variables. These options are illustrated in the following graphs using the `vg_s1m` scheme.

```
graph bar prev_exp tenure ttl_exp hours
```

First, consider this bar chart showing four y-variables. Each y-variable is shown with a different colored bar and with a legend indicating which y-variable corresponds to which bar. See the next example for another way to differentiate these four bars.
Uses nlsw.dta & scheme vg_s1m

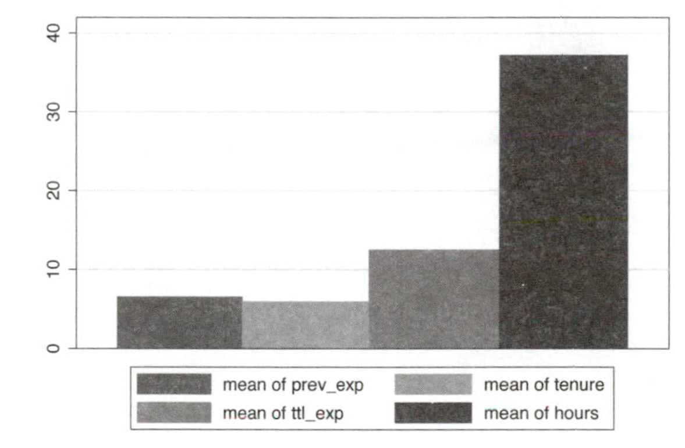

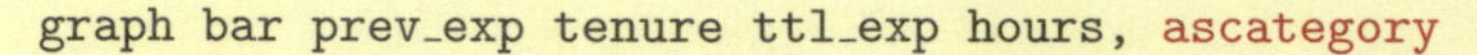

```
graph bar prev_exp tenure ttl_exp hours, ascategory
```

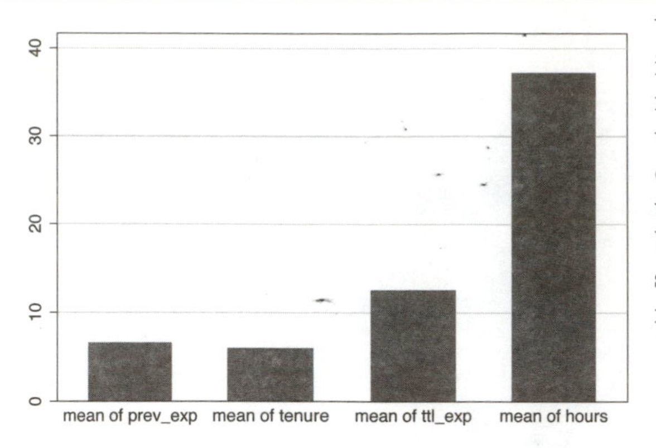

You can use the **ascategory** option to indicate that you want Stata to graph multiple y-variables using the style that would be used for the levels of an `over()` variable. Comparing this graph with the previous graph, note how the bars for the different variables are the same color and labeled on the x-axis rather than using a legend.
Uses nlsw.dta & scheme vg_s1m

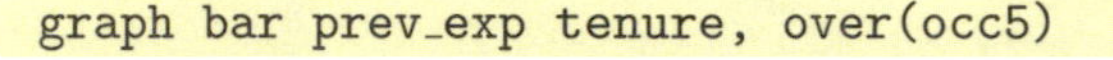

```
graph bar prev_exp tenure, over(occ5)
```

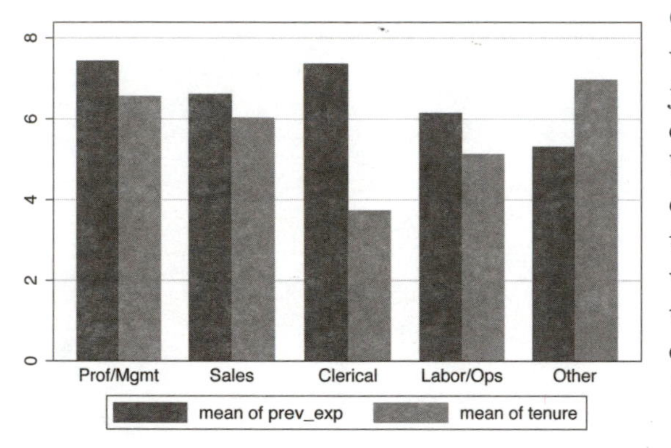

Consider this graph, where we show work experience prior to one's current job (`prev_exp`) and work experience at one's current job (`tenure`) broken down by `occ5`. The total of previous and current work experience represents total work experience, and you might want to show each bar as a percent of total work experience. The next example shows how you can do that.
Uses nlsw.dta & scheme vg_s1m

```
graph bar prev_exp tenure, over(occ5) percentages
```

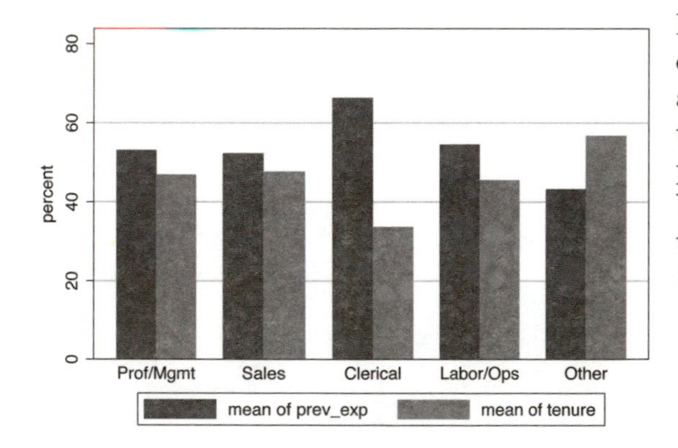

Here, we show the time worked before one's current job, `prev_exp`, and time at the current job, `tenure`, in terms of their percentage of the total (i.e., percentage of total work experience). We can view the bars in this way using the **percentages** option.
Uses nlsw.dta & scheme vg_s1m

```
graph bar prev_exp tenure, over(occ5) stack
```

The stack option shows the y-variables as a stacked bar chart. This allows you to see the mean of each y-variable, as well as the mean of the total y-variables.
Uses nlsw.dta & scheme vg_s1m

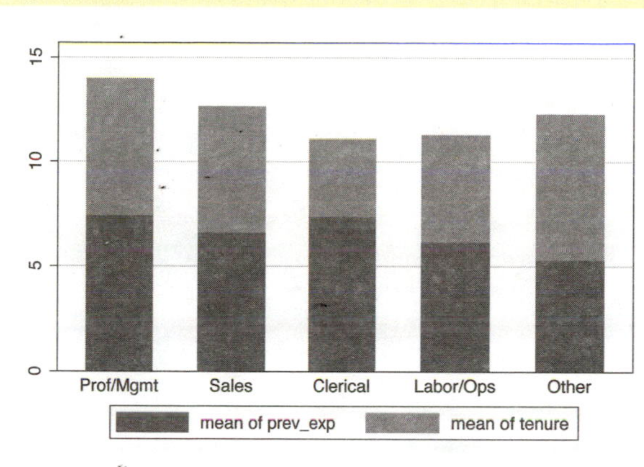


```
graph bar prev_exp tenure, over(occ5) percentages stack
```

We can also combine the stack and percentages options to create a stacked bar chart in terms of percentages.
Uses nlsw.dta & scheme vg_s1m

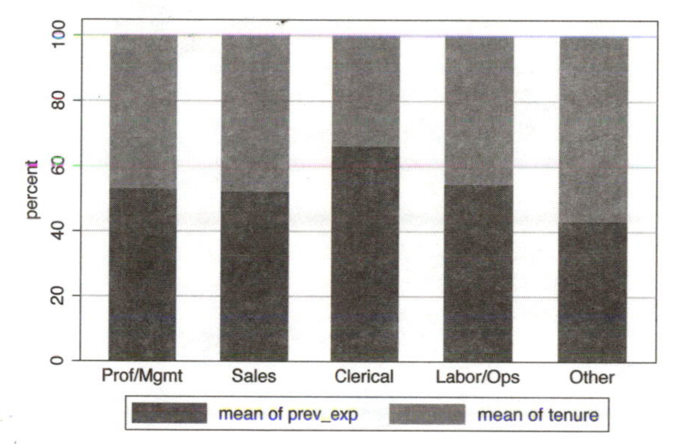

4.2 Graphing bars over groups

This section focuses on the use of the over() option for showing bar charts by one or more categorical variables. It illustrates the use of the over() option with a single y-variable and with multiple y-variables. We also look at some basic options, including options for displaying the over() variable as though its levels were multiple y-variables, including missing values on the over() variable, and suppressing empty combinations of multiple over() variables. See the *group_options* and *over_subopts* tables of [G] **graph bar** for more details.

```
graph hbar wage, over(occ5)
```

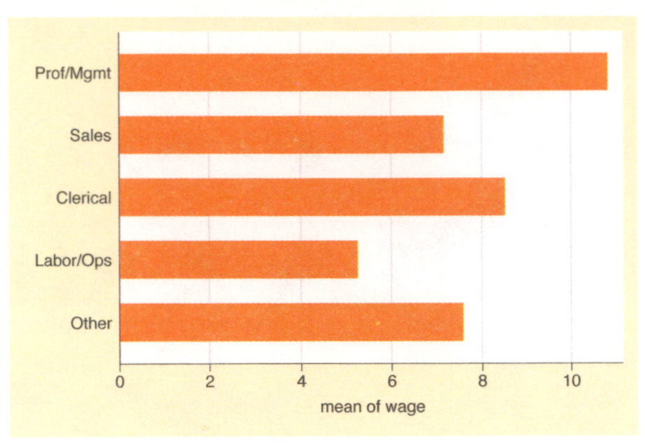

Here, we use the `over()` option to show the average wages broken down by occupation. Note that we are using `graph hbar` to produce horizontal, rather than vertical, bar charts.
Uses nlsw.dta & scheme vg_brite

```
graph hbar wage, over(occ5) over(collgrad)
```

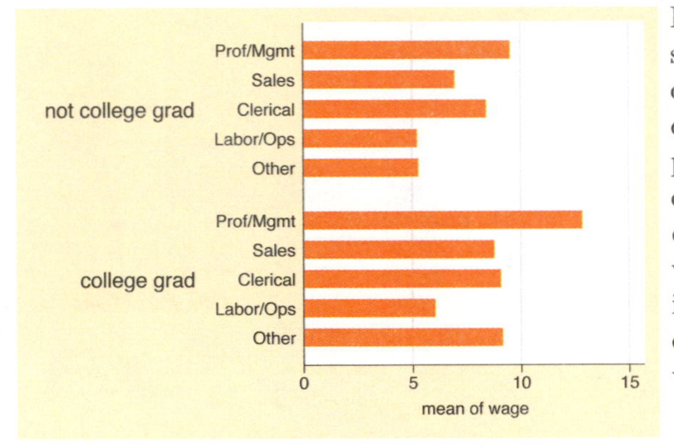

Here, we use the `over()` option twice to show the wages broken down by occupation and whether one graduated college. Note the appropriate way to produce this graph is to use two `over()` options, rather than using a single `over()` option with two variables. As we will see later, each `over()` can have its own options, allowing you to customize the display of each `over()` variable.
Uses nlsw.dta & scheme vg_brite

```
graph hbar wage, over(urban2) over(occ5) over(collgrad)
```

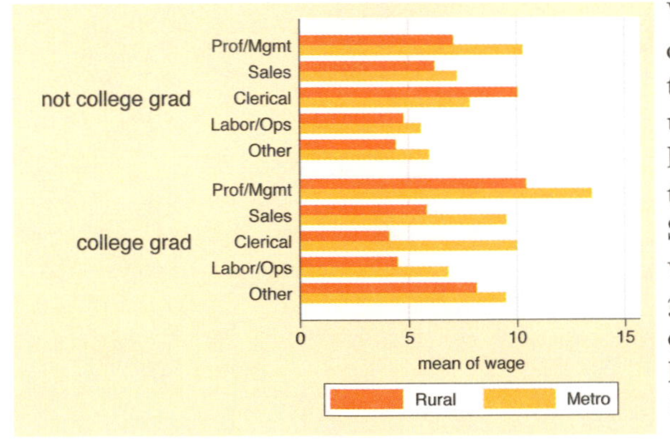

We can even add a third `over()` option, in this case using `over(urban2)` to compare those living in rural versus urban areas. Note the change in the look of the graph when we add the third `over()` variable. This is because Stata is now treating the first `over()` variable as though it were multiple y-variables. Because of this, you can only specify one y-variable when you have three `over()` options.
Uses nlsw.dta & scheme vg_brite

Now, let's look at examples of using multiple y-variables with the `over()` option. We first consider a simple bar graph with multiple y-variables. These examples will use the **vg_lgndc** scheme, which places the legend to the left of the graph and displays it in a single, stacked column.

```
graph hbar prev_exp tenure ttl_exp
```

This graph shows the overall mean of previous, current, and total work experience.
Uses nlsw.dta & scheme vg_lgndc

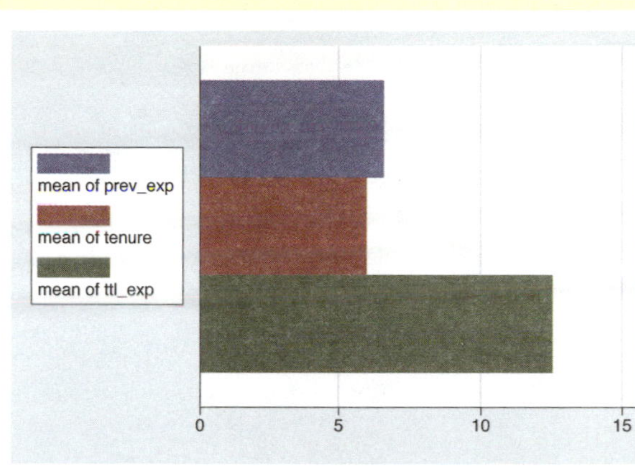


```
graph hbar prev_exp tenure ttl_exp, over(occ5)
```

We can take the graph from above and break the means down by whether one graduated from college by adding the `over(occ5)` option.
Uses nlsw.dta & scheme vg_lgndc

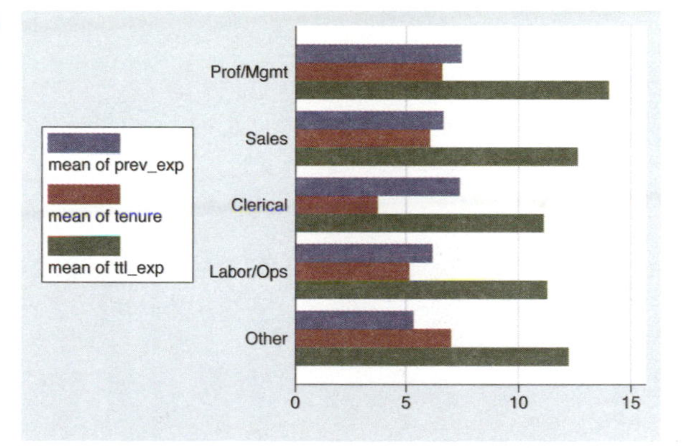

```
graph hbar prev_exp tenure ttl_exp, over(occ5) over(union)
```

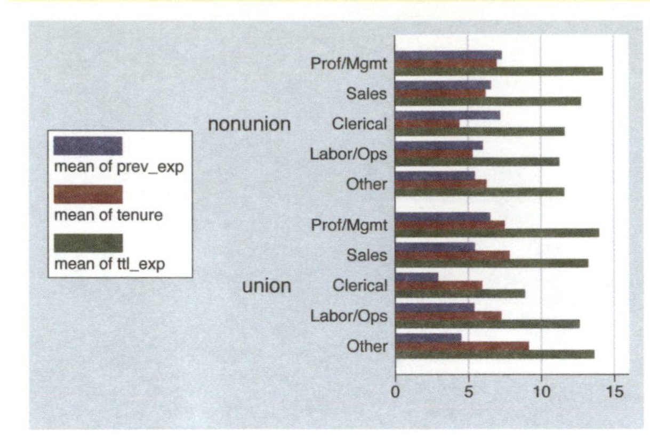

We can take the previous graph and further break the results down by whether one belongs to a union. Note, however, that we cannot add a third `over()` option when we have multiple *y*-variables.
Uses nlsw.dta & scheme vg_lgndc

Now let's consider options that may be used in combination with the `over()` option to customize the behavior of the graphs. We show how you can treat the levels of the variable in the first `over()` option as though they were multiple *y*-variables and can even graph those levels as percentages or stacked bar charts. You can also request that missing values for the levels of the `over()` variables be displayed, and you can suppress empty levels when multiple `over()` options are used. These examples are shown below using the `vg_rose` scheme.

```
graph bar wage, over(occ5) over(union)
```

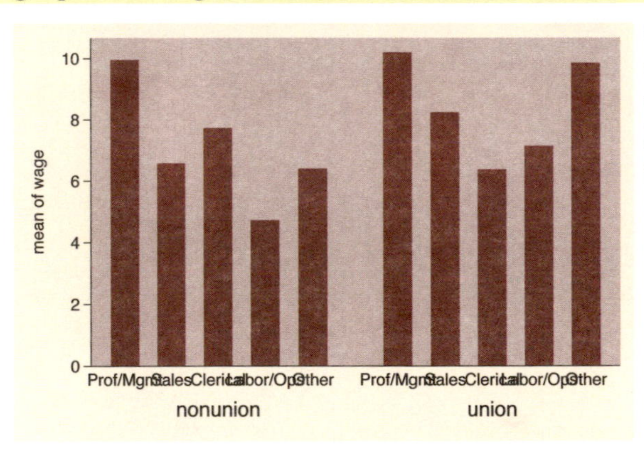

Consider this graph, where we show wages broken down by occupation and whether one belongs to a union. The labels for the levels of `occ5` overlap, but this is mended in the next example.
Uses nlsw.dta & scheme vg_rose

```
graph bar wage, over(occ5) over(union) asyvars
```

If we add the asyvars option, then the first over() variable (occ5) is graphed as if there were five y-variables corresponding to the five levels of occ5. The levels of occ5 are shown as differently colored bars pushed next to each other and labeled using the legend.
Uses nlsw.dta & scheme vg_rose

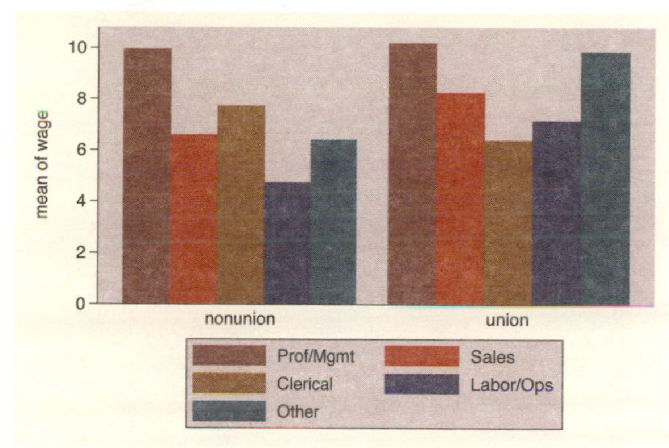


```
graph bar wage, over(occ5) over(union) asyvars percentages
```

With the levels of occ5 considered as y-variables, we can use some of the options that apply when we have multiple y-variables. Here, we request that the values be plotted as percentages.
Uses nlsw.dta & scheme vg_rose

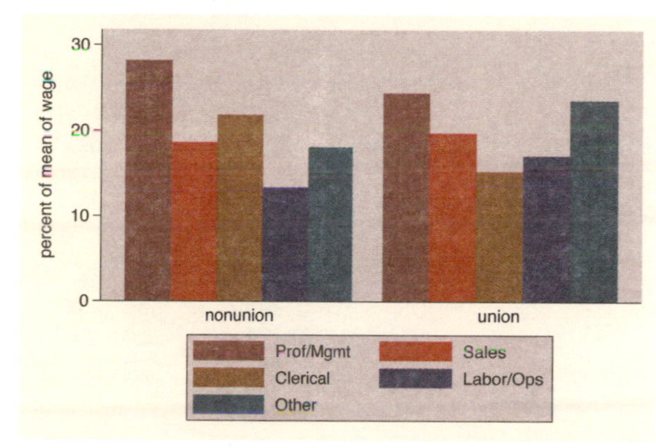


```
graph bar wage, over(occ5) over(union) asyvars percentages stack
```

Again, because we are treating the levels of occ5 as though they were multiple y-variables, we can add the stack option to view the graph as a stacked bar chart.
Uses nlsw.dta & scheme vg_rose

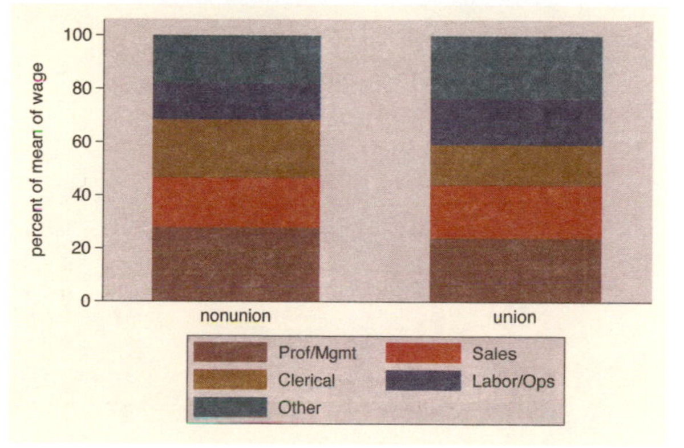

```
graph hbar wage, over(urban3) over(union)
```

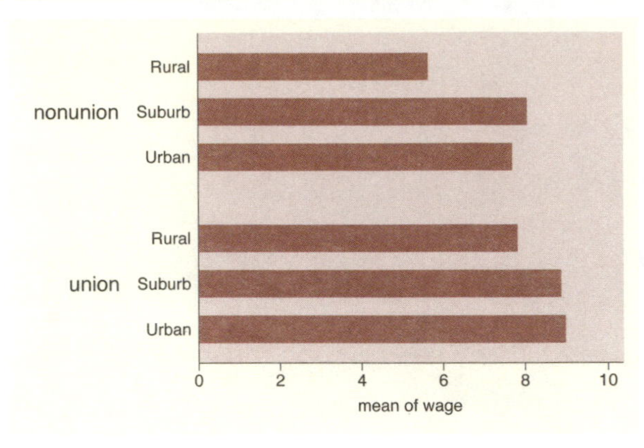

Consider this graph, where we use the `over(union)` option to compare the mean wages of union workers with nonunion workers. One aspect this graph hides is that there are a number of missing values on the variable `union`. *Uses nlsw.dta & scheme vg_rose*

```
graph hbar wage, over(urban3) over(union) missing
```

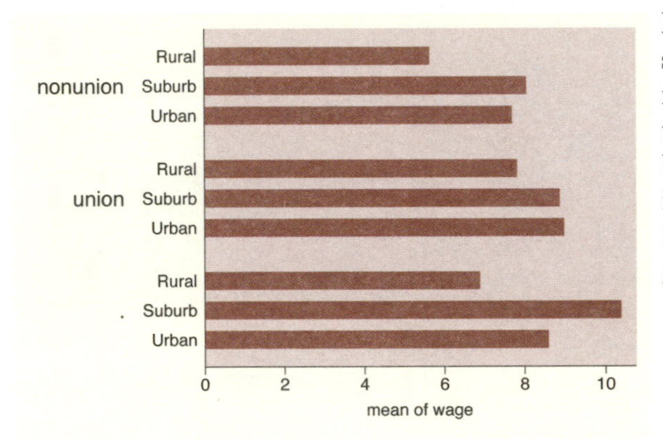

By adding the `missing` option, we then see a category for those who are missing on the `union` variable, shown as the third set of bars. The label for this bar is a single dot, which is the Stata indicator of missing values. The section Bar : Cat axis (123) shows how you can give this bar a more meaningful label. *Uses nlsw.dta & scheme vg_rose*

```
graph bar wage, over(grade) over(collgrad)
```

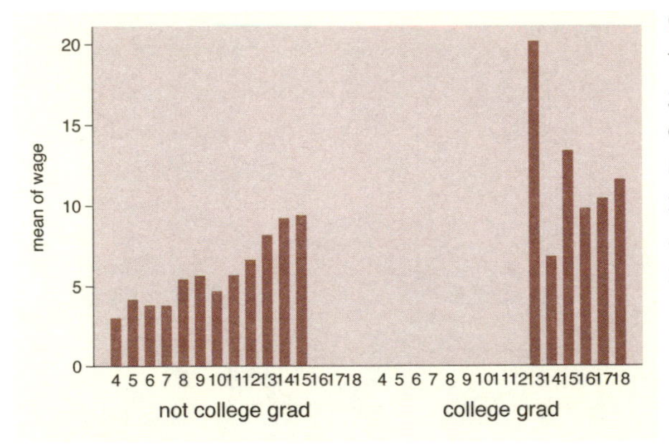

Consider this bar chart, which breaks wages down by two variables: the last grade that one completed and whether one is a college graduate. By default, Stata shows all possible combinations for these two variables. In most cases, all combinations are possible, but not in this case. *Uses nlsw.dta & scheme vg_rose*

```
graph bar wage, over(grade) over(collgrad) nofill
```

If you only want to display the combinations of the over() variables that exist in the data, use the nofill option.
Uses nlsw.dta & scheme vg_rose

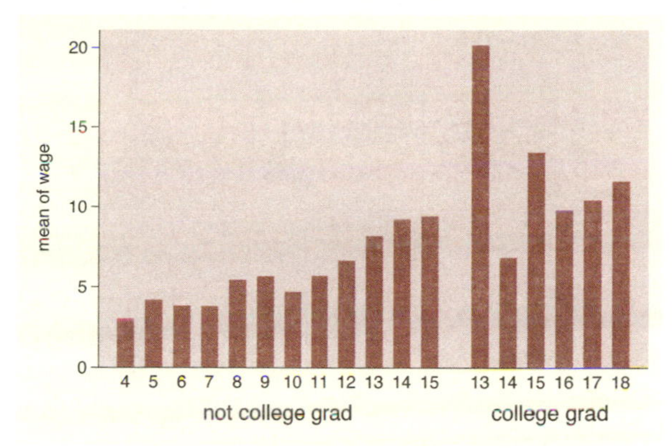

Introduction Twoway Matrix Bar Box Dot Pie Options Standard options Styles Appendix
Y-variables Over Over options Cat axis Legend Y-axis Lookofbar options By

4.3 Options for groups, over options

This section considers some of the options that can be used with the over() and yvaroptions() options for customizing the display of the bars. We will focus on controlling the spacing between the bars and the order in which the bars are displayed. Other options that control the display of the x-axis (such as the labels) are covered in Bar : Cat axis (123). For more information on the over() options covered in this section, see the *over_subopts* table in [G] **graph bar**.

We first consider options that control the spacing among the bars and switch to the vg_s2m scheme.

```
graph hbar wage, over(grade4) over(union)
```

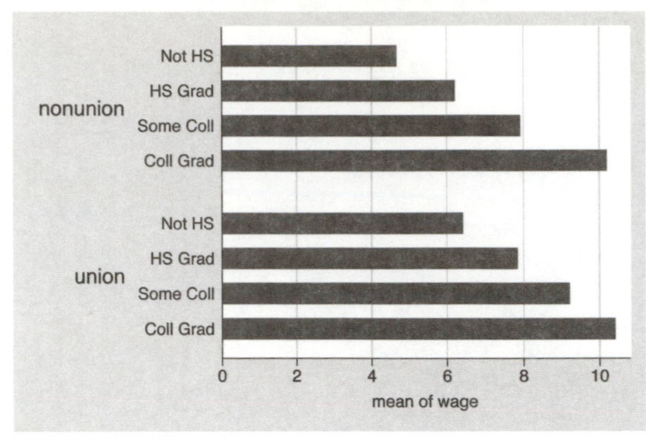

Consider this graph, where we show the mean wages broken down by **grade4** and **union**. Using **graph hbar** displays the chart as a horizontal bar chart, which can be useful when you have many categories to compare.
Uses nlsw.dta & scheme vg_s2m

```
graph hbar wage, over(grade4, gap(*3)) over(union)
```

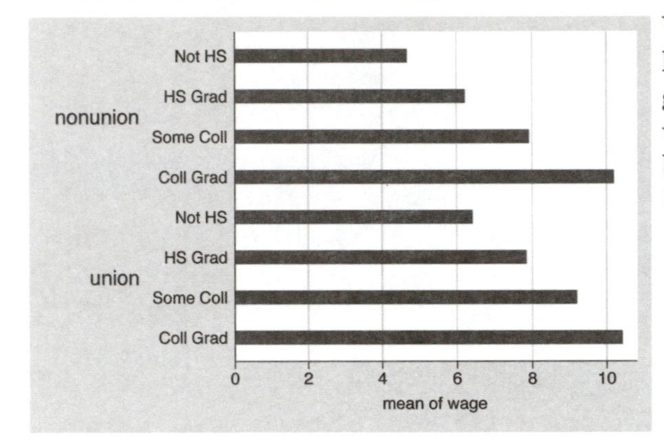

We can change the gap between the levels of **grade4**. Here, we make that gap three times as large as it normally would have been. This leads to thinner bars with a greater gap between them.
Uses nlsw.dta & scheme vg_s2m

```
graph hbar wage, over(grade4, gap(*.3)) over(union)
```

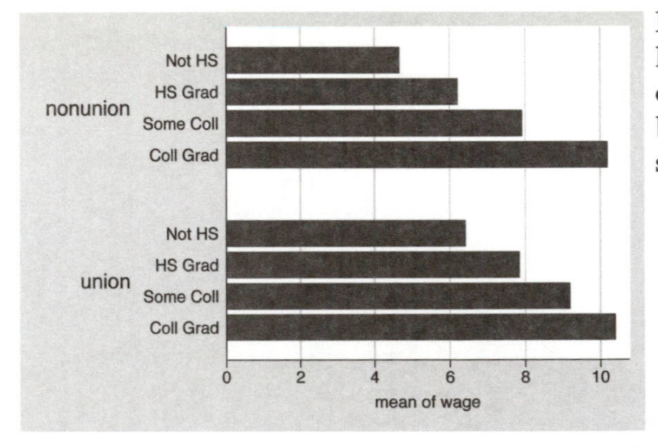

Here, we shrink the gap between the levels of **grade4**, making the gaps 30% of the size they normally would have been. This leads to wider bars with a smaller gap between them.
Uses nlsw.dta & scheme vg_s2m

```
graph hbar wage, over(grade4, gap(*.2)) over(union, gap(*3))
```

We can control the gap with respect to each of the `over()` variables at the same time. In this example, we make the gap among the `grade4` categories smaller (20% their original size) and the gap between the levels of `union` larger (three times the normal size).
Uses nlsw.dta & scheme vg_s2m

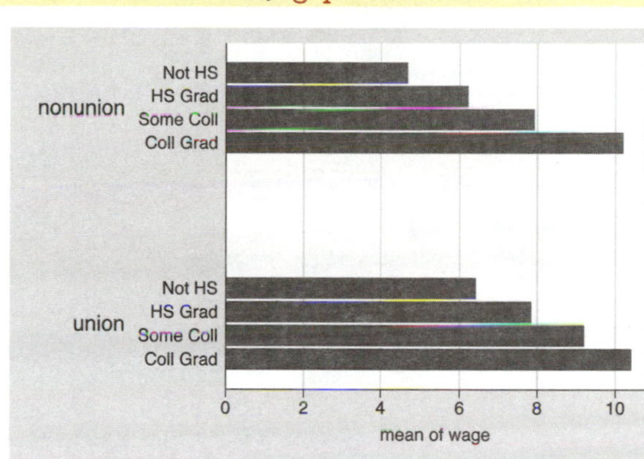

So far, we have let Stata control the order in which the bars are displayed. By default, the bars formed by `over()` variables are ordered in ascending sequence according to the values of the `over()` variable. However, Stata gives you considerable flexibility in the ordering of the bars, as illustrated in the following examples using the `vg_s2c` scheme.

```
graph hbar wage, over(occ7, descending)
```

Consider this graph showing average wages broken down by the seven levels of occupation. The bars are normally ordered by the levels of `occ7`, going from 1 to 7, where 1 is Prof and 7 is Other. Using the `descending` option switches the order of the bars. They still are ordered according to the seven levels of occupation, but the bars are ordered going from 7 to 1.
Uses nlsw.dta & scheme vg_s2c

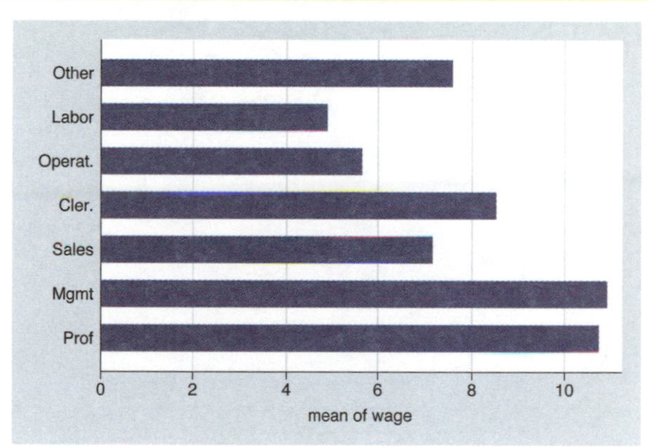

```
graph hbar wage, over(occ7, sort(occ7alpha))
```

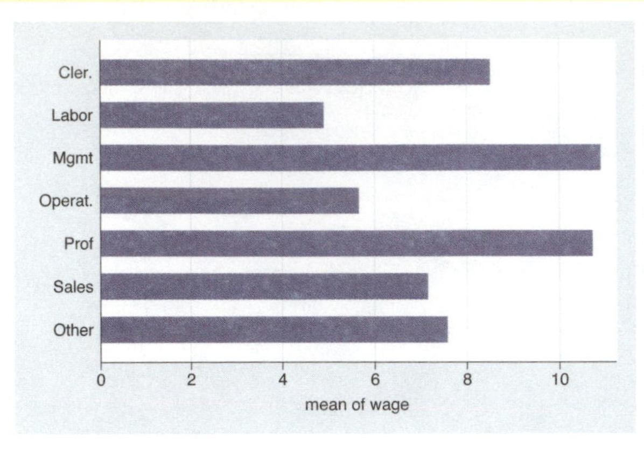

We might want to put these bars in alphabetical order (but with Other still appearing last). We can do this by recoding `occ7` into a new variable (say `occ7alpha`) such that as `occ7alpha` goes from 1 to 7, the occupations are alphabetical. We recoded `occ7` with these assignments: 4 = 1, 6 = 2, 2 = 3, 5 = 4, 1 = 5, 3 = 6, and 7 = 7; see [R] **recode**. Then, the `sort(occ7alpha)` option has the effect of alphabetizing the bars.
Uses nlsw.dta & scheme vg_s2c

```
graph hbar wage, over(occ7, sort(1))
```

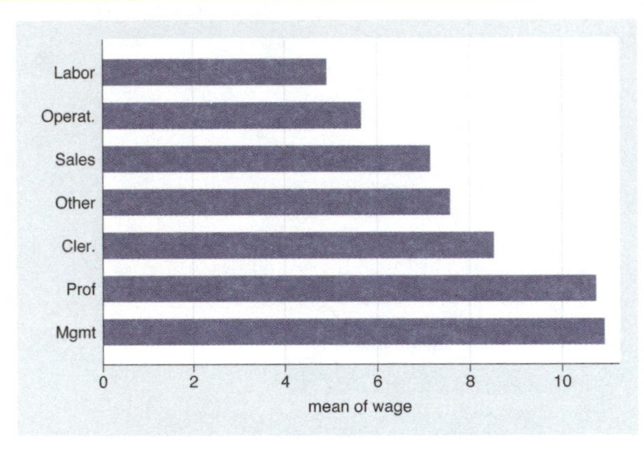

Here, we sort the variables on the height of the bars (in ascending order). The `sort(1)` means to sort the bars according to the height of the first y-variable, in this case, the mean of `wage`.
Uses nlsw.dta & scheme vg_s2c

```
graph hbar wage, over(occ7, sort(1) descending)
```

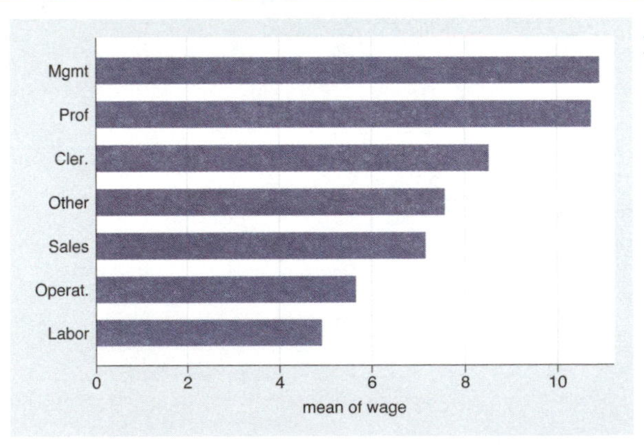

Adding the `descending` option yields bars in descending order.
Uses nlsw.dta & scheme vg_s2c

```
graph hbar wage hours, over(occ7, sort(1))
```

Here, we plot two y-variables. In addition to wages, we also show the average hours worked per week. Including the `sort(1)` option sorts the bars according to the mean of `wage` since that is the first y-variable.
Uses nlsw.dta & scheme vg_s2c

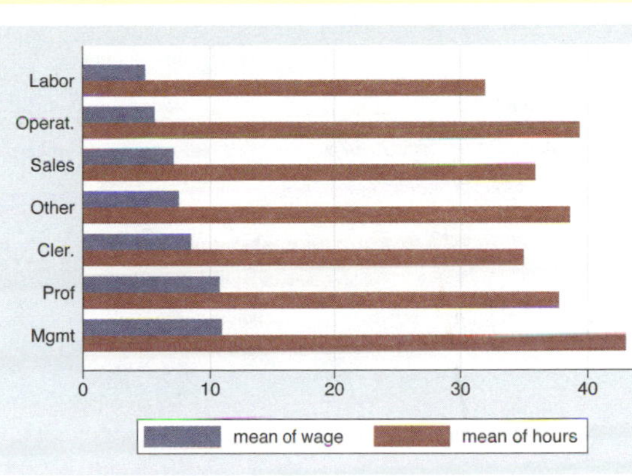


```
graph hbar wage hours, over(occ7, sort(2))
```

Changing `sort(1)` to `sort(2)` sorts the bars according to the second y-variable, the mean of `hours`.
Uses nlsw.dta & scheme vg_s2c

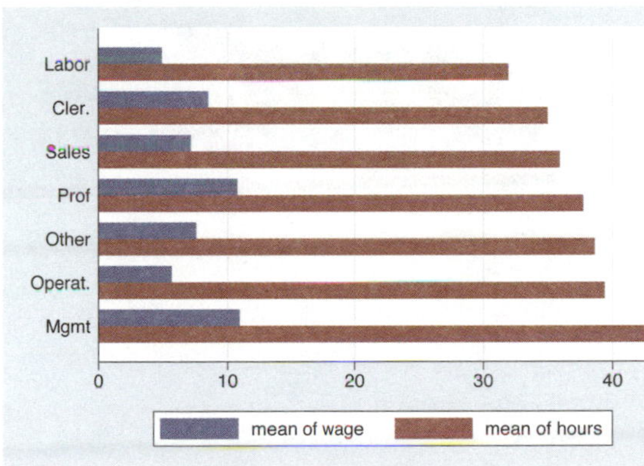


```
graph hbar wage hours, over(occ7, sort(2)) over(married)
```

We can use the `sort()` option when there are additional `over()` variables. Here, the `sort(2)` option orders the bars according to the mean number of hours worked within each level of `married`.
Uses nlsw.dta & scheme vg_s2c

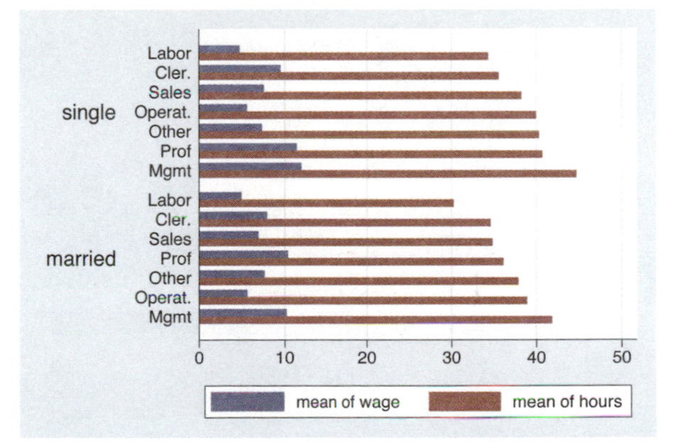

Introduction Twoway Matrix Bar Box Dot Pie Options Standard options Styles Appendix
Y-variables Over Over options Cat axis Legend Y-axis Lookofbar options By

```
graph hbar wage hours, over(occ7, sort(2)) over(married, descending)
```

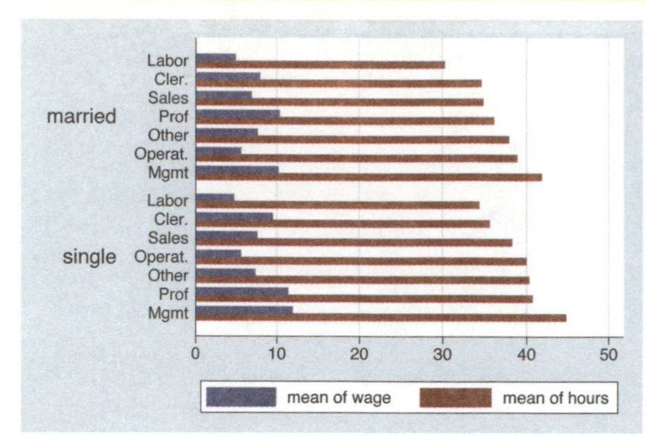

Each `over()` option can have its own separate sorting options. In this example, we add the `descending` option to the second `over()` option, and the levels of `married` are now shown with those who are married appearing first.
Uses nlsw.dta & scheme vg_s2c

```
graph hbar (sum) wage, over(collgrad) over(occ7) asyvars stack
```

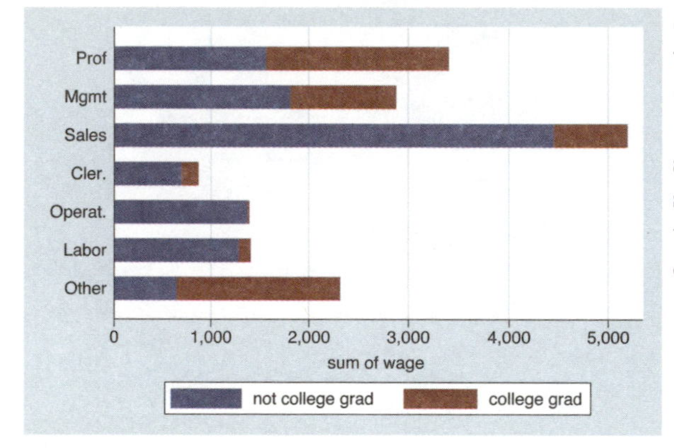

Say that we were to graph the sum of `wage` broken down by `collgrad` and `occ7`. We further treat the levels of `collgrad` as y-variables and form a stacked bar chart. We might want to sort these bars based on the sum of wages for each occupation. See the next example for how we can do that.
Uses nlsw.dta & scheme vg_s2c

```
graph hbar (sum) wage,
   over(collgrad) over(occ7, sort((sum) wage)) asyvars stack
```

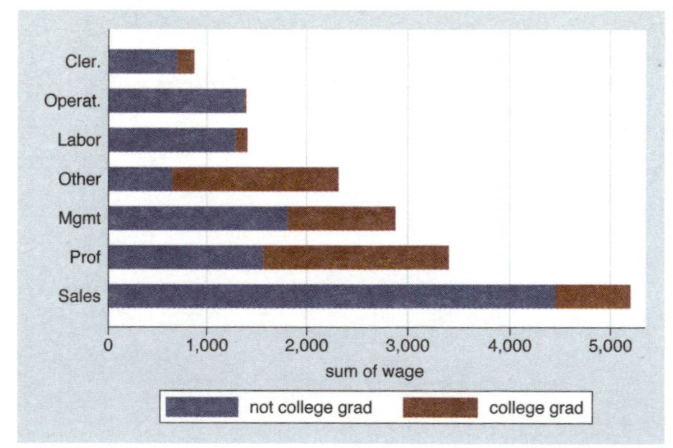

Here, we add `sort((sum) wage)` to the `over()` option for `occ7`, and then the bars are sorted on the sum of wages at each level of `occ7`, sorting the bars on their total height.
Uses nlsw.dta & scheme vg_s2c

```
graph hbar (sum) wage,
   over(collgrad) over(occ7, sort((sum) wage) descending) asyvars stack
```

Here, we add the descending option to change the sort order from highest to lowest. Note the placement of the descending option outside of the sort() option.
Uses nlsw.dta & scheme vg_s2c

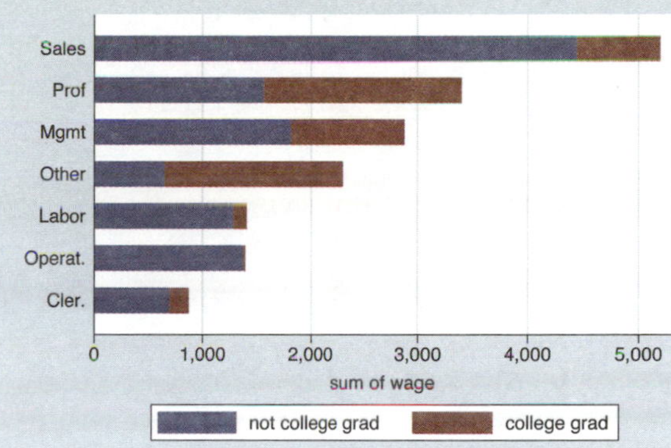

Introduction Twoway Matrix Bar Box Dot Pie Options Standard options Styles Appendix
Y-variables Over Over options Cat axis Legend Y-axis Lookofbar options By

4.4 Controlling the categorical axis

This section describes ways that you can label categorical axes. Bar charts are special since their x-axis is formed by categorical variables. This section describes options you can use to customize these categorical axes. For more details, see [G] ***cat_axis_label_options*** and [G] ***cat_axis_line_options***.

We will start by exploring how you can change the labels for the bars on the x-axis.

```
graph bar wage, over(grade6) over(south) asyvars
```

This bar chart breaks wages down by education level and whether one lives in the South. Adding the asyvars option graphs the levels of education level as differently colored bars, as though they were different y-variables. More importantly, note that the variable south is coded 0/1 and has no labels, leaving the x-axis poorly labeled.
Uses nlsw.dta & scheme vg_s2c

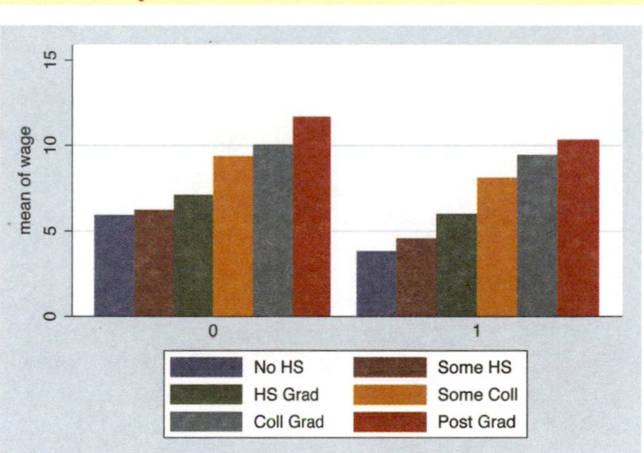

```
graph bar wage, over(grade6) over(south, relabel(1 "N & W" 2 "South"))
   asyvars
```

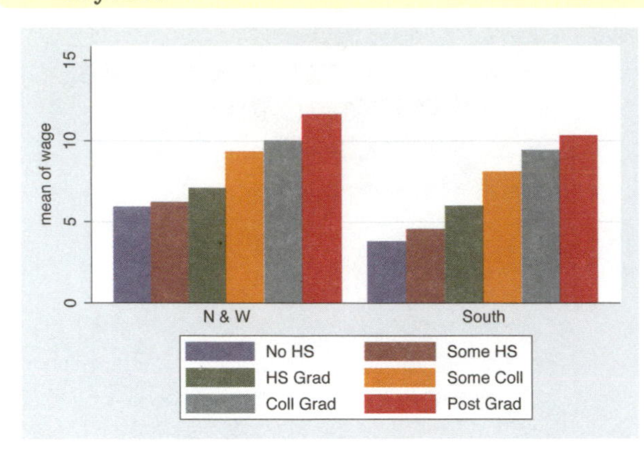

The `relabel()` option is used to change the labels displayed for the levels of **south**, giving the x-axis more meaningful labels. Note that we wrote `relabel(1 "N & W")` and not `relabel(0 "N & W")` since these numbers do not represent the actual levels of **south** but the ordinal position of the levels, i.e., first and second.
Uses nlsw.dta & scheme vg_s2c

```
graph bar wage, over(grade6) over(union, relabel(3 "missing")) missing
   asyvars
```

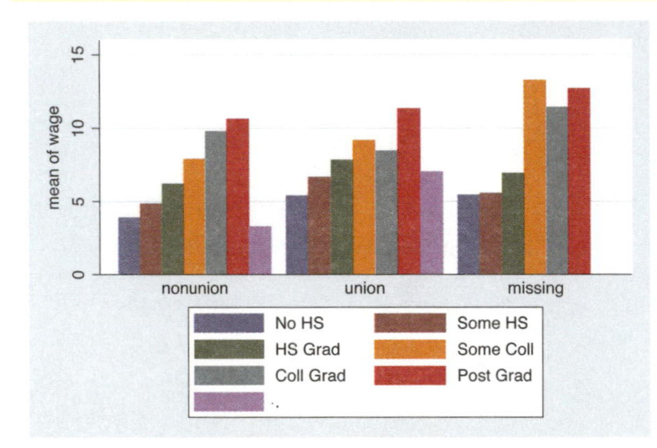

Consider this example, where we show wages broken down by education and union membership with the `missing` option to show a separate category for missing values. Normally, the bar for the missing category would be labeled with a dot, but here we add the `relabel()` option to label that category with the word "missing".
Uses nlsw.dta & scheme vg_s2c

```
graph hbar wage, over(grade6) over(south, relabel(1 "N & W" 2 "South"))
   over(smsa, relabel(1 "Non Metro" 2 "Metro"))
```

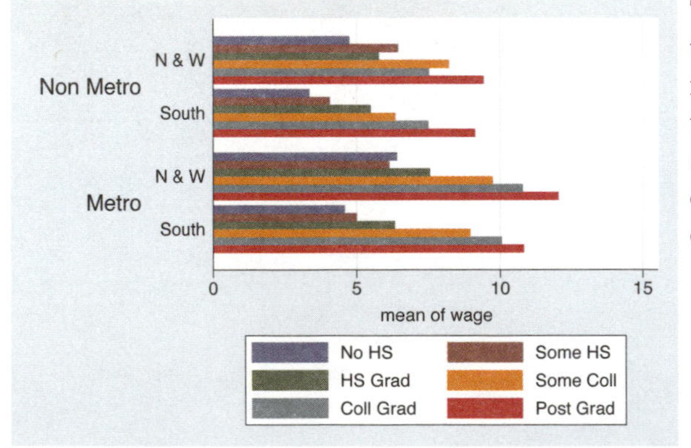

This is an example of a bar chart with three `over()` variables, two of which we relabel. The `relabel()` option is used to change the labels for the levels of **south** and **smsa**. Note each `over()` option can have its own `relabel()` option.
Uses nlsw.dta & scheme vg_s2c

```
graph hbar prev_exp tenure ttl_exp, ascategory over(age3)
```

This bar chart shows three y-variables, but we use the ascategory option to plot the different y-variables as categorical variables on the x-axis. The default labels on the x-axis are not bad, but we might want to change them.
Uses nlsw.dta & scheme vg_s2c

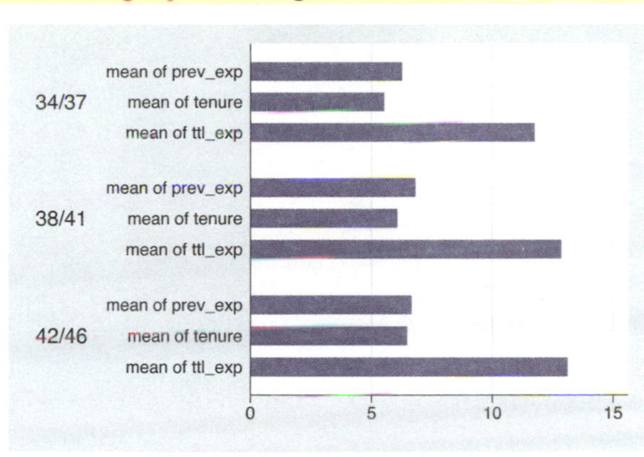


```
graph hbar prev_exp tenure ttl_exp, ascategory over(age3)
   yvaroptions(relabel(1 "Previous Exp" 2 "Current Exp" 3 "Total Exp"))
```

If the three level-of-experience variables were indicated by an `over()` option, we would use the `over(, relabel())` option to change the labels. Instead, since we have treated the multiple y-variables as categories, we then use `yvaroptions(relabel())` to modify the labels on the x-axis.
Uses nlsw.dta & scheme vg_s2c

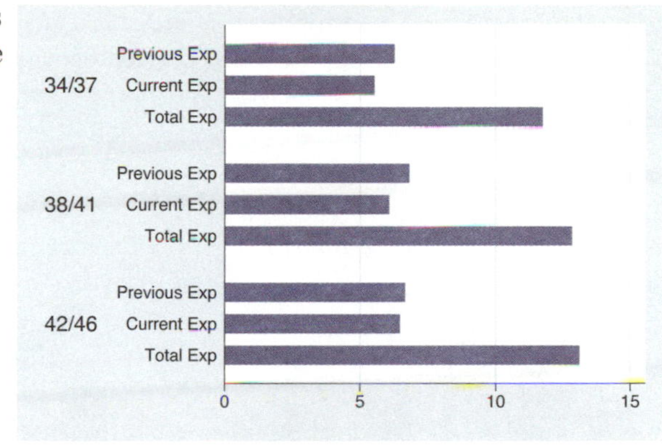


```
graph hbar prev_exp tenure ttl_exp, ascategory
   over(age3, relabel(1 "34-37 yrs" 2 "38-41 yrs" 3 "42-46 yrs"))
   yvaroptions(relabel(1 "Previous Exp" 2 "Current Exp" 3 "Total Exp"))
```

This example is similar to the previous example, but we have added a `relabel()` option to the `over()` variable as well. As before, we use `yvaroptions(relabel())` to modify the labels for the multiple y-variables, and then we also use the `relabel()` option within the `over()` option to change the labels for `age`.
Uses nlsw.dta & scheme vg_s2c

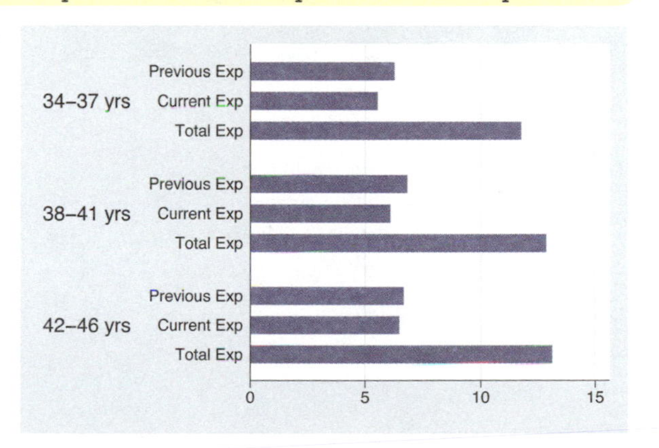

Introduction Twoway Matrix Bar Box Dot Pie Options Standard options Styles Appendix
Y-variables Over Over options Cat axis Legend Y-axis Lookofbar options By

```
graph hbar prev_exp tenure ttl_exp, ascategory xalternate
   over(age3, relabel(1 "34-37 yrs" 2 "38-41 yrs" 3 "42-46 yrs"))
   yvaroptions(relabel(1 "Previous Exp" 2 "Current Exp" 3 "Total Exp"))
```

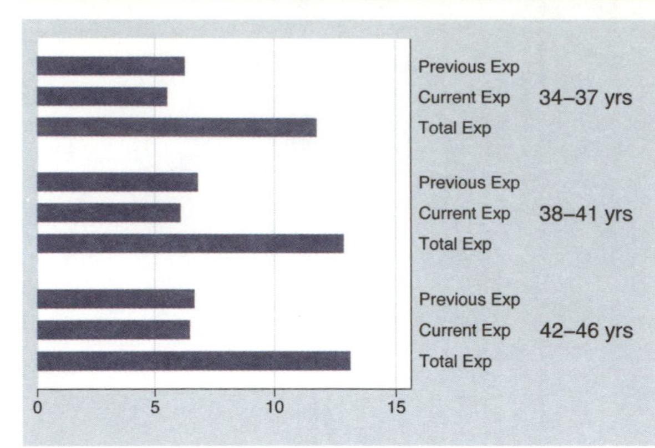

If we wish, we can move the x-axis to the opposite side of the graph. Here, we add the `xalternate` option, which moves the labels for the x-axis to the opposite side, in this case from the left to the right. You can also use the `yalternate` option to move the y-axis to its opposite side.
Uses nlsw.dta & scheme vg_s2c

In the previous examples, we saw that the `relabel` option can be used in the `over()` option to control the labeling of `over()` variables and can be used within `yvaroptions()` to control the labeling of multiple y-variables (provided that the `ascategory` option is used to convert the multiple y-variables into categories). We will further explore other `over()` options, which can be used with either `over()` or `yvaroptions()`.

```
graph bar wage, over(occ7, label(nolabels))
```

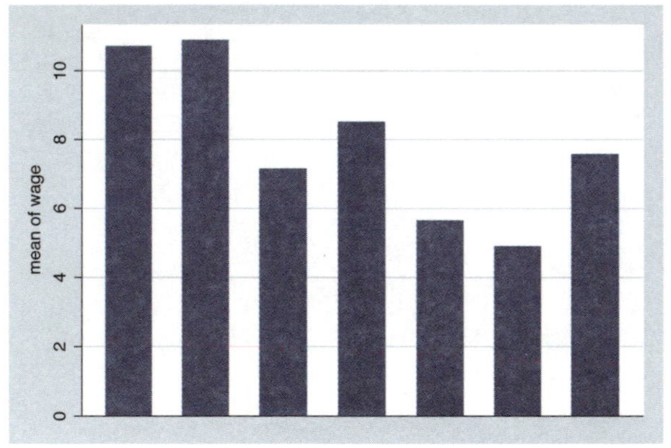

We can use the `label(nolabels)` option to suppress the display of the labels associated with the levels of `occ7`. The `label(nolabels)` option is generally not useful alone but is very useful in combination with other means to label the bars. Consider the next example.
Uses nlsw.dta & scheme vg_s2c

```
graph bar wage, over(occ7, label(nolabels)) blabel(group)
```

By adding the `blabel(group)` (bar label) option, the bars are labeled with the name of the group to which the bar belongs. See Bar : Legend (130) for more about `blabel()`.
Uses nlsw.dta & scheme vg_s2c

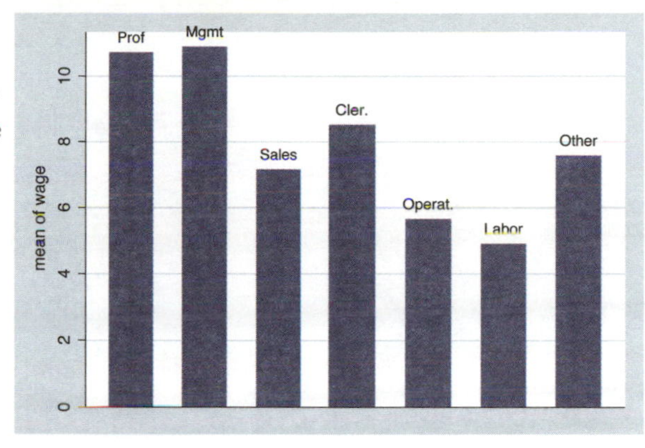


```
graph bar wage, over(occ7, label(angle(45))) over(collgrad)
```

This graph shows wages broken down by occupation and by whether one graduated college. The `label(angle(45))` option is added to rotate the labels for occupation by 45 degrees. If this had been omitted, the labels would have overlapped each other.
Uses nlsw.dta & scheme vg_s2c

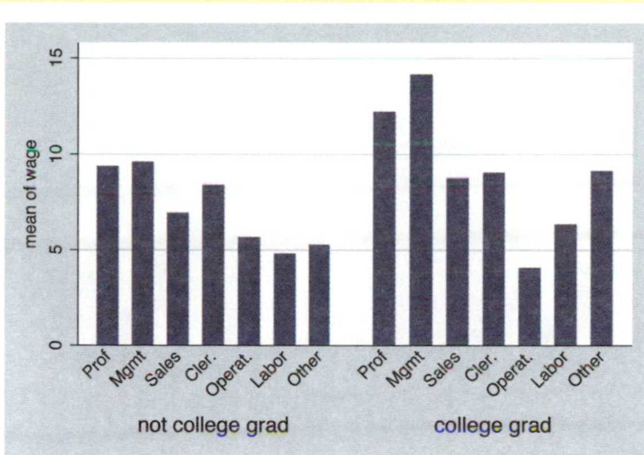


```
graph bar wage, over(occ7, label(alternate)) over(collgrad)
```

Compare this graph with the previous example. This example uses the `label(alternate)` strategy to avoid overlapping by alternating the labels for occupation.
Uses nlsw.dta & scheme vg_s2c

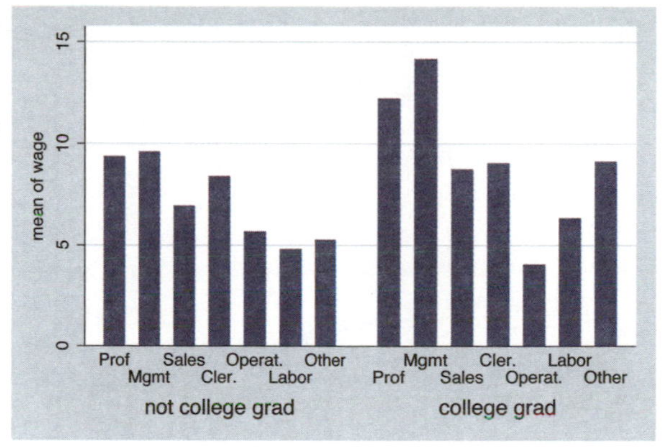

Introduction Twoway Matrix Bar Box Dot Pie Options Standard options Styles Appendix
Y-variables Over Over options Cat axis Legend Y-axis Lookofbar options By

```
graph bar wage hours ttl_exp, ascategory over(collgrad)
   yvaroptions(label(alternate))
```

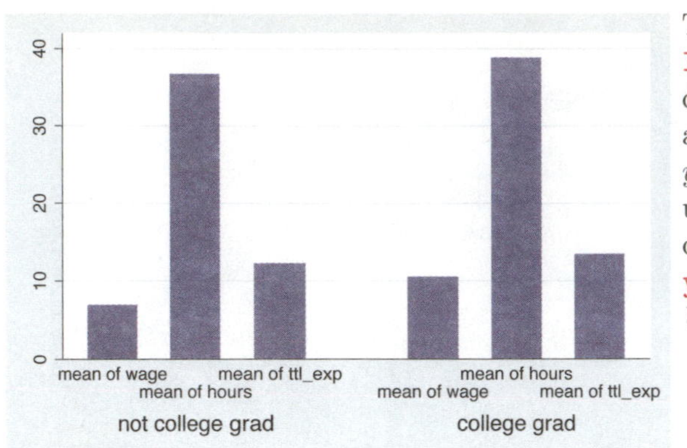

This is another example of using the `label(alternate)` option, but in this case, it is used in the context of alternating labels created by multiple y-variables converted to categories using the `ascategory` option. In such a case, the option is specified as `yvaroptions(label(alternate))`.
Uses nlsw.dta & scheme vg_s2c

```
graph bar wage hours ttl_exp, ascategory over(union) nolabel
```

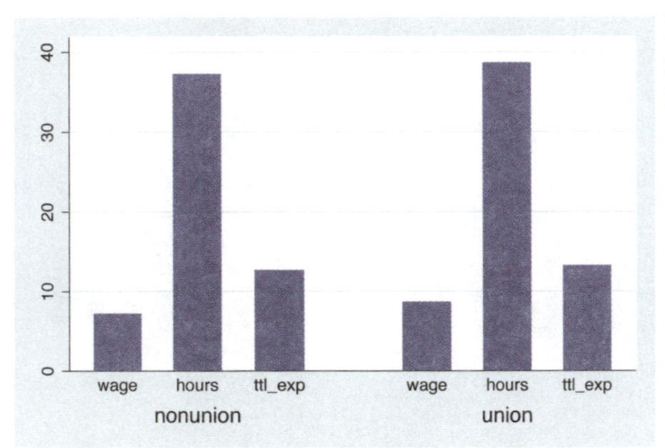

If we add the `nolabel` option, the names of the variables are shown instead of the value labels.
Uses nlsw.dta & scheme vg_s2c

```
graph hbar wage, over(occ5, label(labcolor(green)))
   over(collgrad, label(labcolor(maroon) labsize(small)))
```

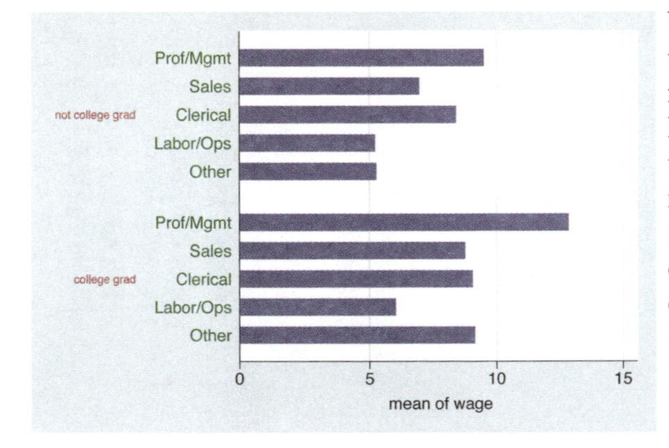

We can change the color of the labels using the `labcolor()` option. Here, we make the label for `occ5` green and the label for `collgrad` maroon. We also use `labsize(small)` to make the labels for `collgrad` small. See Styles : Colors (328) and Styles : Textsize (344) for more details about other values you could choose.
Uses nlsw.dta & scheme vg_s2c

```
graph bar wage,
   over(age3, label(ticks tlwidth(thick) tlength(*2) tposition(crossing)))
   over(collgrad)
```

Stata permits you to add ticks using the `ticks` option. At the same time, we modify the attributes of the ticks, making the tick line width thick, the tick length twice as long as normal, and the tick position crossing the x-axis. See [G] ***cat_axis_label_options*** for more details and other options for controlling ticks.
Uses nlsw.dta & scheme vg_s2c

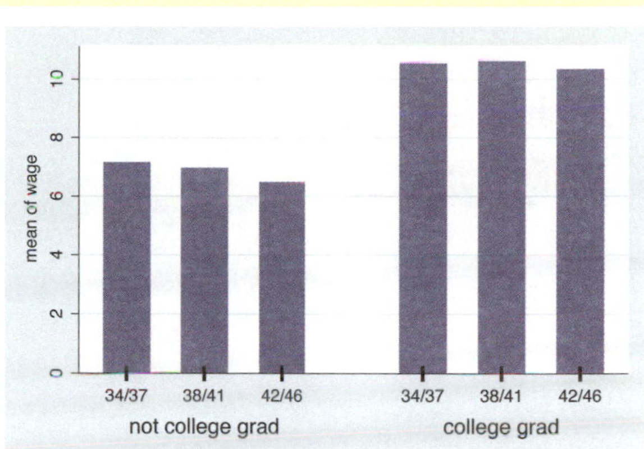


```
graph bar wage, over(age3, label(labgap(*5))) over(collgrad)
```

The `labgap(*5)` option increases the gap between the label and the axis, making the gap between the labels for the levels of `age3` and the axis five times their normal size.
Uses nlsw.dta & scheme vg_s2c

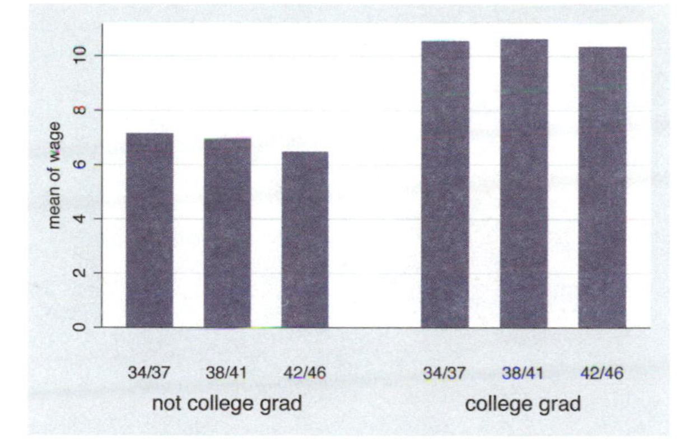


```
graph bar wage, over(age3) over(collgrad, label(labgap(*5)))
```

We use the `label(labgap(*5))` option to control the gap between the labels for `age3` and `collgrad`, making that gap five times the normal size.
Uses nlsw.dta & scheme vg_s2c

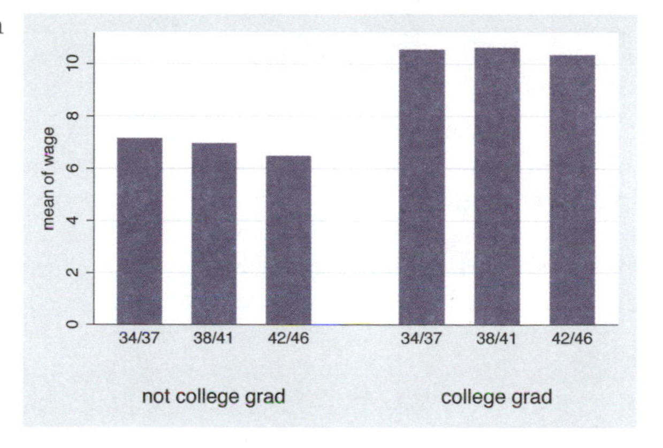

```
graph bar wage, over(age3) over(collgrad, axis(outergap(*20)))
```

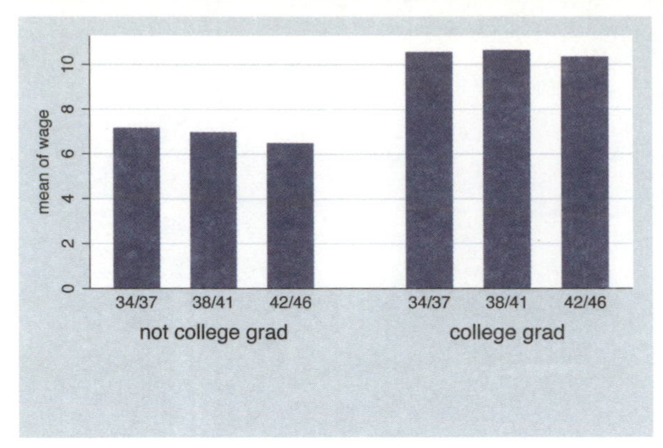

The axis(outergap(*20)) option controls the gap between the labels of the x-axis and the outside of the graph. As you can see, this increases the space below the labels for collgrad and the bottom of the graph.
Uses nlsw.dta & scheme vg_s2c

```
graph bar wage, over(union) over(grade4) asyvars
   b1title("Education Level in Four Categories")
```

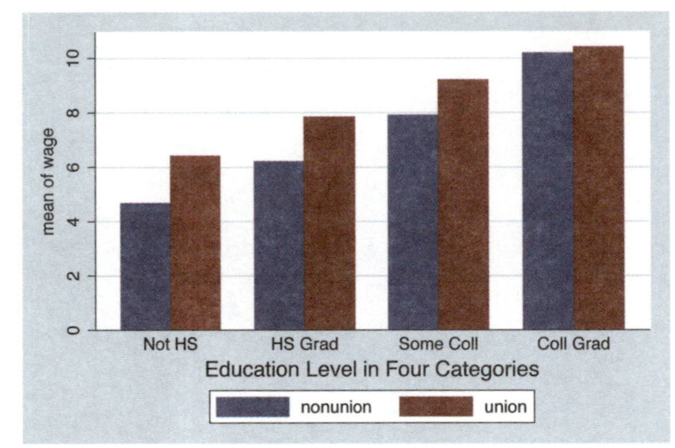

The b1title() option adds a title to the bottom of the graph, in effect labeling the x-axis. We can add a second title below that using the b2title() option. If we used graph hbox, we could label the left axis using the l1title() and l2title() options.
Uses nlsw.dta & scheme vg_s2c

4.5 Controlling legends

This section discusses the use of legends for bar charts, emphasizing the features that are unique to bar charts. The section Options : Legend (287) goes into great detail about legends, as does [G] ***legend_option***. Legends can be used for multiple y-variables or when the first over() variable is treated as a y-variable via the **asyvars** option. See Bar : Y-variables (107) for more information about the use of multiple y-variables and Bar : Over (111) for more examples of treating the first over() variable as a y-variable. Next, we will consider examples that show the different kinds of labels that you can create using the blabel() option. You can create labels that display the name of y-variable, the name of the first over() group, the height of the bar, or the overall height of the bar (when used with the stack option). These examples begin using the vg_s1c scheme.

```
graph bar wage hours tenure ttl_exp age
```

Consider this bar graph of five different y-variables. The bars for the different y-variables are shown with different colors, and a legend is used to identify the y-variables.
Uses nlsw.dta & scheme vg_s1c

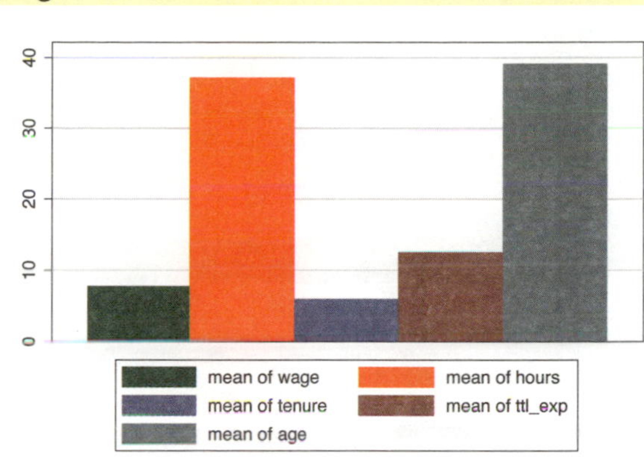


```
graph bar wage, over(occ7) asyvars
```

This is another example of where a legend can arise in a Stata bar graph by specifying the **asyvars** option, which treats an **over()** variable as though the levels were different y-variables.
Uses nlsw.dta & scheme vg_s1c

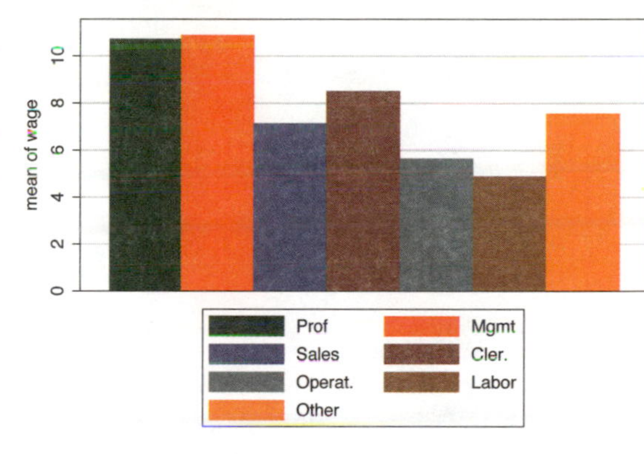

Unless otherwise mentioned, the legend options described below work the same regardless of whether the legend was derived from multiple y-variables or from an **over()** variable that was combined with the **asyvars** option. These next examples use the **vg_s2m** scheme.

```
graph hbar wage hours tenure ttl_exp age, nolabel
```

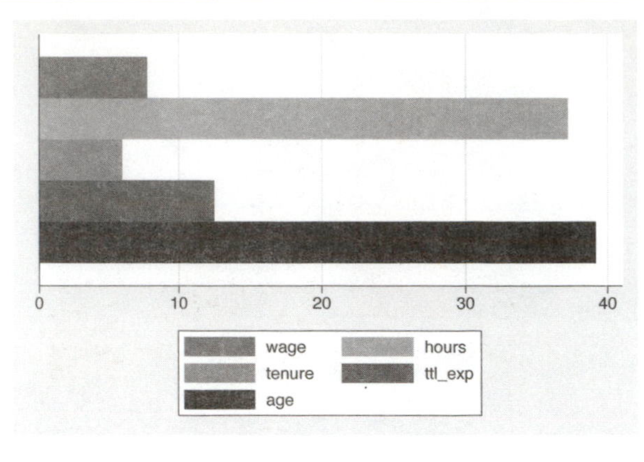

The `nolabel` option only works when you have multiple y-variables. When it is used, the variable names (not the variable labels) are used in the legend. For example, instead of showing the variable label `hourly wage`, it shows the variable name `wage`.
Uses nlsw.dta & scheme vg_s2m

```
graph hbar wage hours tenure ttl_exp age, showyvars
```

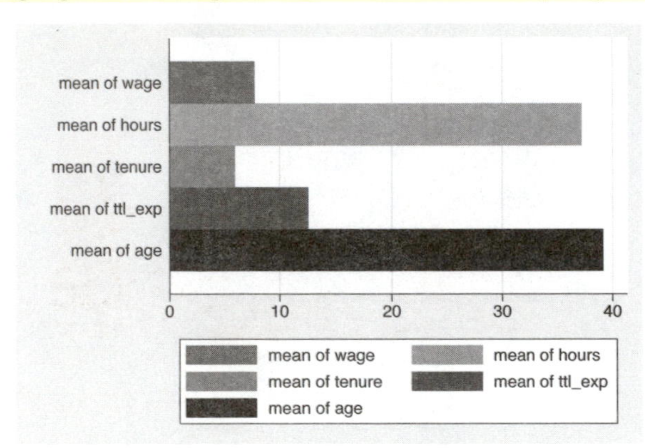

The `showyvars` option puts the labels on the axis, beside or "under" the bars.
Uses nlsw.dta & scheme vg_s2m

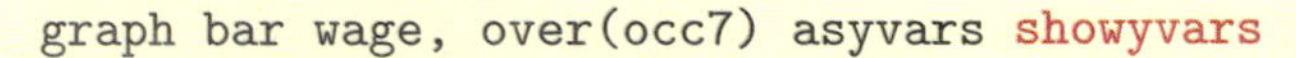

```
graph bar wage, over(occ7) asyvars showyvars
```

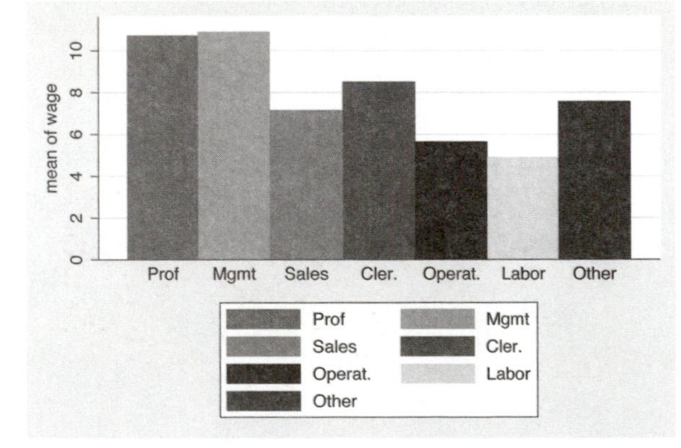

Even though the `showyvars` option sounds like it would work only with multiple y-variables, it also works when you combine the `over()` and `asyvars` options. As you can see, the legend is now redundant and could be suppressed.
Uses nlsw.dta & scheme vg_s2m

```
graph bar wage, over(occ7) asyvars showyvars legend(off)
```

This example is similar to the previous example, but we use the `legend(off)` option to suppress the display of the legend.
Uses nlsw.dta & scheme vg_s2m

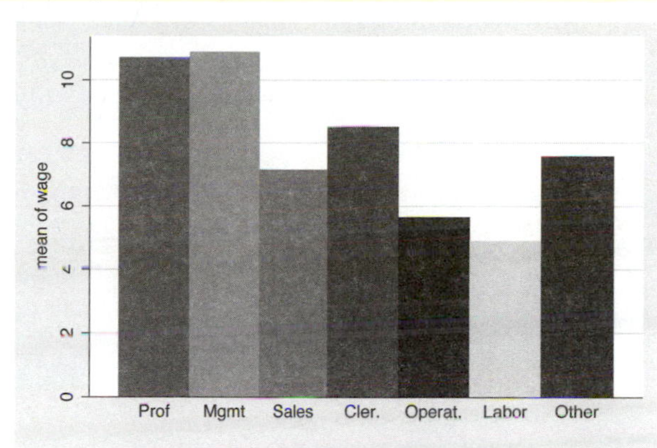


```
graph bar wage, over(occ7) asyvars legend(label(1 "Professional")
   label(2 "Management"))
```

We can use `legend(label())` to change the labels for one or more of the bars in the graph. Here, we change the labels for the first and second bars in the legend. Note that you use a separate `label()` option for each bar. This is in contrast to the `relabel()` option, where all of the label assignments were placed in one `relabel()` option; see Bar : Cat axis (123).
Uses nlsw.dta & scheme vg_s2m

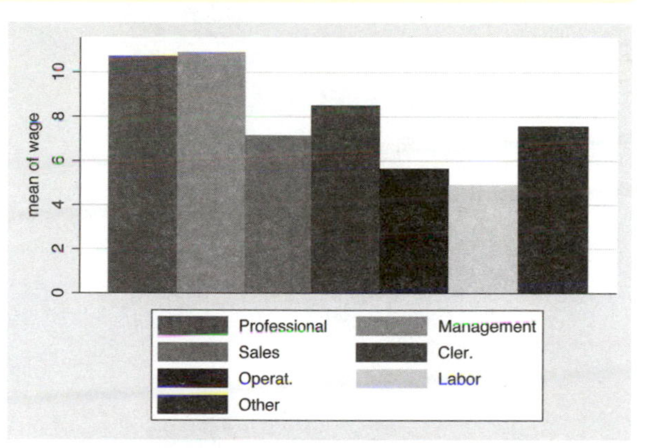


```
graph bar wage, over(occ7) asyvars legend(rows(2) colfirst)
```

In this example, we use the `rows(2)` option combined with `colfirst` to display the legend in two rows and to order the keys by column (instead of the default, which is by row). This yields keys that are more adjacent to the bars that they label.
Uses nlsw.dta & scheme vg_s2m

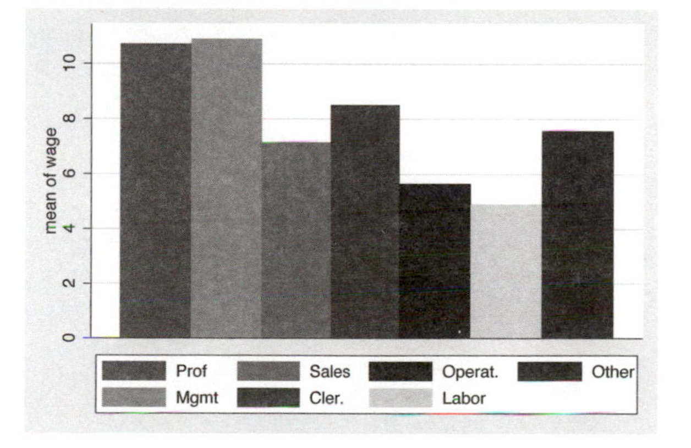

As you can see, the default placement for the legend is below the x-axis. However, Stata gives you tremendous flexibility in the placement of the legend. We now consider options that control the placement of the legend, along with options useful for controlling the placement of the items within the legend. The following examples use the vg_blue scheme.

```
graph bar wage, over(occ7) asyvars legend(position(1))
```

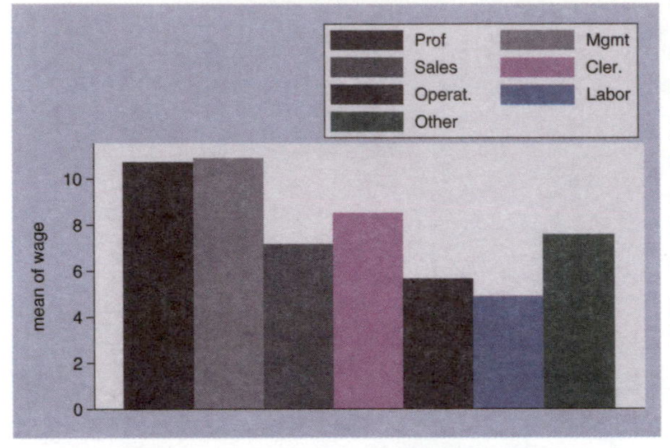

We can use the legend(position(1)) option to place the legend in the top right corner of the graph. The values you supply for position() are like the numbers on a clock face, where 12 o'clock is the top, 6 o'clock is the bottom, and 0 represents the center of the clock face. Specifying 1 o'clock places the legend in the top right; see Styles : Clockpos (330) for more details. *Uses nlsw.dta & scheme vg_blue*

```
graph bar wage, over(occ7) asyvars legend(position(1) ring(0))
```

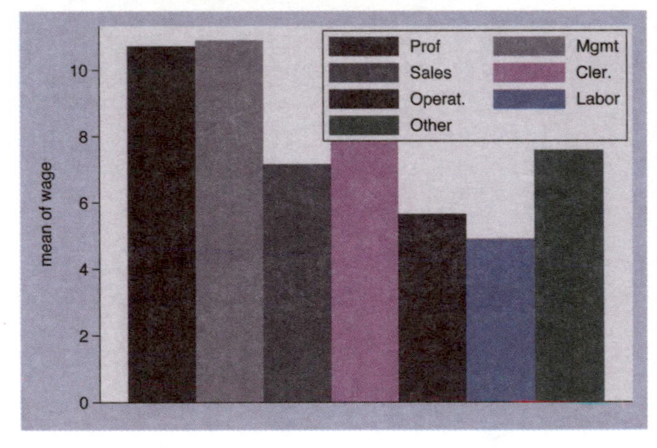

Adding the ring(0) option, we can try to tuck the legend inside the top right corner of the plot area. Think of the ring() option as specifying concentric rings around the graph, where 0 is a position inside the plot region, 1 is just outside the plot region, and increasing values are farther and farther from the center of the plot region. Unfortunately, the legend touches one of the bars, but we will fix that in the next example. *Uses nlsw.dta & scheme vg_blue*

```
graph bar wage, over(occ7) asyvars legend(position(1) ring(0)) exclude0
```

Adding exclude0 no longer forces the y-axis to start at 0 and makes room in the top corner of the plot region for the legend. See Bar : Y-axis (143) for more details about the exclude0 option.
Uses nlsw.dta & scheme vg_blue

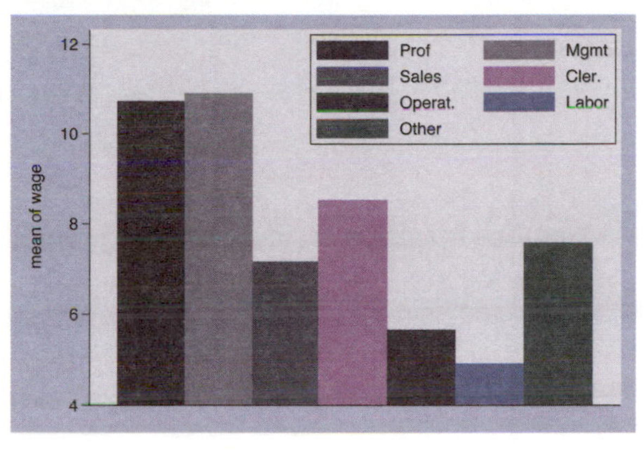


```
graph hbar wage, over(occ7) asyvars legend(cols(1) position(9))
```

We switch to making this a horizontal bar chart and move the legend using the position(9) option to place the legend in the 9 o'clock position. We also use the cols(1) option to display the legend as a single column.
Uses nlsw.dta & scheme vg_blue

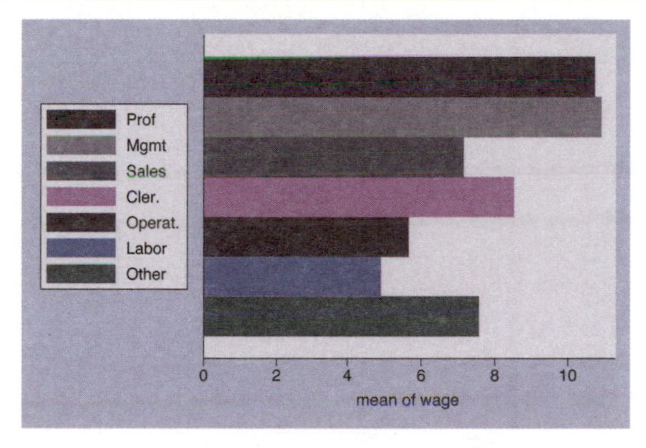


```
graph hbar wage, over(occ7) asyvars legend(cols(1) position(9) textfirst)
```

Adding the textfirst option places the description of the key before the symbol in the legend.
Uses nlsw.dta & scheme vg_blue

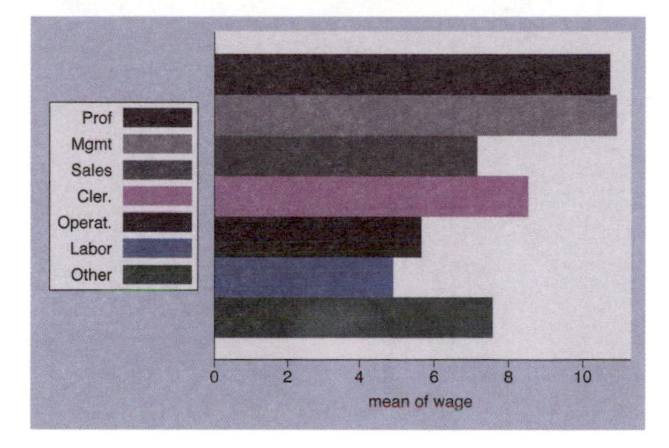

```
graph hbar wage, over(occ7) asyvars legend(cols(1) position(9) stack)
```

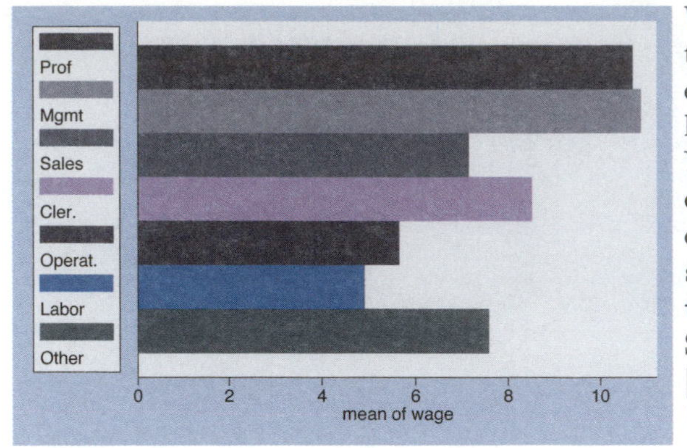

With the stack option, the keys and their labels are placed on top of each other to form an even narrower legend, leaving more room to plot the bars. You have considerable control over the elements within the legend using other options, such as `rowgap()`, `keygap()`, `symxsize()`, `symysize()`, `textwidth()`, and `symplacement()`. See Options : Legend (287) and [G] ***legend_option*** for more details. *Uses nlsw.dta & scheme vg_blue*

```
graph hbar wage, over(occ7) asyvars
```

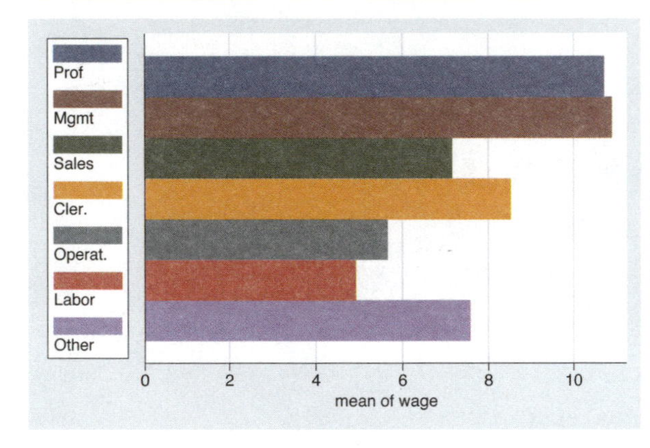

This example uses the vg_lgndc scheme, `set scheme vg_lgndc`. Notice how it positions and customizes the legend, as in the previous example. With this scheme, the legend defaults to the 9 o'clock position, in a single column, with the keys and symbols stacked. *Uses nlsw.dta & scheme vg_lgndc*

Let's now look at how we can use the `blabel()` (bar label) option to add labels to the bars. These labels can show the name of the `over()` option, the name of y-variables, or the height of the bar. These options are illustrated below along with other related options you might use in conjunction with `blabel()` for identifying the bars. These examples begin using the `vg_past` scheme.

```
graph bar wage hours tenure, over(collgrad)
```

Consider this graph, where we look at `wage`, `hours`, and `tenure` broken down by the levels of `collgrad`. The legend identifies the bars for us. In addition to the legend, Stata offers us other ways we can label these bars, as we shall see in the upcoming examples.
Uses nlsw.dta & scheme vg_past

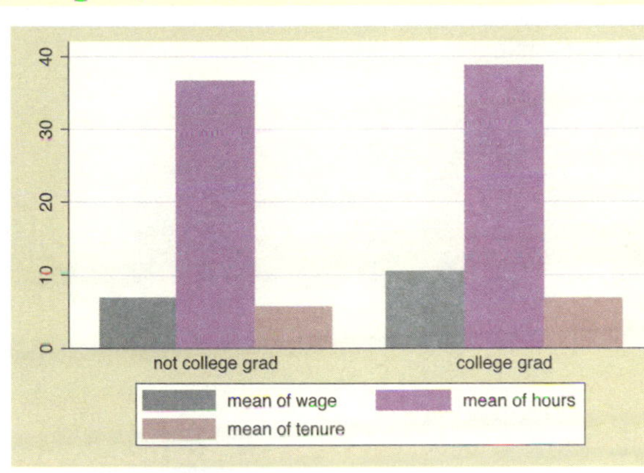


```
graph bar wage hours tenure, over(collgrad) blabel(name)
```

We can add the `blabel(name)` (bar label) option, and it places labels on each of the bars with the name of y-variables. Here, each of these labels is preceded with "mean of" since each bar represents the mean of y-variable.
Uses nlsw.dta & scheme vg_past

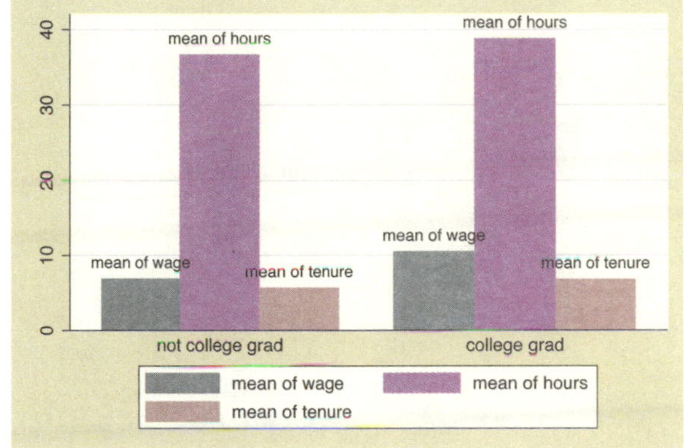


```
graph bar wage hours tenure, over(collgrad) blabel(name) nolabel
```

If we use the `nolabel` option, just the name y-variable is shown. For example, instead of showing the variable label `hourly wage`, it shows the variable name `wage`.
Uses nlsw.dta & scheme vg_past

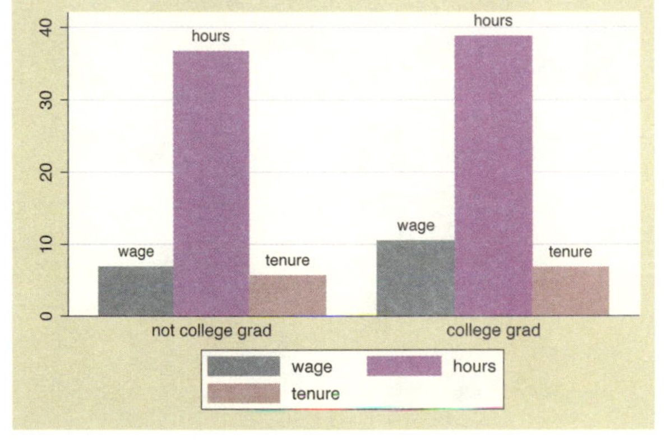

```
graph bar wage hours tenure, over(collgrad) blabel(name) nolabel
   legend(off)
```

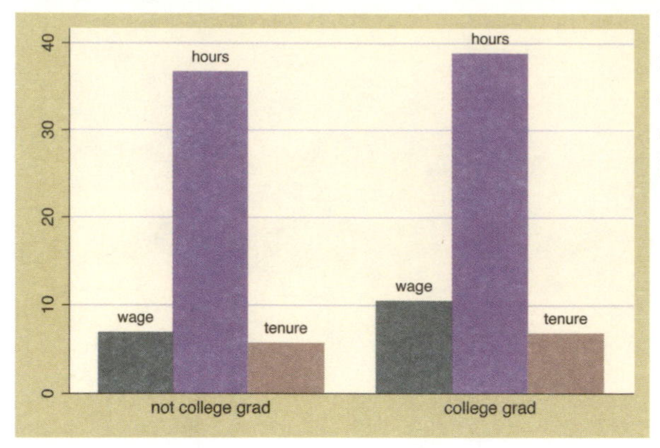

In this case, the legend is no longer needed, so we can suppress the display of the legend with the `legend(off)` option. See Options : Legend (287) for more information about legend options.
Uses nlsw.dta & scheme vg_past

```
graph bar tenure, over(occ7) exclude0 blabel(group)
```

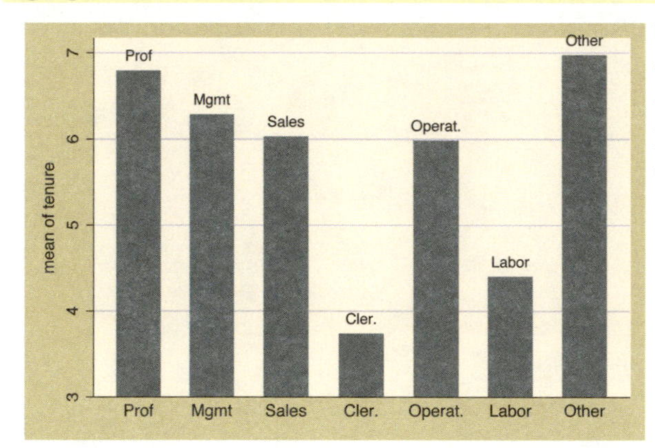

Using the `blabel(group)` option shows the label for the first `over()` group at the top of each bar. In this case, the label at the bottom of the bar becomes unnecessary.
Uses nlsw.dta & scheme vg_past

```
graph bar tenure, over(occ7, label(nolabels)) exclude0
   blabel(group) yscale(range(7.2))
```

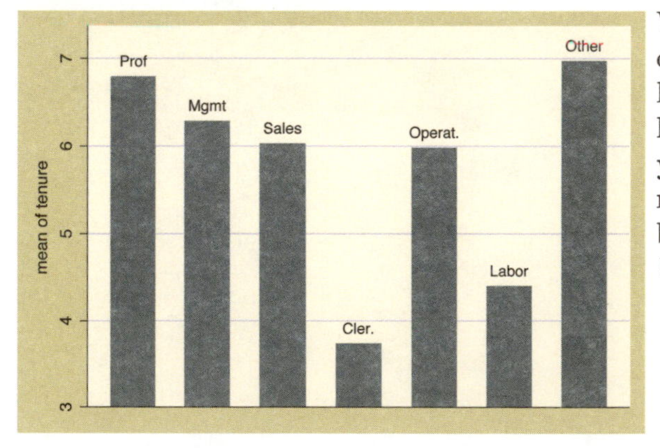

We can add the `label(nolabels)` option to suppress the display of the labels below each bar. Note that we have also used the option `yscale(range(7.2))` to provide more room within the plot area to label the bar for the Other category.
Uses nlsw.dta & scheme vg_past

```
graph bar tenure, over(occ5, label(nolabels)) exclude0 blabel(group)
    yscale(range(7.2)) over(union)
```

Even if we add a second `over()` option, the levels of the first `over()` variable are labeled at the top of each bar due to the `blabel()` option, and the levels of the second `over()` variable are labeled, as usual, at the bottom of the bars. Note that the `blabel()` option does not work this way when you have three `over()` options or multiple y-variables.
Uses nlsw.dta & scheme vg_past

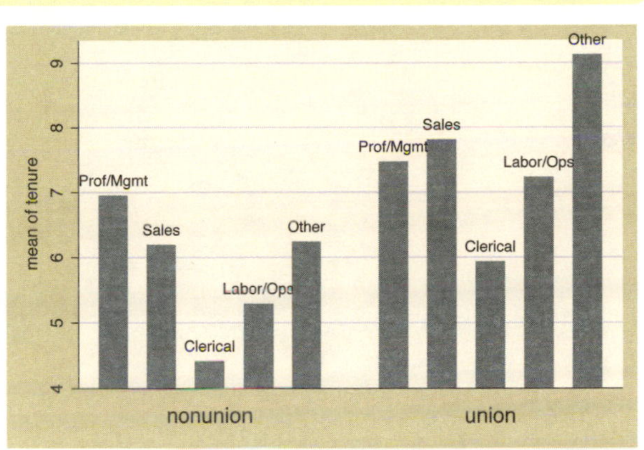


```
graph hbar prev_exp tenure ttl_exp, over(grade4) blabel(bar)
```

Consider this graph showing previous, current, and total work experience broken down by education. In this example, the `blabel(bar)` option is used to display the bar height (in this case, the mean of y-variables).
Uses nlsw.dta & scheme vg_past

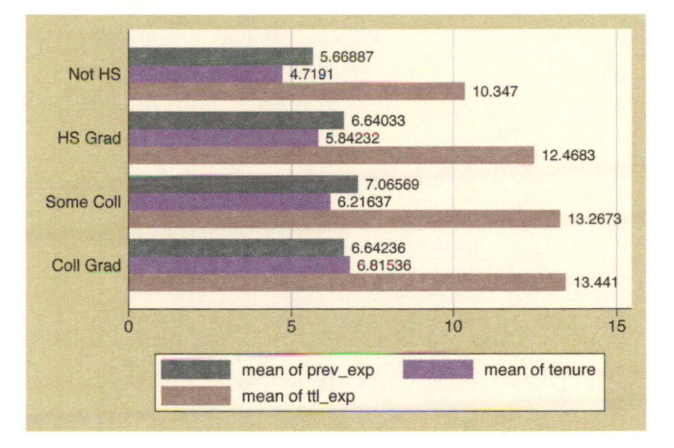


```
graph bar (sum) prev_exp tenure, stack over(grade4) blabel(bar)
```

Using the `(sum)` function, this graph shows the sum of experience for all individuals in a grade level before their current job (`prev_exp`) and the sum of experience for all individuals in a grade level in their current job (`tenure`) and then uses `stack` to stack these two totals. With the `blabel(bar)` option, the bar labels are the sums for each y-variables broken down by `grade4`.
Uses nlsw.dta & scheme vg_past

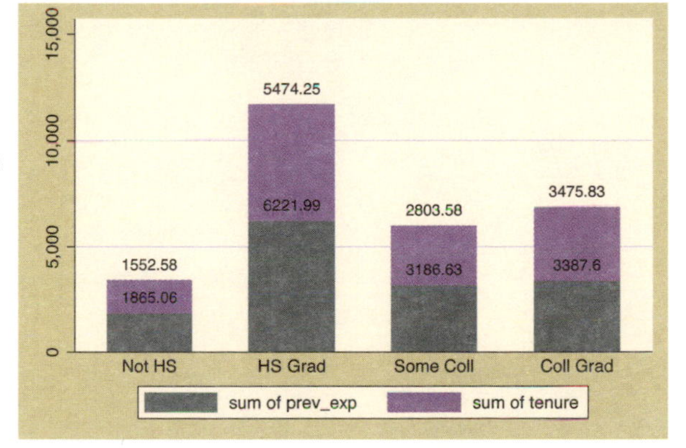

```
graph bar (sum) prev_exp tenure, stack over(grade4) blabel(total)
```

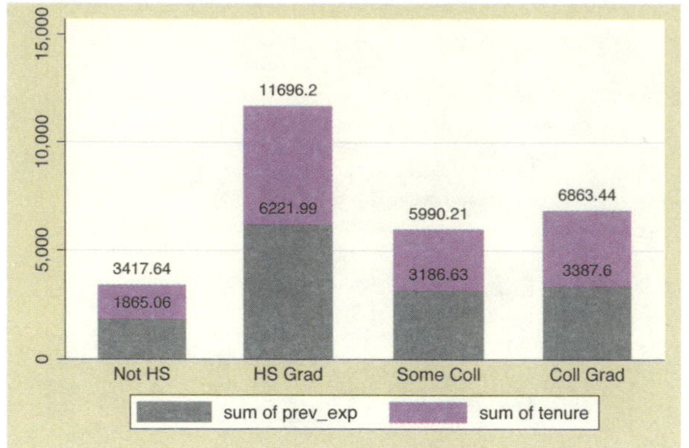

As compared with the prior example, this example uses the blabel(total) option to display the results as totals. Now, the labels represent the cumulative total height of the bar.
Uses nlsw.dta & scheme vg_past

We have seen a variety of ways that you can use the **blabel()** option to label the bars. In addition, Stata offers a variety of options you can use to control the display of these labels. Below, we will consider some of these options that allow you to customize the way these labels are displayed. These example begin using the **vg_palec** scheme.

```
graph hbar hours, over(occ7, label(nolabels)) blabel(group)
```

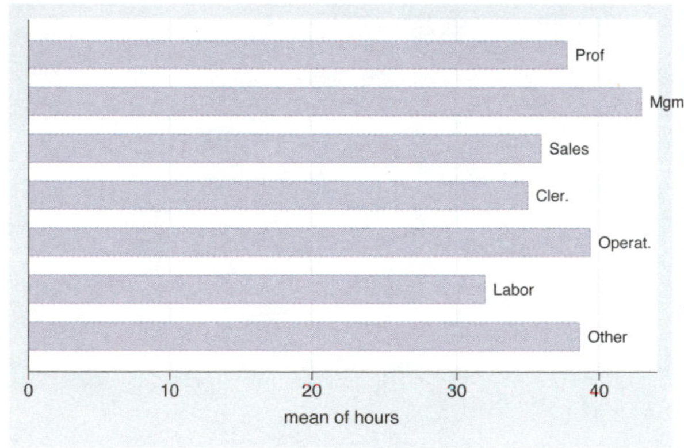

Consider this graph of the average hours worked by occupation. We add labels of the occupation at the top of each bar but suppress the label at the bottom of each bar. The label for the second bar runs off the right of the graph. Fortunately, Stata offers us a number of options to control where these labels are displayed.
Uses nlsw.dta & scheme vg_palec

```
graph hbar hours, over(occ7, label(nolabels))
   blabel(group, position(inside))
```

With the position(inside) option, we can place the group label inside the bar. By default, inside refers to the very "top" of the bar but on the inside of the bar. Note that, because we chose the vg_palec scheme, the bar colors are pale, so the labels within the bars are readable.
Uses nlsw.dta & scheme vg_palec

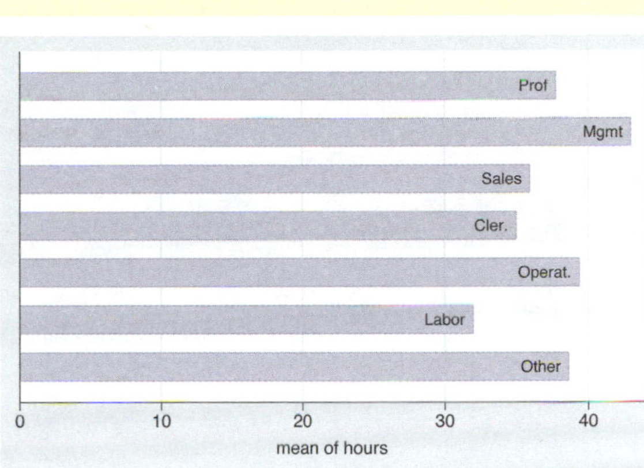


```
graph hbar hours, over(occ7, label(nolabels))
   blabel(group, position(base))
```

With the position(inside) option, we can place the label inside the bar, but at the base of the bar. You can also specify position(center) to place the label in the center of the bar.
Uses nlsw.dta & scheme vg_palec

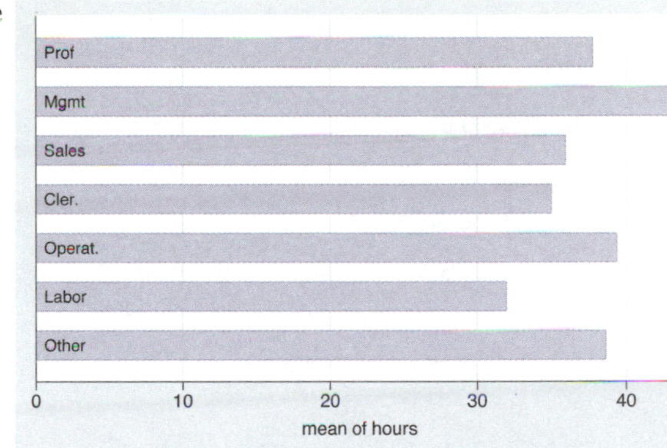


```
graph hbar hours, over(occ7, label(nolabels))
   blabel(group, position(base) gap(*10))
```

The gap() option can be used to fine-tune the placement of the label. Here, we position the label at the base but increase the gap between the label and the base to be 10 times its normal size. You can also use the gap() option with position(inside) to position the label with respect to the top of the bar.
Uses nlsw.dta & scheme vg_palec

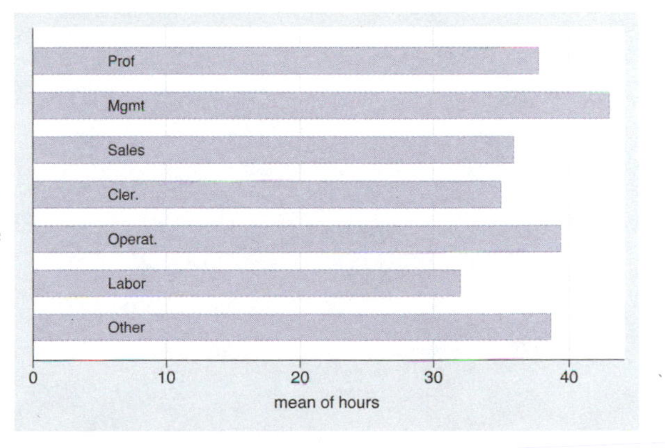

```
graph bar hours, over(occ7) blabel(bar, position(outside)) exclude0
```

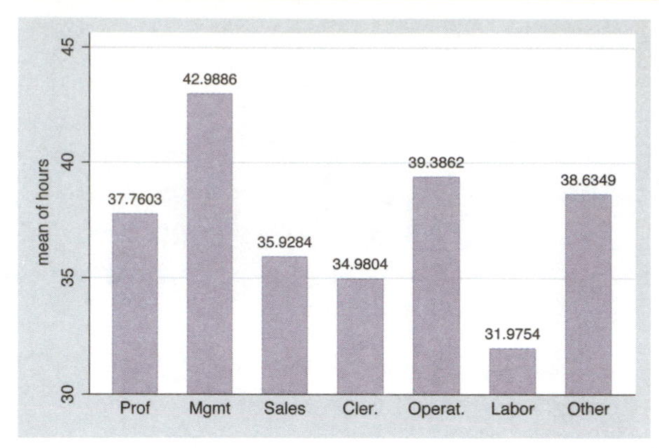

This graph is similar to the previous ones, but the bars are vertical, and we now are labeling the bars with the height of the bar. The label is placed just outside the bar.
Uses nlsw.dta & scheme vg_palec

```
graph bar hours, over(occ7, axis(outergap(*5))) asyvars
   blabel(bar, position(base) gap(-4))
```

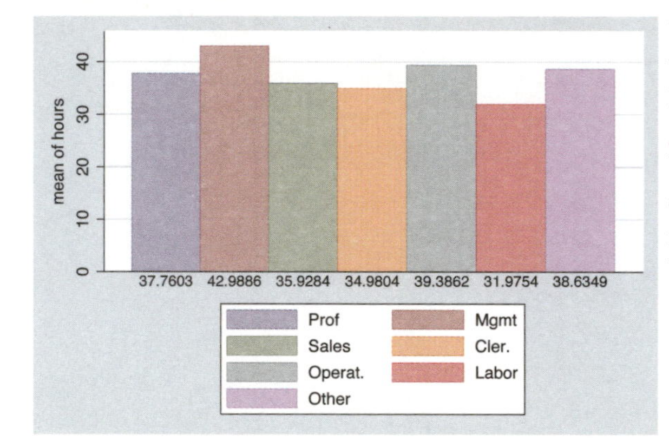

To put the labels just under the bars, we use `position(base)` to put the labels at the base but also specify `gap(-4)` to move the labels below the bars. Adding the `axis(outergap(*5))` option (see Bar : Cat axis (123)), we make enough room so the labels do not bump into the legend.
Uses nlsw.dta & scheme vg_palec

```
graph bar hours, over(occ7) asyvars
   blabel(bar, position(base) box bfcolor(white) size(large) format(%5.2f))
```

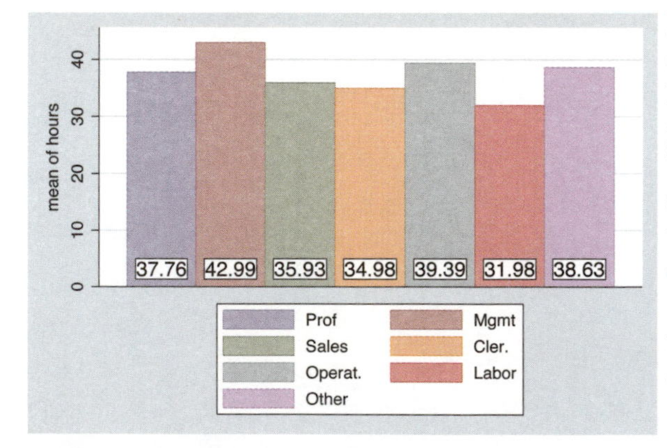

Here, we show more options that you can use to customize the display of the labels. We add a number of options to place a box around the label, make the background fill color white, increase the size of the text to be large, and display the means with a width of 5 and 2 decimal places. See Options : Textboxes (303) for additional examples of how to use textbox options to control the display of text.
Uses nlsw.dta & scheme vg_palec

4.6 Controlling the y-axis

This section describes options you can use to control the y-axis in bar charts. To be precise, when Stata refers to the y-axis on a bar chart, it refers to the axis with the continuous variable, whether the left axis when using **graph bar** or the bottom axis when using **graph hbar**. This section emphasizes the features that are particularly relevant to bar charts. For more details, see Options : Axis titles (254), Options : Axis labels (256), and Options : Axis scales (265). Also see [G] ***axis_title_options***, [G] ***axis_label_options***, and [G] ***axis_scale_options***. This section uses the vg_s2c scheme.

```
graph bar wage, over(occ5) over(married) asyvar
```

Consider this graph showing the mean hourly wage broken down by occupation and marital status.
Uses nlsw.dta & scheme vg_s2c

```
graph bar wage, over(occ5) over(married) asyvar
   ytitle("Years of experience")
```

We can use the ytitle() option to add a title to the y-axis. See Options : Axis titles (254) and [G] ***axis_title_options*** for more details, but please disregard any references to xtitle() since that option is not valid when using **graph bar**.
Uses nlsw.dta & scheme vg_s2c

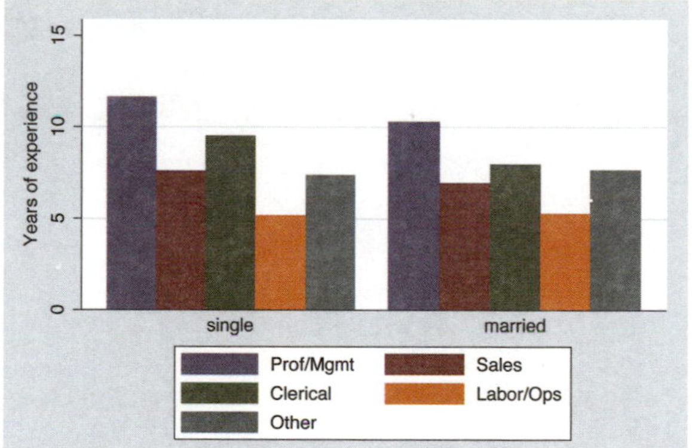

Introduction Twoway Matrix Bar Box Dot Pie Options Standard options Styles Appendix
Y-variables Over Over options Cat axis Legend Y-axis Lookofbar options By

```
graph hbar wage, over(occ5) over(married) asyvar
   ytitle("Years of" "experience")
```

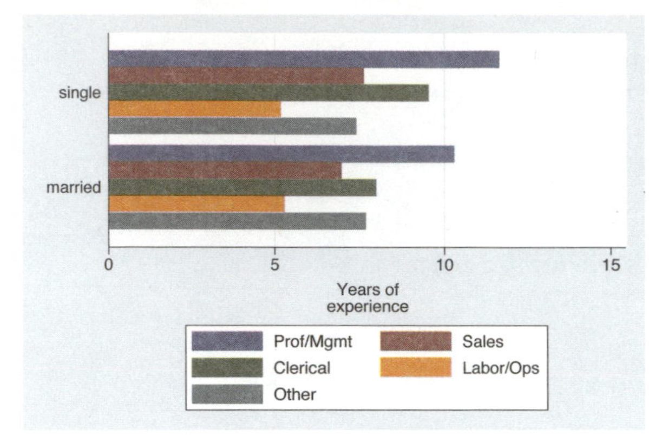

Splitting the title into two separate quoted strings displays the title on separate lines. Note that, when using **graph hbar**, the title of the y-axis now appears at the bottom.
Uses nlsw.dta & scheme vg_s2c

```
graph hbar wage, over(occ5) over(married) asyvar
   ytitle("Years of" "experience", size(vlarge) box bexpand)
```

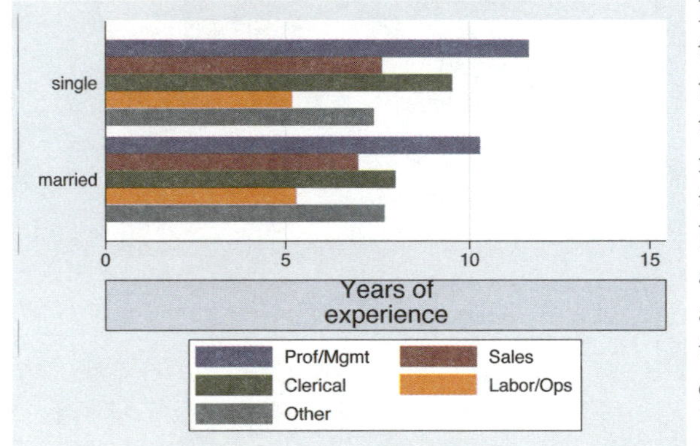

Because this title is considered to be a textbox, you can use a variety of textbox options to control the look of the title. In this example, the title is made large with a box around it, and the `bexpand` (box expand) makes the box expand to fill the width of the plot area. See Options : Textboxes (303) for additional examples of how to use textbox options to control the display of text.
Uses nlsw.dta & scheme vg_s2c

```
graph hbar wage, over(occ5) over(married) asyvar
   yline(8 10, lwidth(thick) lcolor(red) lpattern(dash))
```

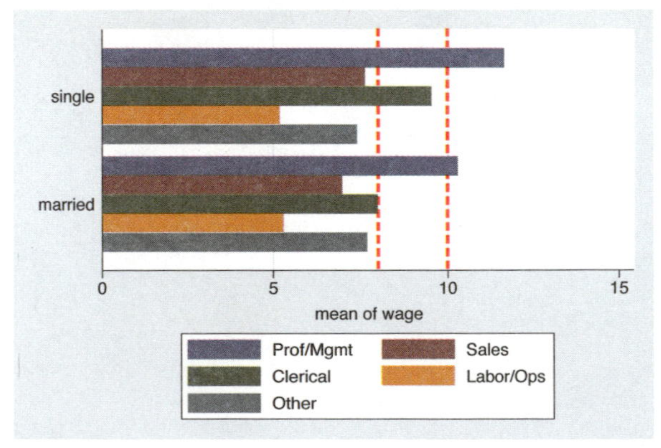

The `yline()` option is used to place a thick, red, dashed line on the graph where y equals 8 and 10. Note that this option is still called `yline()` since the y-axis is the axis with the continuous variable.
Uses nlsw.dta & scheme vg_s2c

```
graph bar hours, over(occ7) asyvar ylabel(30(5)45)
```

We can use the `ylabel()` option to label the y-axis. In this case, we label the y-axis from 30 to 45 by increments of 5. See Options: Axis labels (256) and [G] ***axis_label_options*** for more details. Please disregard any references to `xlabel()` since that option is not valid when using `graph bar`. Note that the y-axis still begins at 0. See the following example to see how you can control that.
Uses nlsw.dta & scheme vg_s2c

```
graph bar hours, over(occ7) asyvar ylabel(30(5)45) exclude0
```

By default, bar charts include 0 on the y-axis, unless you specify the `exclude0` option, as we do here.
Uses nlsw.dta & scheme vg_s2c

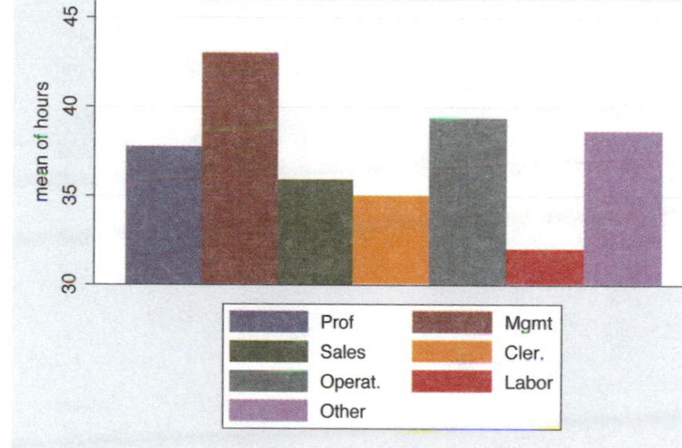


```
graph bar hours, over(occ7) asyvar ylabel(30(5)45, angle(0)) exclude0
```

We can add the `angle()` option to modify the angle of the y-label, making the labels for the y-axis horizontal (zero degrees).
Uses nlsw.dta & scheme vg_s2c

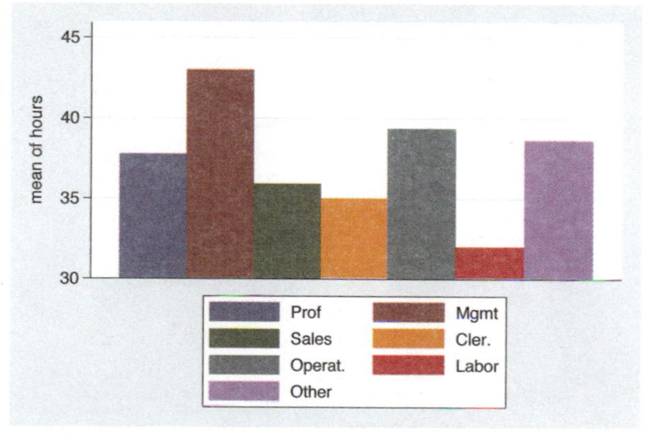

Introduction Twoway Matrix Bar Box Dot Pie Options Standard options Styles Appendix
Y-variables Over Over options Cat axis Legend Y-axis Lookofbar options By

```
graph bar hours, over(occ7) asyvar ylabel(30(5)45, nogrid) exclude0
```

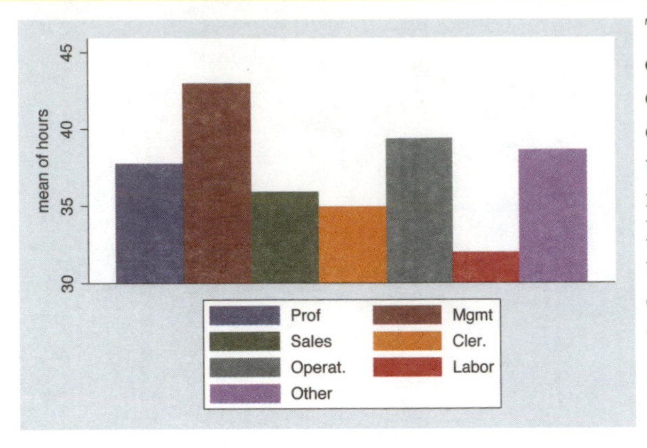

The **nogrid** option suppresses the display of the grid. Note that this option is placed within the `ylabel()` option, thus suppressing the grid for the y-axis. (With bar charts, there is never a grid with respect to the x-axis.) If the grid were absent, and we wanted to include it, we could add the **grid** option. For more details, see Options : Axis labels (256).
Uses nlsw.dta & scheme vg_s2c

```
graph bar prev_exp tenure, over(occ7) yscale(off)
```

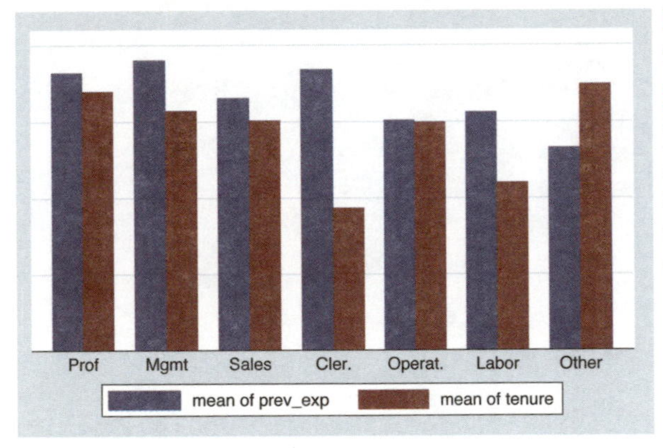

If you want to suppress the display of the y-axis entirely, you can use the `yscale(off)` option. See Options : Axis scales (265) and [G] ***axis_scale_options*** for more details. Please disregard any references to `xscale()` since that option is not valid when using `graph bar`.
Uses nlsw.dta & scheme vg_s2c

```
graph bar prev_exp tenure, over(occ7) yalternate
```

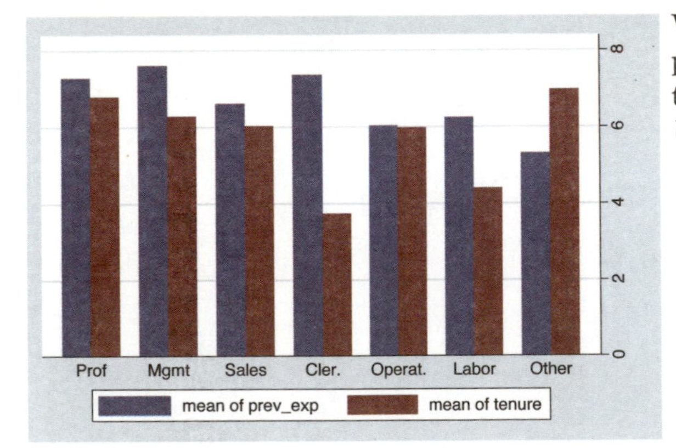

We can use the `yalternate` option to put the y-axis on the opposite side, in this case on the right side of the graph.
Uses nlsw.dta & scheme vg_s2c

```
graph hbar prev_exp tenure, over(occ7) xalternate yreverse
```

You can reverse the direction of the y-axis with the yreverse option. We combine this with the xalternate option to place the labels for the bars on the alternate (right) side of the graph.
Uses nlsw.dta & scheme vg_s2c

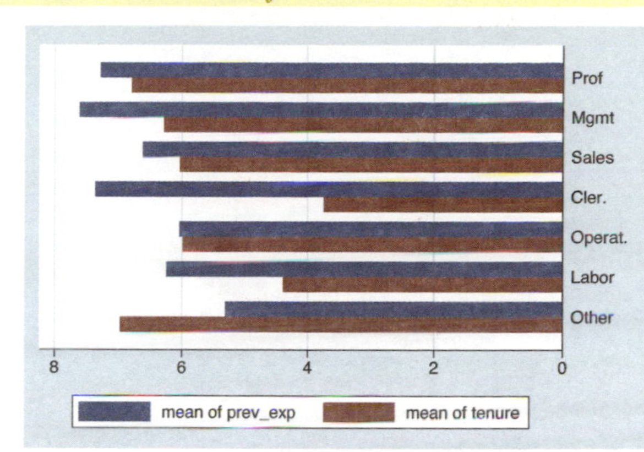

Introduction Twoway Matrix Bar Box Dot Pie Options Standard options Styles Appendix
Y-variables Over Over options Cat axis Legend Y-axis Lookofbar options By

4.7 Changing the look of bars, lookofbar options

This section shows how you can control the look of the bars in your bar charts: the space between the bars, the color of the bars, and the characteristics of the line outlining the bars. For more information, see the *lookofbar_options* table in [G] **graph bar** and [G] ***barlook_options***. This section begins using the vg_rose scheme.

```
graph bar wage hours ttl_exp tenure, over(collgrad)
```

Consider this bar chart. It shows the mean wages, hours worked per week, total experience, and job tenure broken down by whether one graduated college.
Uses nlsw.dta & scheme vg_rose

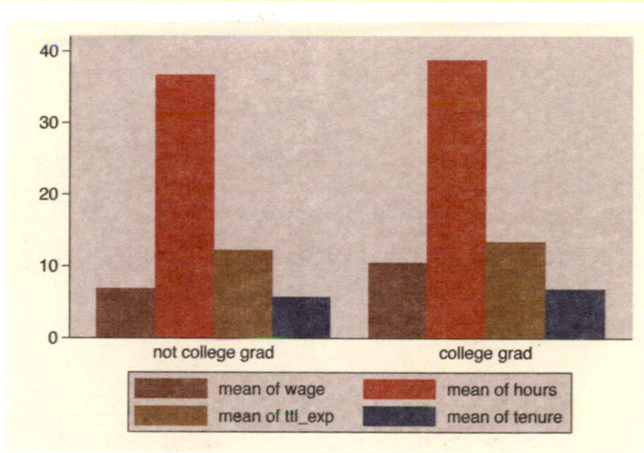

```
graph bar wage hours ttl_exp tenure, over(collgrad)
  outergap(*15)
```

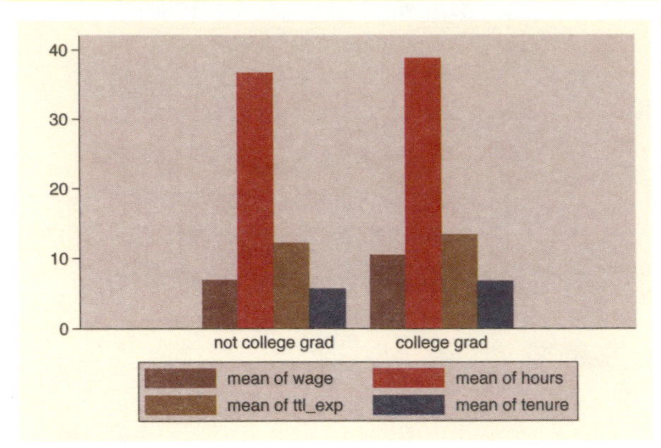

We can change the outer gap between the bars and the edge of the plot area with the `outergap()` option. Here, the gap is fifteen times its normal size. You can also supply values less than 1 to shrink the size of the gap.
Uses nlsw.dta & scheme vg_rose

```
graph bar wage hours ttl_exp tenure, over(collgrad)
  bargap(25)
```

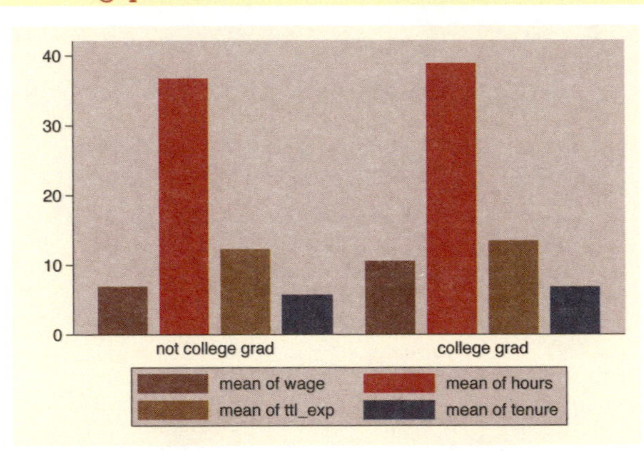

The `bargap()` option controls the size of the gap between the bars. The default value is 0, meaning that the bars touch exactly. Here, we make the gap 25% of the width of the bars.
Uses nlsw.dta & scheme vg_rose

```
graph bar wage hours ttl_exp tenure, over(collgrad)
  bargap(-50)
```

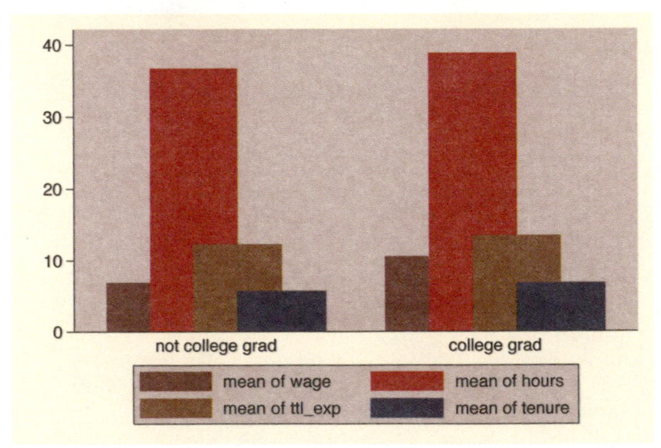

The `bargap()` option permits negative values to indicate that you want the bars to overlap. Here, we make the bars overlap by 50% of the size of the bars.
Uses nlsw.dta & scheme vg_rose

```
graph bar wage hours ttl_exp tenure, over(collgrad)
   intensity(*.5)
```

The **intensity** option is used to control the intensity of the color within the bars. Here, we request that the color be 50% as intense as it normally would be.
Uses nlsw.dta & scheme vg_rose

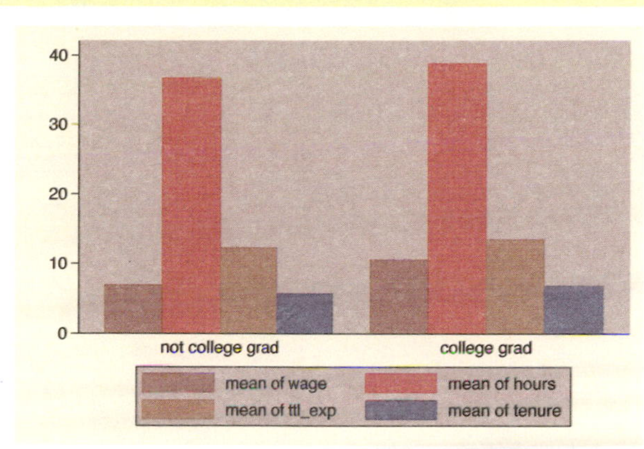


```
graph bar wage hours ttl_exp tenure, over(collgrad)
   intensity(*1.4)
```

In this example, we use the **intensity()** option to make the colors within the bars 1.4 times more intense than they would normally be. Note that Stata also has an option called `lintensity()` that works the same way but controls the intensity of the line surrounding the bar. (This option is not illustrated.)
Uses nlsw.dta & scheme vg_rose

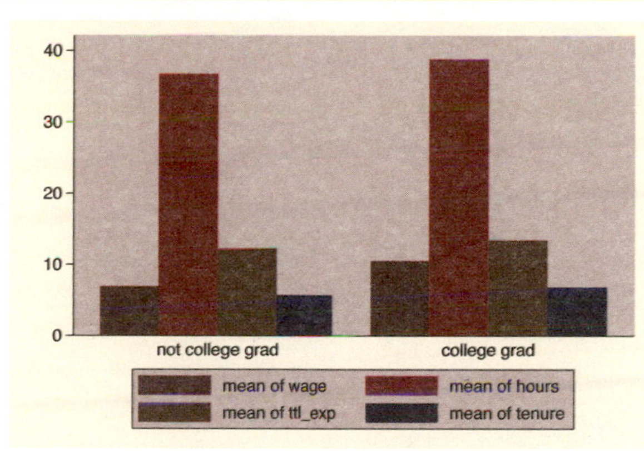

So far, all these options that we have examined determine the overall behavior and look over all of the bars as a group. Using the `bar()` option, you can control the look of the bars for each y-variable, as illustrated below. These graphs use the `vg_s2c` scheme.

```
graph bar wage hours ttl_exp tenure, over(collgrad)
   bar(1, bcolor(dkgreen))
```

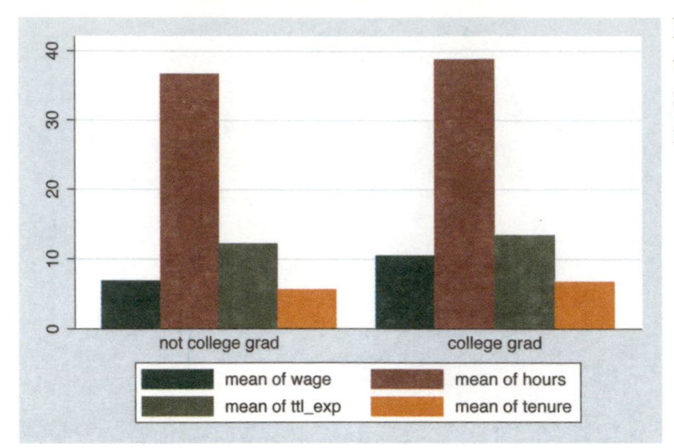

Here, we use the bar() option to make the color of the first bar dark green. See Styles : Colors (328) for more information about colors you can select.
Uses nlsw.dta & scheme vg_s2c

```
graph bar wage hours ttl_exp tenure, over(collgrad)
   bar(1, bfcolor(ltblue) blcolor(blue) blwidth(vthick))
```

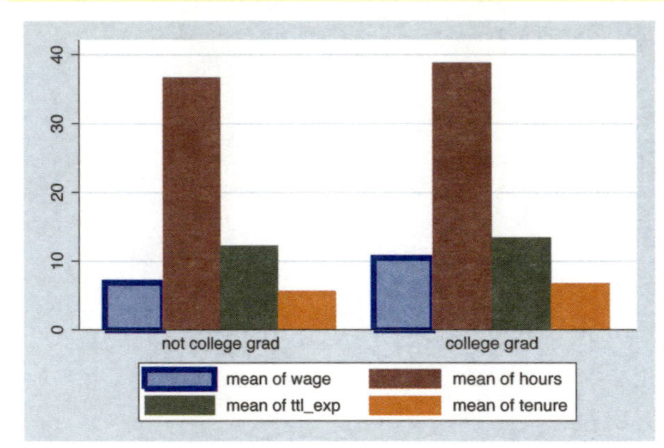

In this example, we make the fill color of the first bar light blue and the outline very thick and blue. See Styles : Linewidth (337) for more details on controlling the thickness of lines. You could also use the `blpattern()` option to control the pattern of the line surrounding the bar; see Styles : Linepatterns (336) for more details.
Uses nlsw.dta & scheme vg_s2c

```
graph bar wage hours ttl_exp tenure, over(collgrad)
```

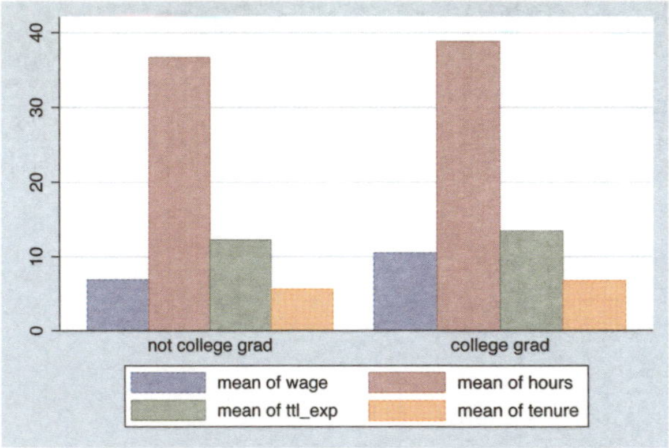

While you can use the `bar()` option to control the look of each bar, selecting a different scheme allows you to control the look of all of the bars. For example, this graph is drawn using the `vg_palec` scheme. See Intro : Schemes (14) for some other schemes you could try and Appendix : Customizing schemes (379) for tips on customizing your own schemes.
Uses nlsw.dta & scheme vg_palec

4.8 Graphing by groups

This section discusses the use of the by() option in combination with graph bar. Normally, you would use the over() option instead of the by() option, but there are cases where the by() option is either necessary or more advantageous. For example, a by() option is useful if you exceed the maximum number of over() options (three if you have a single y-variable or two if you have multiple y-variables). In such cases, the by() option allows you to break your data down by additional categorical variables. Also, by() gives you more flexibility in the placement of the separate panels. For more information about the by() option, see Options : By (272); for more information about the over() option, see Bar : Over (111). These examples are shown using the vg_s1c scheme.

```
graph bar wage, over(urban2) over(married) over(union)
```

Consider this bar graph that breaks wages down by three categorical variables. If we wanted to further break this down by another categorical variable, we could not use another over() option since we can have a maximum of three over() options with a single y-variable.
Uses nlsw.dta & scheme vg_s1c

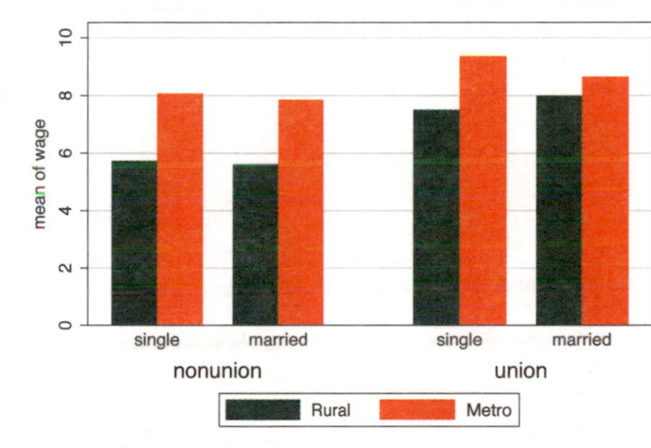


```
graph bar wage, over(urban2) over(married) over(union) by(collgrad)
```

If we want to show the previous graph separately by collgrad, we can use the by() option. This gives us two graphs side by side: one for those who are not college graduates and one for college graduates.
Uses nlsw.dta & scheme vg_s1c

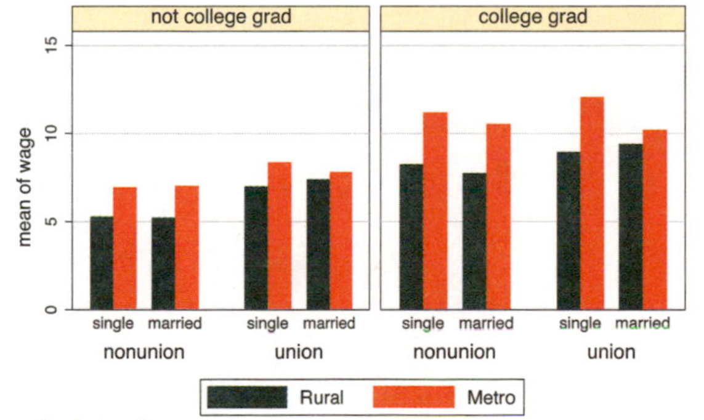

```
graph bar ttl_exp tenure, over(married) over(urban2)
```

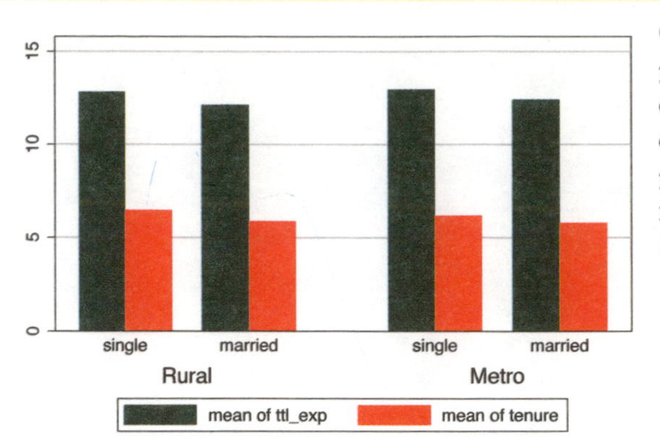

Consider this bar graph with multiple y-variables broken down by two categorical variables using two `over()` options. When you have multiple y-variables, you can only have a maximum of two `over()` options.
Uses nlsw.dta & scheme vg_s1c

```
graph bar ttl_exp tenure, over(married) over(urban2)
   by(union)
```

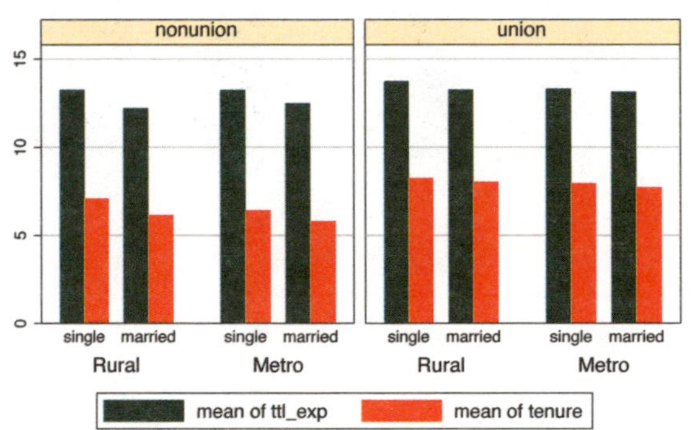

If we want to further show the previous graph by another categorical variable, say `union`, we can use the `by()` option.
Uses nlsw.dta & scheme vg_s1c

```
graph bar ttl_exp tenure, over(married) over(urban2)
   by(union, missing)
```

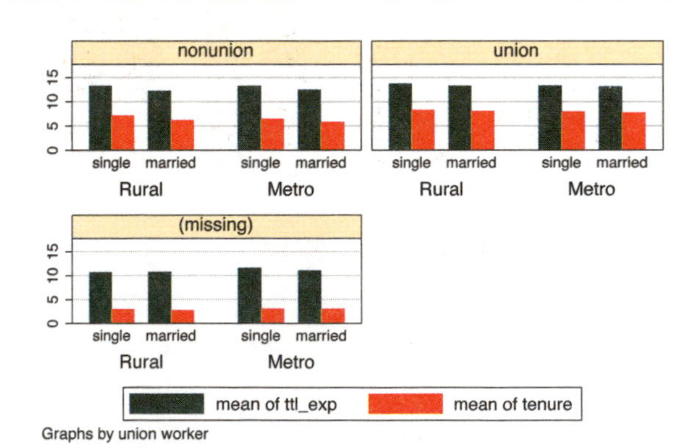

We can add the `missing` option to include a panel for the missing values of `union`.
Uses nlsw.dta & scheme vg_s1c

```
graph bar ttl_exp tenure, over(married) over(urban2)
   by(union, missing total)
```

We can add the total option to include a panel for all observations.
Uses nlsw.dta & scheme vg_s1c

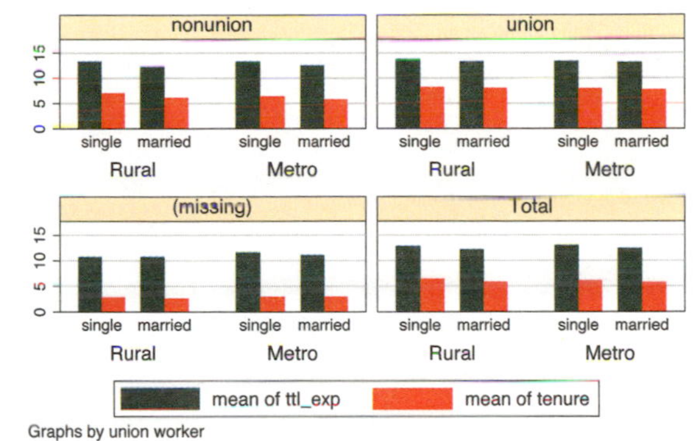


```
graph hbar ttl_exp tenure, over(married) over(urban2)
   by(union, cols(1))
```

We remove the total and missing options and flip the graph to make a horizontal bar chart. We then use the cols(1) option to show these graphs in one column. This makes the graph pretty cramped. Let's explore a number of options we can add to this graph to make it less cramped, adding the options just a small number at a time.
Uses nlsw.dta & scheme vg_s1c

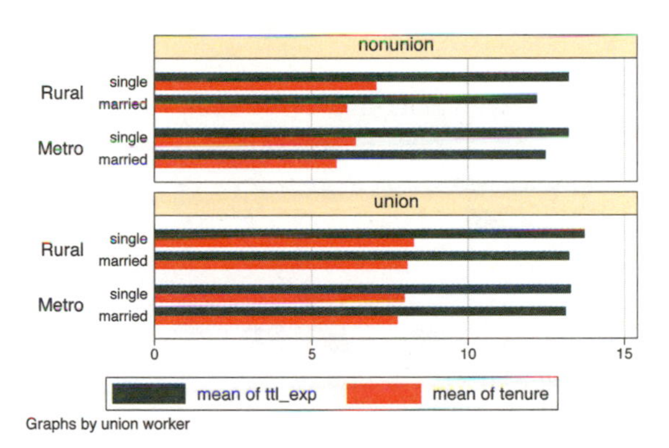


```
graph hbar ttl_exp tenure, over(married) over(urban2)
   by(union, cols(1) note(""))
```

We add the note("") option within the by() option, and that suppresses the note in the left corner, leaving more room for the graph.
Uses nlsw.dta & scheme vg_s1c

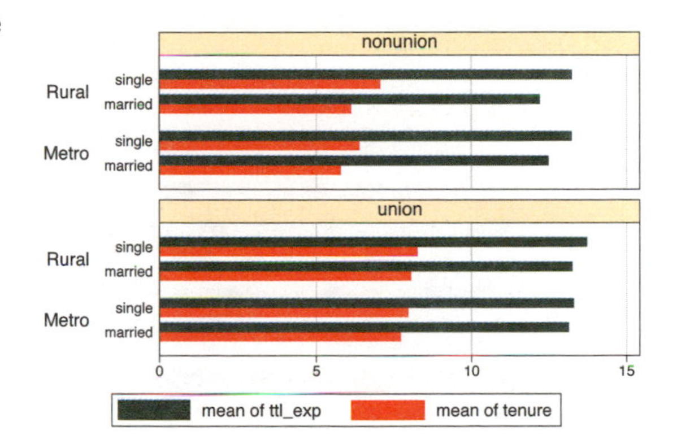

```
graph hbar ttl_exp tenure, over(married) over(urban2)
   by(union, cols(1) note("") legend(position(3)))
```

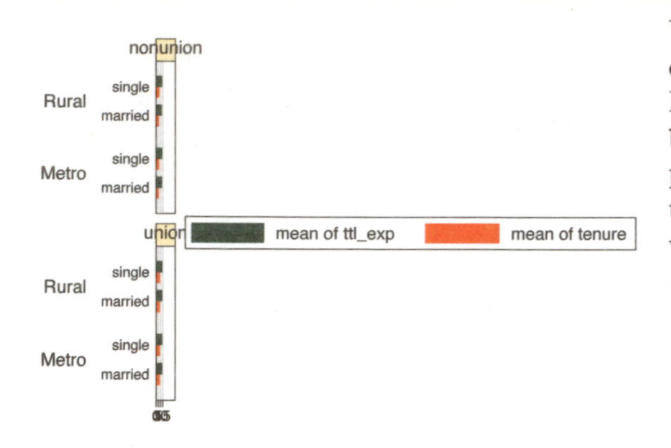

We add the `legend(position(3))` option to put the legend at the right. Note that this is contained within the `by()` option because it changes the position of the legend. If we could make the legend narrow (instead of wide), it would work well in this position.
Uses nlsw.dta & scheme vg_s1c

```
graph hbar ttl_exp tenure, over(married) over(urban2)
   by(union, cols(1) note("") legend(position(3)))
   legend(cols(1) stack label(1 "Tot Exp") label(2 "Curr Exp"))
```

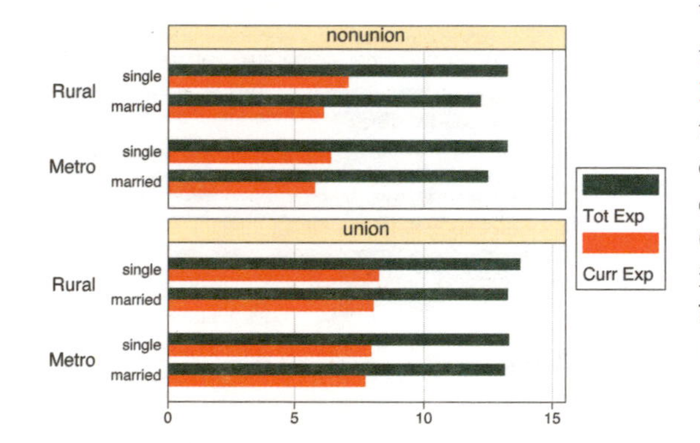

We add the `legend(cols(1) stack)` to make the legend narrow and the `label()` option to change the labels in the legend. Note that this `legend()` option appears outside of the `by()` option. See Options : By (272) and Options : Legend (287) for more information about the interactions of `by()` and `legend()`.
Uses nlsw.dta & scheme vg_s1c

```
graph hbar ttl_exp tenure, over(married) over(urban2)
   by(union, cols(1) note("") legend(position(3)))
   legend(cols(1) stack label(1 "Tot Exp") label(2 "Curr Exp"))
   subtitle(, position(5) ring(0) nobexpand)
```

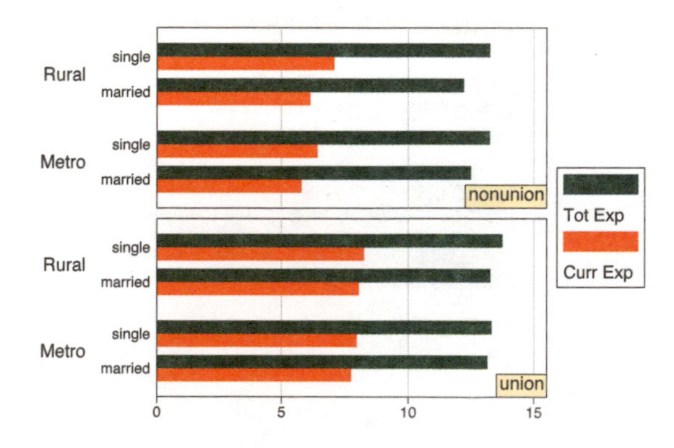

We can add the `subtitle()` option to position the title for each separate graph in the lower right corner. The `position(5)` option puts the title in the 5 o'clock position, and the `ring(0)` option puts the title inside the plot area. The `nobexpand` (no box expand) option keeps the title from expanding to fill the entire plot area.
Uses nlsw.dta & scheme vg_s1c

```
graph bar ttl_exp tenure, over(married) over(urban2)
   by(union collgrad)
```

You can include multiple variables within the by() option. Here, in addition to breaking these variables down by two over() variables, we break them down by two additional variables using the by(union collgrad) option.
Uses nlsw.dta & scheme vg_s1c

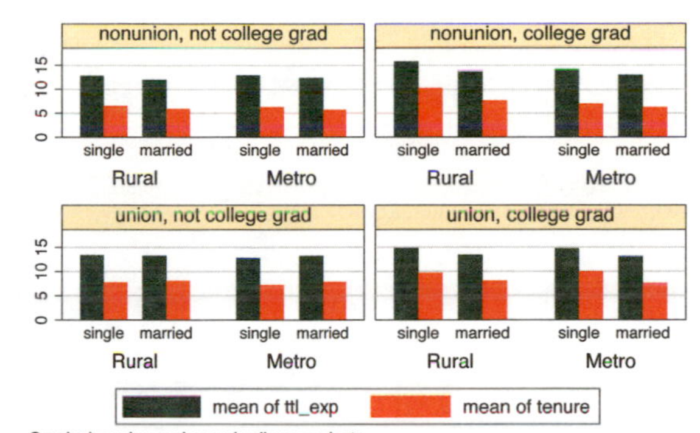

5 Box plots

A box plot displays box(es) bordered at the 25th and 75th percentiles of the y-variable with a *median line* at the 50th percentile. Whiskers extend from the box to the upper and lower adjacent values and are capped with an *adjacent line*. Values exceeding the upper and lower *adjacent values* are called *outside values* and are displayed as markers. This chapter starts by showing the use of the `over()` option to break box plots down by categorical variables and then showing how you can specify multiple y-variables to display plots for multiple variables. Next, we see further options that can be used to customize the display of `over()` option, followed by options that control the display of categorical axes. Next, we discuss options for legends, followed by options that control the display of the y-axis. Finally, we cover options that control the look of boxes and the `by()` option.

5.1 Specifying variables and groups, yvars and over

This section introduces the use of box plots, illustrating the use of the `over()` option for showing box plots by one or more grouping variables. Next, we give examples showing how you can graph multiple variables at once by specifying additional y-variables, followed by some general options for controlling the display of multiple y-variables and the behavior of `over()` options. See the *group_options* table in [G] **graph box** for more details. This section begins with the `vg_s2c` scheme.

```
graph box wage, over(grade4)
```

This is a box plot of wages broken down by education. The `over(grade4)` option breaks down wages by education level (in four categories). By default, the separate levels of `grade4` are graphed using the same color, and the levels are labeled on the x-axis. The graph shows a large number of outside values that are displayed as markers beyond the whiskers. The following example shows how we can suppress the display of the outside values.
Uses nlsw.dta & scheme vg_s2c

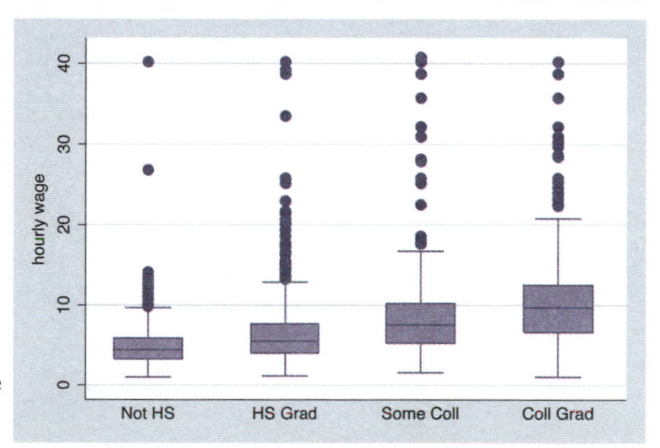

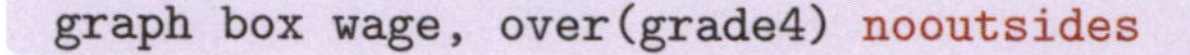

```
graph box wage, over(grade4) nooutsides
```

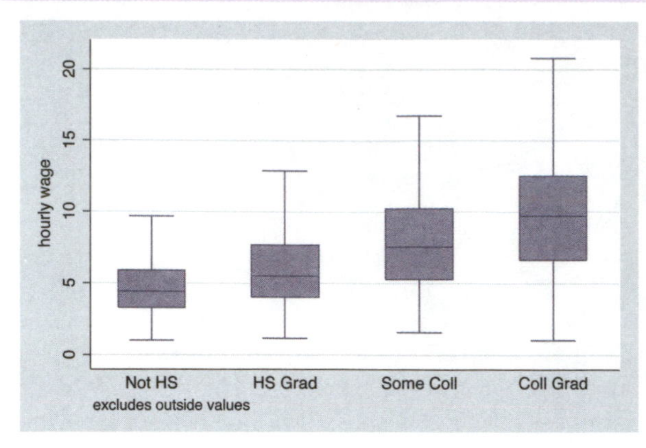

By adding the nooutsides option, we suppress the display of the outside values. Graphs using this option have a note in the bottom left corner indicating that the outside values have been excluded from display in the graph. For most of the graphs in this chapter, there would be a large number of outside values, which would make the graphs very cluttered, so many of the graphs will use the nooutsides option.
Uses nlsw.dta & scheme vg_s2c

```
graph box wage, nooutsides over(grade4) over(union)
```

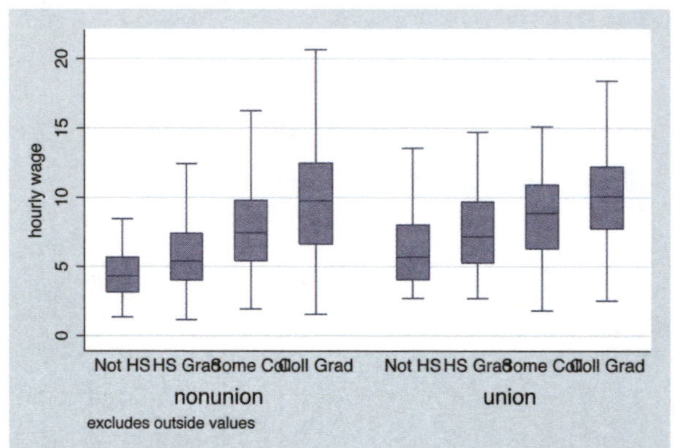

Here, we add the over(union) option to show wages broken down by education and whether one is a member of a union. Note, however, that the labels for grade4 overlap each other. See the next example for one solution.
Uses nlsw.dta & scheme vg_s2c

```
graph hbox wage, nooutsides over(grade4) over(union)
```

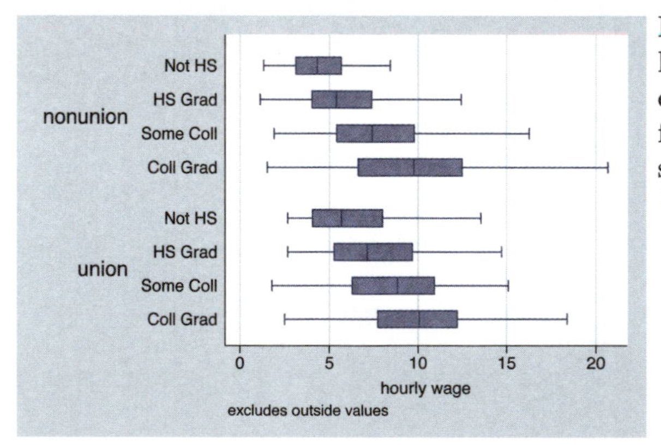

Here, we use graph hbox to make a horizontal box plot. Note that this eliminates the overlapping of the labels for grade4. The next example will show another possible solution.
Uses nlsw.dta & scheme vg_s2c

```
graph box wage, nooutsides over(grade4) over(union) asyvars
```

Using the **asyvars** option, the first **over()** variable, **grade4**, is treated as though it were multiple y-variables. As a result, the levels of **grade4** are shown in multiple colors and labeled via a legend. You can only use **asyvars** when you have a single y-variable.
Uses nlsw.dta & scheme vg_s2c

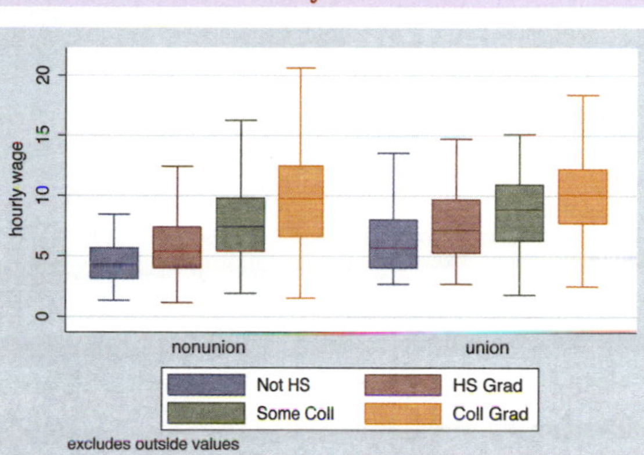


```
graph box wage, nooutsides over(grade4) over(union) over(urban2)
```

In this example, we add a third **over()** option, in this case comparing people who live in rural and metropolitan areas. Note that the first **over()** variable, **grade4**, is now treated as though it were multiple y-variables. Because of this, you can only specify one y-variable when you have three **over()** options.
Uses nlsw.dta & scheme vg_s2c

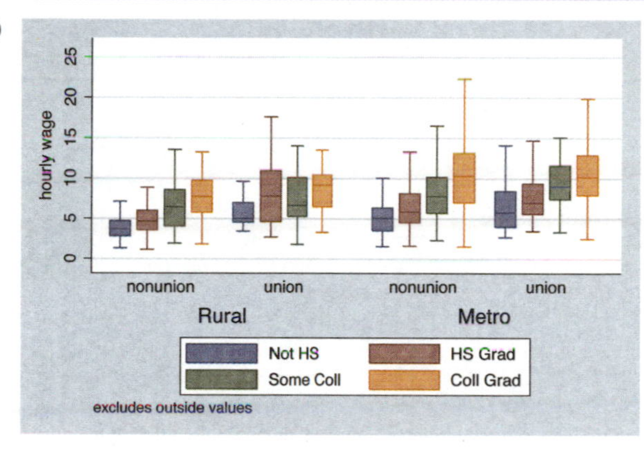

Now, let's look at examples of using multiple y-variables with the **over()** option. We first consider a graph with multiple y-variables. These examples use the **vg_outc** scheme.

```
graph hbox prev_exp tenure, nooutsides
```

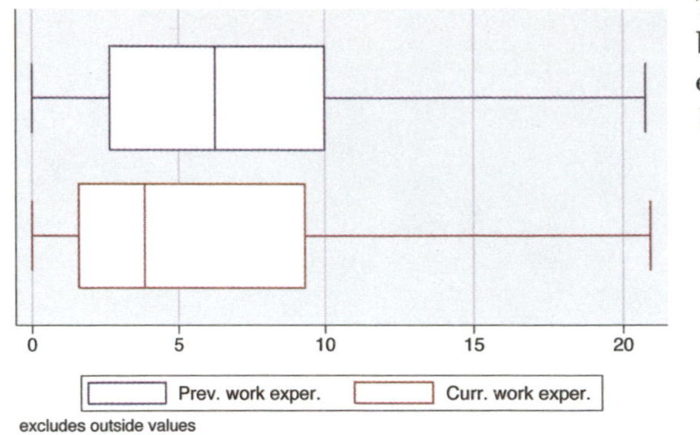

This graph shows work experience before one's current job and work experience at one's current job.
Uses nlsw.dta & scheme vg_outc

```
graph hbox prev_exp tenure, nooutsides over(married)
```

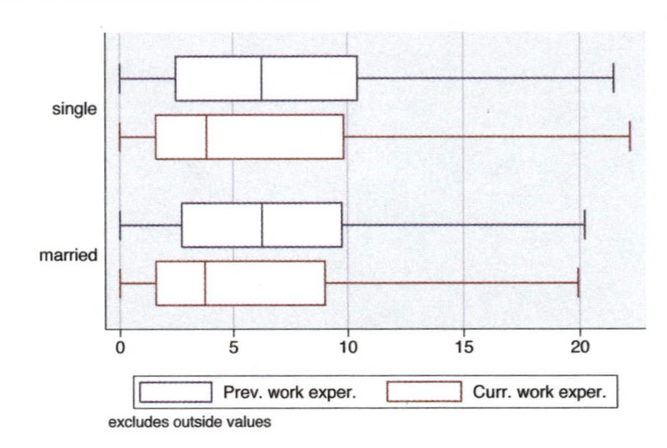

We can further break these variables down by marital status.
Uses nlsw.dta & scheme vg_outc

```
graph hbox prev_exp tenure, nooutsides over(married) over(union)
```

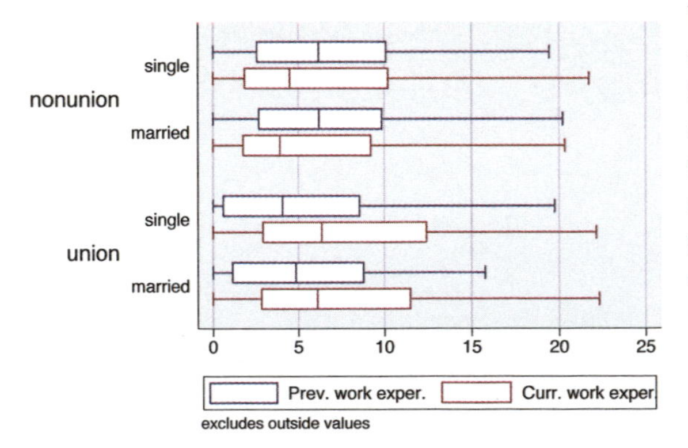

We can take the last graph and add another `over()` option to even further break these variables down by whether one belongs to a union. Note, however, that we cannot add a third `over()` option when we have multiple y-variables, but we could add the `by()` option; see Box : By (189).
Uses nlsw.dta & scheme vg_outc

Now, let's consider options that may be used in combination with the `over()` option to customize the behavior of the graphs. We show how you can treat the levels of the first `over()` option as though they were multiple y-variables. You can also request that missing values for the levels of the `over()` variables be displayed, and you can suppress empty categories when multiple `over()` options are used. These examples are shown below using the `vg_s2m` scheme.

```
graph hbox wage, nooutsides over(grade4) over(union)
```

Consider this graph where we show `wages` broken down by education level and whether one belongs to a union.
Uses nlsw.dta & scheme vg_s2m

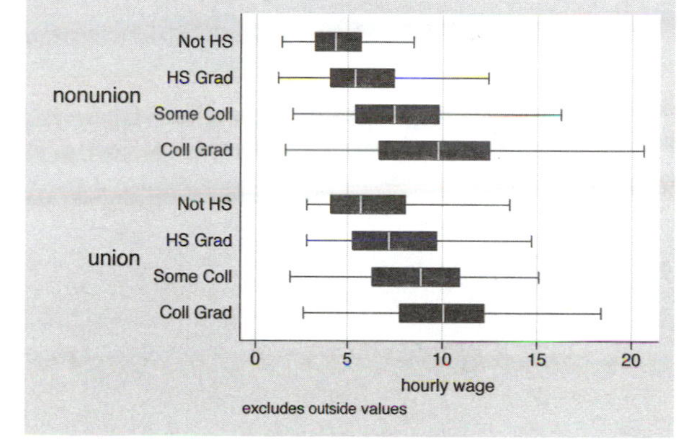


```
graph hbox wage, nooutsides over(grade4) over(union) asyvars
```

If we add the `asyvars` option, then the first `over()` variable (`grade4`) is graphed as if there were four y-variables corresponding to each level of `grade4`. Each level of `grade4` is shown as a differently colored/shaded box and labeled using the legend.
Uses nlsw.dta & scheme vg_s2m

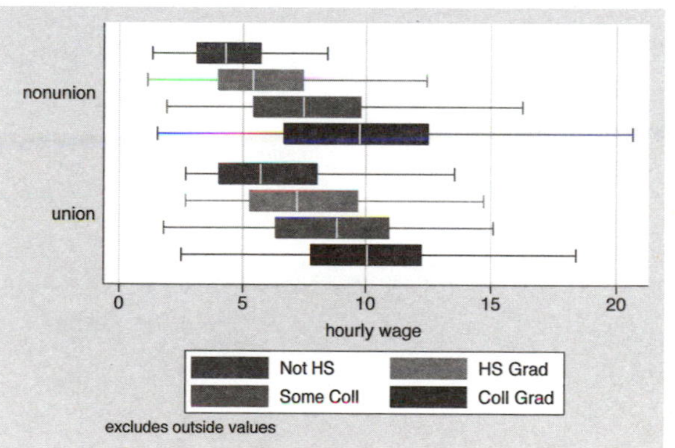

```
graph hbox wage, nooutsides over(grade4) over(union) asyvars missing
```

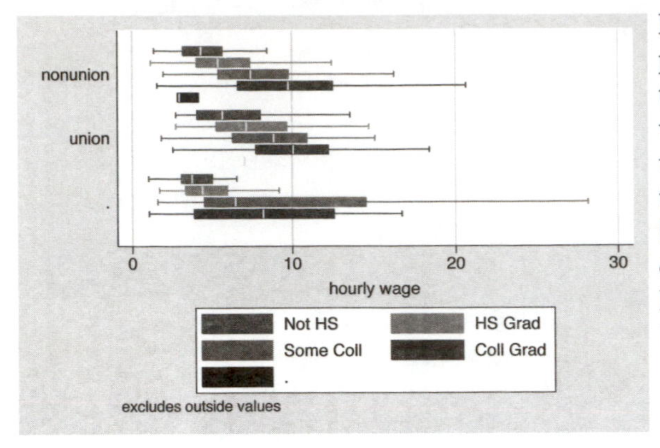

By adding the missing option to the previous graph, we see a category for those who are missing on the union variable, shown as the third group, which is labeled with a dot to indicate that those values are missing; see Box : Cat axis (168) to see how you could label this differently (e.g., labeling it with the word "Missing").
Uses nlsw.dta & scheme vg_s2m

```
graph box wage, nooutsides over(grade) over(collgrad)
```

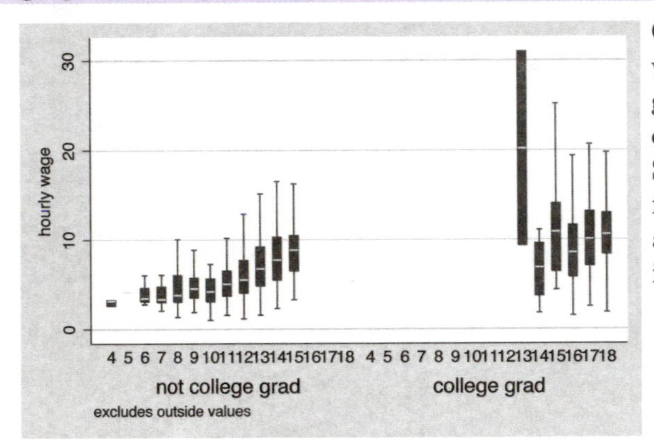

Consider this box chart that breaks wages down by two variables: the last grade that one completed and whether one is a college graduate. By default, Stata shows all possible combinations for these two variables. In most cases, all combinations are possible, but not in this case.
Uses nlsw.dta & scheme vg_s2m

```
graph box wage, nooutsides over(grade) over(collgrad) nofill
```

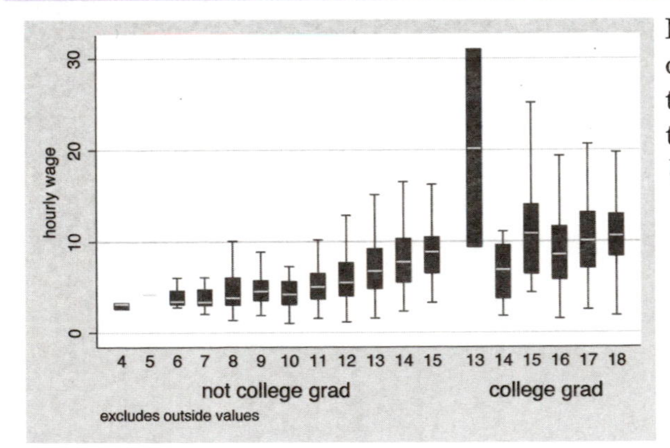

If you only want to display the combinations of the over() variables that exist in the data, then you can use the nofill option.
Uses nlsw.dta & scheme vg_s2m

5.2 Options for groups, over options

Introduction Twoway Matrix Bar Box Dot Pie Options Standard options Styles Appendix
Yvars and over Over options Cat axis Legend Y-axis Boxlook options By

This section considers some of the options that can be used with the `over()` and `yvaroptions()` options for customizing the display of the boxes. We will focus on controlling the spacing between the boxes and the order in which the boxes are displayed. Other options that control the display of the x-axis, such as the labels, are covered in Box : Cat axis (168). For more information on the `over()` options covered in this section, see the *over_subopts* table in [G] **graph box**. We begin by considering options that control the spacing among the boxes and use the `vg_past` scheme.

```
graph hbox tenure, nooutsides over(occ5) over(collgrad)
```

Consider this graph that shows box plots of `tenure` broken down by `occ5` and `collgrad`. We use the `nooutsides` option to suppress the display of outside values. For the rest of the graphs in this section, there would be a large number of outside values, which would make the graphs very cluttered, so we will include the `nooutsides` option for each example.
Uses nlsw.dta & scheme vg_past

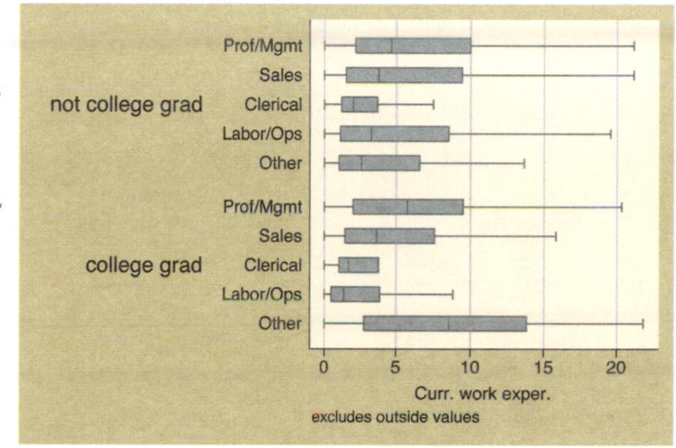


```
graph hbox tenure, nooutsides over(occ5, gap(*3)) over(collgrad)
```

We can change the gap between the levels of `occ5`. Here, we make that gap twice as large as it normally would. This leads to narrow boxes with a sizable gap between them.
Uses nlsw.dta & scheme vg_past

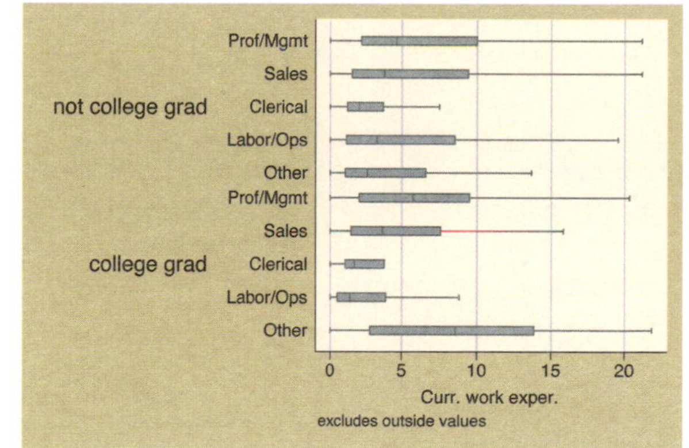

```
graph hbox tenure, nooutsides over(occ5, gap(*.2)) over(collgrad)
```

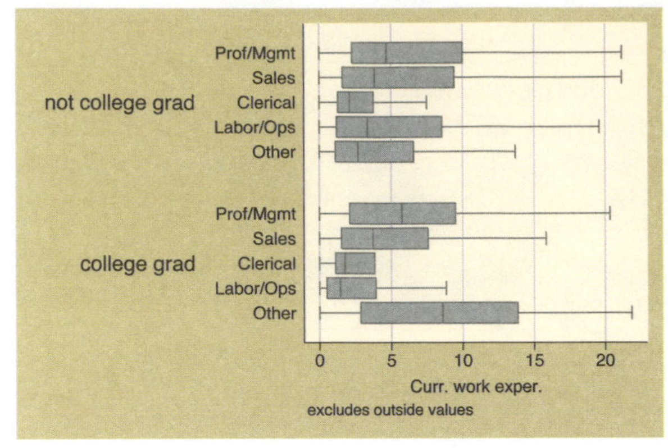

Here, we shrink the gap between the levels of `collgrad`, making the gaps 20% of the size they normally would. This yields boxes that are wider than they normally would.
Uses nlsw.dta & scheme vg_past

```
graph hbox tenure, nooutsides over(occ5, gap(*.4)) over(collgrad, gap(*2))
```

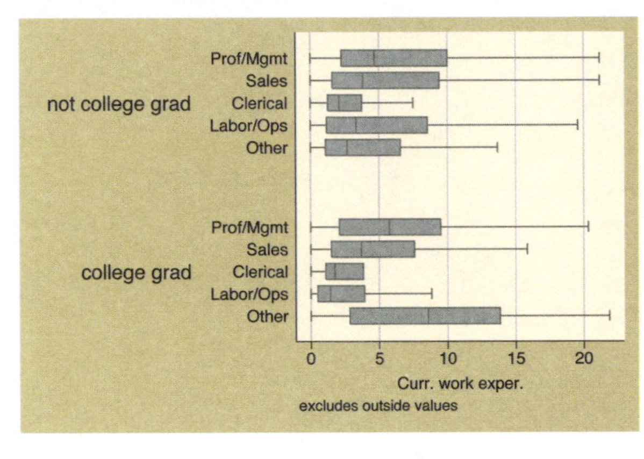

We can control the gap with respect to each of the `over()` variables. In this example, we make the gap among the `occ5` categories small (40% of their original size) and the gap between the levels of `collgrad` larger (two times the normal size).
Uses nlsw.dta & scheme vg_past

By default, the boxes formed by `over()` variables are ordered in ascending sequence according to the values of the `over()` variable. Stata allows us to control the order of the boxes by allowing us to put them in descending order, order them according to the values of another variable, or sort the boxes according to their medians. These options are illustrated in the following examples.

```
graph hbox tenure, nooutsides over(occ7, descending)
```

Consider this graph showing **tenure** broken down by the seven levels of occupation. The boxes would normally be ordered by levels of **occ7**, going from 1 to 7. The **descending** option switches the order of the boxes. They still are ordered according to the seven levels of occupation, but the boxes are ordered going from 7 to 1.
Uses nlsw.dta & scheme vg_past

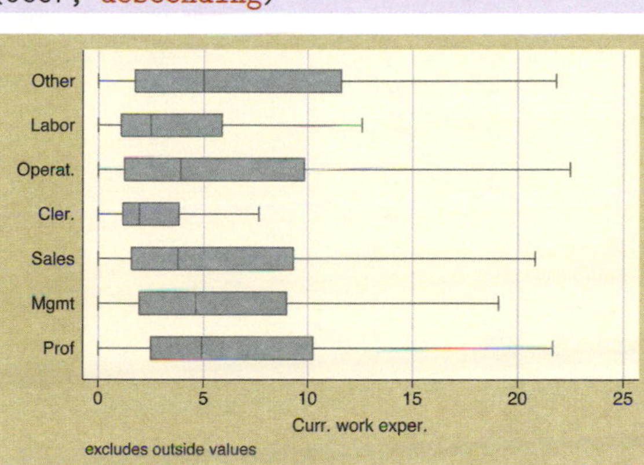


```
graph hbox tenure, nooutsides over(occ7, sort(occ7alpha))
```

We might want to put these boxes in alphabetical order, but with Other still appearing last. We can do this by recoding **occ7** into a new variable (say **occ7alpha**) such that, as **occ7alpha** goes from 1 to 7, the occupations alphabetically ordered. We recoded **occ7** with these assignments: 4 = 1, 6 = 2, 2 = 3, 5 = 4, 1 = 5, 3 = 6, and 7 = 7. Then, the **sort(occ7alpha)** option alphabetizes the boxes (but with Other still appearing last).
Uses nlsw.dta & scheme vg_past

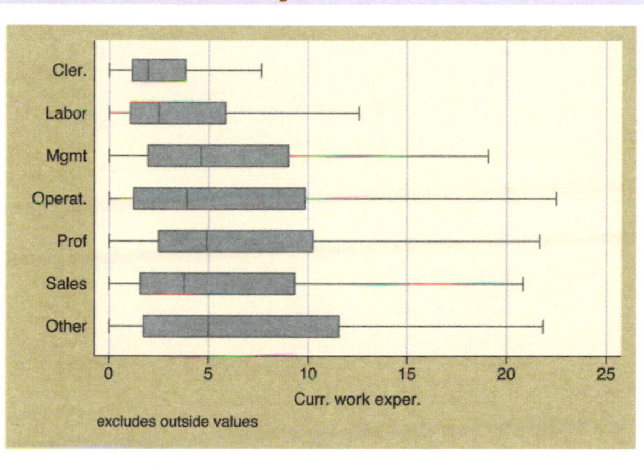


```
graph hbox tenure, nooutsides over(occ7, sort(1))
```

Here, we sort the variables based on the median of **tenure**, yielding boxes with medians in ascending order. The **sort(1)** option sorts the boxes according to the median of the first y-variable, meaning to sort on the median of **tenure**.
Uses nlsw.dta & scheme vg_past

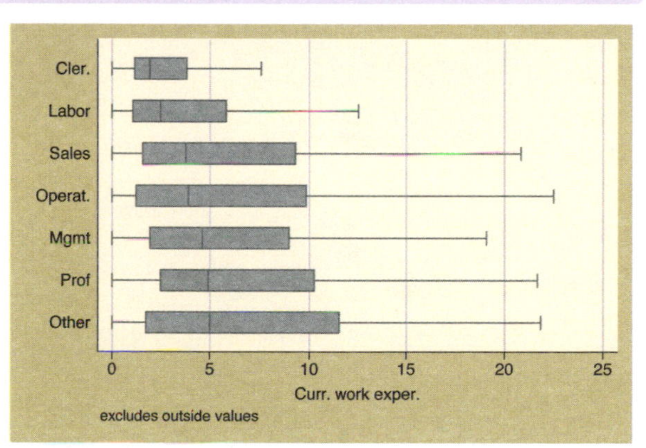

```
graph hbox tenure, nooutsides over(occ7, sort(1) descending)
```

Adding the `descending` option yields boxes in descending order, going from highest median `tenure` to lowest median `tenure`.
Uses nlsw.dta & scheme vg_past

```
graph hbox prev_exp tenure, nooutsides over(occ7)
```

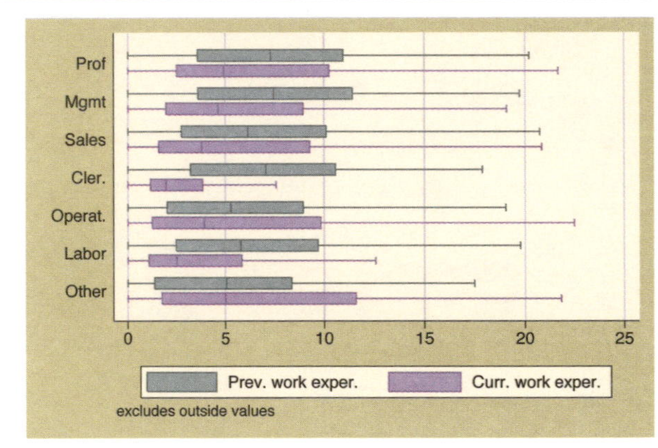

Here, we plot two y-variables: the number of years of work experience before one's current job and the years in one's current job. Since we have removed any `sort()` options, the boxes are sorted according to the values of `occ7`.
Uses nlsw.dta & scheme vg_past

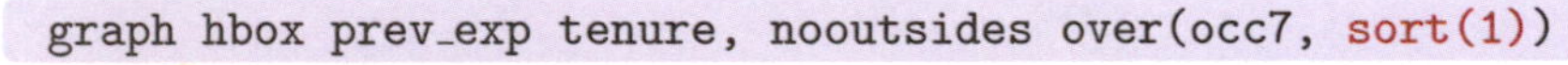

```
graph hbox prev_exp tenure, nooutsides over(occ7, sort(1))
```

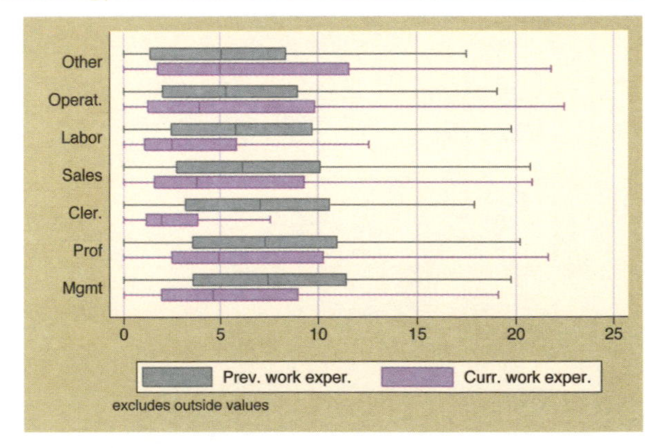

Adding the `sort(1)` option now sorts the boxes according to the median of `prev_exp` since that is the first y-variable.
Uses nlsw.dta & scheme vg_past

```
graph hbox prev_exp tenure, nooutsides over(occ7, sort(2))
```

Changing `sort(1)` to `sort(2)` then sorts the boxes according to the median of the second *y*-variable, `tenure`.
Uses nlsw.dta & scheme vg_past

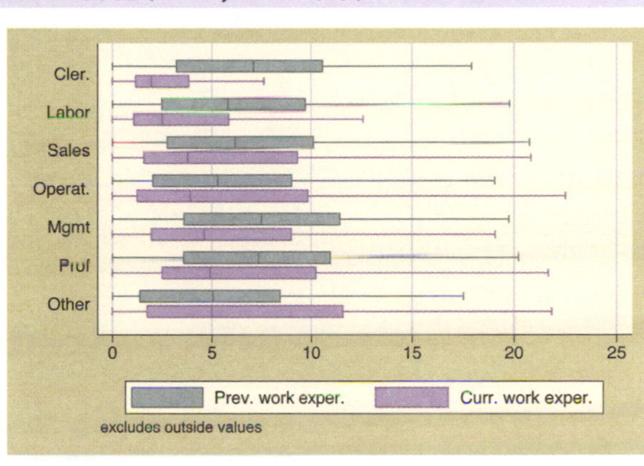


```
graph hbox tenure, nooutsides over(occ7, sort(1)) over(collgrad)
```

We can use the `sort()` option when there are additional `over()` variables. Here, the boxes are ordered according to the median of `tenure` across `occ7` but within each level of `collgrad`.
Uses nlsw.dta & scheme vg_past

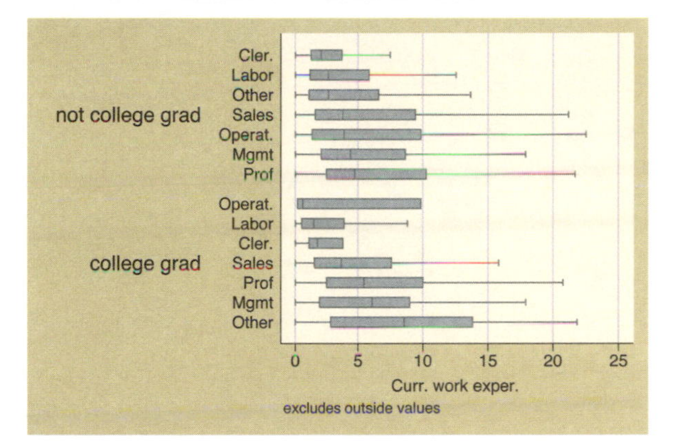


```
graph hbox tenure, nooutsides over(occ7, sort(1)) over(collgrad, descending)
```

We add the `descending` option to the second `over()` option, and the levels of `collgrad` are now shown with college graduates appearing first.
Uses nlsw.dta & scheme vg_past

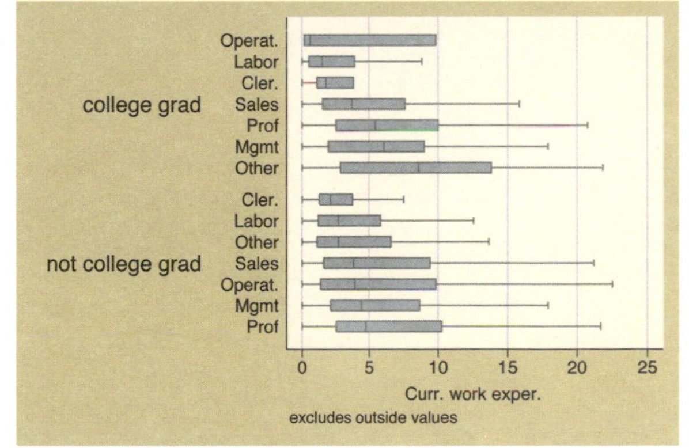

5.3 Controlling the categorical axis

This section describes ways that you can label categorical axes. Box plots are similar to bar charts, but they are different from other graphs because their x-axes are represented by categorical variables. This section describes options you can use to customize these categorical axes. For more details on this, see [G] ***cat_axis_label_options*** and [G] ***cat_axis_line_options***.

We will start by showing examples of how you can change the labels for the x-axis for these categorical variables. The next set of examples will use the `vg_teal` scheme.

```
graph box wage, nooutsides over(south)
```

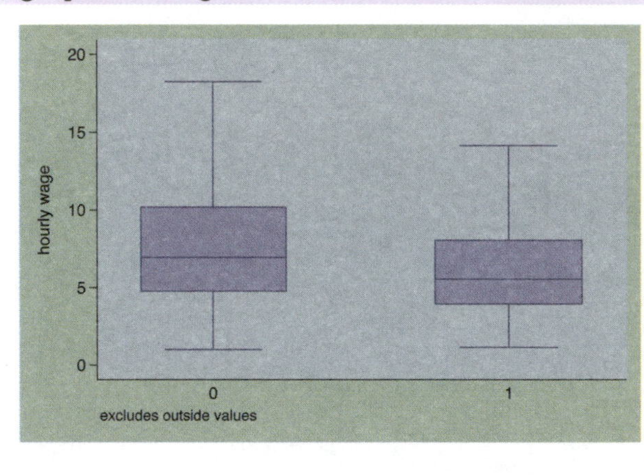

This is an example of a box plot with one `over()` variable graphing wages broken down by whether one lives in the South. The variable `south` is a dummy variable that does not have any value labels, so the x-axis is not labeled very well. We use the `nooutsides` option to suppress the display of outside values. For the rest of the graphs in this section, there would be a large number of outside values, which would make the graphs very cluttered, so we will include the `nooutsides` option for each example.
Uses nlsw.dta & scheme vg_teal

```
graph box wage, nooutsides over(south, relabel(1 "N & W" 2 "South"))
```

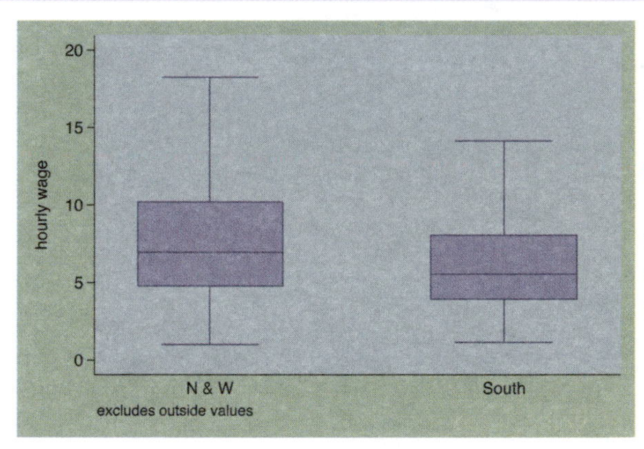

We can use the `relabel()` option to change the labels displayed for the levels of `south`, giving the x-axis more meaningful labels. Note that we wrote `relabel(1 "N & W")`, not `relabel(0 "N & W")`, since these numbers do not represent the actual levels of `south` but the ordinal position of the levels, i.e., first and second.
Uses nlsw.dta & scheme vg_teal

```
graph box wage, nooutsides over(south, relabel(1 "N & W" 2 "South"))
   over(smsa, relabel(1 "Non Metro" 2 "Metro"))
```

This is an example of a box plot with two `over()` variables. Here, we use the `relabel()` option to change the labels displayed for the levels of `south` and `smsa`.
Uses nlsw.dta & scheme vg_teal

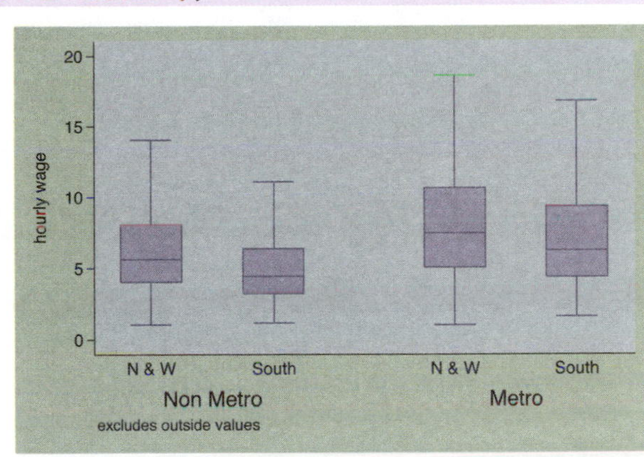


```
graph box prev_exp tenure ttl_exp, nooutsides ascategory
```

This shows a box plot with multiple y-variables but uses the `ascategory` option to plot the different y-variables as if they were categorical variables. The boxes for the different variables are the same color, and the categories are labeled on the x-axis rather than with a legend. The default labels on the x-axis are not bad, but we might want to change them.
Uses nlsw.dta & scheme vg_teal

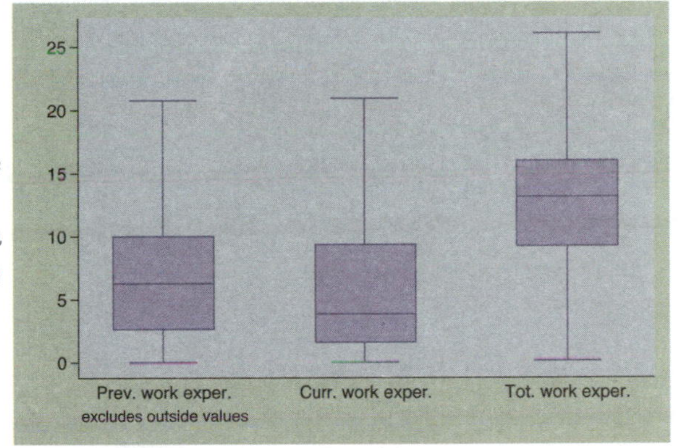


```
graph box prev_exp tenure ttl_exp, nooutsides ascategory
   yvaroptions(relabel(1 "Prev Exp" 2 "Curr Exp" 3 "Tot Exp"))
```

If we had an `over()` option, we would use the `relabel()` option to change the labels on the x-axis. But since we had multiple y-variables that we have treated as categories, we then use the `yvaroptions(relabel())` option to modify the labels on the x-axis.
Uses nlsw.dta & scheme vg_teal

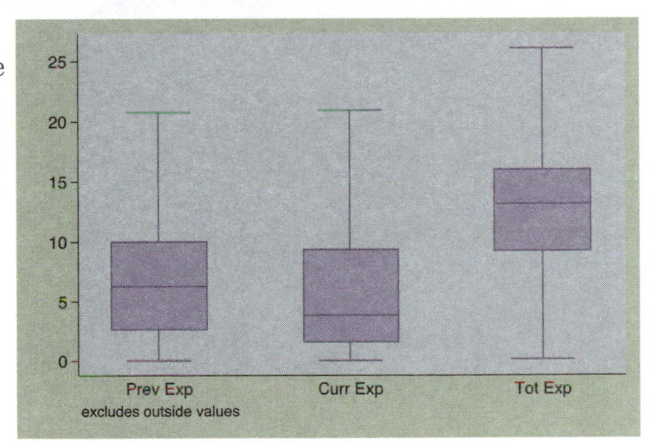

```
graph box prev_exp tenure ttl_exp, nooutsides ascategory
   over(south, relabel(1 "N & W" 2 "South"))
   yvaroptions(relabel(1 "Prev Exp" 2 "Curr Exp" 3 "Tot Exp"))
```

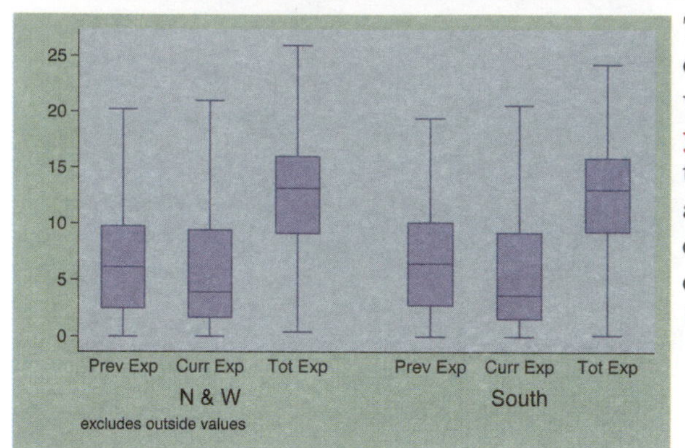

This example is similar to the previous example, but we have added an **over()** variable as well. As before, we use **yvaroptions(relabel())** to modify the labels for the multiple y-variables, and then we also use the **relabel()** option within the **over()** option to change the labels for **south**.
Uses nlsw.dta & scheme vg_teal

```
graph box prev_exp tenure ttl_exp, nooutsides ascategory xalternate
   over(south, relabel(1 "N & W" 2 "South"))
   yvaroptions(relabel(1 "Prev Exp" 2 "Curr Exp" 3 "Tot Exp"))
```

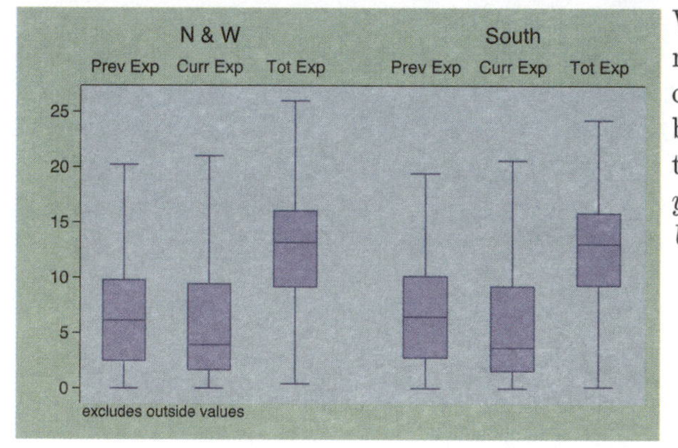

We add the **xalternate** option, which moves the labels for the x-axis to the opposite side, in this case from the bottom to the top. You can also use the **yalternate** option to move the y-axis to its opposite side.
Uses nlsw.dta & scheme vg_teal

In the examples above, we have seen that, even though the **relabel()** option is called an **over()** option, it can be used within **yvaroptions()** to control the labeling of multiple y-variables (provided that the **ascategory** option is used to convert the multiple y-variables into categories). We will next explore other **over()** options, which also can be used with either **over()** or **yvaroptions()**. These examples will use the **vg_rose** scheme.

```
graph box wage, nooutsides over(occ7, label(angle(45))) over(collgrad)
```

In this example, the levels of occ7 might overlap each other. Using the label(angle(45)) option makes the angle of the labels for occ7 45 degrees, and they do not overlap.
Uses nlsw.dta & scheme vg_rose

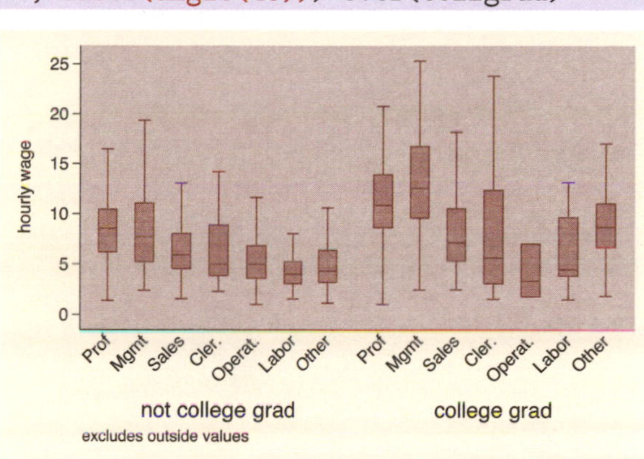


```
graph box wage, nooutsides over(occ7, label(alternate)) over(collgrad)
```

Another way we can avoid overlapping is by adding the label(alternate) option. As you can see, the labels alternate in height, avoiding overlapping.
Uses nlsw.dta & scheme vg_rose

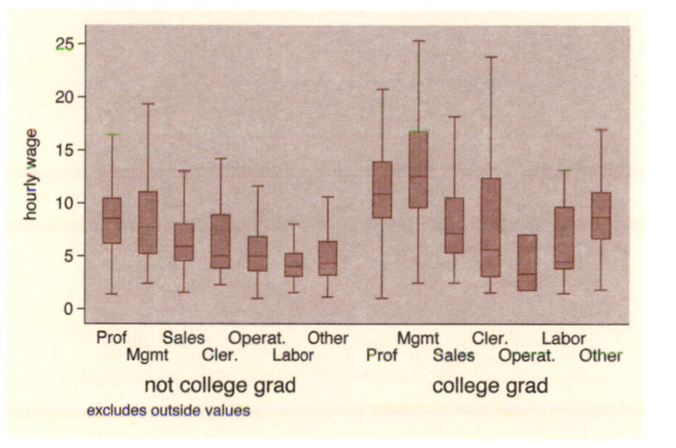


```
graph box wage, nooutsides over(occ7, label(labsize(small))) over(collgrad)
```

We can instead make the size of the labels smaller to make them fit without overlapping. Here, we make the label size small using the label(labsize(small)) option. See Styles: Textsize (344) for other values you could choose for labsize().
Uses nlsw.dta & scheme vg_rose

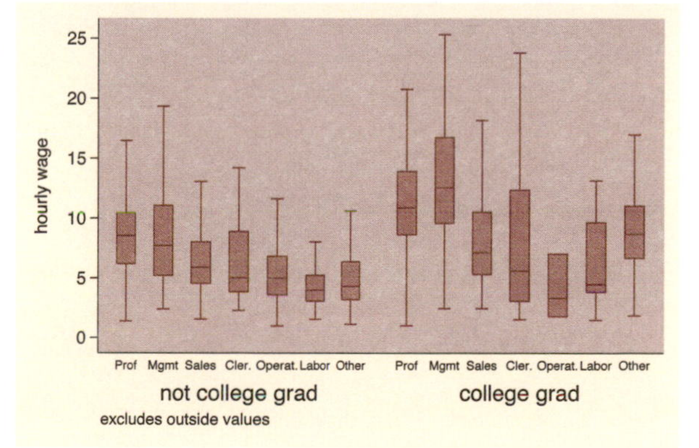

```
graph hbox wage, nooutsides over(occ5, label(labcolor(maroon)))
   over(collgrad)
```

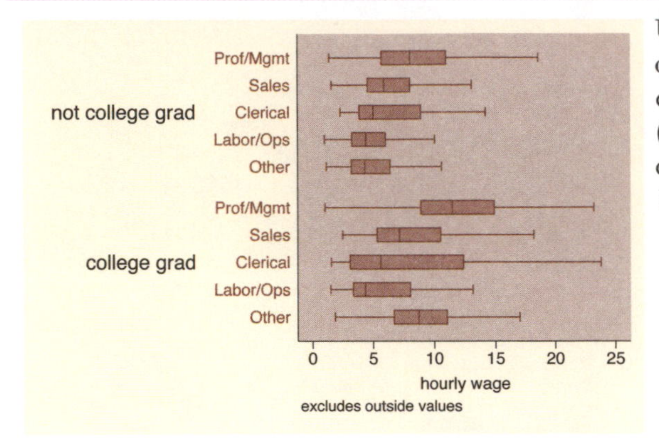

Using the `label(labcolor(maroon))` option, we change the label color for `occ5` to maroon. See Styles : Colors (328) for more details about other colors you could choose.
Uses nlsw.dta & scheme vg_rose

```
graph hbox wage, nooutsides
   over(occ5, label(ticks tlwidth(thick) tlength(*2) tposition(crossing)))
   over(collgrad)
```

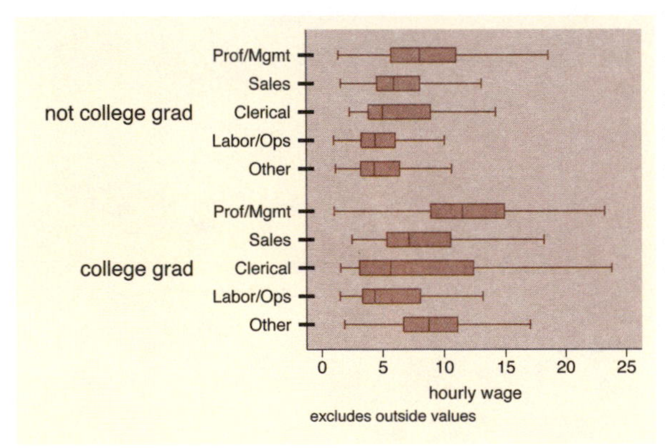

We can use the `label(ticks)` option to place ticks under each box. We also modify the attributes of the ticks, making the tick thick, twice as long as normal, and crossing the x-axis. See [G] ***cat_axis_label_options*** for more details and other options for controlling ticks.
Uses nlsw.dta & scheme vg_rose

```
graph hbox wage, nooutsides over(occ5, label(labgap(*5))) over(collgrad)
```

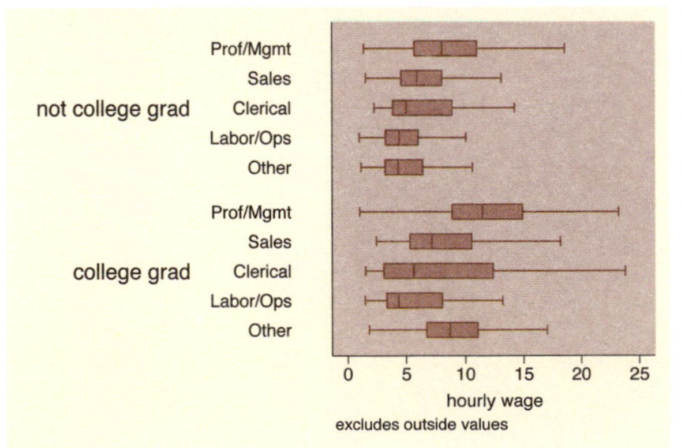

The `label(labgap(*5))` option controls the gap between the label and the ticks. Here, we increase the gap between the label for the levels of `occ5` and the axis line to five times its normal size.
Uses nlsw.dta & scheme vg_rose

```
graph hbox wage, nooutsides over(occ5) over(collgrad, label(labgap(*7)))
```

Using the `label(labgap(*7))` option, we increase the gap associated with `collgrad`. This example makes the gap between `collgrad` and `occ5` seven times its normal size.
Uses nlsw.dta & scheme vg_rose

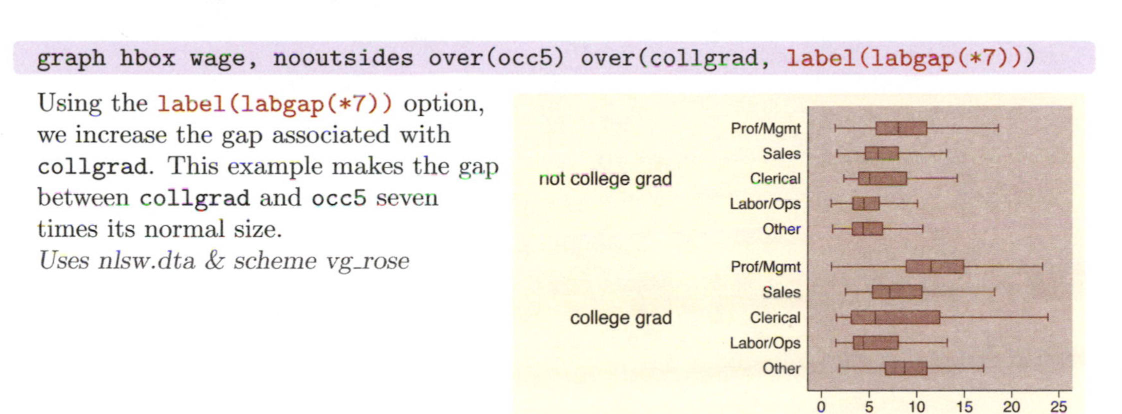


```
graph box wage, nooutsides over(occ7, axis(outergap(*20)))
```

We use the `axis(outergap())` option to increase the gap between the labels of the x-axis and the outside of the graph. As you can see, this increases the space between the labels for `occ7` and the bottom of the graph.
Uses nlsw.dta & scheme vg_rose

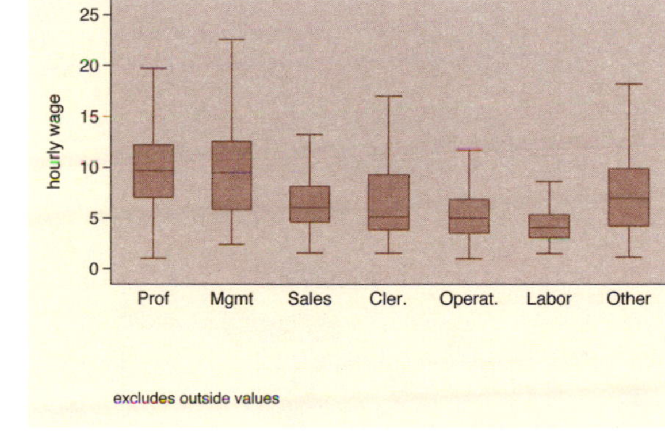

So far, we have focused on labeling the values on the categorical x-axis, but we have not yet looked at how to add a title to that axis. We might be tempted to use `xtitle()`, but that option is not valid for a categorical axis. Instead, we can use other means for giving titles to these axes, as illustrated in the examples below using the `vg_s1c` scheme.

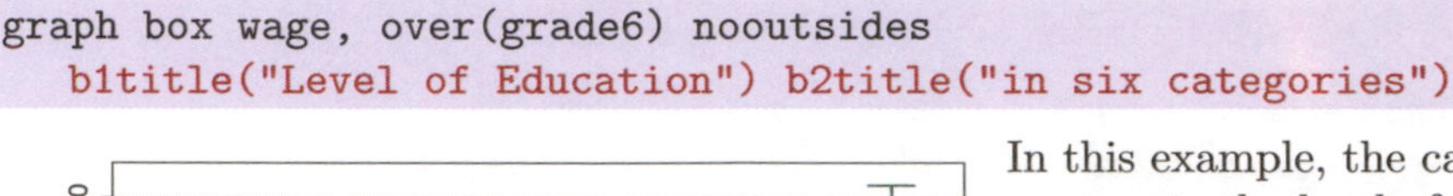

```
graph box wage, over(grade6) nooutsides
  b1title("Level of Education") b2title("in six categories")
```

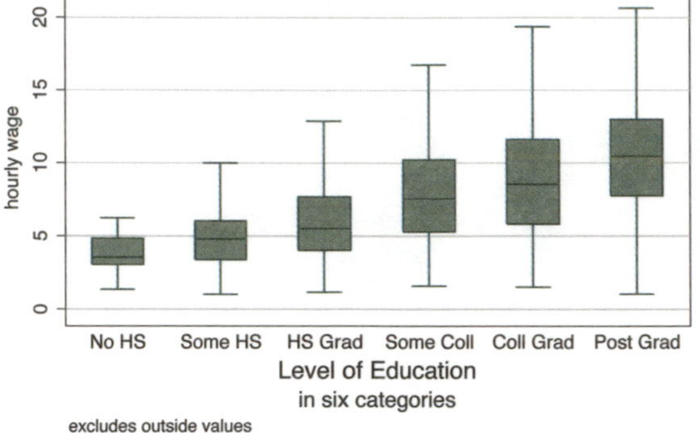

In this example, the categorical axis represents the level of education, and we can use the `b1title()` and `b2title()` options to add titles to the bottom of the graph. See Standard options : Titles (313) for more details.
Uses nlsw.dta & scheme vg_s1c

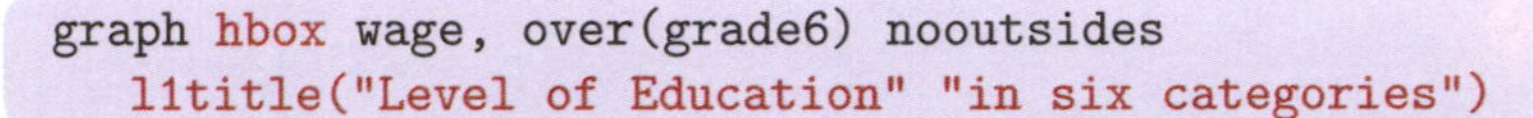

```
graph hbox wage, over(grade6) nooutsides
  l1title("Level of Education" "in six categories")
```

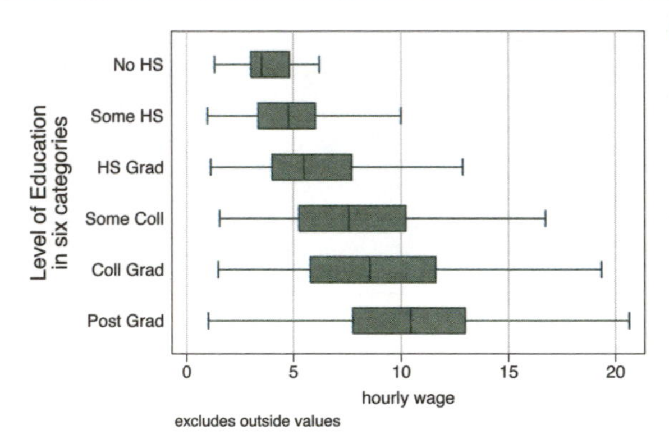

By using `graph hbox`, the categorical axis is now on the left axis, so we then use the `l1title()` to add a title to the x-axis. We could also use the `l2title()` to add a second title as well.
Uses nlsw.dta & scheme vg_s1c

5.4 Controlling legends

This section discusses the use of legends for box charts, emphasizing the features that are unique to box charts. The section Options : Legend (287) goes into great detail about legends, as does [G] ***legend_option***. Legends can be used for multiple y-variables or when the first `over()` variable is treated as a y-variable via the `asyvars` option. See Box : Yvars and over (157) for more information about using multiple y-variables and more examples of treating the first `over()` variable as a y-variable. These first examples use the `vg_brite` scheme.

```
graph box prev_exp tenure ttl_exp, nooutsides
```

Consider this box plot of three different variables. These variables are shown with different colors, and a legend is used to identify the variables. We use the `nooutsides` option to suppress the display of outside values. For the rest of the graphs in this section, there would be a large number of outside values, which would make the graphs very cluttered, so we will include the `nooutsides` option for each example.
Uses nlsw.dta & scheme vg_brite

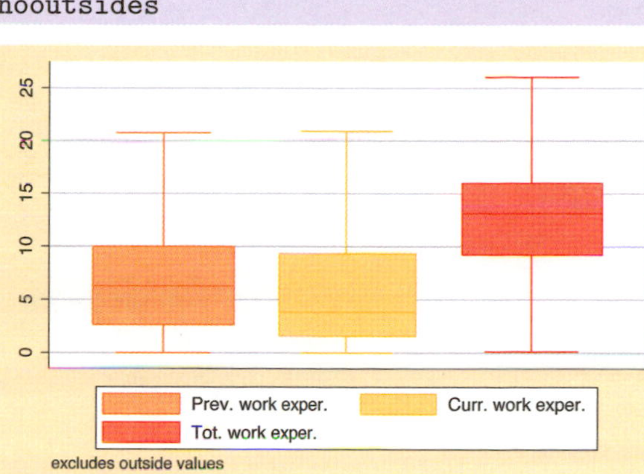


```
graph box wage, nooutsides over(occ7) asyvars
```

This is another example of where a legend can arise in a Stata box plot by using the `asyvars` option, which treats an `over()` variable as though the levels were different y-variables.
Uses nlsw.dta & scheme vg_brite

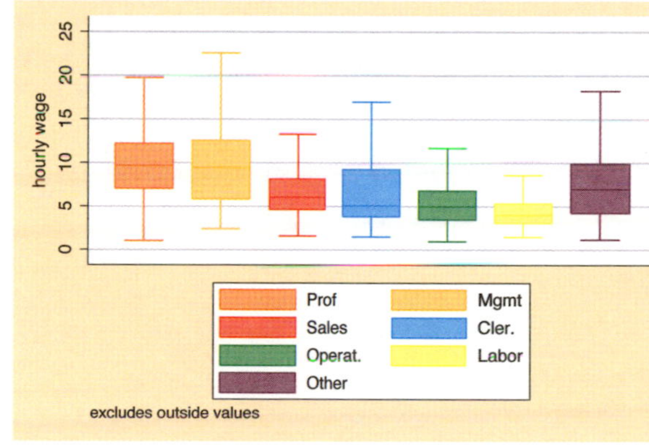

Unless otherwise mentioned, the `legend()` options described below work the same whether the legend was derived from multiple y-variables or from an `over()` option that was combined with the `asyvars` option. These examples use the `vg_teal` scheme.

```
graph box prev_exp tenure ttl_exp, nooutsides nolabel
```

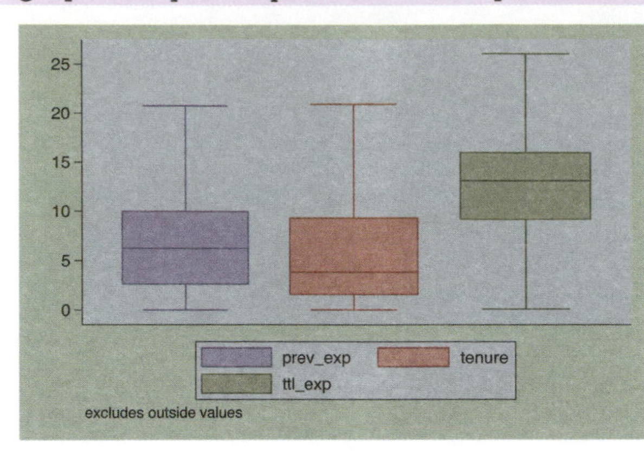

The `nolabel` option only works when you have multiple y-variables. When this option is used, the variable names (not the variable labels) are used in the legend. For example, instead of showing the variable label `Prev. work exper.`, it shows the variable name `prev_exp`.
Uses nlsw.dta & scheme vg_teal

```
graph box prev_exp tenure ttl_exp, nooutsides showyvars
```

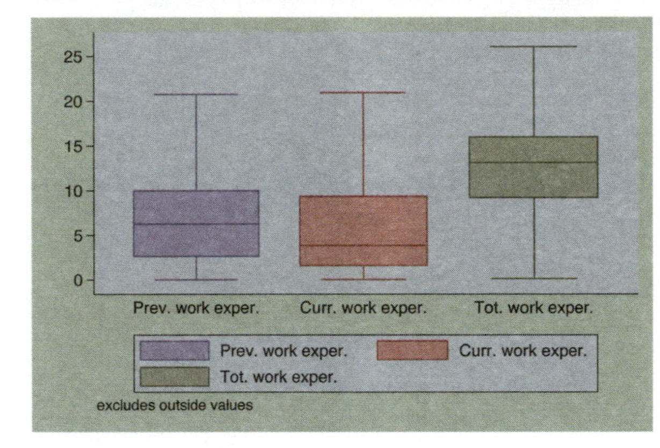

The `showyvars` option puts the labels under the boxes.
Uses nlsw.dta & scheme vg_teal

```
graph box prev_exp tenure ttl_exp, nooutsides showyvars legend(off)
```

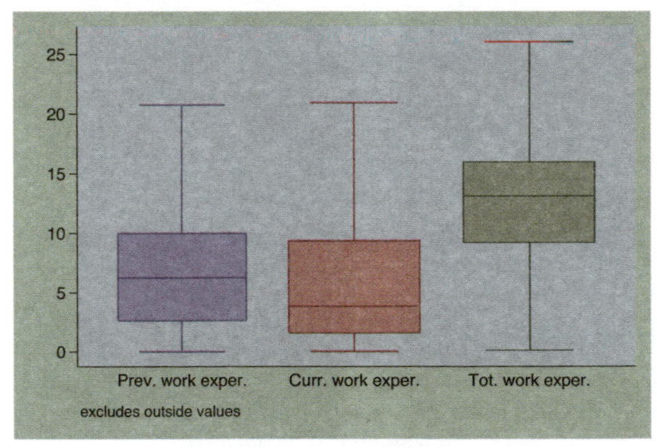

One instance when the `showyvars` option would be useful is when you want separately colored boxes labeled at the bottom. Here, we use `showyvars` to show the labels at the bottom of the boxes and the `legend(off)` option to suppress the display of the legend.
Uses nlsw.dta & scheme vg_teal

```
graph box wage, nooutsides over(occ7) asyvars showyvars legend(off)
```

Even though the `showyvars` option sounds like it would work only with multiple y-variables, it also works when you combine the `over()` and `asyvars` options. As before, we suppress the legend in this example using the `legend(off)` option.
Uses nlsw.dta & scheme vg_teal

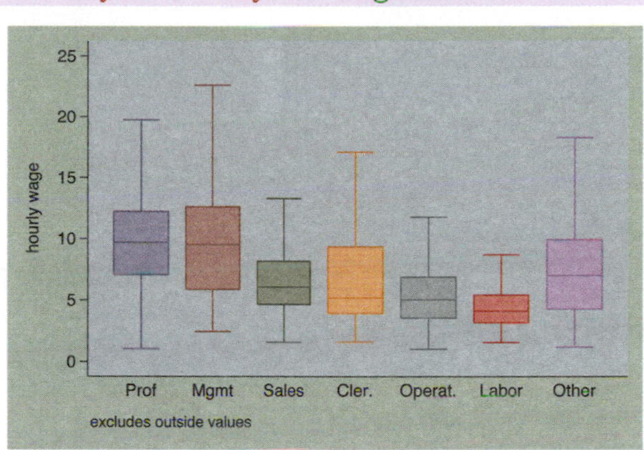


```
graph box wage, nooutsides over(occ7) asyvars
   legend(label(1 "Professional") label(2 "Management"))
```

We use the `legend(label())` option to change the labels for the first and second variables in the legend. Note that you use a separate `label()` option for each bar. This is in contrast to the `relabel()` option, where all the label assignments were placed in one `relabel()` option; see Box : Cat axis (168).
Uses nlsw.dta & scheme vg_teal

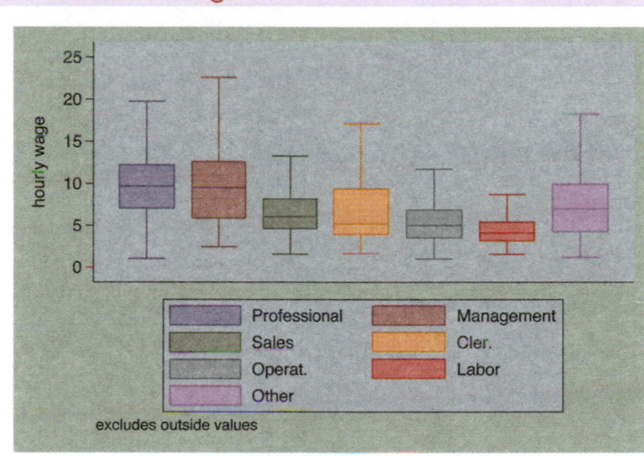


```
graph box wage, nooutsides over(occ7) asyvars legend(rows(2) colfirst)
```

In this example, we use the `legend(rows(2) colfirst)` options to display the legend in two rows and to order the keys by column (instead of the default, which is by row). This yields keys that are more adjacent to the boxes that they label.
Uses nlsw.dta & scheme vg_teal

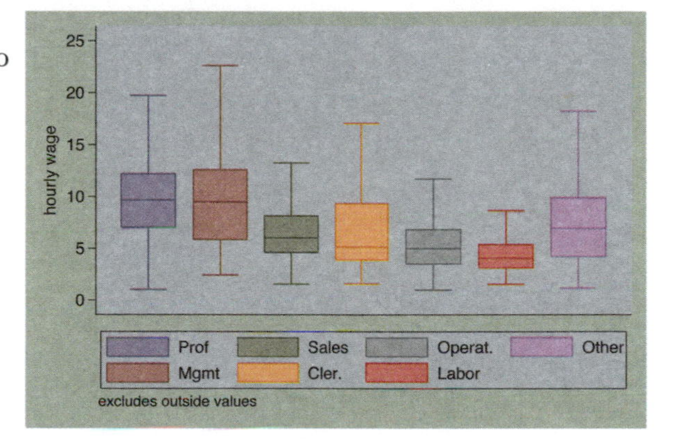

```
graph box wage, nooutsides over(occ7) asyvars
   legend(position(1))
```

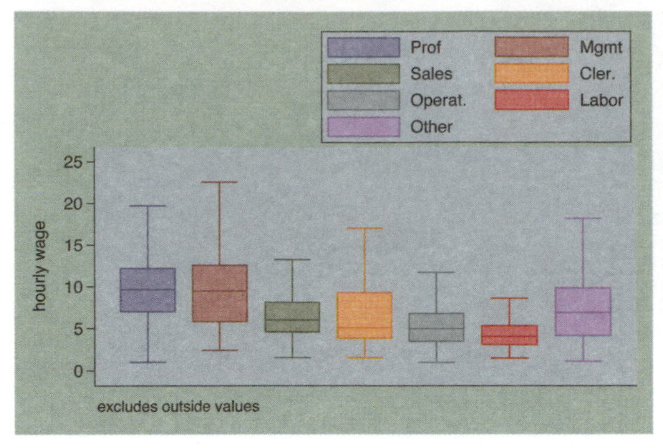

We can put the legend up in the top right corner of the graph with the legend(position(1)) option. The values you supply for position() are like the numbers on a clock face, where 12 o'clock is the top, 6 o'clock is the bottom, and 0 represents the center of the clock face; see Styles : Clockpos (330) for more details.
Uses nlsw.dta & scheme vg_teal

```
graph hbox wage, nooutsides over(occ7) asyvars
   legend(cols(1) position(9))
```

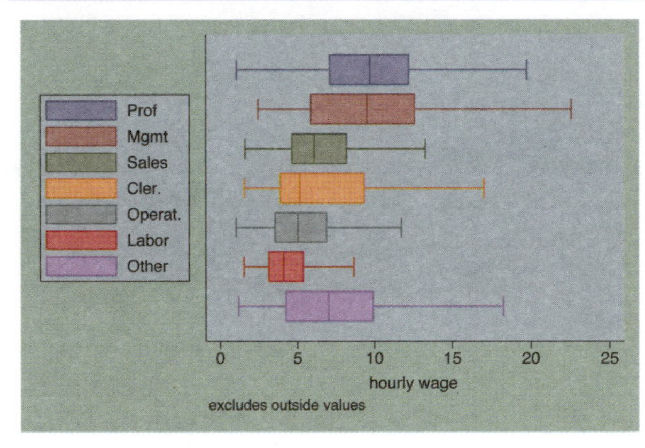

We switch to making this a horizontal box chart and then move the legend using the legend(position(9)) option. The legend is now placed in the 9 o'clock position and is displayed as a single column.
Uses nlsw.dta & scheme vg_teal

```
graph hbox wage, nooutsides over(occ7) asyvars
   legend(cols(1) position(9) textfirst)
```

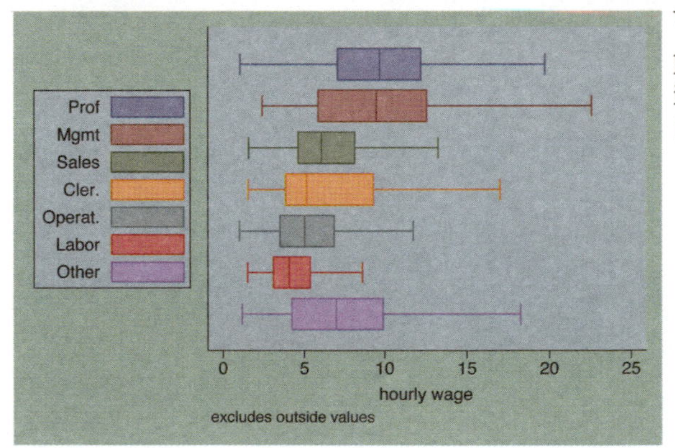

We can add the textfirst option to put the key description before the key in the legend.
Uses nlsw.dta & scheme vg_teal

```
graph hbox wage, nooutsides over(occ7) asyvars
   legend(cols(1) position(9) stack)
```

With the stack option, we can place the keys and their labels on top of each other to form an even more compact column. You have considerable control over the elements within the legend using other options like `rowgap()`, `keygap()`, `symxsize()`, `symysize()`, `textwidth()`, and `symplacement()`. See Options : Legend (287) and [G] ***legend_option*** for more details.
Uses nlsw.dta & scheme vg_teal

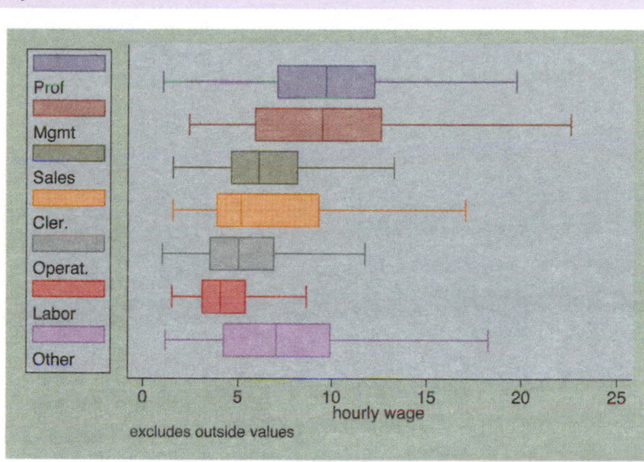


```
graph hbox wage, nooutsides over(occ7) asyvars
```

Switching to the vg_lgndc scheme, by typing `set scheme vg_lgndc`, positions the legend at the left in a single column, by default, without the need to specify options.
Uses nlsw.dta & scheme vg_lgndc

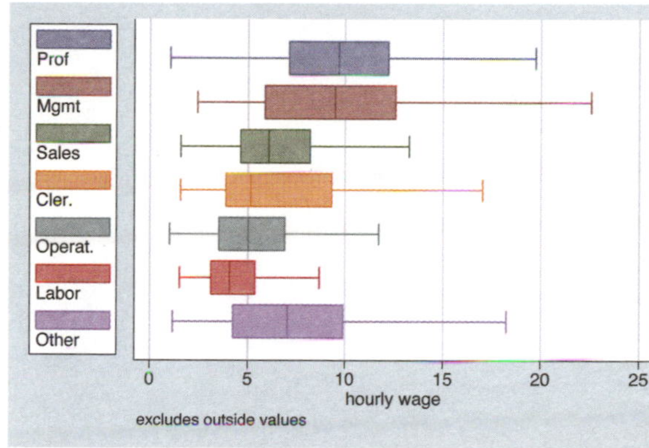

5.5 Controlling the y-axis

This section describes options you can use with respect to the y-axis with box charts. To be precise, when Stata refers to the y-axis on a box chart, it refers to the axis with the continuous variable, whether the left axis when using `graph box` or the bottom axis when using `graph hbox`. This section emphasizes the features that are particularly relevant to box charts. For more details, see Options : Axis titles (254), Options : Axis labels (256), and Options : Axis scales (265). See also [G] ***axis_title_options***, [G] ***axis_label_options***, and [G] ***axis_scale_options***. These examples are shown using the `vg_lgndc` scheme, which places the legend to the left in a single column.

```
graph box wage, nooutside over(occ5)
```

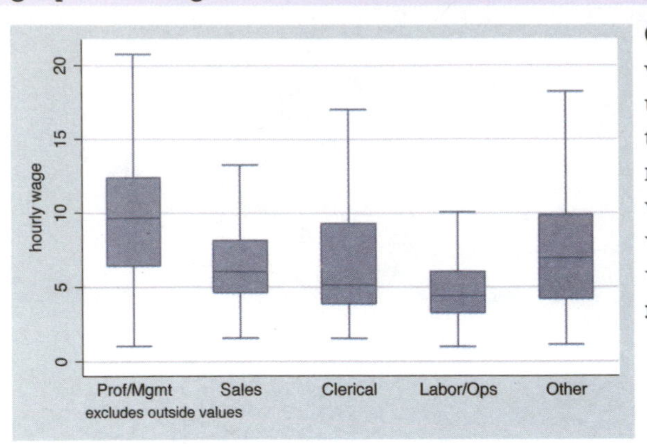

Consider this graph showing the hourly wages broken down by occupation. We use the **nooutsides** option to suppress the display of outside values. For the rest of the graphs in this section, there would be a large number of outside values, which would make the graphs very cluttered, so we will include the **nooutsides** option for each example.
Uses nlsw.dta & scheme vg_lgndc

```
graph box prev_exp tenure, nooutside over(occ5)
   ytitle("Years of experience")
```

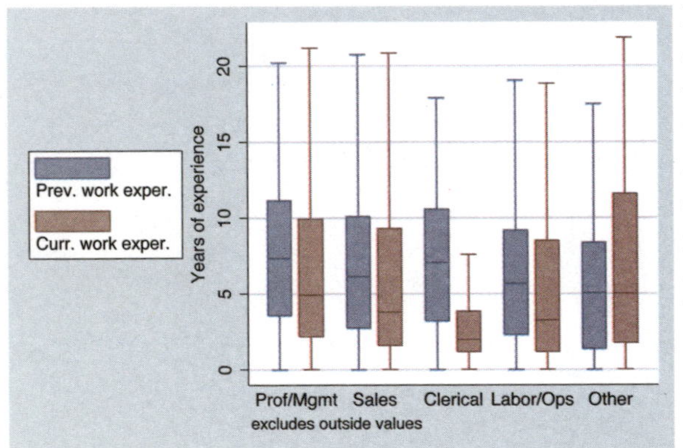

Looking at previous and current work experience over occupations, we can use the **ytitle()** option to add a title to the *y*-axis. See Options : Axis titles (254) and [G] ***axis_title_options*** for more details, but please disregard any references to **xtitle()** there since that option is not valid when using **graph box**.
Uses nlsw.dta & scheme vg_lgndc

```
graph hbox prev_exp tenure, nooutside over(occ5)
   ytitle("Years of" "experience")
```

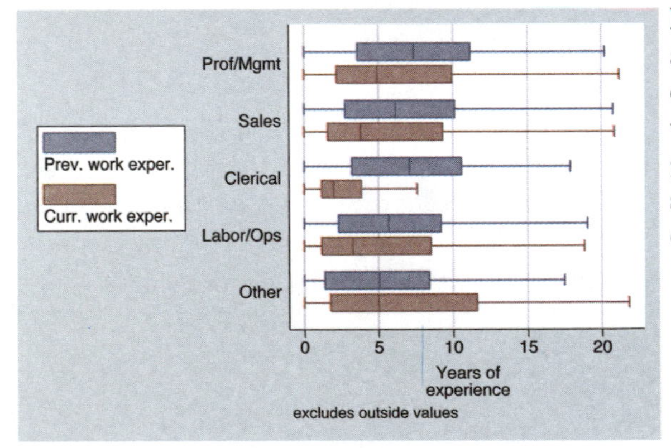

In this example, we place the title across two lines by using two separate quoted strings. Note that, even though we have used **graph hbox** to place the *y*-axis on the bottom axis, we still should use **ytitle()** to change the title of that axis.
Uses nlsw.dta & scheme vg_lgndc

```
graph hbox prev_exp tenure, nooutside over(occ5)
   ytitle("Years of experience", size(vlarge) box bexpand)
```

Because this title is considered to be a textbox, you can use a variety of textbox options to control the look of the title. This example makes the title very large, surrounds it with a box, and uses the bexpand (box expand) option to stretch the box to fill the width of the plot area. See Options : Textboxes (303) for additional examples of how to use textbox options to control the display of text.
Uses nlsw.dta & scheme vg_lgndc

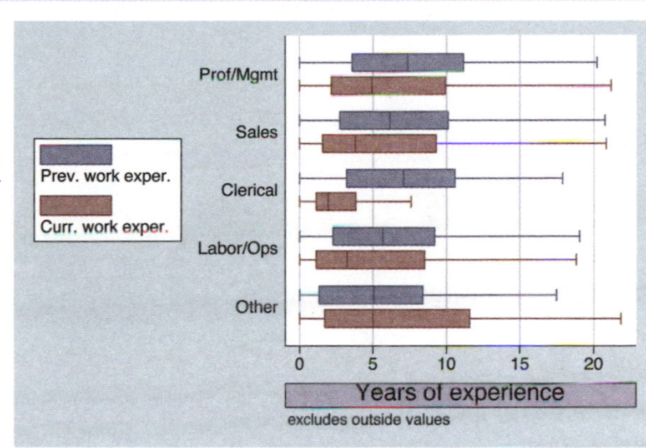


```
graph box wage, nooutside over(occ5) over(collgrad) asyvar
   yline(4 12, lwidth(medthick) lcolor(maroon) lpattern(dash))
```

In this example, we use the yline() option to add a medium-thick, maroon, dashed line to the points in the graph where wages equal 4 and 12. Note that we would still use yline(), even if we used graph hbox, placing the y-axis at the bottom.
Uses nlsw.dta & scheme vg_lgndc

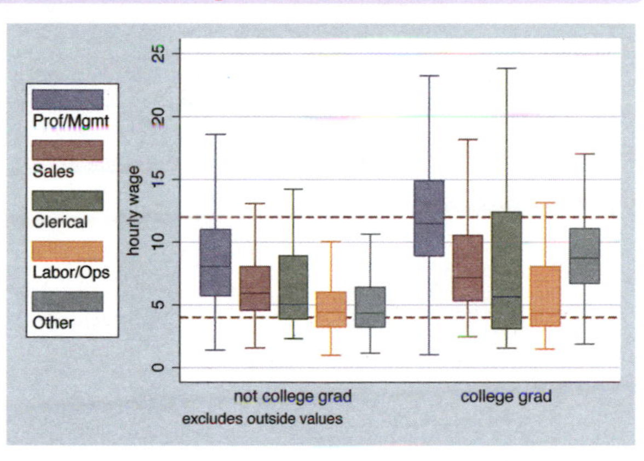


```
graph box wage, nooutside over(occ5) over(collgrad) asyvar
   ylabel(5(10)25)
```

We can use the ylabel() option to label the y-axis. In this case, we use the labels going from 5 to 25 by increments of 10. Note that the y-axis still starts at 0, and we would have to supply the exclude0 option, so 0 is not necessarily the starting point for the y-axis. See Options : Axis labels (256) and [G] ***axis_label_options*** for more details. Please disregard any references to xlabel() since that option is not valid when using graph box.
Uses nlsw.dta & scheme vg_lgndc

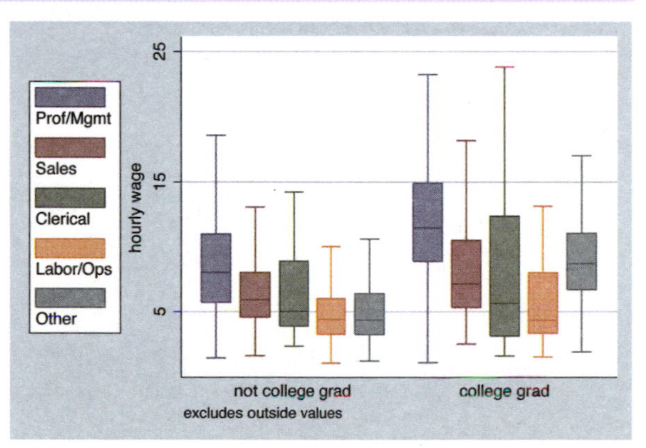

Introduction Twoway Matrix Bar Box Dot Pie Options Standard options Styles Appendix
Yvars and over Over options Cat axis Legend Y-axis Boxlook options By

```
graph box wage, nooutside over(occ5) over(collgrad) asyvar
    ylabel(5(10)25, angle(0))
```

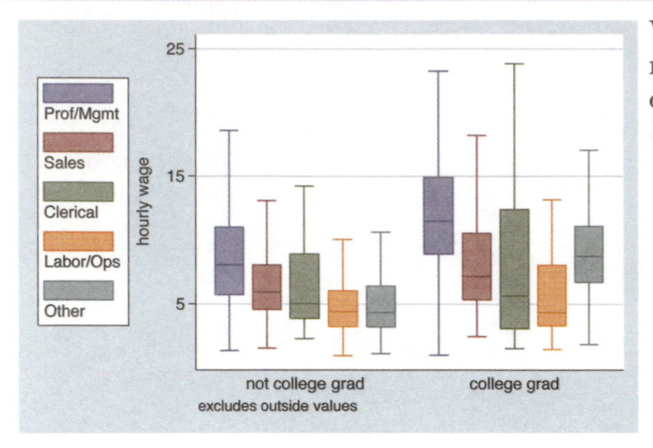

We can add the angle(0) option to modify the angle of the y-labels, in this case making them display horizontally. *Uses nlsw.dta & scheme vg_lgndc*

```
graph box wage, nooutside over(occ5) over(collgrad) asyvar
    ylabel(5(10)25, nogrid)
```

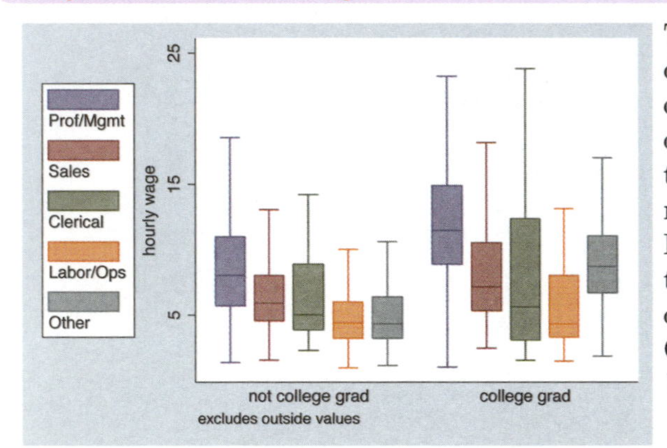

The nogrid option suppresses the display of the grid. Note that this option is placed within the `ylabel()` option, thus suppressing the grid for the y-axis. (With box plots, there is never a grid with respect to the x-axis.) If the grid were absent and we wanted to include it, we could add the `grid` option. For more details, see Options : Axis labels (256). *Uses nlsw.dta & scheme vg_lgndc*

```
graph box wage, nooutside over(occ5) over(collgrad) asyvar yscale(off)
```

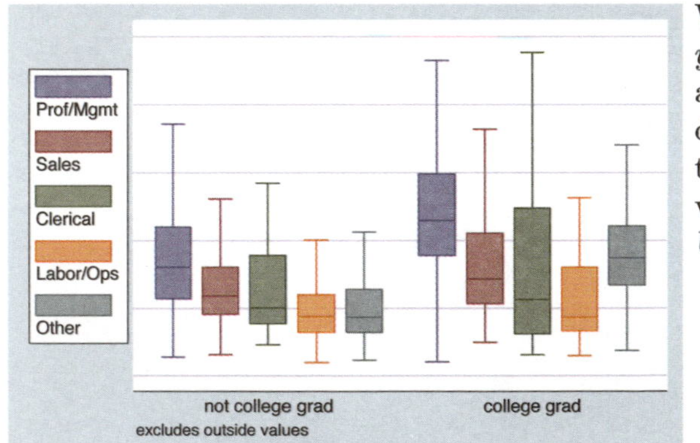

We can use yscale(off) to turn off the y-axis. See Options : Axis scales (265) and [G] ***axis_scale_options*** for more details. Please disregard any references to `xscale()`, since that option is not valid when using `graph box`. *Uses nlsw.dta & scheme vg_lgndc*

```
graph box wage, nooutside over(occ5) over(collgrad) asyvar yalternate
```

We can put the y-axis on the opposite side, in this case on the right side of the graph, using the **yalternate** option.
Uses nlsw.dta & scheme vg_lgndc

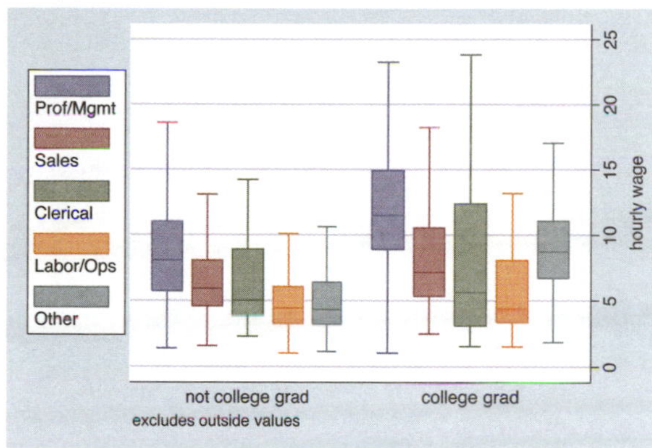


```
graph box wage, nooutside over(occ5) over(collgrad) asyvar yreverse
```

You can reverse the direction of the y-axis, in effect turning your boxes upside down, with the **yreverse** option.
Uses nlsw.dta & scheme vg_lgndc

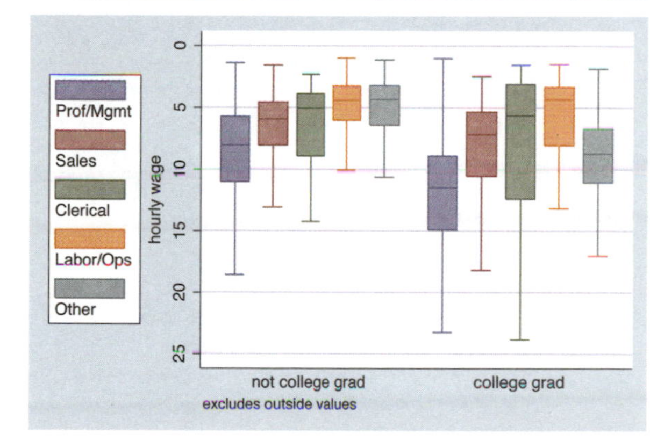

5.6 Changing the look of boxes, boxlook options

This section shows how you can control the look of the boxes in your box charts: control the space between the boxes, the color of the boxes, and the characteristics of the line outlining the boxes. For more information, see the *boxlook_options* table in [G] **graph box**. These examples begin with the `vg_blue` scheme.

```
graph box prev_exp tenure ttl_exp, over(collgrad)
```

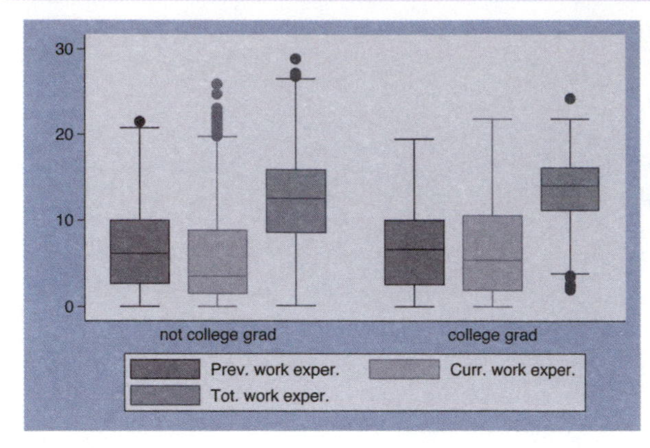

Consider this box chart, which shows the distribution of previous work experience, current work experience, and total work experience. These three variables are broken down by whether one graduated college.
Uses nlsw.dta & scheme vg_blue

```
graph box prev_exp tenure ttl_exp, nooutsides over(collgrad)
```

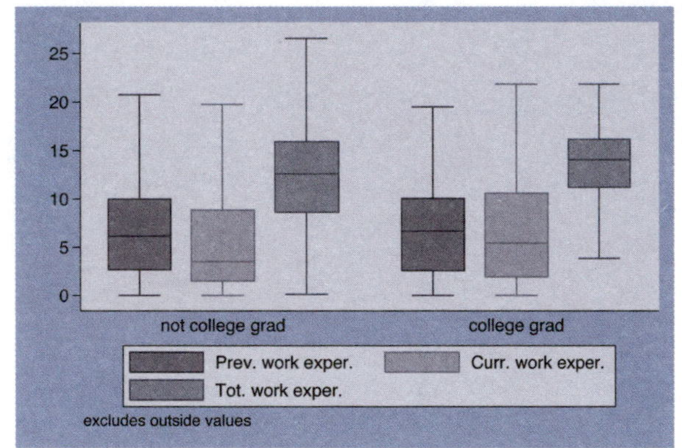

We add the `nooutsides` option to suppress the display of outside values. We will use this option for most of the graphs in this section.
Uses nlsw.dta & scheme vg_blue

```
graph box prev_exp tenure ttl_exp, nooutsides over(collgrad)
    outergap(*5)
```

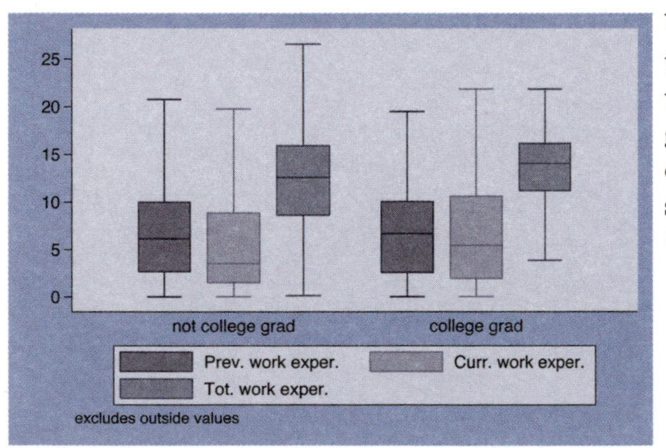

We can change the outer gap between the boxes and the edge of the plot area with the `outergap()` option. Here, the gap is five times its normal size. You could also supply a value less than 1 to shrink the size of the outer gap.
Uses nlsw.dta & scheme vg_blue

```
graph box prev_exp tenure ttl_exp, nooutsides over(collgrad)
   boxgap(10)
```

The boxgap() option controls the size of the gap among the boxes formed by the multiple y-variables. The default value is 33, meaning that the distance between the boxes is 33% of the width of the boxes. Here, we make the gap smaller, making the boxes for the y-variables closer to each other.
Uses nlsw.dta & scheme vg_blue

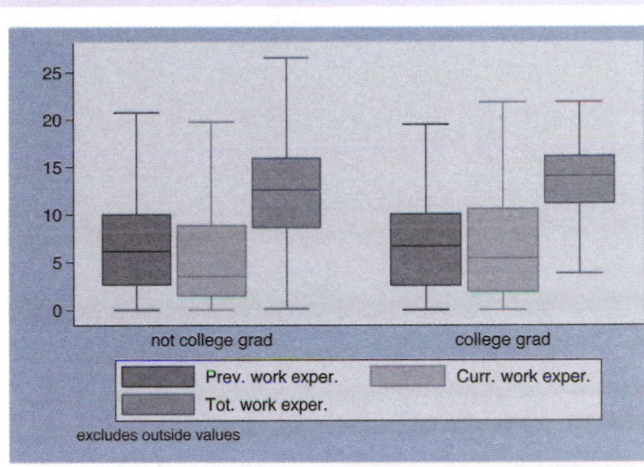


```
graph box prev_exp tenure ttl_exp, nooutsides over(collgrad, gap(*3))
```

Here, we use the gap() option to control the gap between the college graduate group and the noncollege graduate group. Here, we make the gap three times the width of a box. See Box : Over options (163) for more information about controlling the gap among boxes created by the over() option.
Uses nlsw.dta & scheme vg_blue

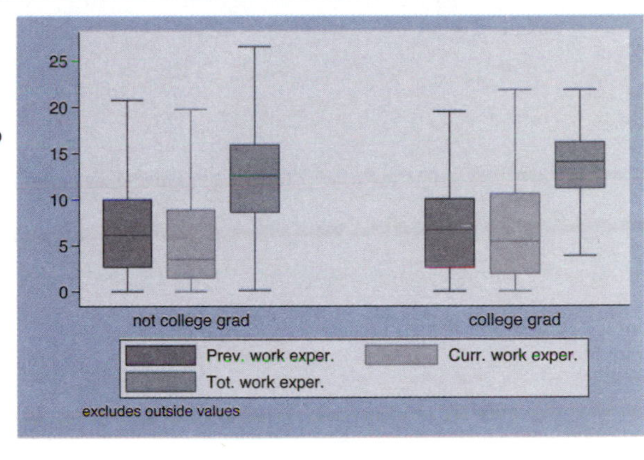

Let's now look at options that allow us to control the color of the boxes. We will first look at options that control the overall intensity of the color for all the boxes and then show how you can control the color of each box. We will use the vg_s2c scheme for the following examples.

```
graph box wage, over(occ5) over(collgrad) asyvars nooutsides intensity(*.5)
```

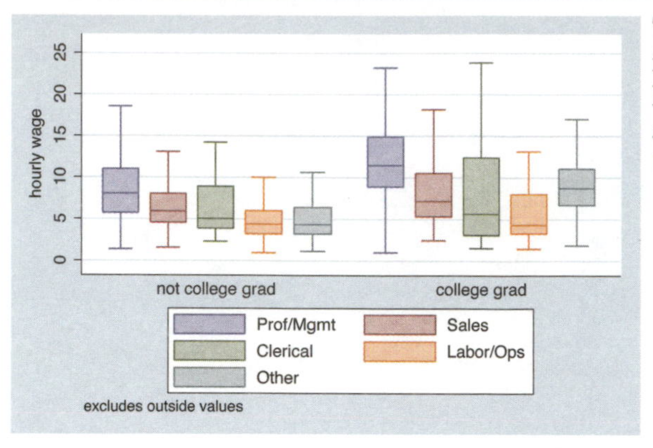

The **intensity** option controls the intensity of the color within the boxes. Here, we request that the color be 50% as intense as it normally would be.
Uses nlsw.dta & scheme vg_s2c

```
graph box wage, over(occ5) over(collgrad) asyvars nooutsides intensity(*1.5)
```

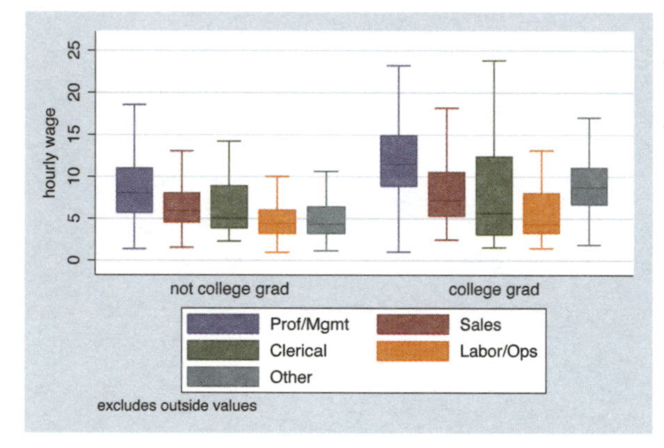

In this example, we use the **intensity** option to make the colors within the boxes 1.5 times more intense than they would normally.
Uses nlsw.dta & scheme vg_s2c

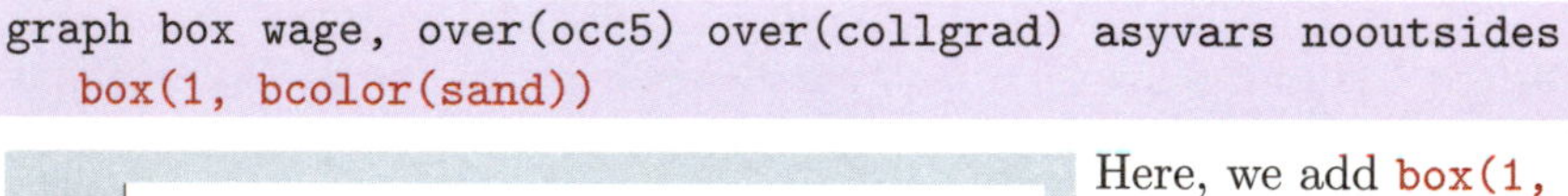

```
graph box wage, over(occ5) over(collgrad) asyvars nooutsides
   box(1, bcolor(sand))
```

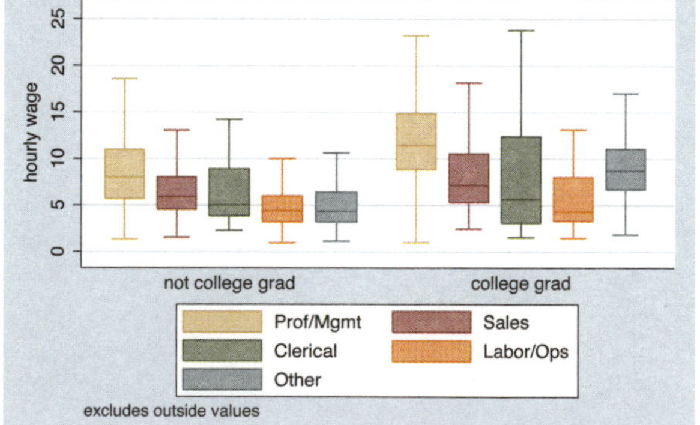

Here, we add **box(1, bcolor(sand))** to make the box color for the first bar a sand color. See Styles : Colors (328) for more information about colors you can select.
Uses nlsw.dta & scheme vg_s2c

```
graph box wage, over(occ5) over(collgrad) asyvars nooutsides
   box(1, bcolor(sand) blcolor(brown) blwidth(thick))
```

We add the `blcolor()` (box line color) and `blwidth()` (box line width) options to make the outline for the first box brown and thick. Note that, while you can control the color of the boxes and outline characteristics via the `box()` option, if you want to extensively change these characteristics for many graphs, you might consider making your own scheme. See Intro : Schemes (14) and Appendix : Customizing schemes (379).
Uses nlsw.dta & scheme vg_s2c

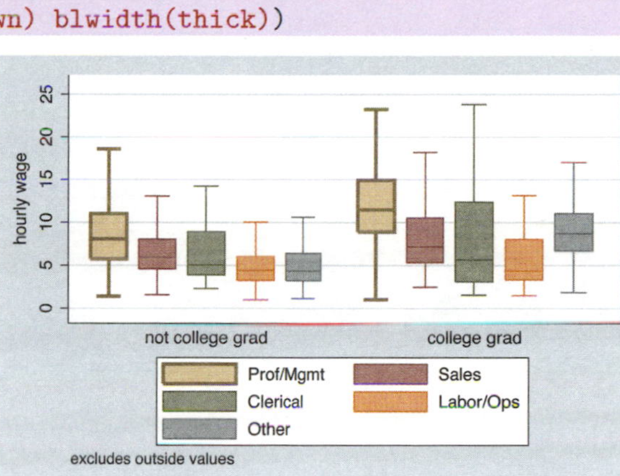

Now, let's consider options that allow us to control the display of the median, whiskers, caps, and outside markers. These examples use the `vg_s1m` scheme.

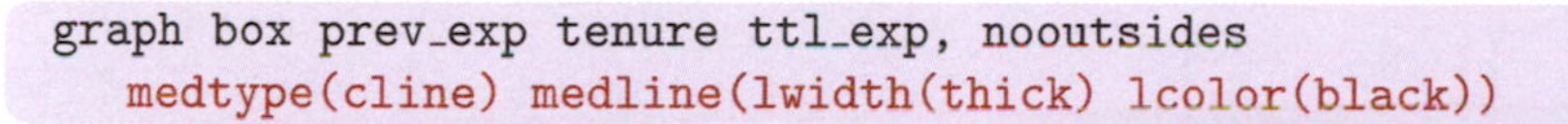

```
graph box prev_exp tenure ttl_exp, nooutsides
   medtype(cline) medline(lwidth(thick) lcolor(black))
```

The `medtype(cline)` option sets the median type to be a custom line. We then customize the median line using the `medline()` option to specify that the line width be thick and the line color be black.
Uses nlsw.dta & scheme vg_s1m

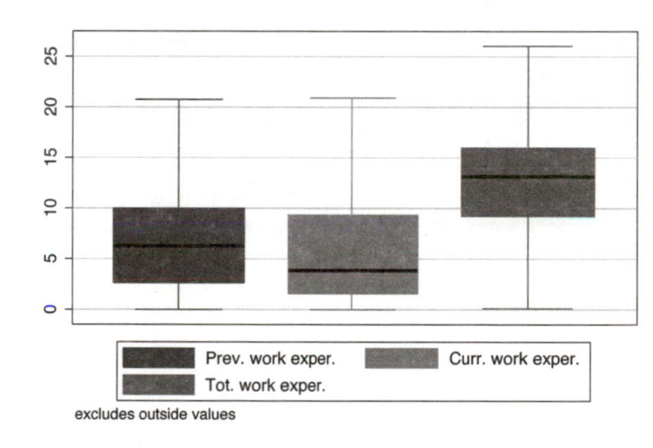

```
graph box prev_exp tenure ttl_exp, nooutsides
   medtype(marker) medmarker(msymbol(+) msize(large))
```

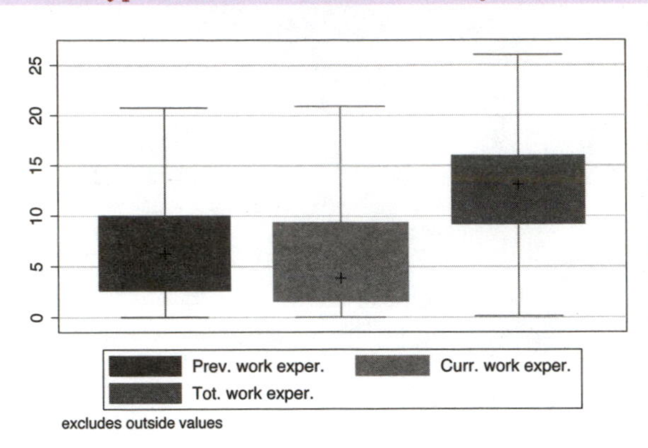

We can use the medtype(marker) option to tell Stata that we want to use a marker symbol to label the median and then use the medmarker() option to control the display of the median marker. In this case, we make the marker symbol a plus sign and make the marker size large.
Uses nlsw.dta & scheme vg_s1m

```
graph box prev_exp tenure ttl_exp, nooutsides
   cwhiskers lines(lwidth(thick) lcolor(black))
```

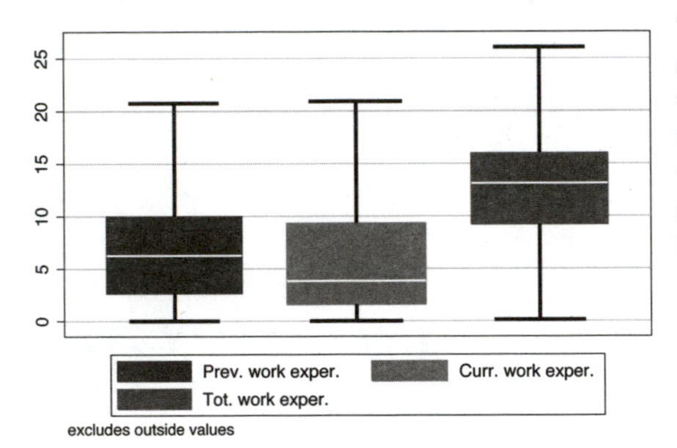

To customize the whiskers, we need to specify the cwhiskers (customize whiskers) option, and then we can add the lines() option to specify how we want the whiskers customized. In this case, we make the whiskers thick and black.
Uses nlsw.dta & scheme vg_s1m

```
graph box prev_exp tenure ttl_exp, nooutsides alsize(20)
```

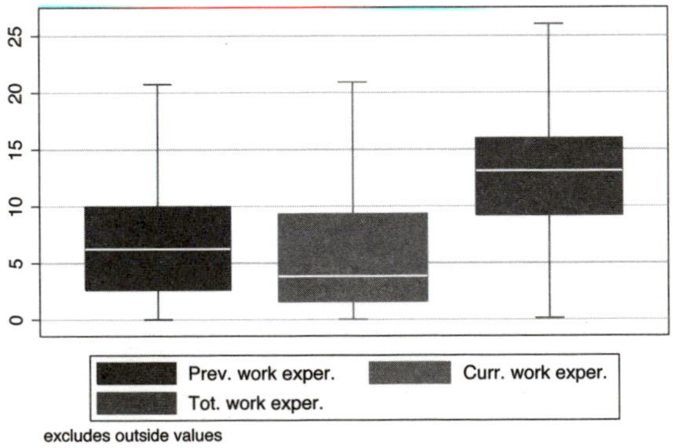

The alsize() (adjacent line size) option allows you to control the size (width) of the adjacent line. By default, the adjacent line is 67% of the width of the box. Here, we make the adjacent line much smaller, 20% of the width of the box.
Uses nlsw.dta & scheme vg_s1m

```
graph box prev_exp tenure ttl_exp, nooutsides capsize(5)
```

The capsize() option allows you to specify the size of the caps (if any) on the adjacent line. The default value is 0, meaning that no cap is displayed. Here, we add a small cap to the adjacent line.
Uses nlsw.dta & scheme vg_s1m

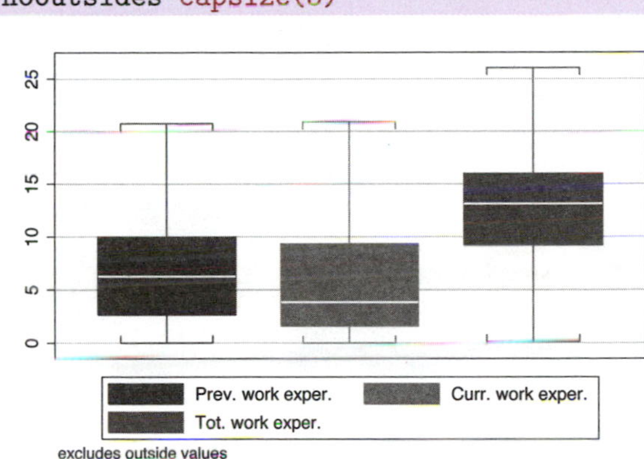


```
graph box prev_exp tenure ttl_exp, marker(2, msymbol(Oh) msize(vlarge))
```

The marker() option allows you to control the markers used to display the outside values. You can control this separately for each y-variable. Here, we make the outside value for tenure display as large, hollow circles.
Uses nlsw.dta & scheme vg_s1m

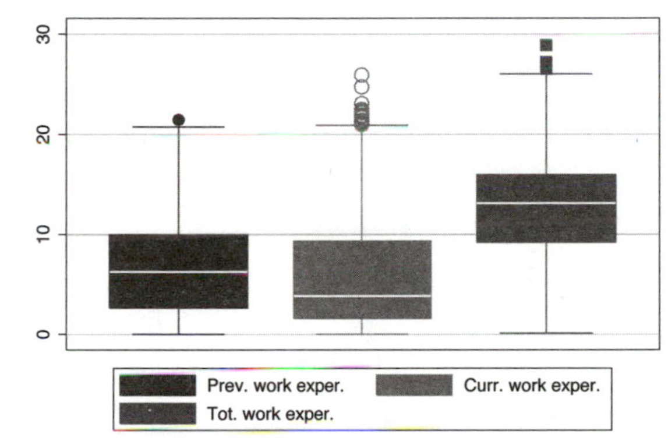

5.7 Graphing by groups

This section discusses the use of the by() option in combination with graph box. Normally, you would use the over() option instead of the by() option, but in some cases the by() option is either necessary or more advantageous. For example, a by() option is useful if you exceed the maximum number of over() options (three if you have a single y-variable or two if you have multiple y-variables). In such cases, the by() option allows you to break your data down by additional categorical variables. Also, by() gives you more flexibility in the placement of the separate panels. For more information about the by() option, see Options : By (272); for more information about the over() option, see Box : Yvars and over (157).

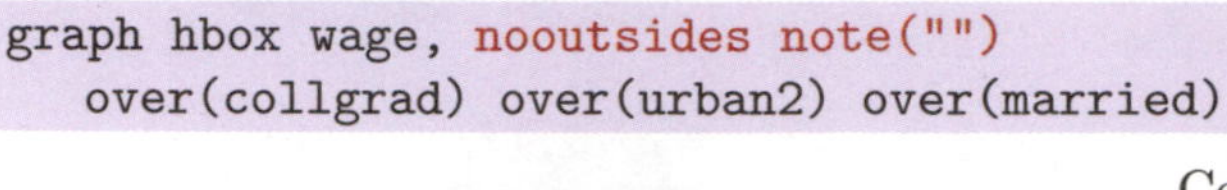

```
graph hbox wage, nooutsides note("")
   over(collgrad) over(urban2) over(married)
```

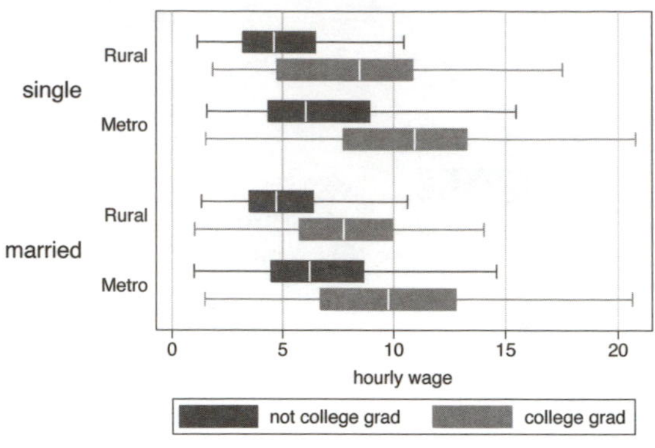

Consider this box graph, which breaks wages down by three categorical variables. If we wanted to further break this down by another categorical variable, we could not use another `over()` option since we can have a maximum of three `over()` options with a single y-variable. We use the `nooutsides` option to suppress the display of outside values for this graph and the rest of the graphs in this section.
Uses nlsw.dta & scheme vg_s1m

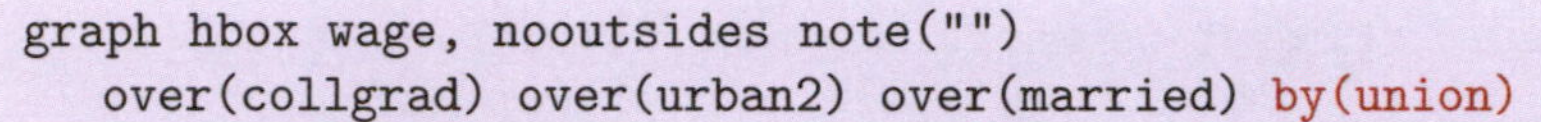

```
graph hbox wage, nooutsides note("")
   over(collgrad) over(urban2) over(married) by(union)
```

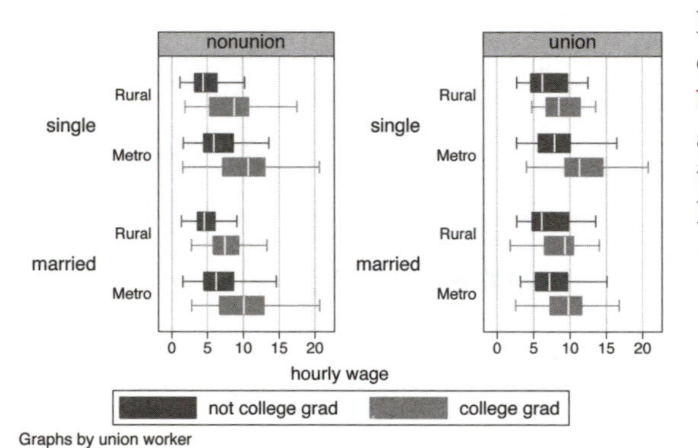

If we want to further break `prev_exp` down by `union`, we can use the `by(union)` option to do this. We also add the `note("")` option to suppress the note saying that the outside values have been omitted.
Uses nlsw.dta & scheme vg_s1m

```
graph hbox prev_exp tenure, nooutsides note("")
   over(urban2) over(married)
```

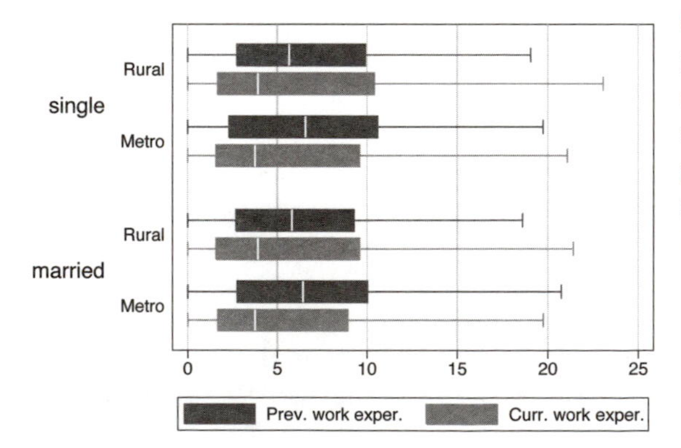

Consider this box graph with multiple y-variables breaking them down by two categorical variables using two `over()` options. When you have multiple y-variables, you can only have a maximum of two `over()` options.
Uses nlsw.dta & scheme vg_s1m

```
graph box prev_exp tenure, nooutsides note("")
   over(urban2) over(married) by(union)
```

If we want to further break prev_exp down by another categorical variable, say union, we can use the by(union) option. We can include multiple variables within by(), although this can make some very small graphs.
Uses nlsw.dta & scheme vg_s1m

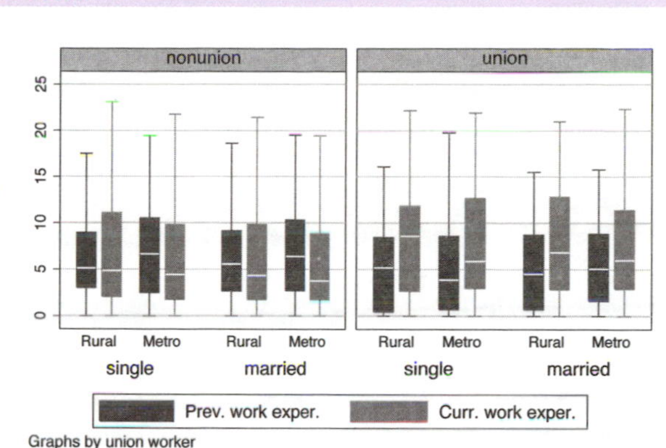


```
graph hbox ttl_exp tenure, nooutsides note("")
   over(urban2) over(married) by(union, missing)
```

We can use the missing option to include a panel for the missing values of union.
Uses nlsw.dta & scheme vg_s1m

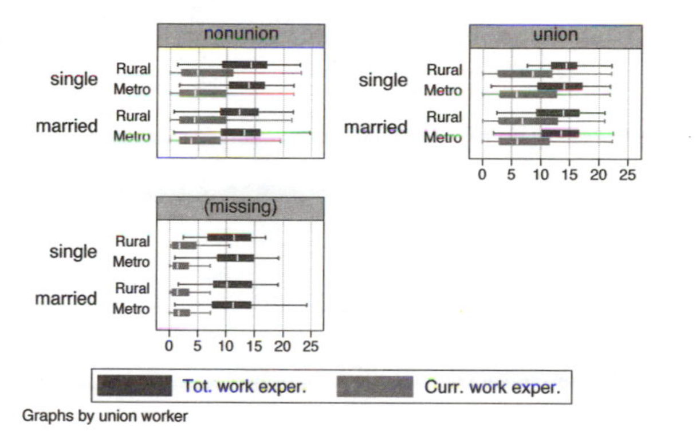


```
graph hbox ttl_exp tenure, nooutsides note("")
   over(urban2) over(married) by(union, total)
```

We can add the total option to include a panel for all observations.
Uses nlsw.dta & scheme vg_s1m

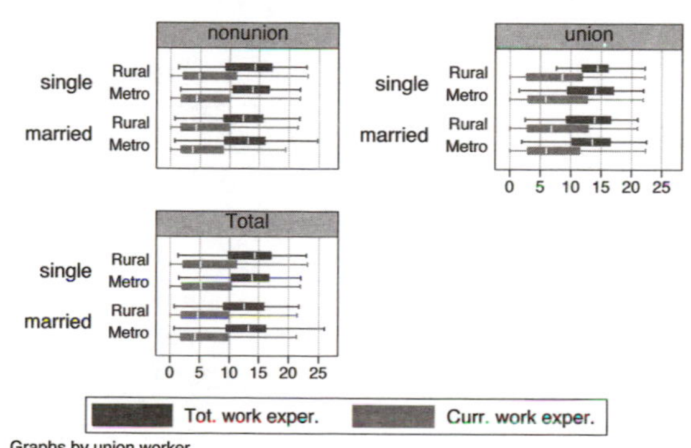

```
graph hbox ttl_exp tenure, nooutsides note("")
   over(urban2) over(married) by(union, total row(1))
```

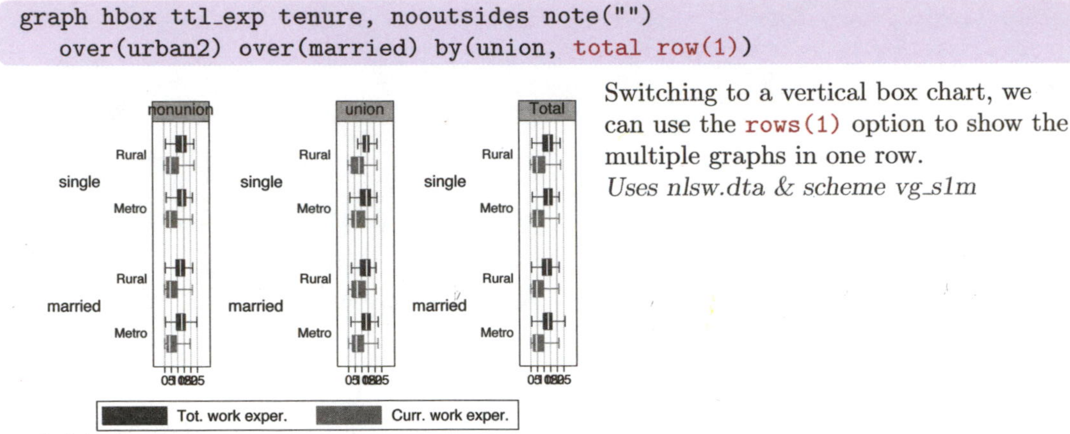

Switching to a vertical box chart, we can use the rows(1) option to show the multiple graphs in one row.
Uses nlsw.dta & scheme vg_s1m

```
graph hbox ttl_exp tenure, nooutsides note("")
   over(urban2) over(married) by(union, cols(1))
```

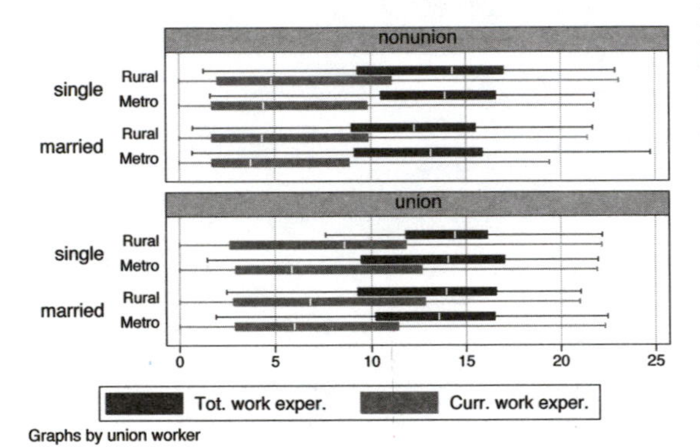

Here, we flip the graph back to a horizontal box chart and use the cols(1) option to show both graphs in one column.
Uses nlsw.dta & scheme vg_s1m

```
graph hbox ttl_exp tenure, nooutsides note("")
   over(urban2) over(married) by(union, cols(1) legend(position(9)))
   legend(cols(1) stack)
```

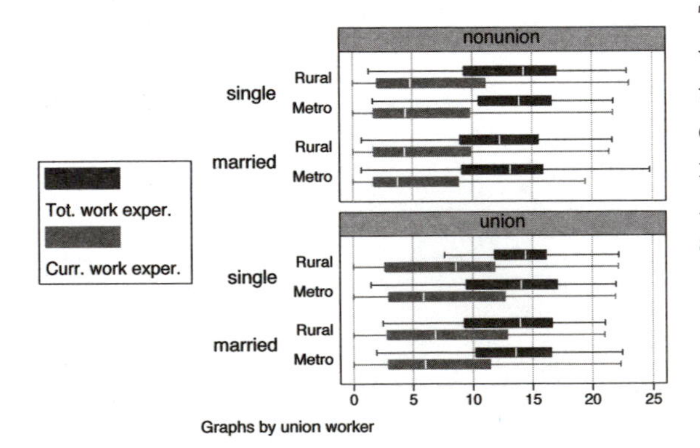

To make the last graph more readable, we can add the legend(pos(9)) within the by() option to put the legend at 9 o'clock and legend(cols(1) stack) to make the legend one stacked column. Adding note("") suppresses the note about outside values being omitted.
Uses nlsw.dta & scheme vg_s1m

6 Dot plots

This chapter discusses the use of dot plots in Stata. We start by showing how you can specify multiple y-variables to display plots for multiple variables and how you can use the `over()` option to break dot plots down by categorical variables. Then, we discuss `over()` options that can be used to customize the display of these categorical variables, followed by options concerning the display of display of categorical axes. Next, we cover options that control legends, followed by options that control the y-axis. Finally, we discuss options that control the look of the lines and dots that form the dot plot and, lastly, the `by()` option.

6.1 Specifying variables and groups, yvars and over

This section introduces the use of dot plots. It shows how you can use the `over()` option for displaying dot plots by one or more grouping variables. It then shows how you can specify one or more y-variables in a plot and control the summary statistic used for collapsing the y-variable(s). See the *group_options* table in [G] **graph dot** for more details. The graphs in this section begin using the `vg_s1c` scheme.

```
graph dot tenure, over(occ7)
```

Here, we use the `over()` option to show the average current work experience broken down by occupation. By default, the y-variable (`tenure`) is placed on the bottom axis and is considered to be the y-axis. Likewise, the levels of `occ7` are placed on the left axis and are considered to form the x-axis, or categorical axis.
Uses nlsw.dta & scheme vg_s1c

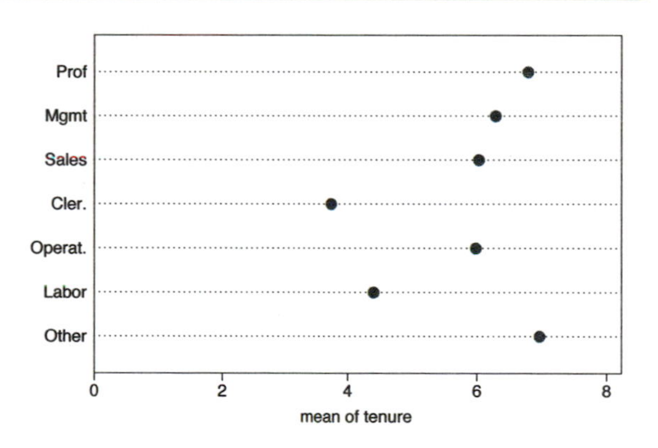

```
graph dot tenure, over(occ7) over(collgrad)
```

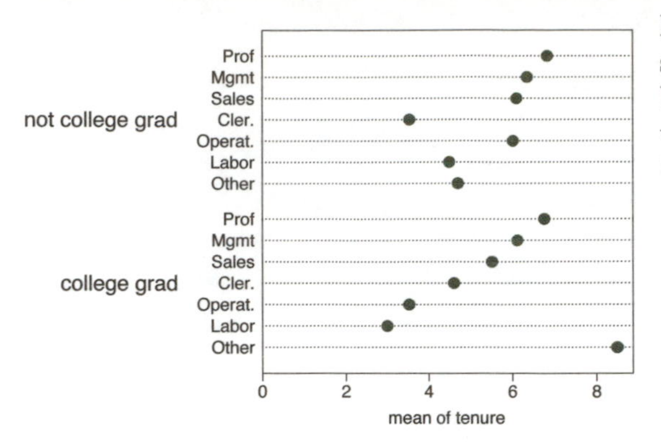

Here, we use a second `over()` option to show the mean of work experience broken down by occupation and whether one graduated college.
Uses nlsw.dta & scheme vg_s1c

```
graph dot tenure, over(occ7) over(collgrad) over(married)
```

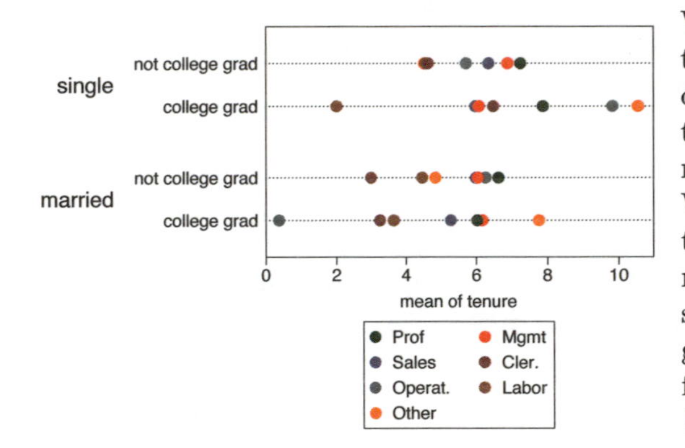

We can add a third `over()` option, in this case further breaking the `tenure` down by whether one is married. Note that the first `over()` variable (`occ7`) is now treated as multiple y-variables. When you use three `over()` options, the first variable is then treated as multiple y-variables, as though you had specified the **`asyvars`** option. This graph can be difficult to read with `occ7` forming the multiple y-variables.
Uses nlsw.dta & scheme vg_s1c

```
graph dot tenure, over(married) over(occ7) over(collgrad)
```

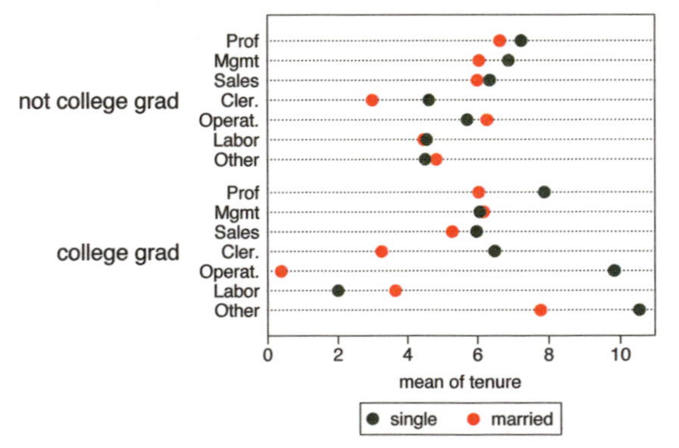

This graph shows the same data as the last one, except we have switched the order of the `over()` options, making `over(married)` come first and thus forming the multiple y-variables. This might be easier to read than the previous graph.
Uses nlsw.dta & scheme vg_s1c

Let's now consider examples with multiple y-variables. These examples are shown using the `vg_outc` scheme.

```
graph dot prev_exp tenure, over(occ7)
```

This graph shows the average previous experience and average current tenure broken down by occupation. While you do not need to use the `over()` option, omitting it may make a fairly boring graph.
Uses nlsw.dta & scheme vg_outc

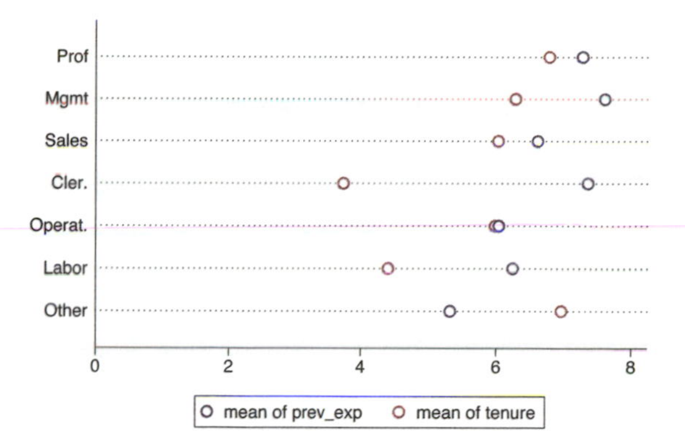


```
graph dot prev_exp tenure, over(occ7) over(collgrad)
```

This graph adds whether one is a college graduate as an additional grouping level. Because the command has multiple y-variables, we cannot include another `over()` option since dot plots support three levels of nesting and the multiple y-variables account for a level.
Uses nlsw.dta & scheme vg_outc

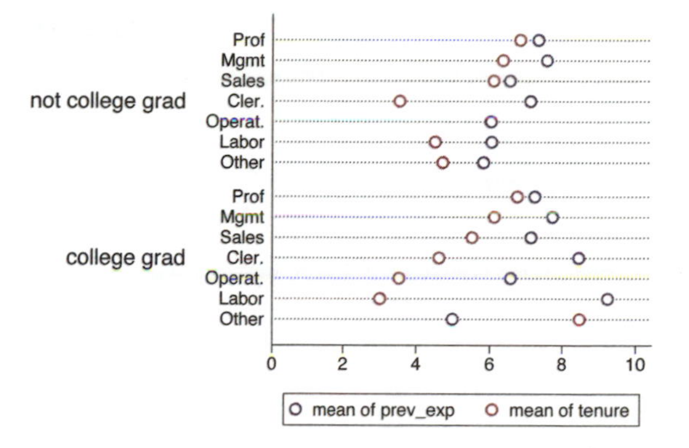

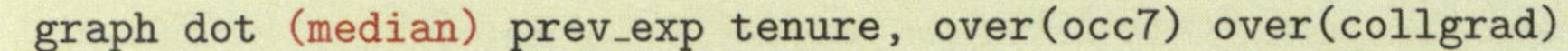

```
graph dot (median) prev_exp tenure, over(occ7) over(collgrad)
```

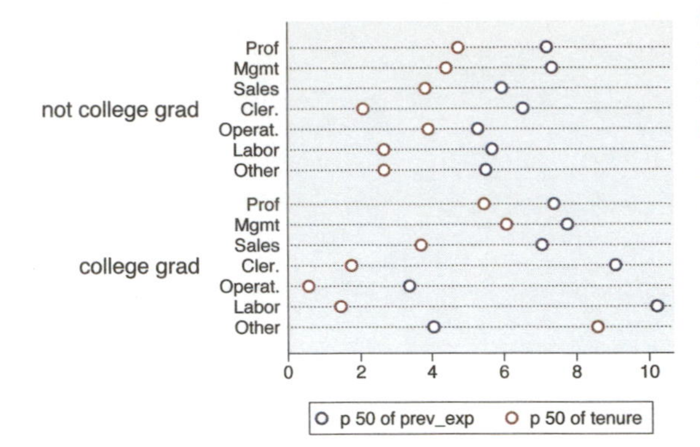

So far, all the examples we have seen have graphed the mean of y-variable(s). Here, we preface the y-variables with `(median)`, plotting the median for each y-variable.
Uses nlsw.dta & scheme vg_outc

```
graph dot (p10) wage_p10=wage (p25) wage_p25=wage
   (p50) wage_p50=wage (p75) wage_p75=wage (p90) wage_p90=wage,
   over(occ7)
```

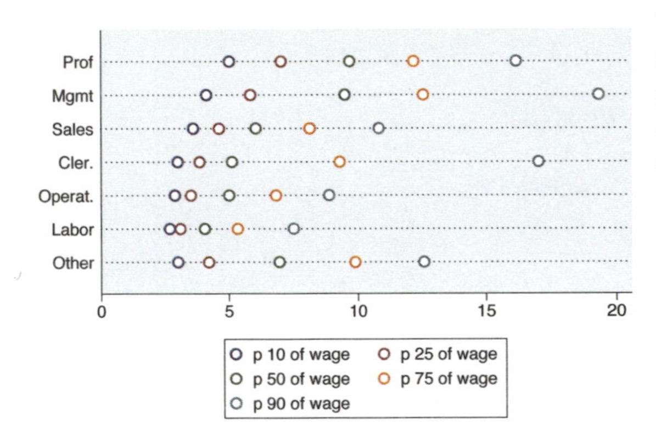

You can request different statistics for the same variable, such as in this example, which shows the 10th, 25th, 50th, 75th, and 90th percentiles of wages broken down by occupation.
Uses nlsw.dta & scheme vg_outc

Now, let's consider options that can be used in combination with the `over()` option to customize the behavior of the graphs. We show how you can treat the levels of the first `over()` option as though they were multiple y-variables. You can also request that missing values for the levels of the `over()` variables be displayed, and you can suppress empty categories when multiple `over()` options are used. These examples are shown below using the `vg_s2m` scheme.

```
graph dot tenure, over(collgrad) over(occ7)
```

Consider this graph, which shows the average current work experience broken down by whether one is a college graduate and by occupation.
Uses nlsw.dta & scheme vg_s2m

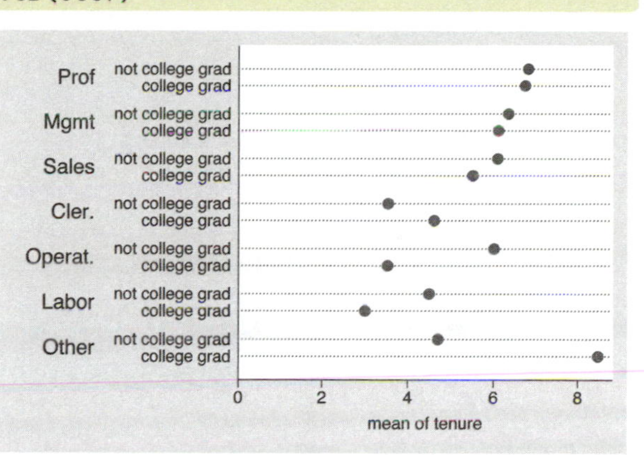


```
graph dot tenure, over(collgrad) over(occ7) asyvars
```

If we add the asyvars option, the first over() variable (collgrad) is graphed as if there were two y-variables. The two levels of collgrad are shown as different markers on the same line, and they are labeled using the legend.
Uses nlsw.dta & scheme vg_s2m

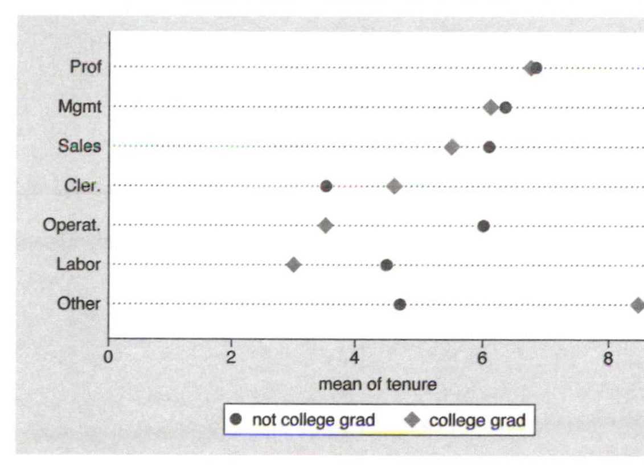


```
graph dot tenure, over(occ5) over(union) missing
```

Consider this graph in which we use the over() option to show tenure broken down by occ5 and union. By including the missing option, we then see the category for those who are missing on the union variable, shown as the third group labeled with a dot. See Dot : Cat axis (202) for examples showing how you could change the label (.) to something more meaningful, e.g., “Missing”.
Uses nlsw.dta & scheme vg_s2m

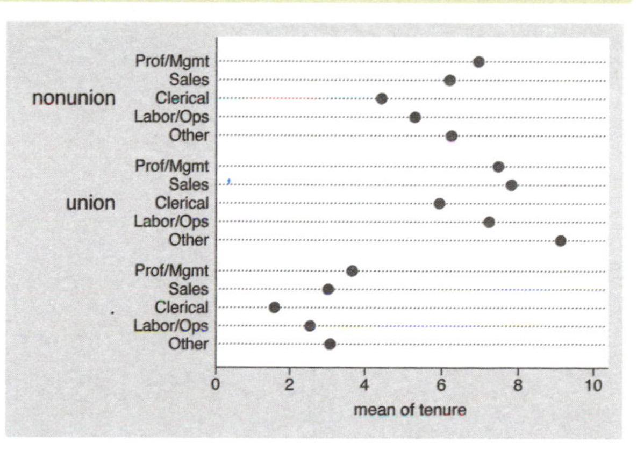

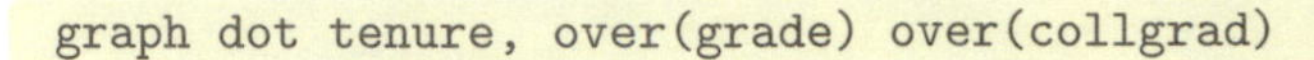

```
graph dot tenure, over(grade) over(collgrad)
```

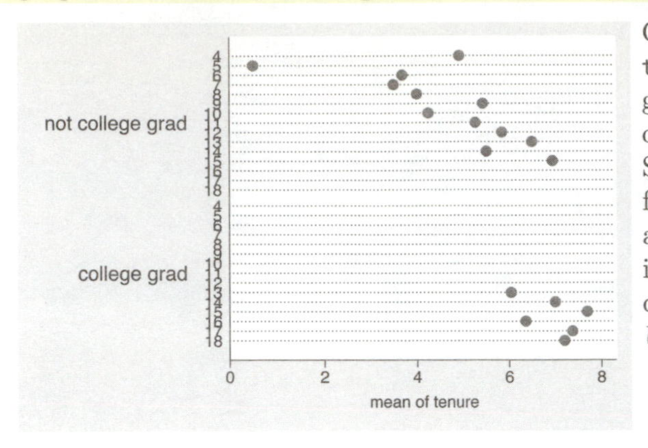

Consider this dot plot, which breaks `tenure` down by two variables: the last grade that one completed and whether one is a college graduate. By default, Stata shows all possible combinations for these two variables. In most cases, all combinations are possible, but not in this case, and including them has caused the labels for grade to overlap.
Uses nlsw.dta & scheme vg_s2m

```
graph dot tenure, over(grade) over(collgrad) nofill
```

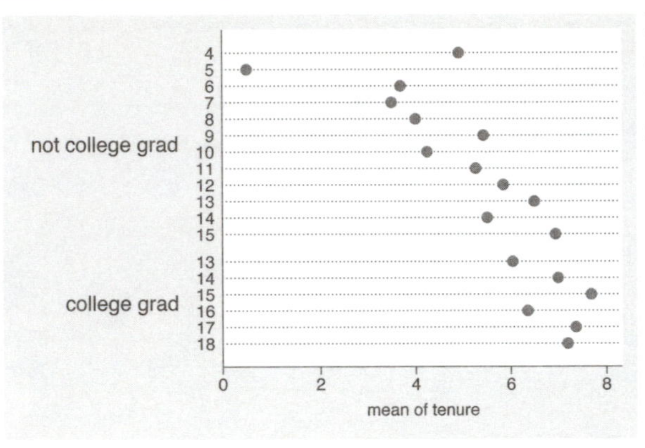

If you only want to display only the combinations of the `over()` variables that exist in the data, you can use the `nofill` option.
Uses nlsw.dta & scheme vg_s2m

6.2 Options for groups, over options

This section considers some of the options that can be used with the `over()` and `yvaroptions()` options for customizing the display of the markers. We will focus on controlling the spacing between the markers and the order in which the markers are displayed. Other options that control the display of the x-axis (such as the labels) are covered in Dot : Cat axis (202). For more information on the `over()` options covered in this section, see the *over_subopts* table in [G] **graph dot**. We first consider options that control the spacing among the markers and then options that change the order in which the markers are sorted. These examples begin with the `vg_blue` scheme.

```
graph dot tenure, over(occ5) over(collgrad)
```

Consider this graph in which we show a dot plot of `tenure` broken down by `occ5` and `collgrad`.
Uses nlsw.dta & scheme vg_blue

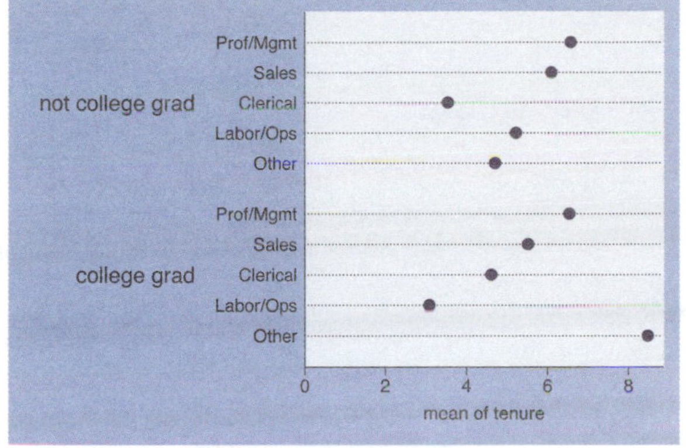


```
graph dot tenure, over(occ7) over(collgrad, gap(*5))
```

Suppose that we wanted to make the gap between the levels of `collgrad` larger. Here, we use the `gap(*5)` option to make this gap five times as large as it normally would be.
Uses nlsw.dta & scheme vg_blue

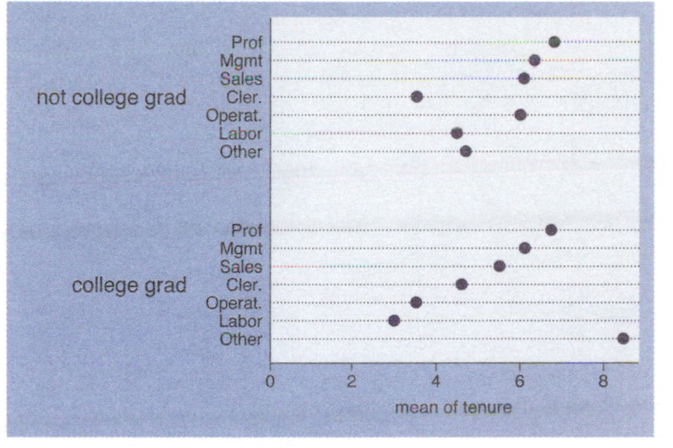


```
graph dot tenure, over(occ7)
```

Consider this graph showing `tenure` broken down by the seven levels of occupation. The markers are ordered by levels of `occ7`, going from 1 to 7.
Uses nlsw.dta & scheme vg_blue

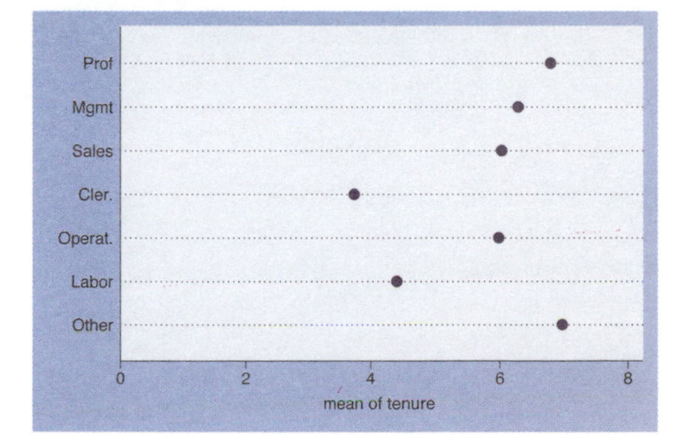

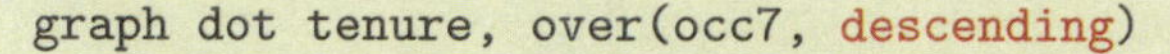

```
graph dot tenure, over(occ7, descending)
```

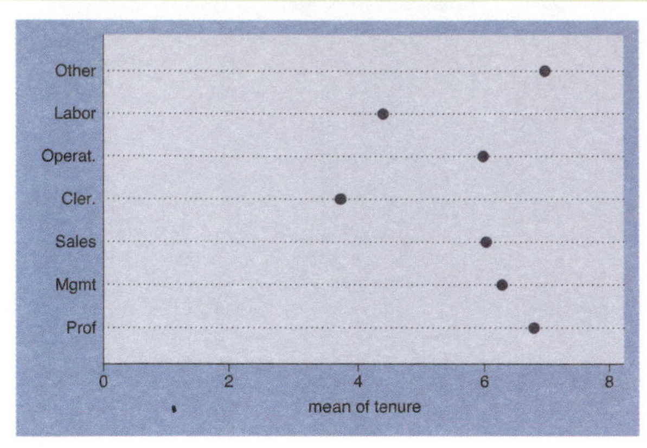

The `descending` option switches the order of the markers. They still are ordered according to the seven levels of occupation, but the markers are ordered from 7 to 1.
Uses nlsw.dta & scheme vg_blue

```
graph dot tenure, over(occ7, sort(occ7alpha))
```

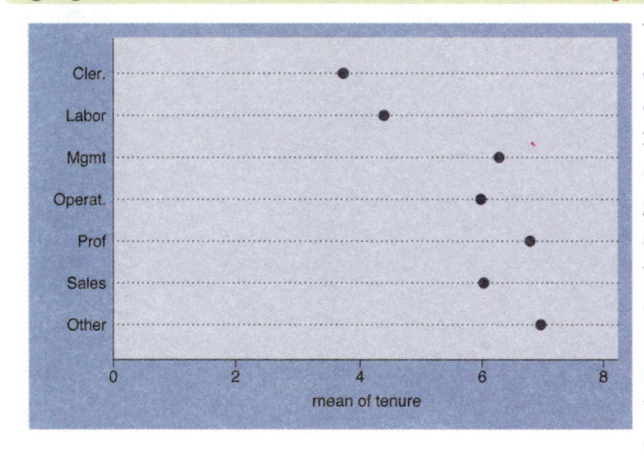

We might want to put these markers in alphabetical order (but with Other appearing last). We can do this by recoding `occ7` into a new variable (say `occ7alpha`), such that, as `occ7alpha` goes from 1 to 7, the occupations are alphabetical. We recoded `occ7` with these assignments: $4 = 1$, $6 = 2$, $2 = 3$, $5 = 4$, $1 = 5$, $3 = 6$, and $7 = 7$; see [R] **recode**. Then, the `sort(occ7alpha)` option alphabetizes the markers (but with Other still appearing last).
Uses nlsw.dta & scheme vg_blue

```
graph dot tenure, over(occ7, sort(1))
```

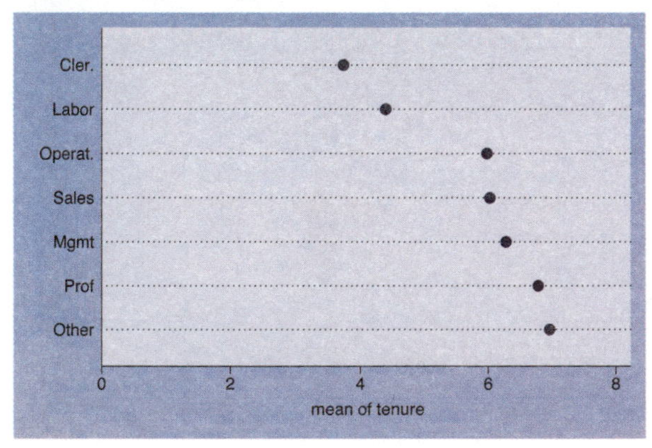

Here, we sort the variables based on the mean of `tenure`, yielding markers with means in ascending order. The `sort(1)` option sorts the markers according to the mean of the first y-variable, the mean of `tenure`. In this case, there is only one variable.
Uses nlsw.dta & scheme vg_blue

```
graph dot tenure, over(occ7, sort(1) descending)
```

Adding the `descending` option yields markers in descending order, going from highest mean `tenure` to lowest mean `tenure`.
Uses nlsw.dta & scheme vg_blue

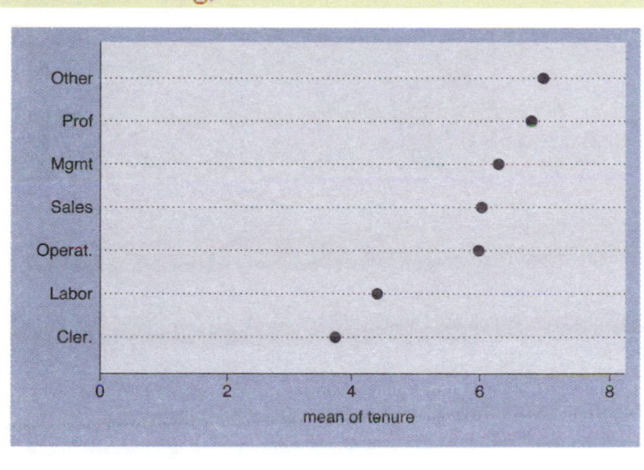


```
graph dot tenure prev_exp, over(occ7, sort(2))
```

Adding a second y-variable and changing `sort(1)` to `sort(2)` sorts the markers according to the second y-variable, the mean of `prev_exp`.
Uses nlsw.dta & scheme vg_blue

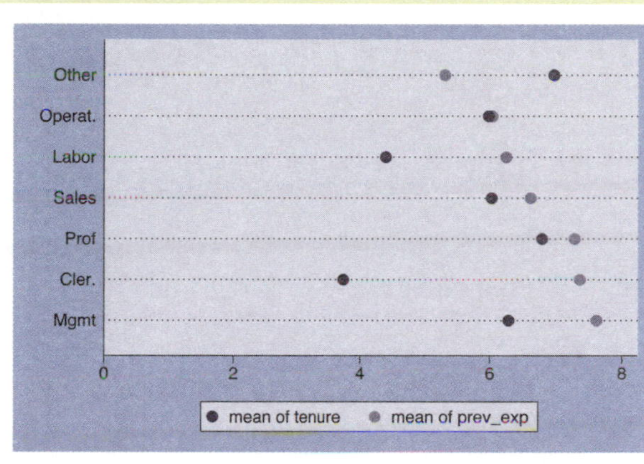


```
graph dot tenure prev_exp, over(occ7, sort(1)) over(collgrad)
```

We can use the `sort()` option when there are additional `over()` variables. Here, the markers are ordered according to the mean of `tenure` within each level of `collgrad`.
Uses nlsw.dta & scheme vg_blue

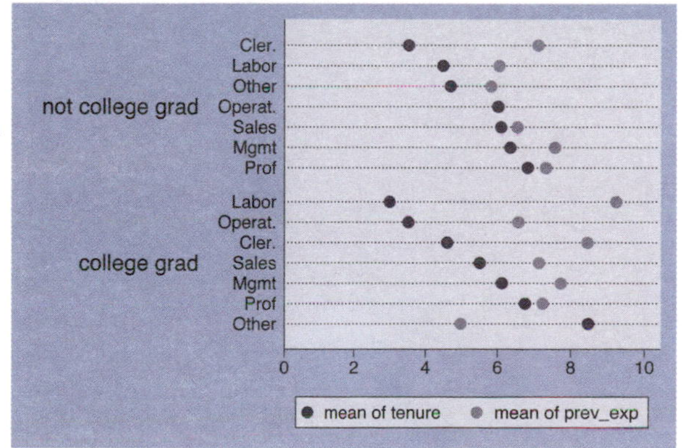

```
graph dot tenure prev_exp, over(occ7, sort(1)) over(collgrad, descending)
```

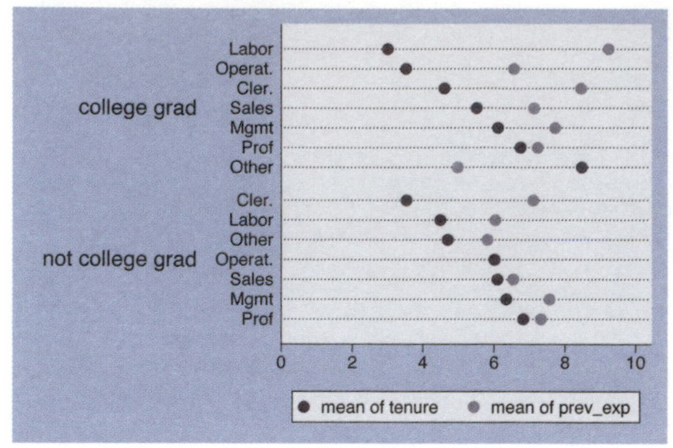

We add the **descending** option to the second **over()** option, and the levels of **collgrad** are now shown with college graduates appearing first.
Uses nlsw.dta & scheme vg_blue

6.3 Controlling the categorical axis

This section describes ways that you can label the categorical axis in dot plots. Dot plots, like bar and box plots, are different from other plots since their x-axis is formed by categorical variables. (Remember that Stata calls the axis with the categorical variable(s) the x-axis, even though it may be placed on the left axis.) This section describes options you can use to customize the categorical axis. For more details on this, see [G] ***cat_axis_label_options*** and [G] ***cat_axis_line_options***. We will start by showing how you can change the labels used for the categorical axis. These examples use the `vg_past` scheme.

```
graph dot tenure, over(occ7) over(south)
```

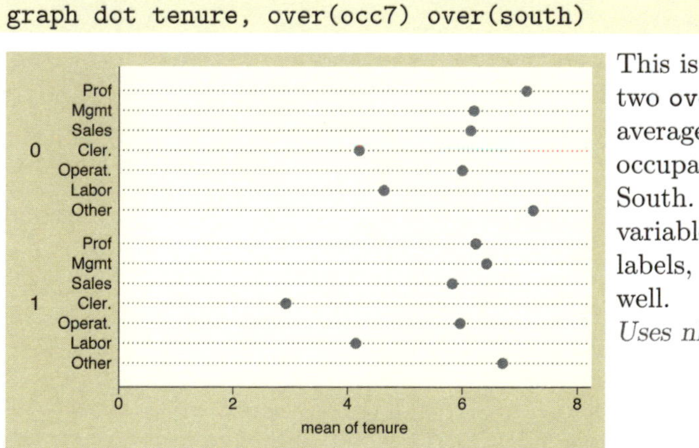

This is an example of a dot plot with two **over()** variables graphing the average tenure broken down by occupation and whether one lives in the South. The variable **south** is a dummy variable that does not have any value labels, so the x-axis is not labeled very well.
Uses nlsw.dta & scheme vg_past

```
graph dot wage, over(occ7) over(south, relabel(1 "N & W" 2 "South"))
```

We can use the `relabel()` option to change the labels displayed for the levels of **south**, giving the x-axis more meaningful labels. Note that we wrote `relabel(1 "N & W")`, not `relabel(0 "N & W")`, since these numbers do not represent the actual levels of **south** but the ordinal position of the levels, i.e., first and second.
Uses nlsw.dta & scheme vg_past


```
graph dot prev_exp tenure ttl_exp, over(occ5) ascategory
```

This **graph dot** command has multiple y-variables but uses the `ascategory` option to plot the different y-variables as if they were categorical variables. The dots for the different y-variables are plotted on different lines using the same symbol, and each line is labeled on the x-axis rather than using a legend. The default labels on the x-axis are not bad, but we might want to change them.
Uses nlsw.dta & scheme vg_past

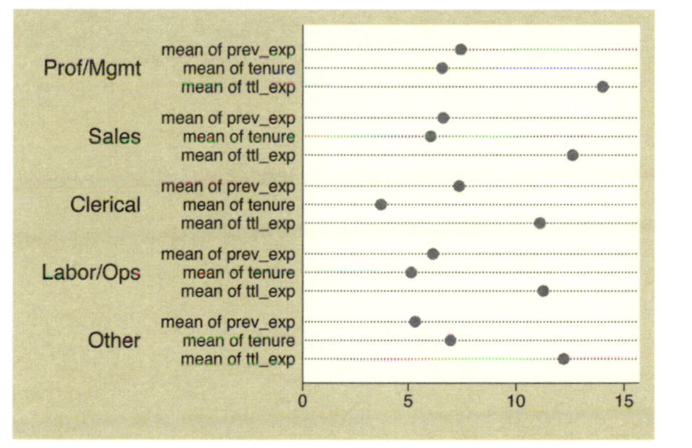


```
graph dot prev_exp tenure ttl_exp, over(occ5) ascategory
   yvaroptions(relabel(1 "Previous" 2 "Current" 3 "Total"))
```

If we had an **over()** option, we could use the **relabel()** option to change the labels on the x-axis. But since we have multiple y-variables that we have treated as categories, we then use the `yvaroptions(relabel())` option to modify the labels on the x-axis.
Uses nlsw.dta & scheme vg_past

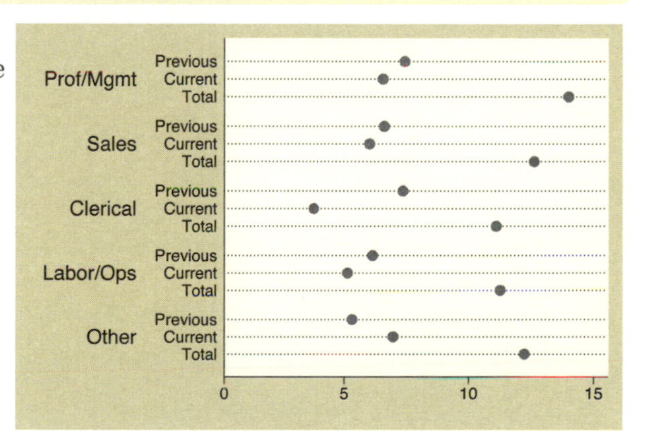

Introduction Twoway Matrix Bar Box Dot Pie Options Standard options Styles Appendix
Yvars and over Over options Cat axis Legend Y-axis Dotlook options By

```
graph dot prev_exp tenure ttl_exp, ascategory
   over(south, relabel(1 "N & W" 2 "South"))
   yvaroptions(relabel(1 "Previous" 2 "Current" 3 "Total"))
```

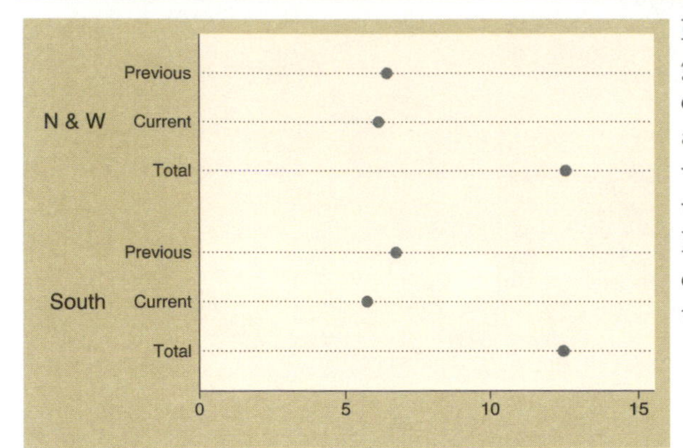

In this example, we have multiple y-variables that are converted into categorical variables via the `ascategory` option, and an `over()` variable, as well. The `relabel()` option within the `over()` option changes the labels for `south`, and the `relabel()` option within `yvaroptions()` changes the labels for the multiple y-variables.
Uses nlsw.dta & scheme vg_past

```
graph dot prev_exp tenure ttl_exp, ascategory xalternate
   over(south, relabel(1 "N & W" 2 "South"))
   yvaroptions(relabel(1 "Previous" 2 "Current" 3 "Total"))
```

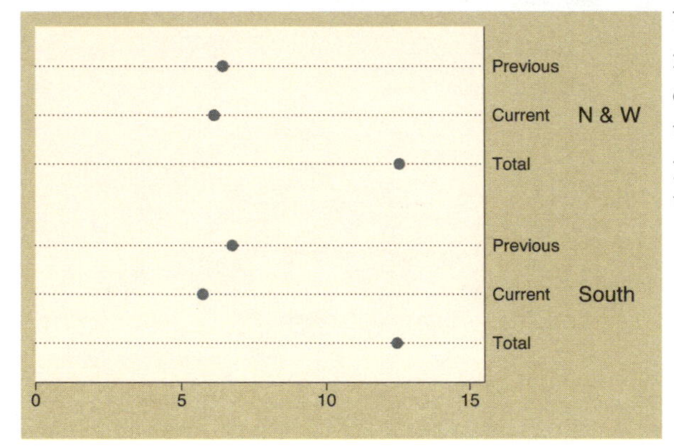

We add the `xalternate` option, which moves the labels for the x-axis to the opposite side, in this case, from the left to the right. We could also use the `yalternate` option to move the y-axis to its opposite side.
Uses nlsw.dta & scheme vg_past

```
graph dot wage, over(occ7)
   l1title("Occupations recoded" "into seven categories")
```

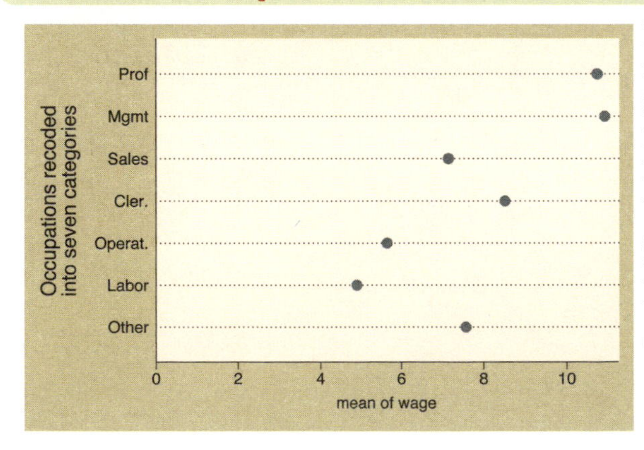

In this example, the categorical axis represents the occupation after recoding it into seven categories. We can use the `l1title()` option to add a title to the left of the graph labeling this axis. Note that we broke the title into two quoted strings that appear on the graph as two lines. We could also add a second title to the left with `l2title()`; see Standard options : Titles (313) for more details.
Uses nlsw.dta & scheme vg_past

6.4 Controlling legends

This section discusses the use of legends for dot plots, emphasizing the features that are unique to dot plots. The section Options : Legend (287) goes into great detail about legends, as does [G] ***legend_option***. Legends can be used for multiple y-variables or when the first `over()` variable is treated as a y-variable via the `asyvars` option. See Dot : Yvars and over (193) for more information about using multiple y-variables and more examples of treating the first `over()` variable as a y-variable. These following examples use the `vg_rose` scheme.

```
graph dot prev_exp tenure ttl_exp, over(occ7)
```

Consider this dot plot of three different variables. These variables are shown with different markers, and a legend is used to identify the variables.
Uses nlsw.dta & scheme vg_rose

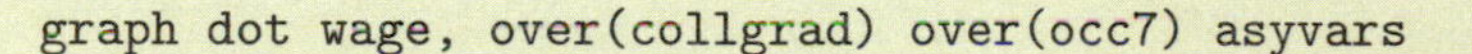

```
graph dot wage, over(collgrad) over(occ7) asyvars
```

This is another example of how a legend can arise in a Stata dot plot if you use the `over()` variable with the `asyvars` option. Stata treats the levels of the `over()` variable as if they were really multiple y-variables.
Uses nlsw.dta & scheme vg_rose

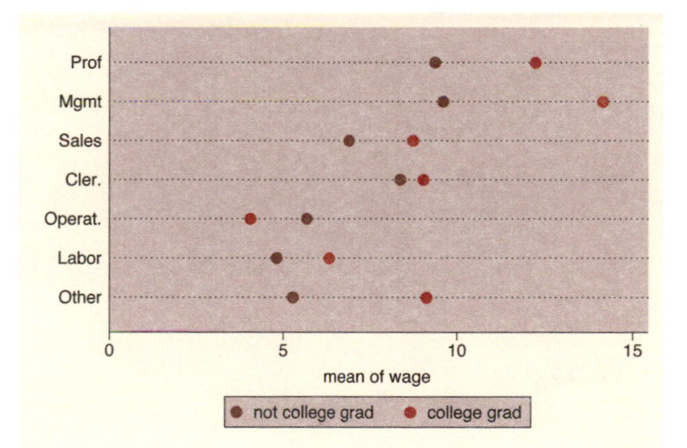

Unless otherwise mentioned, the legend options described below work the same whether the legend was derived from multiple y-variables or from an over() variable that was combined with the asyvars option.

```
graph dot prev_exp tenure ttl_exp, over(occ7) nolabel
```

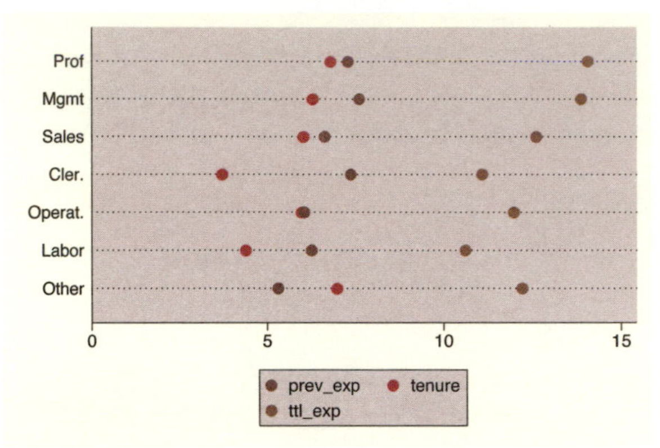

The nolabel option only works when you have multiple y-variables. When this option is used, the variable names (not the variable labels) are used in the legend. For example, instead of showing the variable label Prev. work exper., this option shows the variable name prev_exp.
Uses nlsw.dta & scheme vg_rose

```
graph dot prev_exp tenure ttl_exp, over(occ7)
   legend(label(1 "Previous") label(2 "Current") label(3 "Total")
   title("Work Experience"))
```

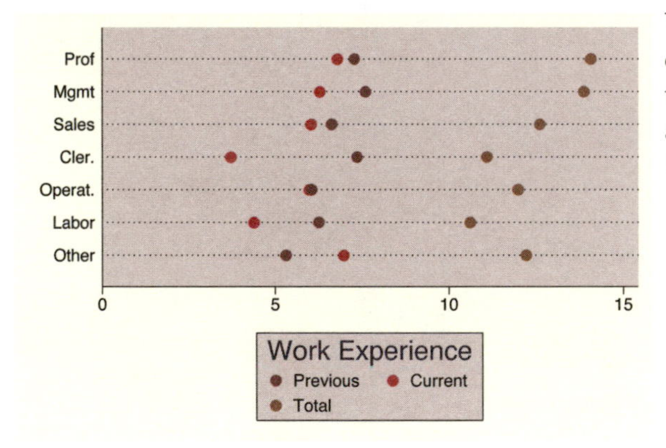

We use the legend(label()) option to change the labels for the variables in the legend and the title() option to add a title to the legend.
Uses nlsw.dta & scheme vg_rose

```
graph dot prev_exp tenure ttl_exp, over(occ7)
   legend(position(12) rows(1))
```

We can put the legend at the top of the graph with the `legend(position(12))` option. The values you supply for `position()` are similar to the numbers on a clock face, where 12 o'clock is the top, 6 o'clock is the bottom, and 0 represents the center of the clock face; see Styles : Clockpos (330) for more details. We also add the `rows(1)` option to make the legend display as one row.
Uses nlsw.dta & scheme vg_rose

```
graph dot prev_exp tenure ttl_exp, over(occ7)
   legend(cols(1) position(9))
```

Here, the legend is moved to the left and displayed in a single column using the `legend(cols(1) position(9))` options.
Uses nlsw.dta & scheme vg_rose

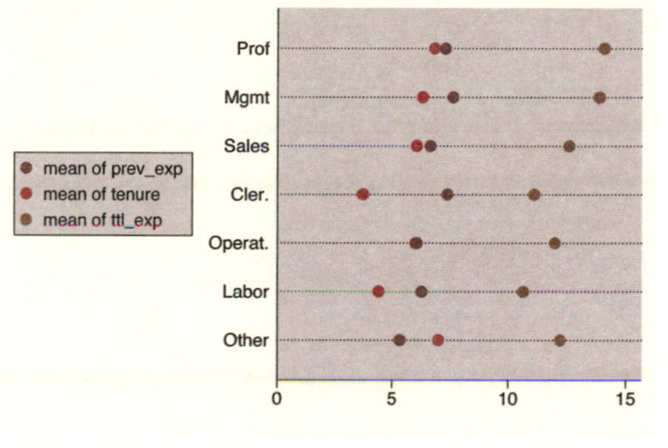

6.5 Controlling the y-axis

This section describes options to customize the y-axis with dot plots. To be precise, when Stata refers to the y-axis on a dot plot, it refers to the axis with the continuous variable, which is placed on the bottom (where the x-axis would traditionally be placed). This section emphasizes the features that are particularly relevant to dot plots. For more details, see Options : Axis titles (254), Options : Axis labels (256), and Options : Axis scales (265). Also, see [G] ***axis_title_options***, [G] ***axis_label_options***, and [G] ***axis_scale_options***. These examples use the `vg_teal` scheme.

```
graph dot hours, over(occ7) ytitle("Hours Worked" "Per Week")
```

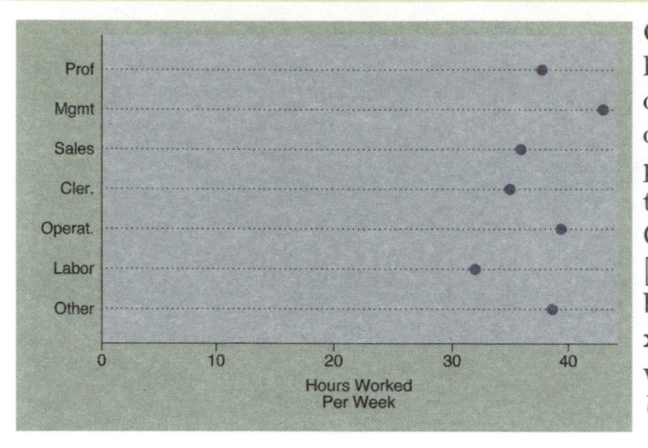

Consider this graph showing the mean hourly wage broken down by occupation. We use the `ytitle()` option to add a title to the y-axis. We place the title across two lines by using two separate, quoted strings. See Options : Axis titles (254) and [G] ***axis_title_options*** for more details, but please disregard any references to `xtitle()`, since that option is not valid when using **`graph dot`**.
Uses nlsw.dta & scheme vg_teal

```
graph dot hours, over(occ7)
   ytitle("Hours Worked" "Per Week", bfcolor(eggshell) box bexpand)
```

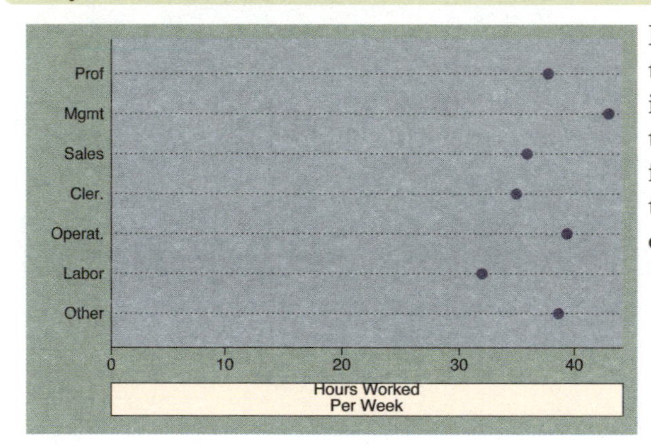

Because the title is considered to be a textbox, you can use textbox options as illustrated here to control the look of the title. See Options : Textboxes (303) for additional examples of how to use textbox options to control the display of text.
Uses nlsw.dta & scheme vg_teal

```
graph dot hours, over(occ7)
   yline(35 40, lwidth(thin) lcolor(navy) lpattern(dash))
```

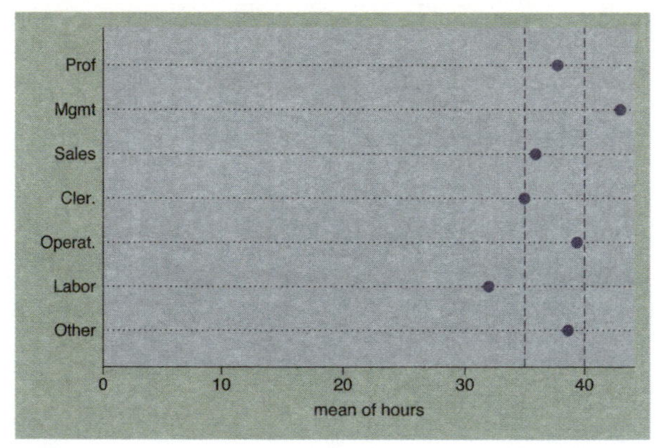

This example uses the `yline()` option to add a thin, navy, dashed line to the graph where the hours worked equal 35 and 40.
Uses nlsw.dta & scheme vg_teal

```
graph dot hours, over(occ7) ylabel(30(5)45)
```

We use the `ylabel()` option to label the y-axis from 30 to 45 by increments of 5. See Options : Axis labels (256) and [G] ***axis_label_options*** for more details. Please disregard any references to `xlabel()` since that option is not valid when using `graph dot`. Note that the y-axis still begins at 0, but see the next example for how you can override this.
Uses nlsw.dta & scheme vg_teal

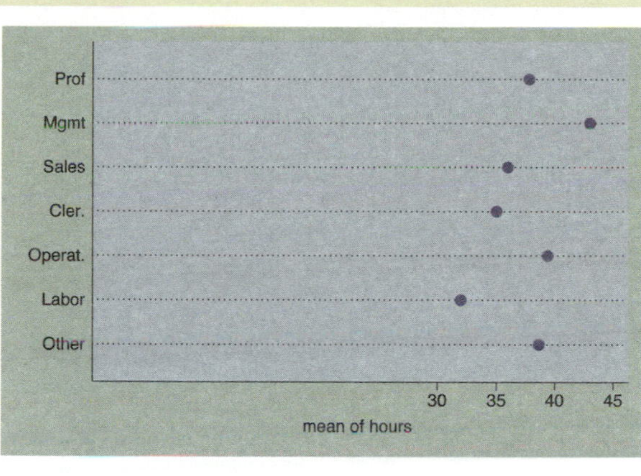


```
graph dot hours, over(occ7) ylabel(30(5)45) exclude0
```

When we add the `exclude0` option, the dot plot does not automatically begin at 0. In this case, it starts at 30 since that is the value we specified as the starting point on the `ylabel()` option.
Uses nlsw.dta & scheme vg_teal

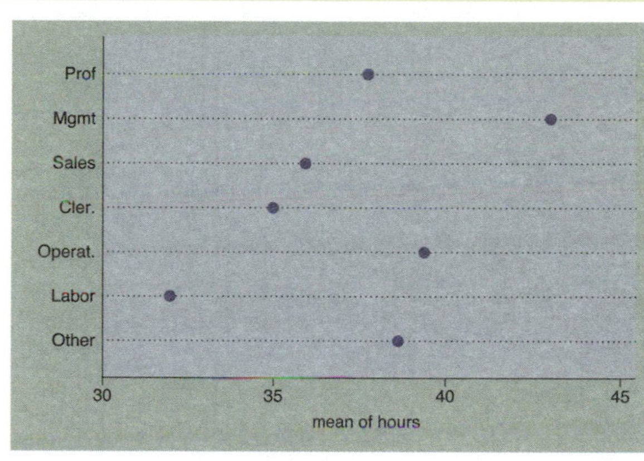


```
graph dot hours, over(occ7) yscale(off)
```

We can use the `yscale(off)` option to turn off the y-axis. See Options : Axis scales (265) and [G] ***axis_scale_options*** for more details. Please disregard any references to `xscale()` since that option is not valid when using `graph dot`.
Uses nlsw.dta & scheme vg_teal

```
graph dot hours, over(occ7) yalternate
```

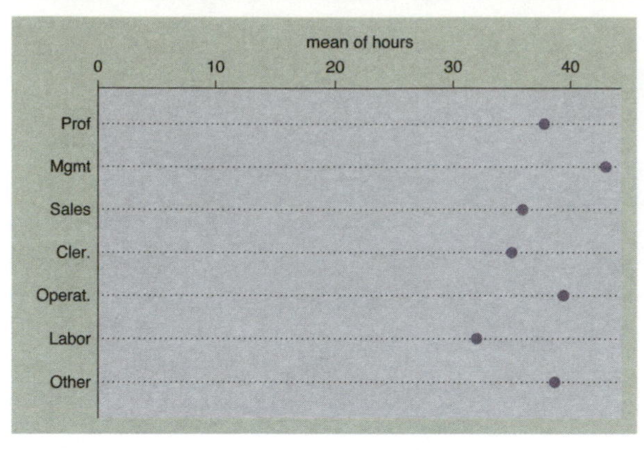

The yalternate option puts the y-axis on the opposite side, in this case on the top side of the graph.
Uses nlsw.dta & scheme vg_teal

```
graph dot hours, over(occ7) yreverse
```

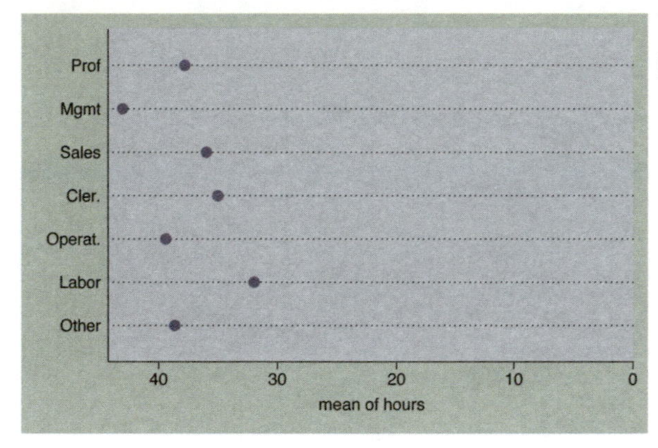

You can reverse the direction of the y-axis with the yreverse option.
Uses nlsw.dta & scheme vg_teal

6.6 Changing the look of dot rulers, dotlook options

This section shows how you can control the look of the lines in your dot plots. We show how you can control the space between the lines, the color of the lines, and other characteristics of the line. For more information, see the *linelook_options* table in [G] **graph dot**. These graphs are shown using the vg_s2c scheme.

```
graph dot prev_exp tenure, over(occ7)
```

Consider this dot plot showing previous and current work experience broken down by occupation. Each dot plot has a series of small dots that forms a line on which the symbols are plotted.
Uses nlsw.dta & scheme vg_s2c

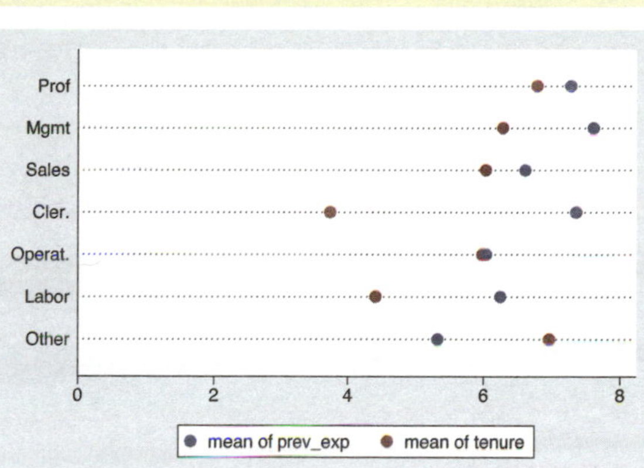


```
graph dot prev_exp tenure, over(occ7)
   ndots(50) dots(msymbol(Oh) msize(medium) mcolor(dkgreen))
```

By default, each line would be composed of 100 small dots, but here we use the `ndots(50)` option to display 50 small dots. Further, using the `dots()` option, the small dots are displayed as medium-sized, dark green, hollow circles. See Styles : Symbols (342), Styles : Markersize (340), and Styles : Colors (328) for more information.
Uses nlsw.dta & scheme vg_s2c

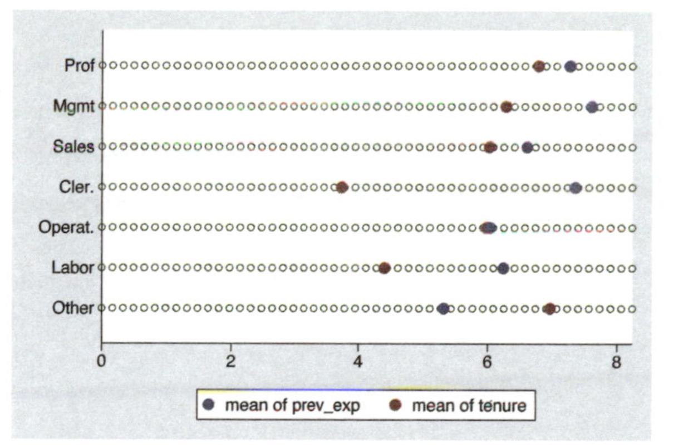


```
graph dot prev_exp tenure, over(occ7) linetype(line)
   lines(lwidth(thick) lcolor(erose))
```

Using the `linetype(line)` option, the dots are instead displayed as lines. Further, we use the `lines()` option to make the line width thick and the line color rose. We could also add the `lpattern()` option to control the line pattern. See Styles : Linewidth (337), Styles : Colors (328), and Styles : Linepatterns (336) for more information.
Uses nlsw.dta & scheme vg_s2c

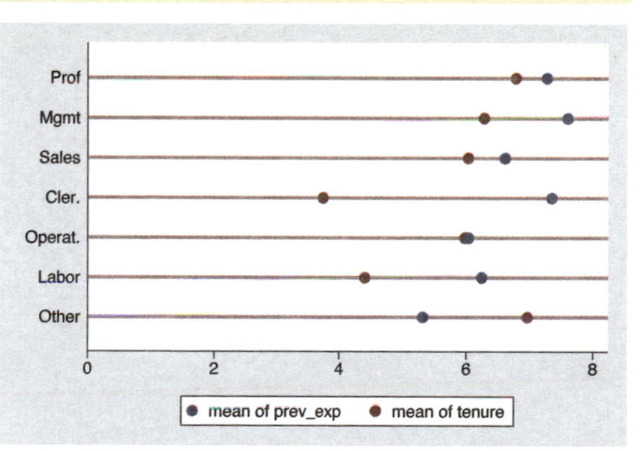

```
graph dot prev_exp tenure, over(occ7) linetype(rectangle)
  rwidth(3) rectangles(fcolor(erose) lcolor(maroon))
```

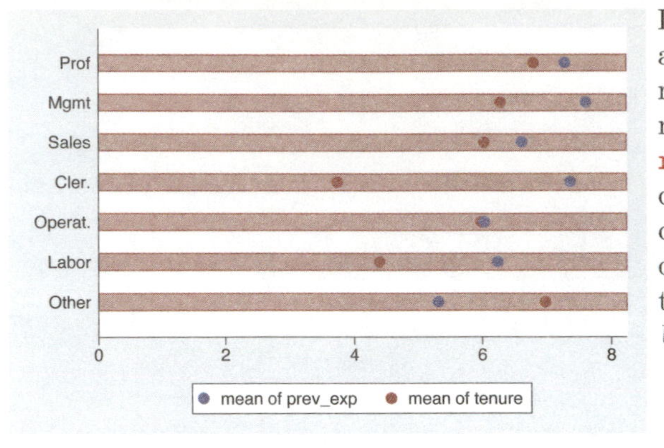

Here, we change the `linetype()` to be a rectangle. The `rwidth(3)` sets the rectangle width to be three times its normal width. In addition, the `rectangle()` option is used to customize it, using the `fcolor()` (fill color) and `lcolor()` (line color) options to make the rectangle rose on the inside with a maroon outline. *Uses nlsw.dta & scheme vg_s2c*

Let's now look at options that allow us to control the markers and whether the markers are displayed on the same line.

```
graph dot prev_exp tenure, over(occ7)
  marker(1, msymbol(D) mcolor(teal) msize(large))
```

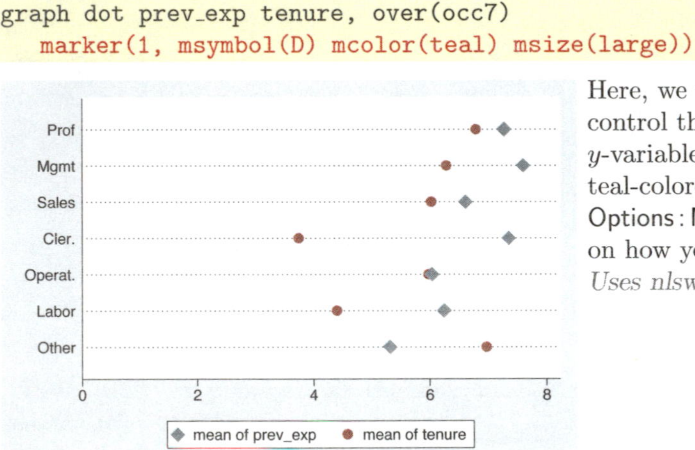

Here, we use the `marker()` option to control the marker used for the first y-variable, making it a large teal-colored diamond. See Options : Markers (235) for more details on how you can control markers. *Uses nlsw.dta & scheme vg_s2c*

```
graph dot prev_exp tenure, over(occ7)
   marker(1, msymbol(d) mfcolor(teal) mlcolor(dkgreen) mlwidth(thick))
   marker(2, msymbol(S) mfcolor(ltblue) mlcolor(blue) mlwidth(thick))
```

In this example, we use two `marker()` options, so we can control both markers. The first marker is now a diamond with a teal fill and a thick, dark green outline. The second marker is a square, light blue on the inside with a thick blue outline. The section Options : Markers (235) has more details on controlling markers.
Uses nlsw.dta & scheme vg_s2c

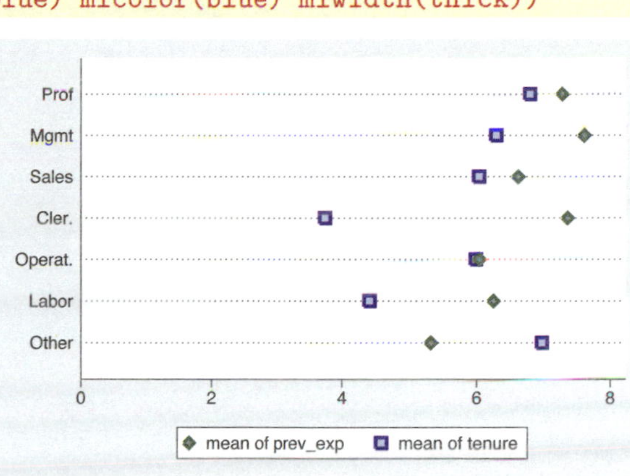


```
graph dot prev_exp tenure, over(occ7) linegap(45)
```

We can use the `linegap()` option to display the y-variables on different lines and specify the gap between these lines. The default value is 0, meaning that all y-variables are displayed on the same line.
Uses nlsw.dta & scheme vg_s2c

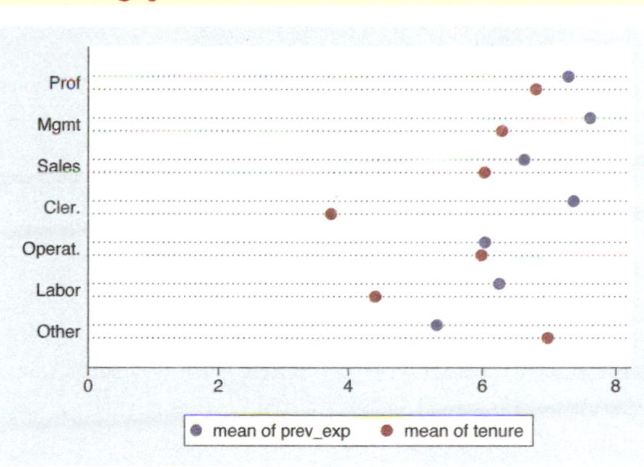


```
graph dot tenure, over(occ5) over(collgrad) over(married)
```

Consider this graph. Since we have used three `over()` options, the levels of the first `over()` variable are displayed as though they were different y-variables. We may want to use the `linegap()` option to display the different y-variables on different lines to make the graph more readable; see the next example.
Uses nlsw.dta & scheme vg_s2c

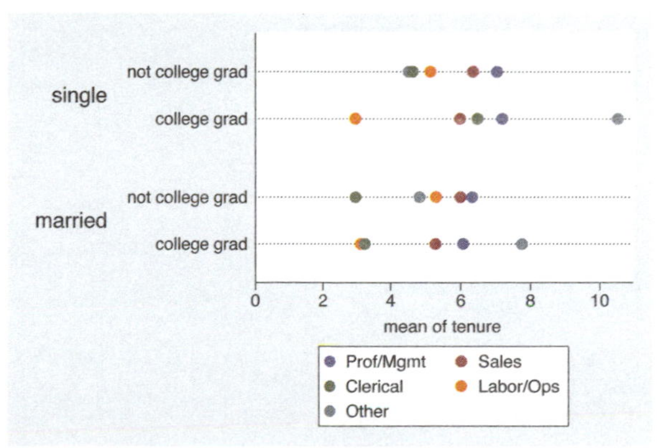

```
graph dot tenure, over(occ5) over(collgrad) over(married) linegap(30)
   legend(rows(1) span)
```

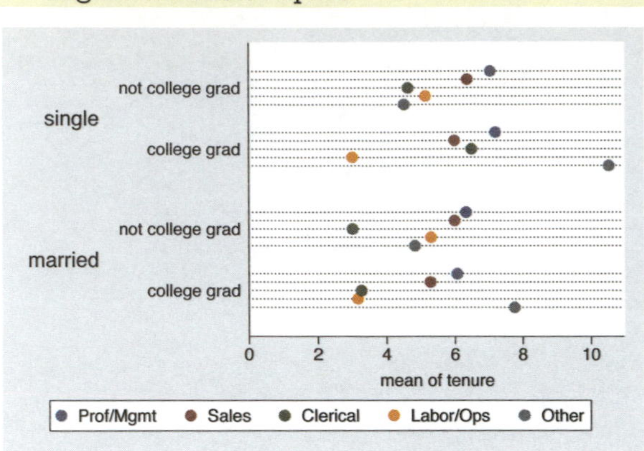

This example is the similar to the previous one, but we have added the linegap(30) option to make the levels of occ5 display on separate lines, making the results more readable. We have also added a legend() option to make the legend display in one line and span the width of the graph.
Uses nlsw.dta & scheme vg_s2c

6.7 Graphing by groups

This section discusses the use of the by() option in combination with graph dot. Normally, you would use the over() option instead of the by() option, but in some cases, the by() option is either necessary or more advantageous. For example, a by() option is useful if you exceed the maximum number of over() options (three if you have a single y-variable or two if you have multiple y-variables). In such cases, the by() option allows you to break your data down by additional categorical variables. by() also gives you more flexibility in the placement of the separate panels. For more information about the by() option, see Options: By (272), and for more information about the over() option, see Dot: Yvars and over (193). The examples in this section use the vg_s1m scheme.

```
graph dot wage, over(collgrad) over(occ5) over(urban2)
```

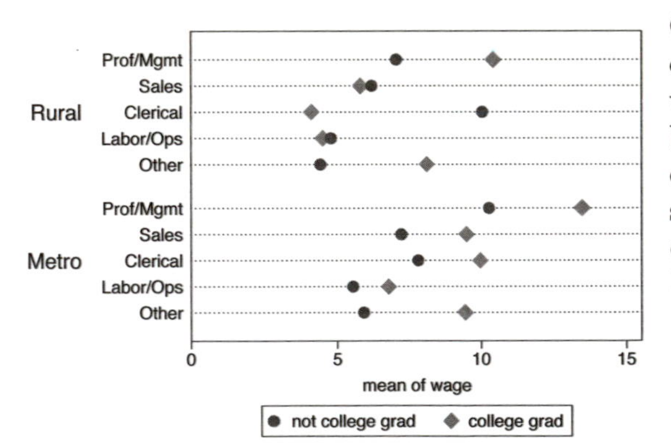

Consider this dot graph breaking wages down by three categorical variables. If we wanted to break this down further by another categorical variable, we could not use another over() option since we can have a maximum of three over() options with a single y-variable.
Uses nlsw.dta & scheme vg_s1m

```
graph dot wage, over(collgrad) over(occ5) over(urban2) by(union)
```

If we want to break wage down further by union, we can use the by(union) option.
Uses nlsw.dta & scheme vg_s1m

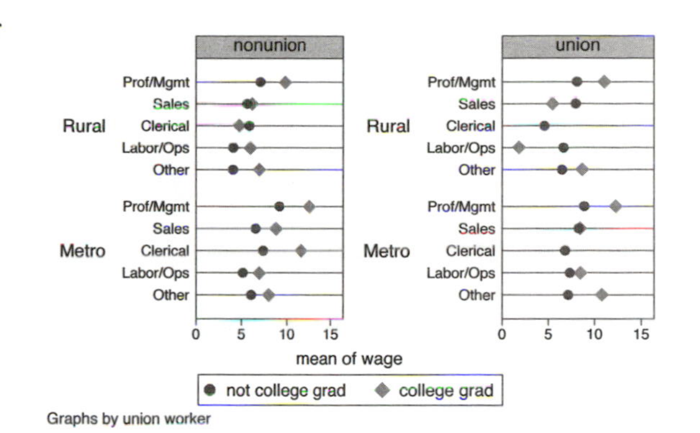


```
graph dot tenure ttl_exp, over(occ5) over(urban2)
```

Consider this dot graph with multiple y-variables breaking them down by two categorical variables using two over() options. When you have multiple y-variables, you can have a maximum of two over() options.
Uses nlsw.dta & scheme vg_s1m

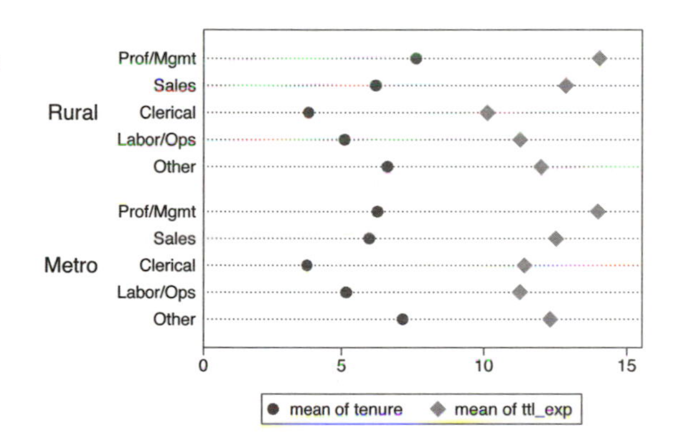


```
graph dot tenure ttl_exp, over(occ5) over(urban2)
    by(union)
```

If we want to break tenure down further by another categorical variable, say union, we can use the by(union) option. Although this example shows only a single variable in the by() option, you can specify multiple variables.
Uses nlsw.dta & scheme vg_s1m

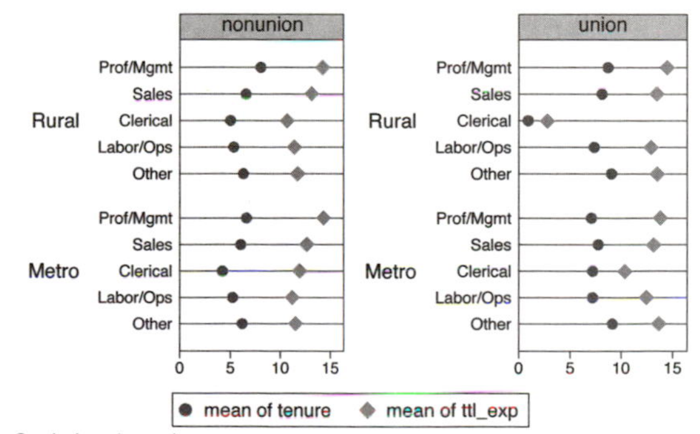

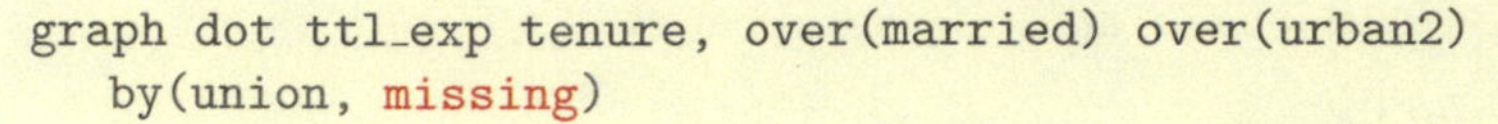

```
graph dot ttl_exp tenure, over(married) over(urban2)
   by(union, missing)
```

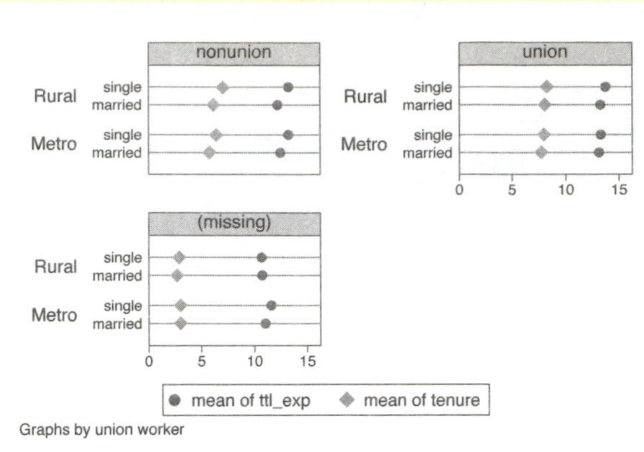

We can use the `missing` option to include a panel for the missing values of `union`. Note that we changed the first `over()` variable to be `over(married)` to make an example that was more readable.
Uses nlsw.dta & scheme vg_s1m

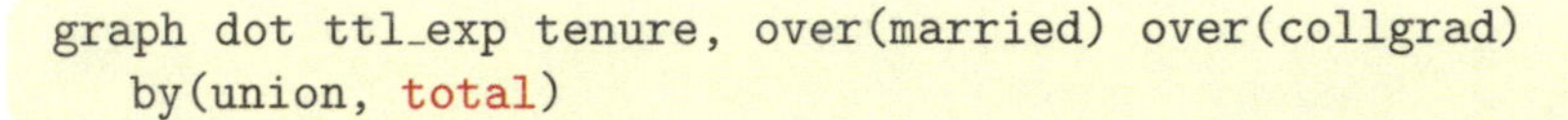

```
graph dot ttl_exp tenure, over(married) over(collgrad)
   by(union, total)
```

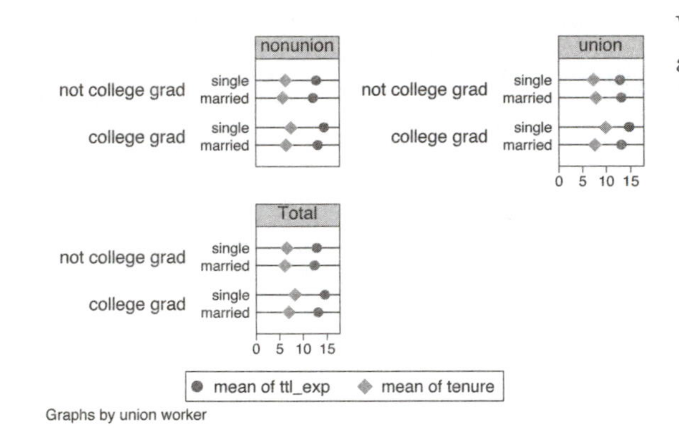

We can add the `total` option to include a panel for all observations.
Uses nlsw.dta & scheme vg_s1m

```
graph dot ttl_exp tenure, over(married) over(collgrad)
   by(union, total cols(1))
```

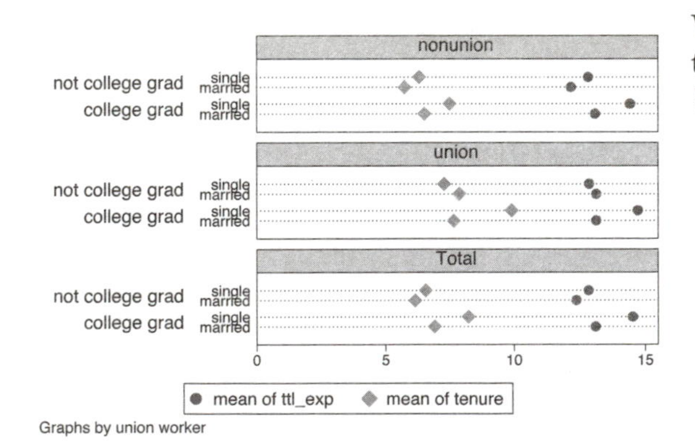

We can use the `cols(1)` option to show the graphs in one column.
Uses nlsw.dta & scheme vg_s1m

7 Pie graphs

This chapter discusses the use of pie charts in Stata. We start by illustrating the different kinds of ways that you can create pie charts in Stata, followed by showing how you can sort the slices in your pie charts. Next, we show how you can customize the display of individual slices, as well as control the colors of the pie chart. Then, we demonstrate different ways you can label the pie slices and then how you can control the legends for pie charts. Finally, we discuss how to use the `by()` option.

7.1 Types of pie graphs

This section describes different ways to produce pie charts using Stata. Stata allows you to produce pie charts based on multiple y-variables, with each y-variable corresponding to a slice. You can also create a pie chart based on a single y-variable broken down by a single `over()` variable. Finally, you can create a pie chart with no y-variables broken down by an `over()` variable, which counts the number of observations by each level of the `over()` variable. For more details, see [G] **graph pie**. This section uses the `vg_s1c` scheme.

```
graph pie poplt5 pop5_17 pop18_64 pop65p
```

In this syntax, you supply multiple y-variables, and each y-variable corresponds to a slice in the pie. The first y-variable is the population in the state that is younger than 5 years old, the next the population 5 to 17 years old, the next 18 to 64 years old, and the last 65 years and older. The entire pie would correspond to the sum of all of these variables across all states. The first slice then corresponds to the percentage of the total population that is younger than 5 years old.
Uses allstates.dta & scheme vg_s1c

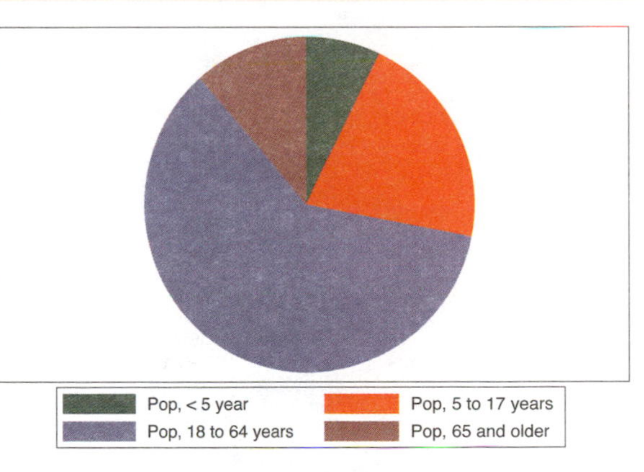

```
graph pie pop, over(division)
```

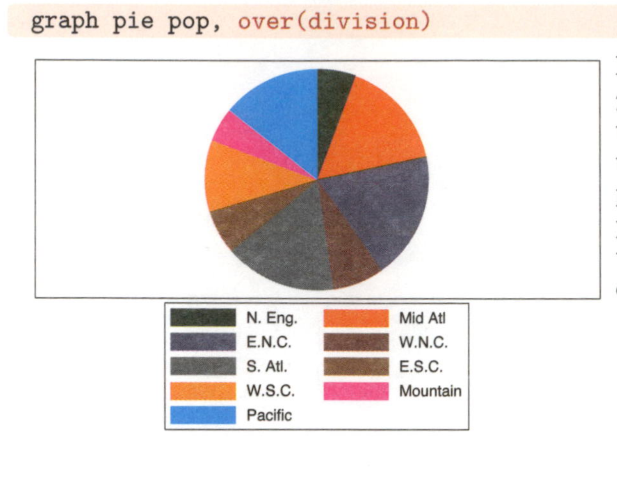

In this syntax, you supply a single y-variable and an `over()` option. In this case, the y-variable corresponds to the population of the state, the entire pie corresponds to the entire population, and each slice corresponds to the percentage of the population for each level of `division`.
Uses allstates.dta & scheme vg_s1c

```
graph pie, over(occ7)
```

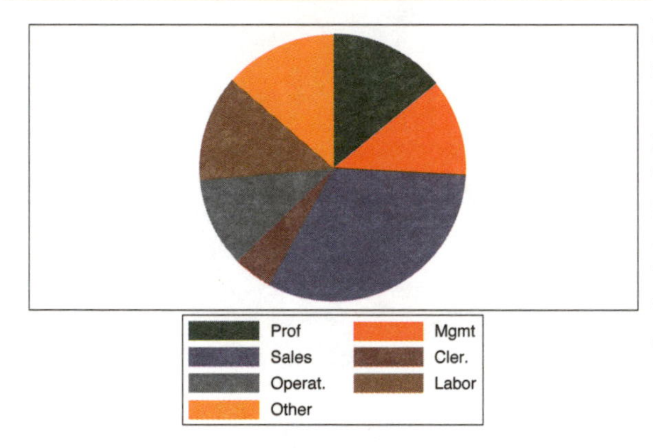

For this third example, we switch to the `nlsw` data file. In this syntax, an `over()` option is supplied, but no y-variable is supplied (in a sense, the observation itself serves as the y-variable). This pie chart is much like a visual frequency distribution of `occ7`, where the size of each slice corresponds to the proportion of women in each occupation.
Uses nlsw.dta & scheme vg_s1c

```
graph pie, over(union) missing
```

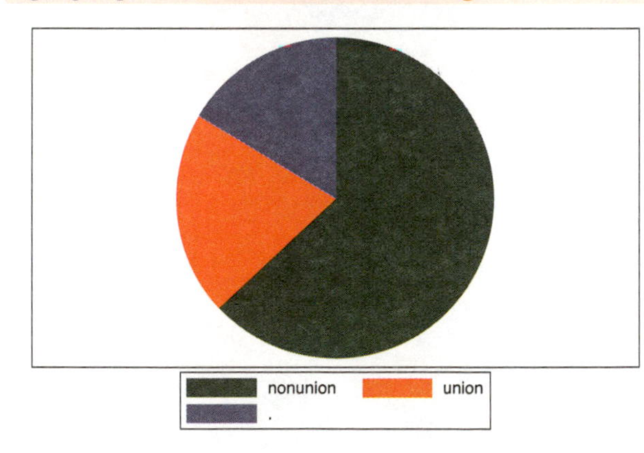

This example shows the proportion of women in union and nonunion jobs. We add the `missing` option, and another pie slice is added for the observations in which `union` is missing.
Uses nlsw.dta & scheme vg_s1c

7.2 Sorting pie slices

This section describes how you can sort and arrange slices in pie charts. For more details, see [G] **graph pie**. This section uses the vg_lgndc scheme, which places the legend at the left in a single column.

```
graph pie, over(occ7)
```

Consider this pie chart showing the number of women who work in these seven different occupations. The slices are ordered according to the levels of occ7 from 1 to 7, rotating clockwise, starting with the first slice, which is positioned at 90 degrees.
Uses nlsw.dta & scheme vg_lgndc

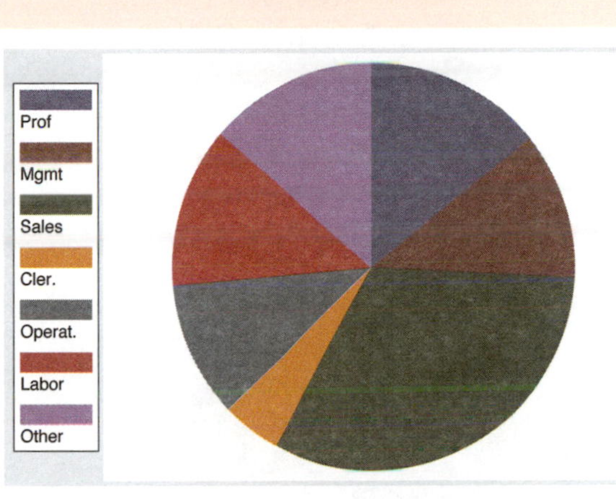


```
graph pie, over(occ7) noclockwise
```

With the noclockwise option, you can display the slices in counterclockwise order.
Uses nlsw.dta & scheme vg_lgndc

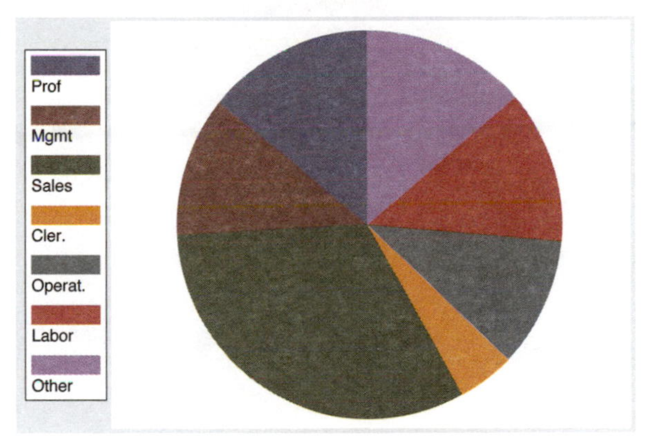

```
graph pie, over(occ7) angle0(0)
```

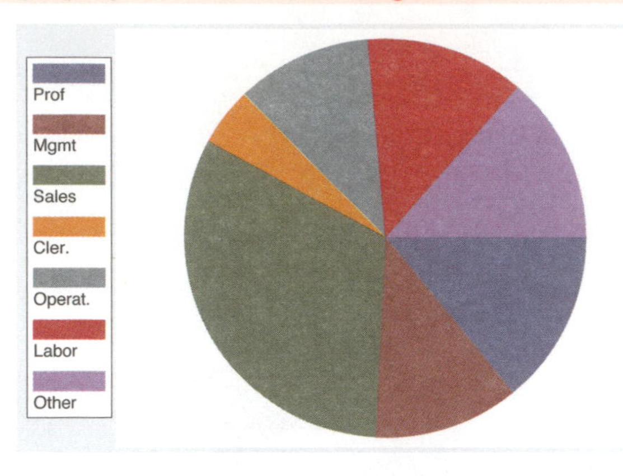

With the `angle0()` option, you can set the angle of the line that begins the first pie slice. Here, we make the first pie slice begin at 0 degrees.
Uses nlsw.dta & scheme vg_lgndc

```
graph pie, over(occ7) sort
```

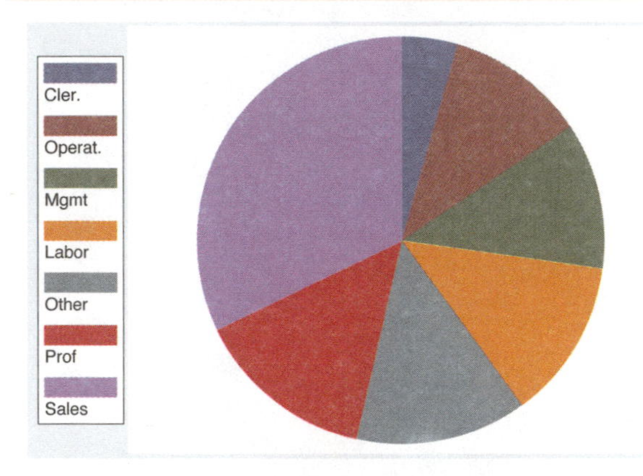

The `sort` option sorts the slices according to their size, from smallest to largest.
Uses nlsw.dta & scheme vg_lgndc

```
graph pie, over(occ7) sort descending
```

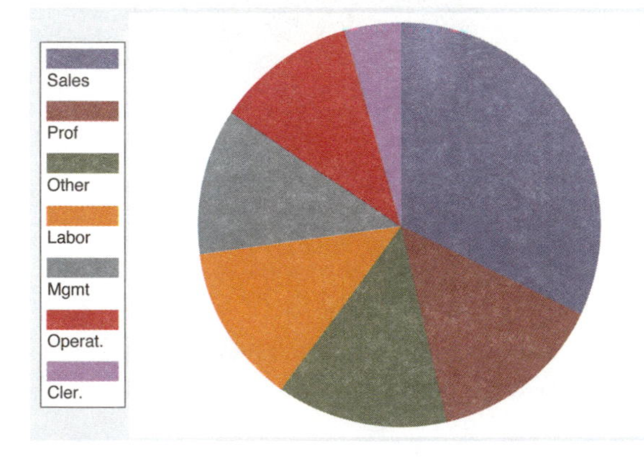

Adding the `descending` option to the `sort` option orders the slices from largest to smallest.
Uses nlsw.dta & scheme vg_lgndc

```
graph pie, over(occ7) sort(occ7alpha)
```

Say that we wanted to sort the slices (alphabetically) by occupation name. We have created a new variable, `occ7alpha`, that is a recoded version of `occ7`. It is recoded such that, as `occ7alpha` goes from 1 to 7, the occupations are alphabetized (except for Other, which is placed last). We add `sort(occ7alpha)`, and the slices are ordered alphabetically.
Uses nlsw.dta & scheme vg_lgndc

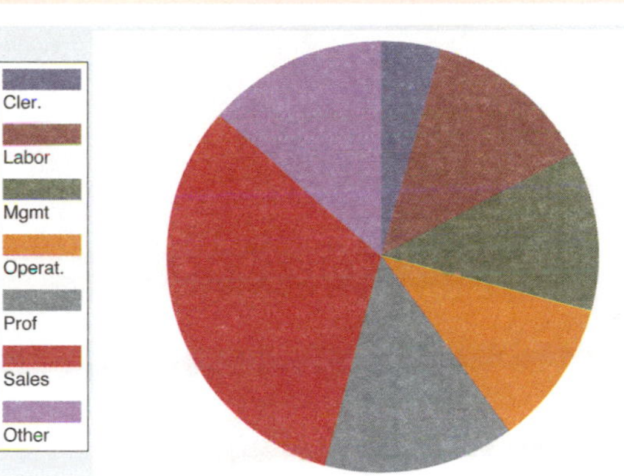

Introduction Twoway Matrix Bar Box Dot Pie Options Standard options Styles Appendix
Types of pie graphs Sorting Colors and exploding Labels Legend By

7.3 Changing the look of pie slices, colors, and exploding

This section describes how to change the color of pie slices, explode pie slices, control the overall intensity of colors, and control the characteristics of lines surrounding the pie slices. For more details, see [G] **graph pie**. This section uses the `vg_rose` scheme.

```
graph pie, over(occ7)
```

Consider this pie chart showing the number of women who work in these seven different occupations. The slices are colored using the colors indicated by the scheme. None of the slices are exploded, and no lines surround the slices.
Uses nlsw.dta & scheme vg_rose

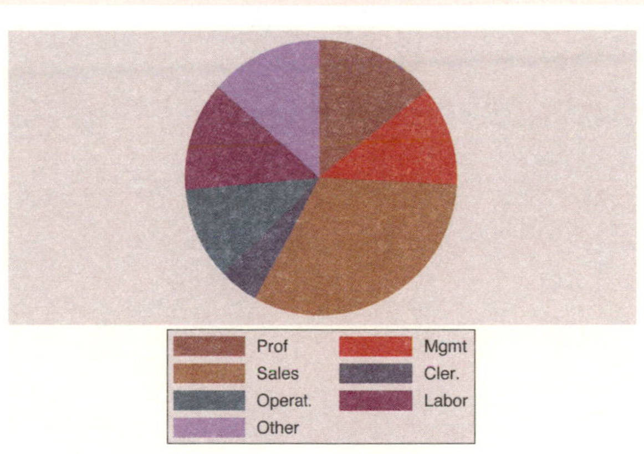

```
graph pie, over(occ7) pie(3, explode)
```

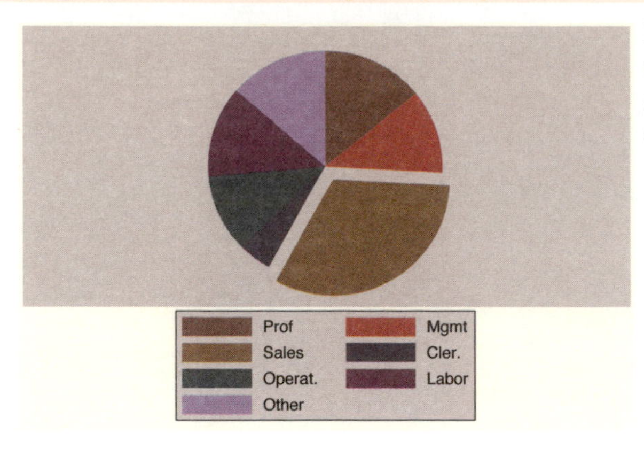

In this example, we use the `pie()` option to explode the third pie slice, calling attention to this slice. By default, it is exploded by 3.8 units.
Uses nlsw.dta & scheme vg_rose

```
graph pie, over(occ7) pie(3, explode(5) color(cyan))
```

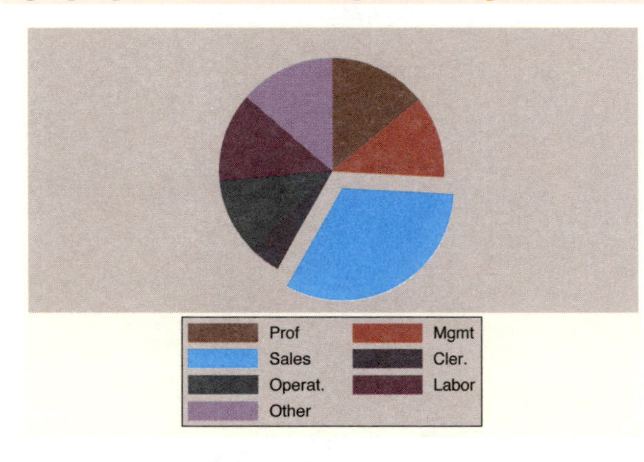

Here, we specify `explode(5)` to increase the distance this slice is exploded to 5 units. We also make the third slice cyan to make it more noticeable. See Styles : Colors (328) for other colors you could choose.
Uses nlsw.dta & scheme vg_rose

```
graph pie, over(occ7) pie(3, color(cyan) explode(5))
   pie(1, color(gold) explode(2.5))
```

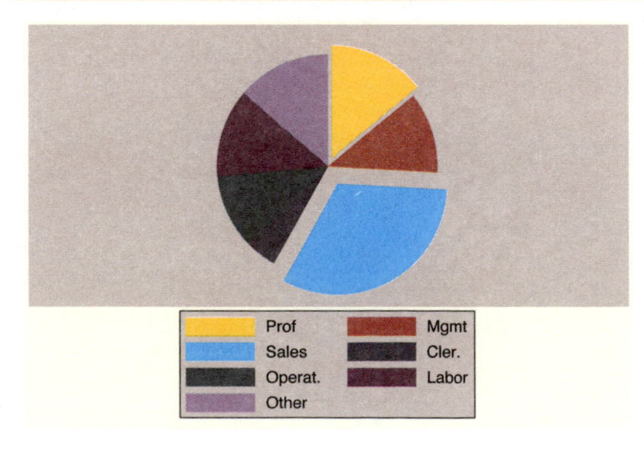

You can use the `pie()` option repeatedly. Here, we change the color and explode slices 1 and 3.
Uses nlsw.dta & scheme vg_rose

```
graph pie, over(occ7) intensity(*1.5)
```

Using the `intensity()` option, we make the colors of all of the slices 1.5 times their normal intensity.
Uses nlsw.dta & scheme vg_rose

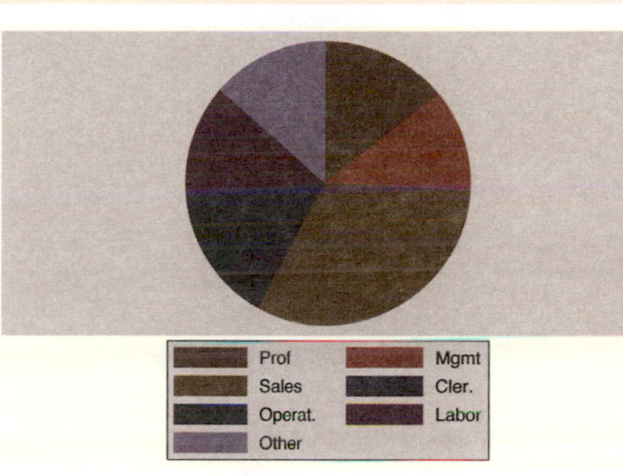


```
graph pie, over(occ7) intensity(*.6)
```

In this example, we make the intensity of the colors 60% of the normal color.
Uses nlsw.dta & scheme vg_rose

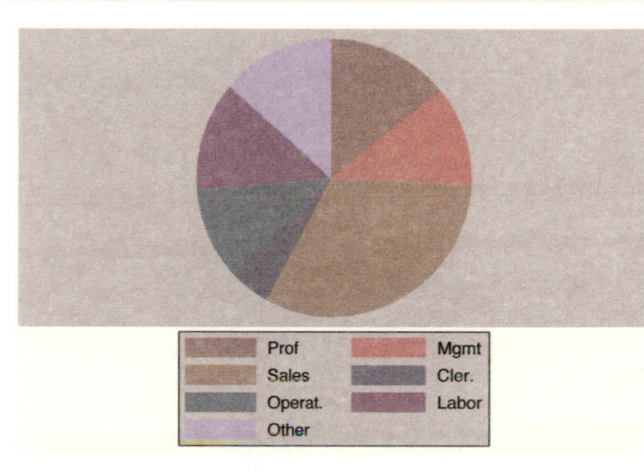


```
graph pie, over(occ7) line(lcolor(sienna) lwidth(thick))
```

The `line()` option can be used to change the characteristics of the lines surrounding the pie slices. Here, we add the `lcolor()` (line color) and `lwidth()` (line width) options to make the line sienna and thick.
Uses nlsw.dta & scheme vg_rose

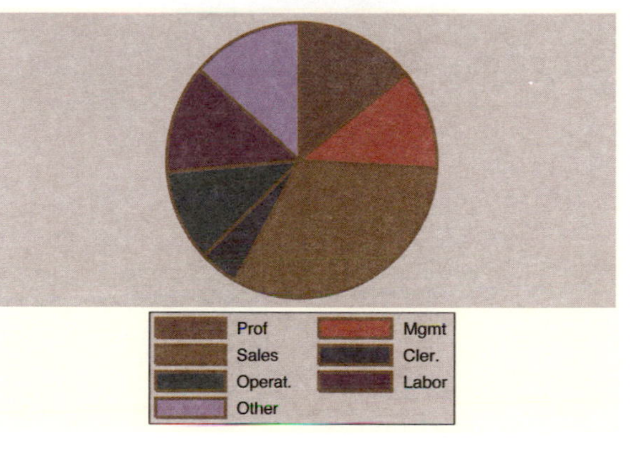

7.4 Slice labels

This section describes how you can label the pie slices. For more details, see [G] **graph pie**. For this section, we will use the `economist` scheme.

```
graph pie, over(occ7) plabel(_all sum)
```

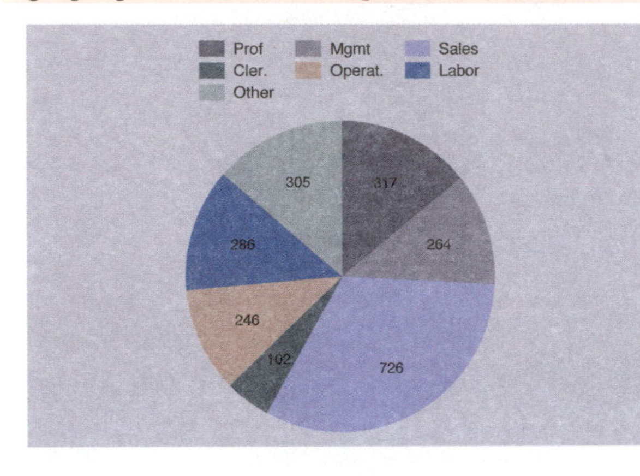

Consider this pie chart showing the number of women who work in these seven different occupations. Here, we use the `plabel()` (pie label) option to label all slices with the sum, in this case the frequency of women who work in each occupation. Notice how readable the labels are because of the pale colors of the pie slices selected by the `vg_past` scheme. Other schemes with more intense colors would have made these labels hard to read.
Uses nlsw.dta & scheme economist

```
graph pie, over(occ7) plabel(_all percent)
```

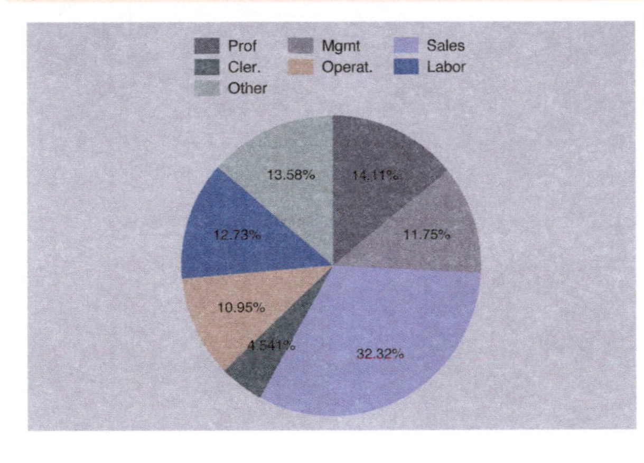

Using the `percent` option, we can show the percent of women who work in each occupation.
Uses nlsw.dta & scheme economist

```
graph pie, over(occ7) plabel(_all name)
```

The name option adds a label that is the name of the occupation.
Uses nlsw.dta & scheme economist

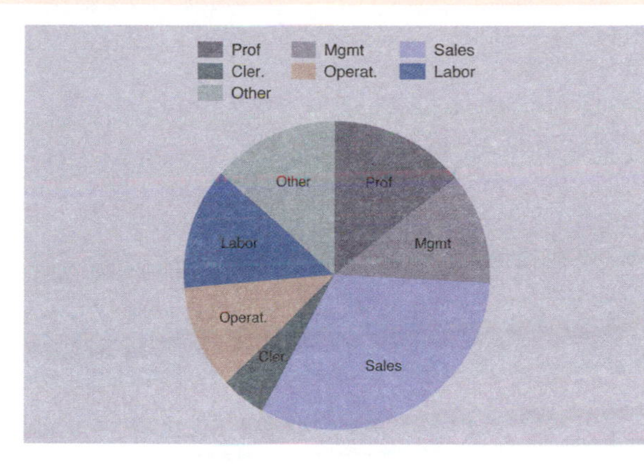


```
graph pie, over(occ7) plabel(_all name) legend(off)
```

When the name option is used, the legend is not as necessary and can be suppressed using the legend(off) option.
Uses nlsw.dta & scheme economist

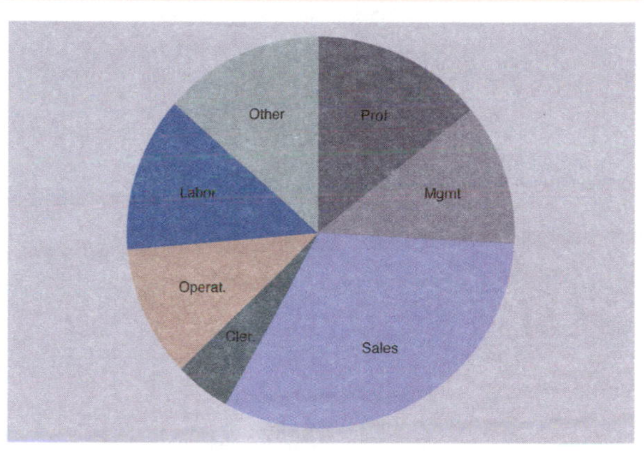


```
graph pie, over(occ7) plabel(1 "Prof=14.11")
   plabel(3 "Sales=32.32%")
```

The plabel() option can also be used to put any text that you want into all slices or into individual slices. Here, we add text to the first and third slices.
Uses nlsw.dta & scheme economist

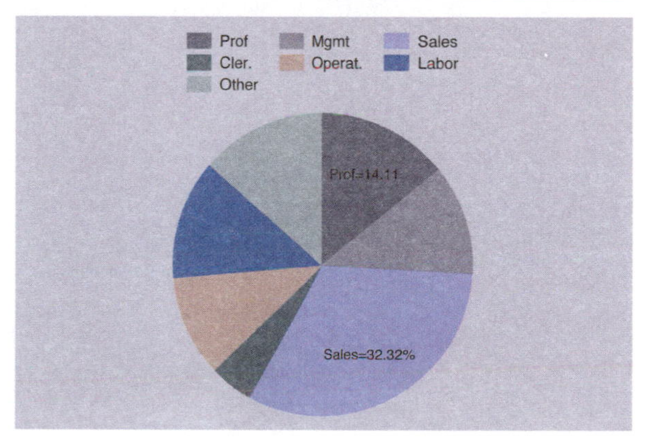

```
graph pie, over(occ7) plabel(_all percent, format("%2.0f"))
```

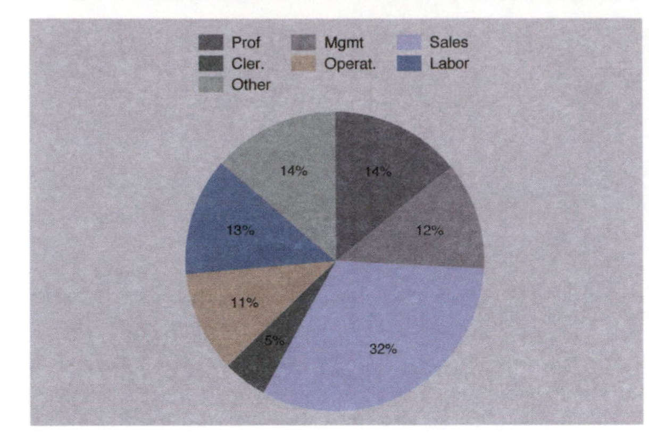

When you use `plabel` to label slices with a `sum` or `percent`, you can use the `format()` option to control the format of the numeric values displayed. Here, we display the percentages as whole numbers.
Uses nlsw.dta & scheme economist

```
graph pie, over(occ7) plabel(_all percent, gap(-5))
```

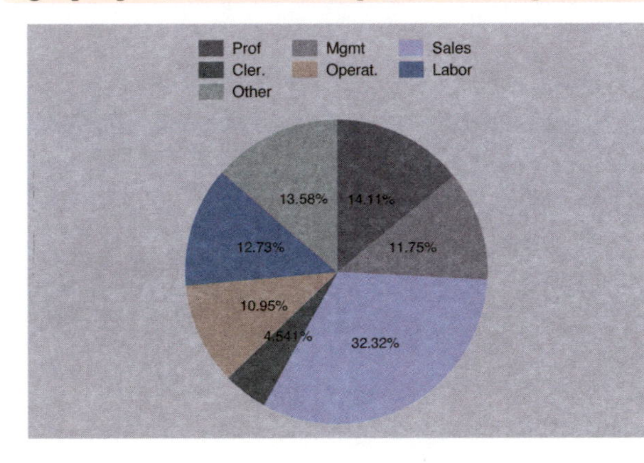

You can use the `gap()` option to adjust the position of the label with respect to the center of the pie. A positive number pushes the label away from the center of the pie, and a negative value pushes the label closer to the center of the pie.
Uses nlsw.dta & scheme economist

```
graph pie, over(occ7) plabel(_all percent, size(large) color(maroon))
```

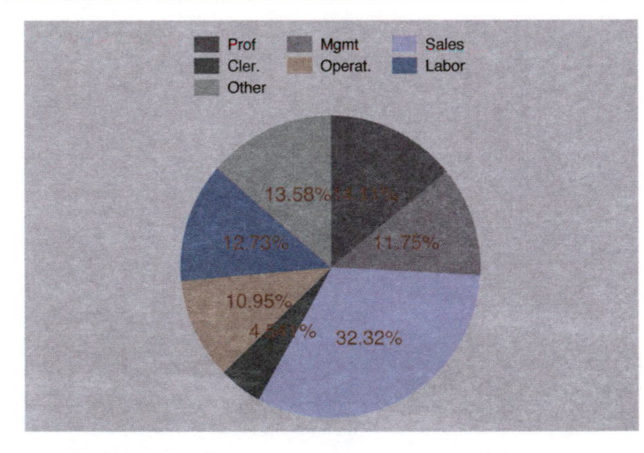

You can use textbox options to modify the display of the text labeling the pie slices. Here, we increase the size of the text and change its color to maroon. See Options : Textboxes (303) for more options you can use.
Uses nlsw.dta & scheme economist

```
graph pie, over(occ7) plabel(_all name, gap(-5))
   plabel(1 "32%", gap(5)) legend(off)
```

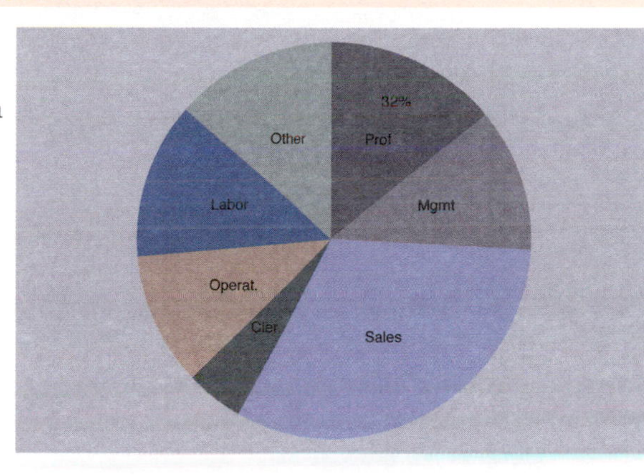

We can include multiple plabel() options. In this example, the first plabel() option assigns the occupation names to all the slices and moves the names 5 units inward. The second plabel() option assigns text to the second slice and displays it 5 more units from the center. Since the legend was not needed, we suppressed it with the legend(off) option.
Uses nlsw.dta & scheme economist

```
graph pie, over(occ7) plabel(_all name, gap(-5))
   plabel(_all percent, gap(5) format("%2.0f"))
   legend(off)
```

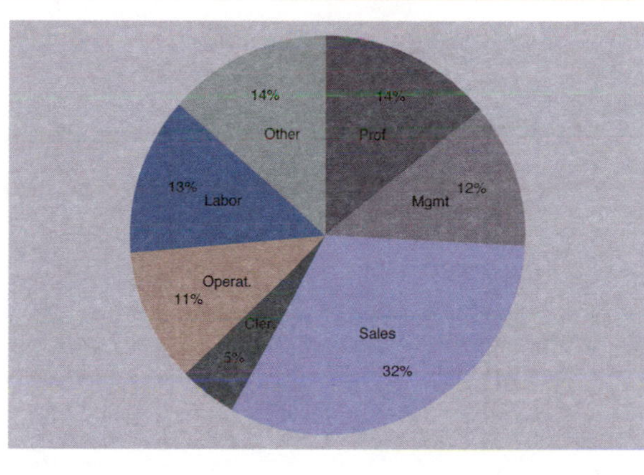

Here, we use the plabel() option twice to label the slices with the occupation name and with the percentage. We use the gap() option to move the names closer to the center by 5 extra units and move the percentage 5 extra units from the center.
Uses nlsw.dta & scheme economist

```
graph pie, over(occ7) ptext(0 30 "This is some text")
```

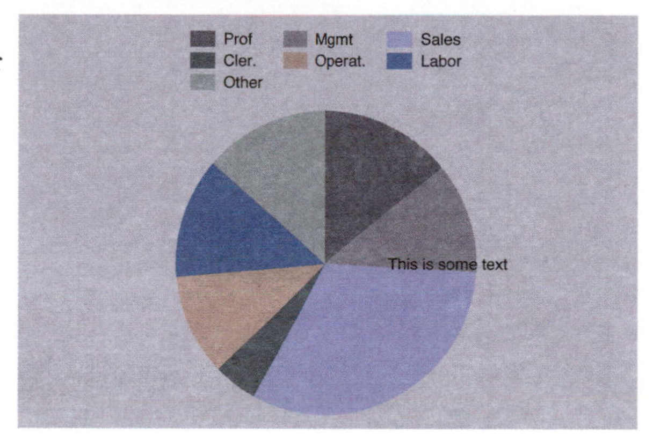

The ptext() (pie text) option can be used to add text to the pie chart. Polar coordinates are used to determine the location of the text by specifying the angle and distance from the center. Here, the angle is 0, and the distance from the center is 30.
Uses nlsw.dta & scheme economist

```
graph pie, over(occ7) ptext(-10 10 "This is some text")
```

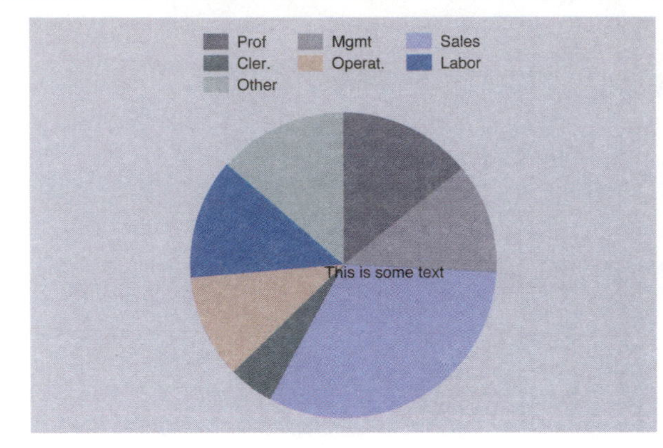

Here, we choose an angle of −10 (putting it 10 degrees below 0) and a distance of 10 units from the center. Note that the angle determines only the position of the text but not its actual angle of display, which is controlled with the `orientation()` option. See the next example for more details.
Uses nlsw.dta & scheme economist

```
graph pie, over(occ7)
   ptext(-10 10 "This is some text", orientation(rvertical)
   placement(s) box margin(medsmall) bfcolor(sand))
```

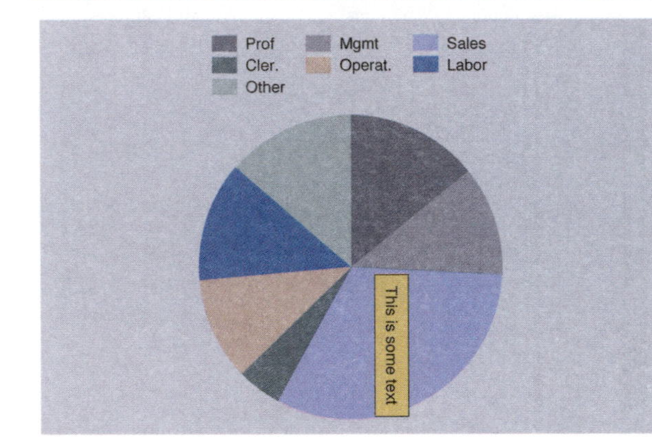

Here, we choose an angle of −10 degrees and a distance of 10 units. We also add a number of textbox options to make the text reverse vertical, meaning that it is placed to the south of the given coordinates, within a box that has with a medium-small margin and is filled with a sand color. For more information on these kinds of textbox options, see Options : Textboxes (303).
Uses nlsw.dta & scheme economist

7.5 Controlling legends

This section illustrates some of the options that you can use to control the display of legends with pie charts. While this section illustrates the use of legends, it emphasizes options that may be particularly useful with pie charts. See Options : Legend (287) for more details about legends; those details apply well to pie charts, even if the examples use other kinds of graphs. Also, see [G] ***legend_option*** for more details. We begin this section using the `vg_brite` scheme.

```
graph pie, over(occ7)
```

Consider this pie graph showing the frequencies of women in these seven occupational categories.
Uses nlsw.dta & scheme vg_brite

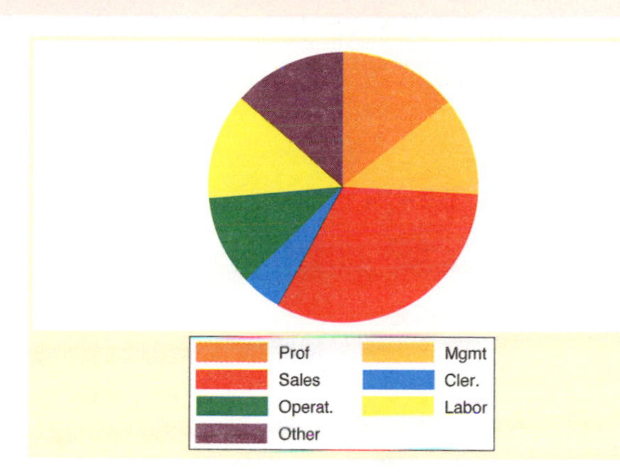


```
graph pie, over(occ7) legend(label(1 "Professional"))
```

We can use the `legend(label())` option to change the label for the first occupation.
Uses nlsw.dta & scheme vg_brite

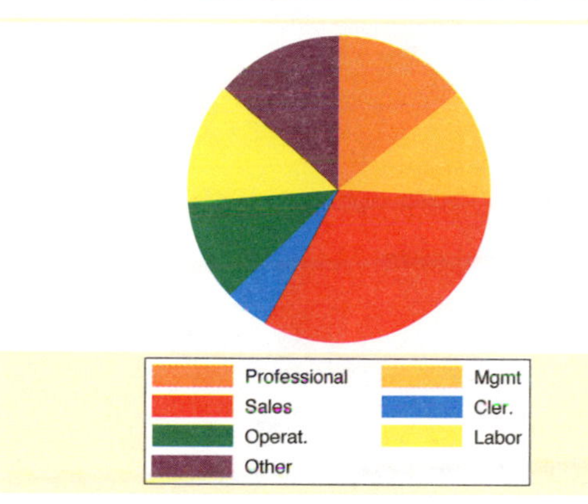


```
graph pie, over(occ7) legend(title(Occupation))
```

We can add the `title()` option to the `legend()` option to add a title to the legend. In fact, we can also use `subtitle()`, `note()`, and `caption()` options as well, much as we would for adding titles to a graph; see Standard options : Titles (313) for more details.
Uses nlsw.dta & scheme vg_brite

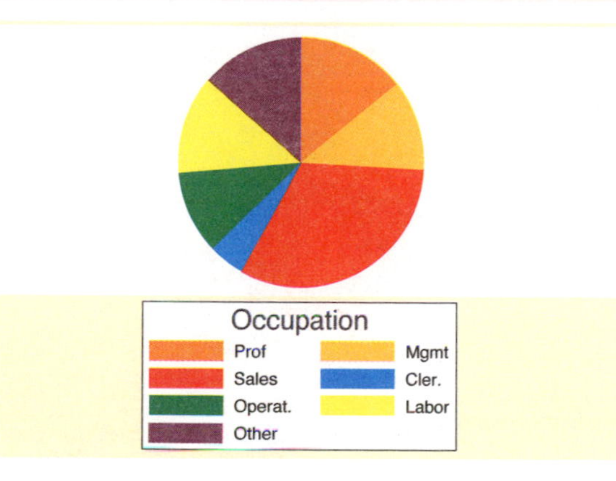

```
graph pie, over(occ7) legend(title(Occupation, position(6)))
```

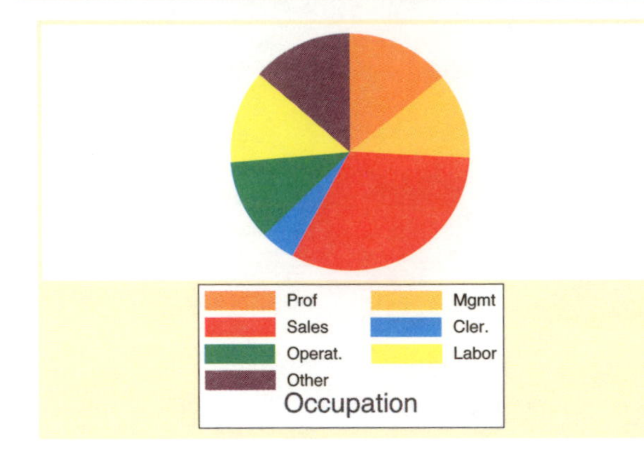

We can use the `position()` option within the `title()` option to control the position of the title. Here, we put the title in the 6 o'clock position, placing it at the bottom of the legend.
Uses nlsw.dta & scheme vg_brite

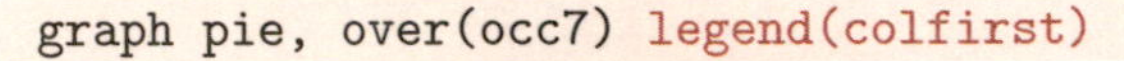

```
graph pie, over(occ7) legend(colfirst)
```

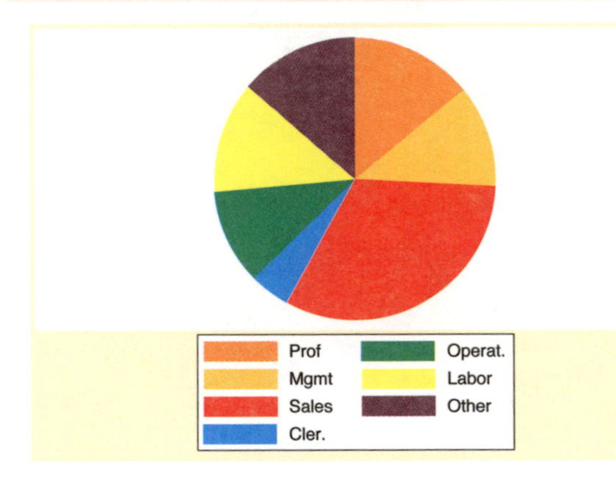

We can use the `legend(colfirst)` option to order the items in the legend by columns instead of rows.
Uses nlsw.dta & scheme vg_brite

```
graph pie, over(occ7) legend(colfirst order(7 6 5 1 2 3 4) holes(1))
```

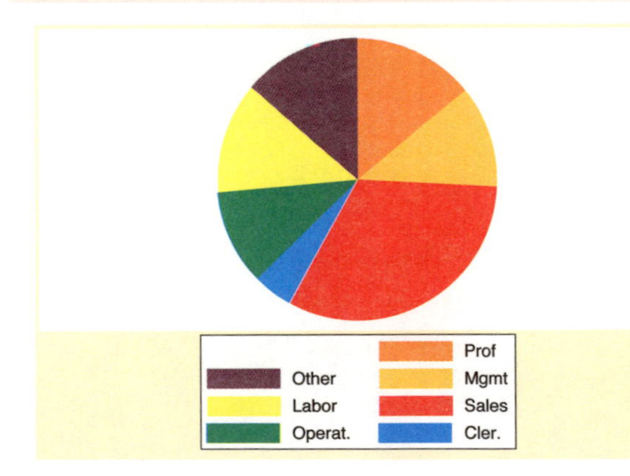

The pie wedges rotate clockwise, and here we make the items within the legend rotate in a similar clockwise fashion, starting from the top right. The `order()` option puts the items in the legend in a clockwise order, and the `holes(1)` option leaves the first position empty.
Uses nlsw.dta & scheme vg_brite

```
graph pie, over(occ7) legend(position(12) rows(2))
```

We can use the position() option to control the position of the legend, indicating its position like the numbers on a clock face; see Styles : Clockpos (330). Here, we put the legend at the 12 o'clock position, placing it at the top of the chart, and also add the rows(2) option to make the legend display in two rows.
Uses nlsw.dta & scheme vg_brite

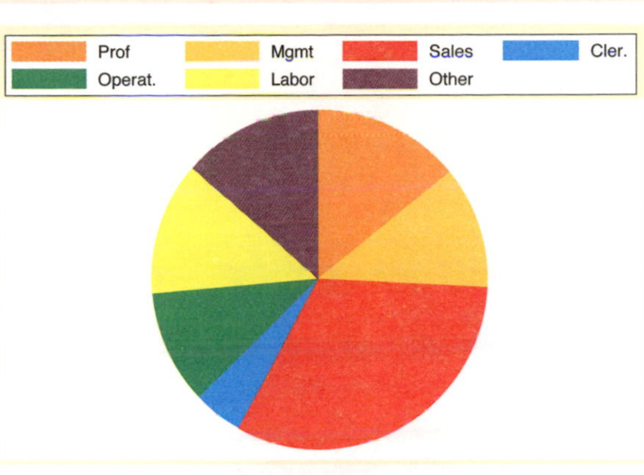


```
graph pie, over(occ7) legend(position(9) cols(1) stack)
```

Here, we use the same options as those in the last example but use them to place the legend to the left of the graph (in the 9 o'clock position) and make the legend display in a single column. We also add the stack option to the previous example to stack the symbol and descriptive text above each other. This makes an even narrower column, leaving more room for the pie chart.
Uses nlsw.dta & scheme vg_brite

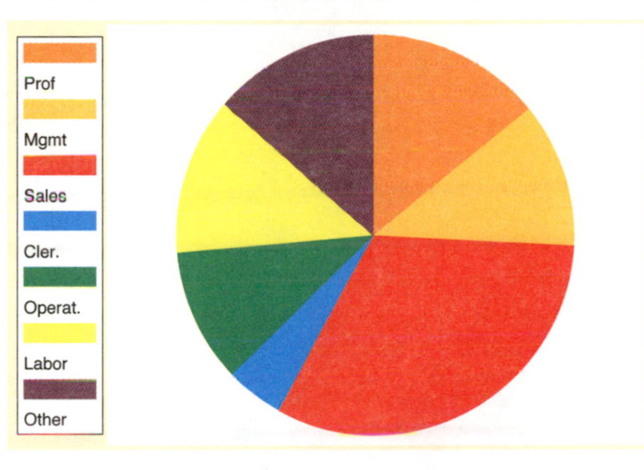


```
graph pie, over(occ7)
```

Here, we use the vg_lgndc scheme. Using this scheme places the legend to the left in a single column with the symbol stacked above the description.
Uses nlsw.dta & scheme vg_lgndc

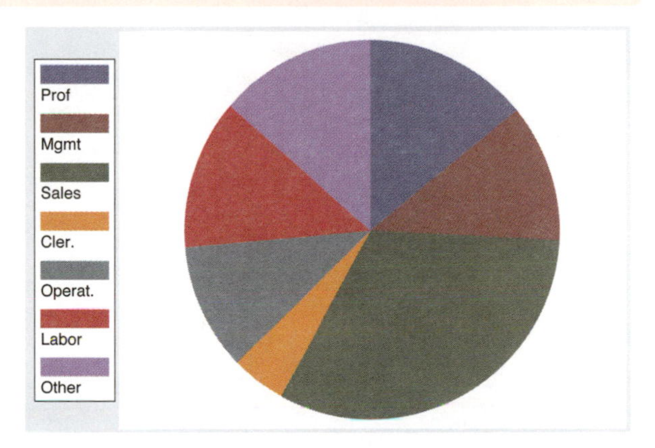

Introduction Twoway Matrix Bar Box Dot Pie Options Standard options Styles Appendix
Types of pie graphs Sorting Colors and exploding Labels Legend By

7.6 Graphing by groups

This section describes the use of the by() option with pie charts, focusing on features that are specifically relevant to pie charts. For more details, see Options : By (272) and [G] ***by_option***.

```
graph pie, over(occ7)
```

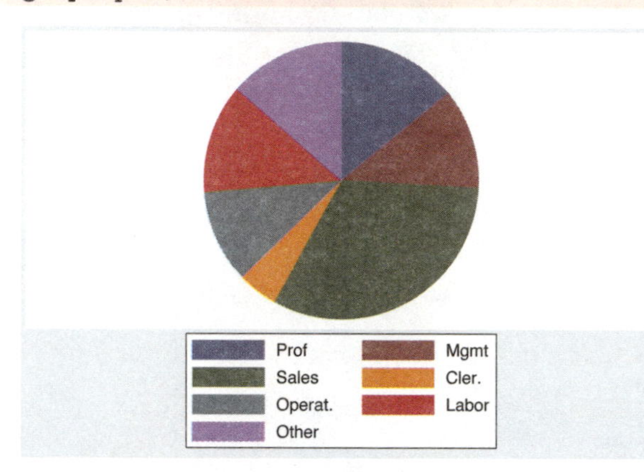

Here, we see a basic pie chart showing the distribution of occupations.
Uses nlsw.dta & scheme vg_s2c

```
graph pie, over(occ7) by(union)
```

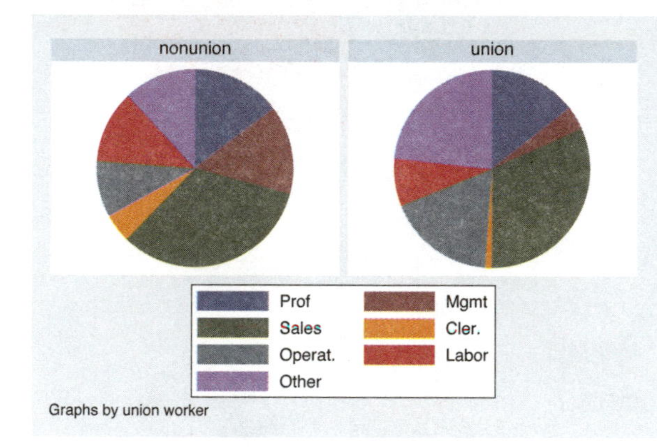

In this graph, the occupations are broken down by whether one belongs to a union.
Uses nlsw.dta & scheme vg_s2c

```
graph pie, over(occ7) by(union) pie(2, explode)
```

If we add the `pie(2, explode)` option, the second slice is exploded in both graphs.
Uses nlsw.dta & scheme vg_s2c

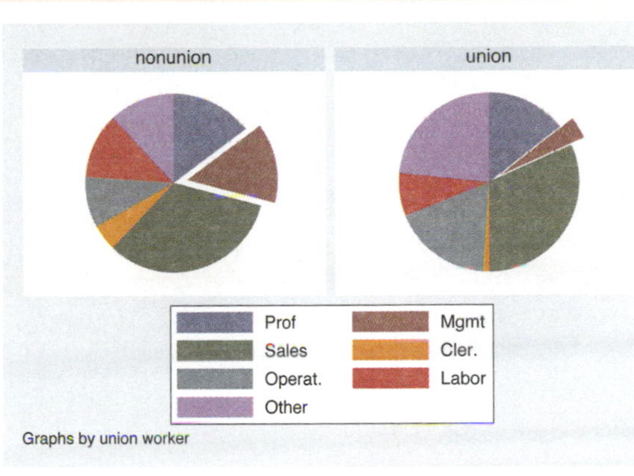


```
graph pie, over(occ7) by(union) sort
```

Here, we sort the slices from least frequent to most frequent. Note that separate legends are shown for each chart. This is because the slices can be ordered differently in the two different graphs when sorted. Thus, when you use the `sort` option for pie charts, Stata shows two separate legends to assure proper labeling of the slices.
Uses nlsw.dta & scheme vg_s2c

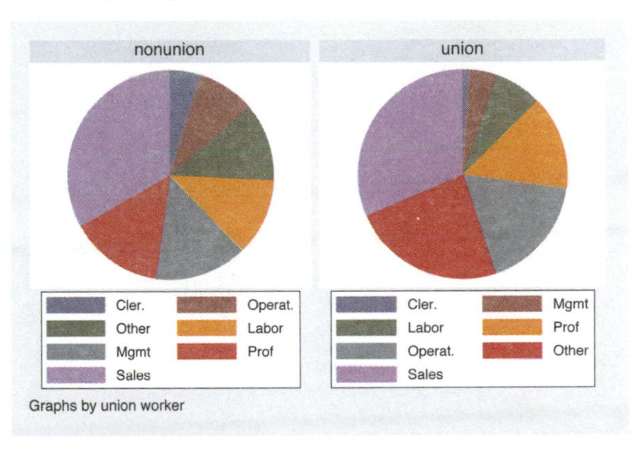


```
graph pie, over(occ7) by(union, legend(off))
    plabel(_all name)
```

Here, we add the `plabel()` option to label the inside of each slice with the name of the slice, so the legend is no longer needed. We suppress the legend with the `legend(off)` option, which is placed within the `by()` option because it, in a way, determines the placement of the legend by turning it off.
Uses nlsw.dta & scheme vg_s2c

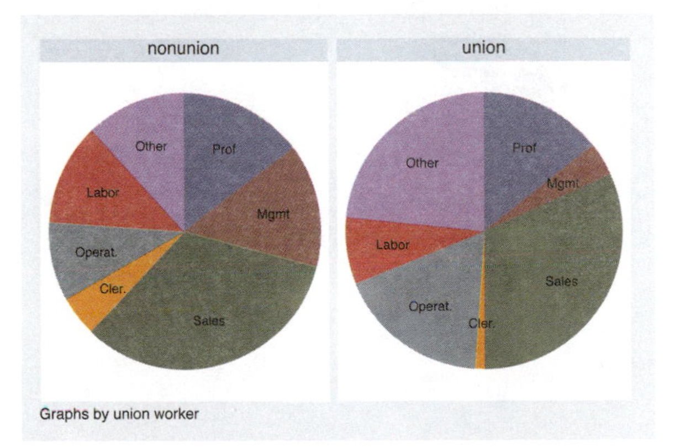

```
graph pie, over(occ7) by(union, legend(pos(3))) legend(cols(1) stack)
```

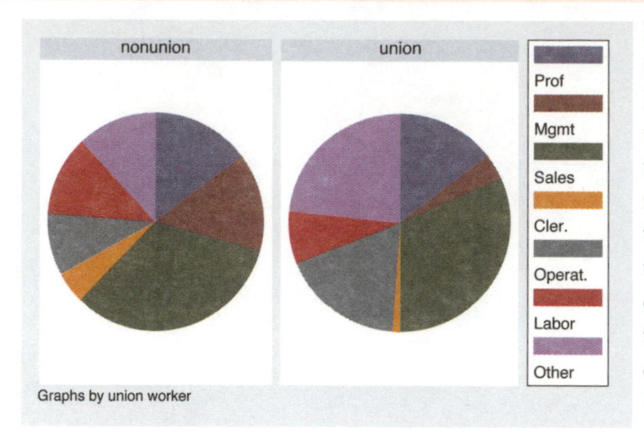

Here, we place the legend to the right using the `legend(pos(3))` option. Note that this option is contained within the `by()` option because it alters the position of the legend. We also make the legend a single column with the legend symbols and labels stacked with the `legend(cols(1) stack)` option. Note this option is outside of the `by()` option since it does not determine the position of the legend.
Uses nlsw.dta & scheme vg_s2c

```
graph pie, over(occ7) by(union) legend(pos(3) cols(1) stack) sort
```

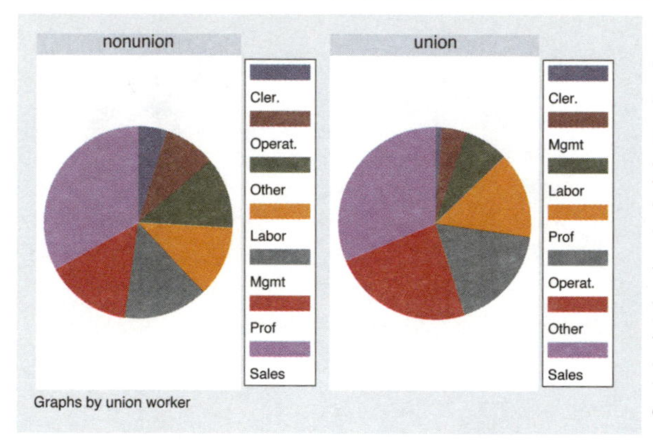

This example is similar to the previous example, but we have added the `sort` option. Note that, when we add the `sort` option, we need to move the `pos()` option from within the `by()` option to outside of the `by()` option. This is an exception to the general rule that legend options that control the position of the legend are placed within the `by()` option. Here, we get the legends that we desire, each to the right of the pie.
Uses nlsw.dta & scheme vg_s2c

```
graph pie, over(occ7) by(urban3, legend(at(4)))
```

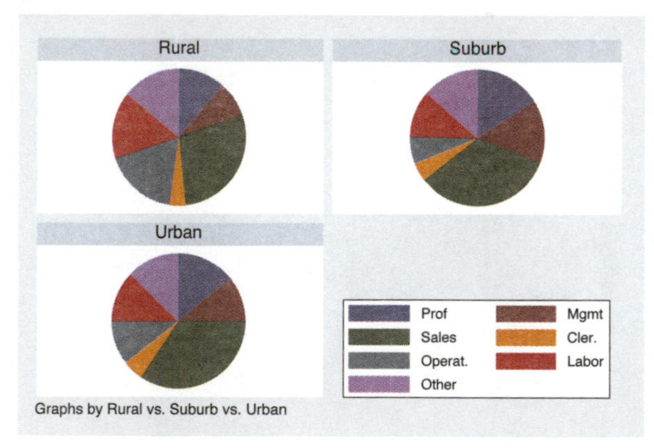

Here, we break down occupation by a three-level variable, leaving a fourth position open. We can specify the `legend(at(4))` option within the `by()` option to place the legend in the space in the fourth position, conserving space on the graph.
Uses nlsw.dta & scheme vg_s2c

8 Options available for most graphs

This chapter discusses options that are used in many, but not all, kinds of graphs in Stata, as compared with the Standard options (313) chapter, which covers options that are standard in all Stata graphs. This chapter goes into greater detail about how to use these options to customize your graphs. As you can see from the *Visual Table of Contents* at the right, this chapter covers markers, connecting, axis titles, labels, scales, selection, using the `by()` option, legends, added text, and textboxes. For further details, the examples will frequently refer to sections of Styles (327) and to [G] **graph**.

8.1 Changing the look of markers

This section looks at options that we can use for controlling markers. While the examples in this section focus on `twoway scatter`, these options apply to any graph where you have markers and can control them. This section will show how to change the marker symbol, marker size, and color (both fill and outline color). For more information, see [G] ***marker_options***. We will start this section using the `vg_s2c` scheme.

```
twoway scatter ownhome borninstate
```

Consider this scatterplot showing the relationship between the percentage of people in a state who own their home and the percentage of people born in their state of residence. The markers used in this plot are filled circles.
Uses allstates.dta & scheme vg_s2c

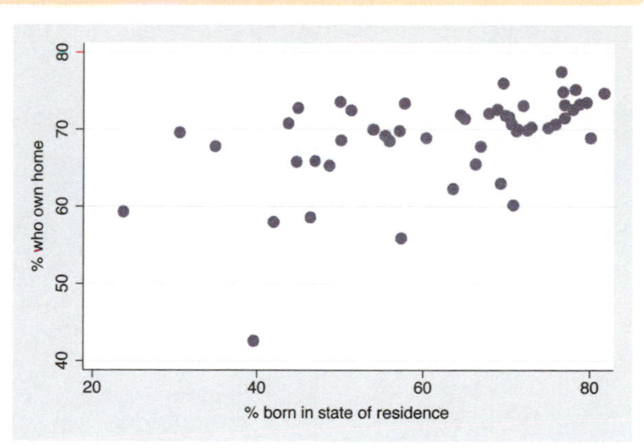

```
twoway scatter ownhome borninstate, msymbol(S)
```

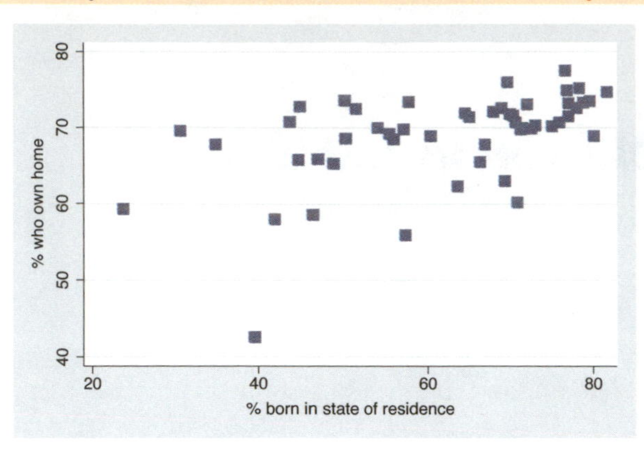

We can control the shape of the marker with the `msymbol()` (marker symbol) option. Here, we make the symbols large squares.
Uses allstates.dta & scheme vg_s2c

```
twoway scatter ownhome borninstate, msymbol(s)
```

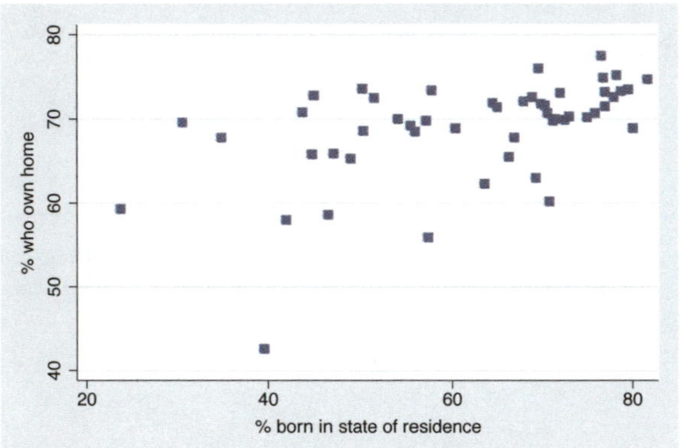

Specifying `msymbol(s)`, which uses a lowercase `s`, displays smaller squares.
Uses allstates.dta & scheme vg_s2c

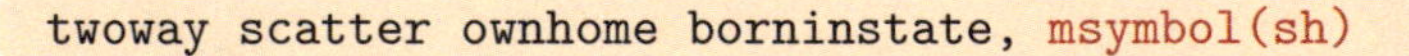

```
twoway scatter ownhome borninstate, msymbol(sh)
```

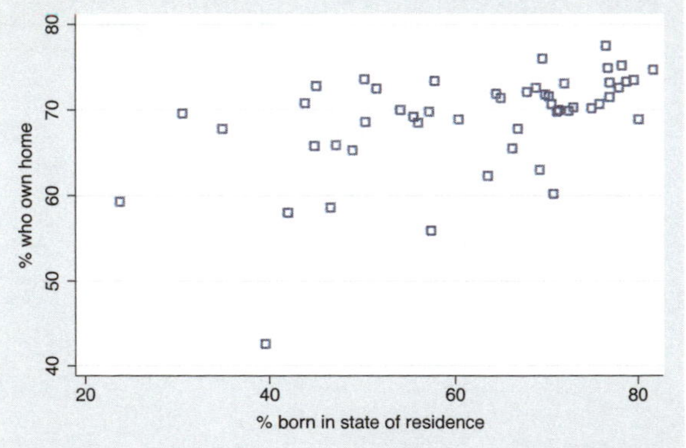

We can append an `h` (i.e., `msymbol(sh)`) to yield hollow squares. In addition to choosing `S` for larger squares and `s` for small squares, we can specify `D` (large diamond), `T` (large triangle), and `O` (large circles). We can specify lowercase letters to get smaller versions of these symbols and append the `h` for hollow versions.
Uses allstates.dta & scheme vg_s2c

```
twoway scatter ownhome borninstate, msymbol(X)
```

We can also specify `msymbol(X)` to use a large X shape for the markers. We could also use a lowercase `x` for smaller markers. We cannot append an `h` since we cannot make a hollow X.
Uses allstates.dta & scheme vg_s2c

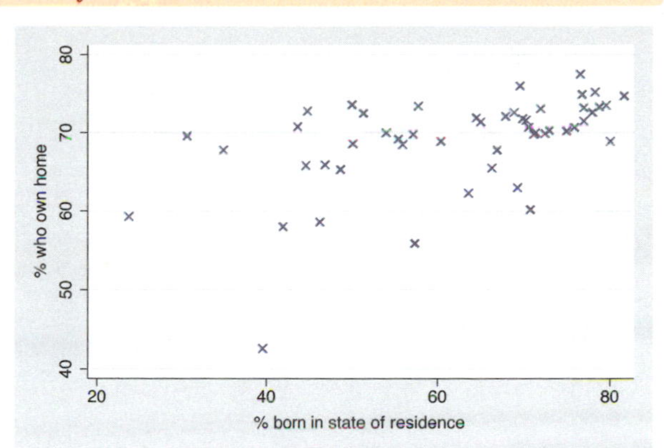


```
twoway scatter ownhome borninstate, msymbol(+)
```

Specifying `msymbol(+)` yields a plus sign shape for the markers. As with the `X`, we cannot make these hollow, nor is there a symbol for a smaller version of plus signs.
Uses allstates.dta & scheme vg_s2c

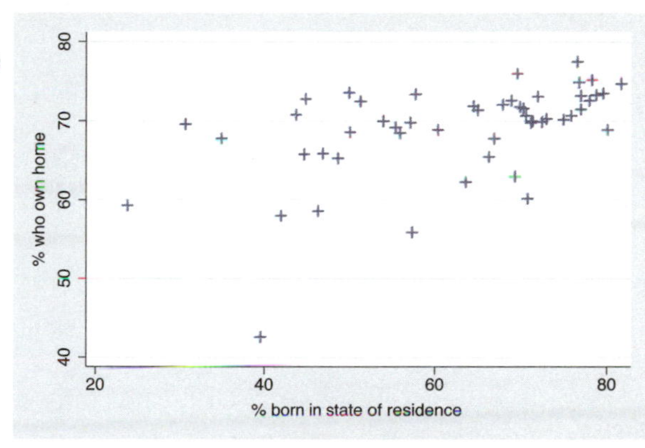


```
twoway scatter heatdd cooldd, msymbol(p)
```

Here, we switch to the `citytemp` data file to illustrate the use of the `msymbol(p)` option to plot very small points. Although each point is hard to see because they are so small, we can see the overall pattern of the data because of the large number of points and the strong trend in the data. See Styles : Symbols (342) for more information about symbols.
Uses citytemp.dta & scheme vg_s2c

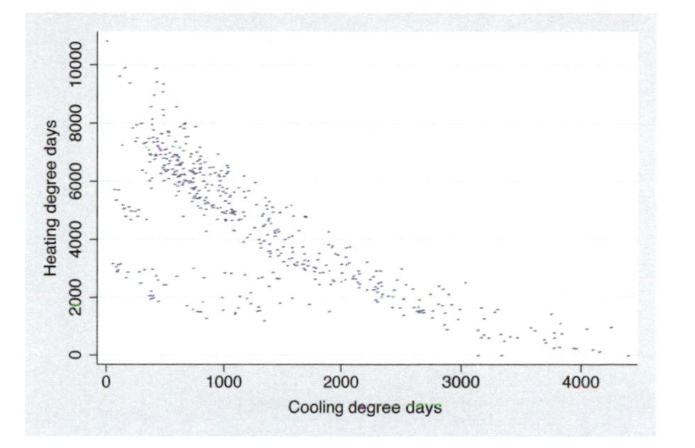

```
twoway scatter ownhome propval100 borninstate
```

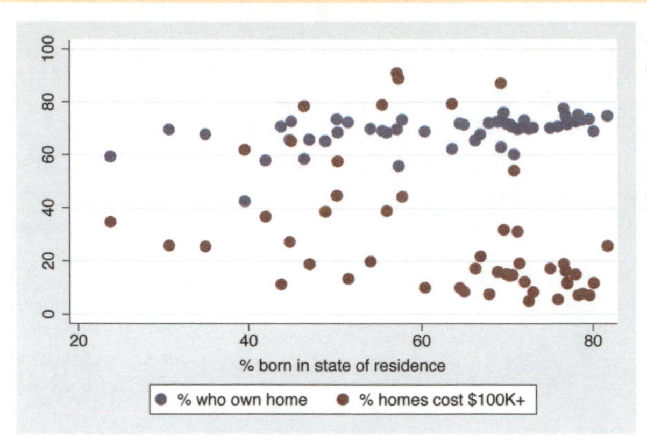

Aside from aesthetics, choosing different marker symbols is useful to differentiate multiple markers displayed in the same plot. In this example, we plot two y-variables, and Stata displays both as solid circles, differing in color.
Uses allstates.dta & scheme vg_s2c

```
twoway scatter ownhome propval100 borninstate,
   msymbol(t Oh)
```

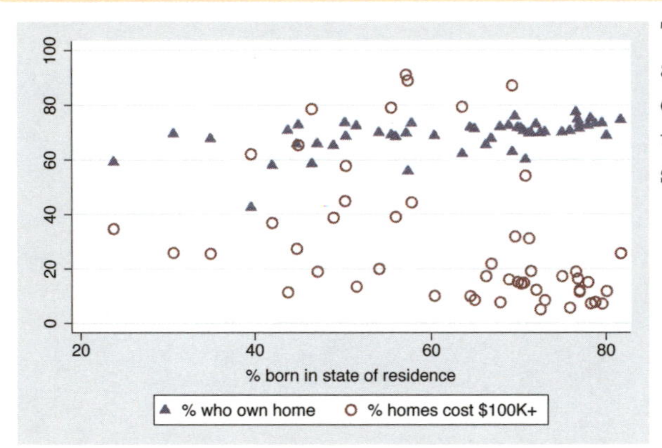

To further differentiate the symbols, we add the `msymbol(t Oh)` option to control both markers. Here, we make the first marker a small triangle and the second a larger hollow circle.
Uses allstates.dta & scheme vg_s2c

```
twoway scatter ownhome propval100 borninstate,
   msymbol(.  Oh)
```

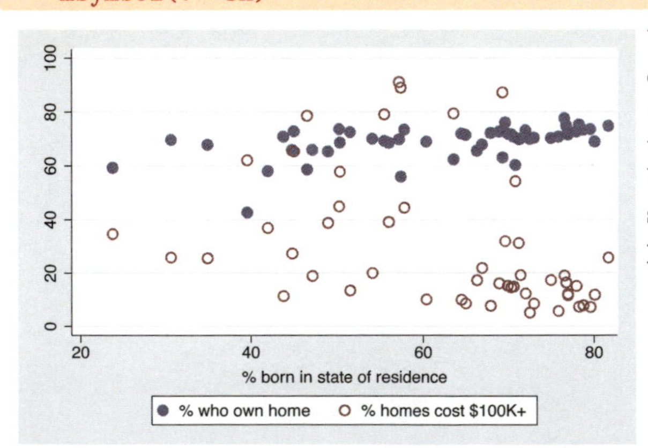

Using the `msymbol(.  Oh)` option, we can leave the first symbol unchanged (as indicated by the dot) and change the second symbol to a hollow circle. We might think that the dot indicates a small point, but that is indicated by the `p` option.
Uses allstates.dta & scheme vg_s2c

```
twoway scatter ownhome borninstate, mlabel(stateab) mlabpos(center)
```

One last marker symbol is `i` for invisible, allowing us to hide the marker symbol. In this example, we use the `mlabel(stateab)` (marker label) option to display a marker label with the state abbreviation for each observation and the `mlabpos(center)` (marker label position) option to center the marker label. However, the marker symbol (the circle) and the marker label (the abbreviation) are right on top of each other.
Uses allstates.dta & scheme vg_s2c

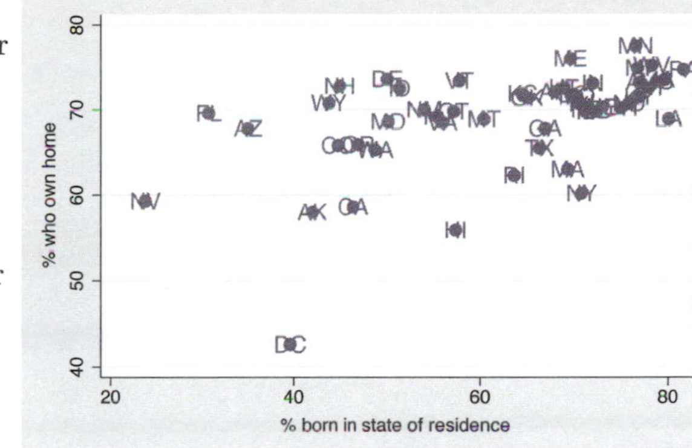


```
twoway scatter ownhome borninstate, mlabel(stateab) mlabpos(center)
   msymbol(i)
```

If we use `msymbol(i)` to make the marker symbol invisible, the marker label (the state abbreviation) can be displayed without being obscured by the marker symbol. See Styles : Symbols (342) for more information about symbols.
Uses allstates.dta & scheme vg_s2c

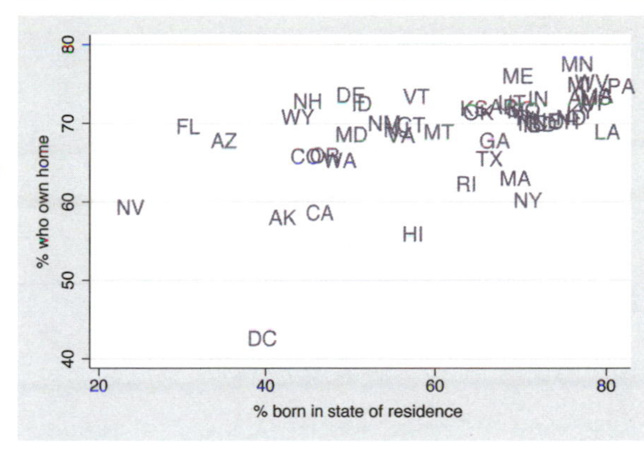

So far, we have seen that the `msymbol()` option can be used to control the marker symbol and, to a certain extent, can be used to control the marker size (e.g., using `O` yields large circles, and using `o` yields small circles). As the following examples show, the `msize()` option can be used to exert more flexible control over the size of the markers. The following examples will use the `vg_s1m` scheme.

```
twoway scatter ownhome borninstate, msymbol(+) msize(small)
```

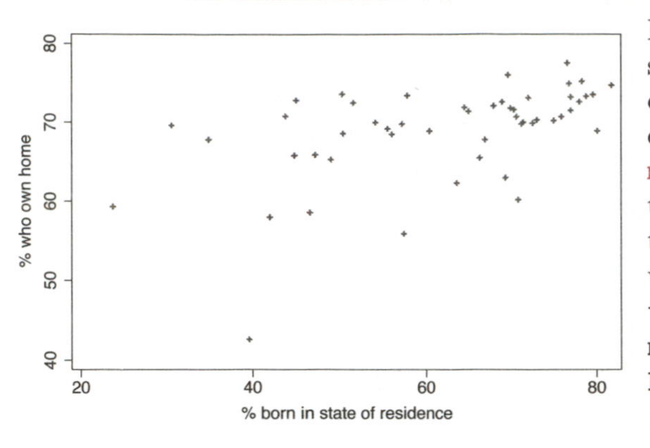

Previously, we saw that the size of the symbols created using `O`, `D`, `S`, and `T` could be modified using an uppercase or lowercase letter. Here, we use the `msize()` (marker size) option to control the size of the marker symbol, making the marker symbol small. Other values we could have chosen include `vtiny`, `tiny`, `vsmall`, `small`, `medsmall`, `medium`, `medlarge`, `large`, `vlarge`, `huge`, `vhuge`, and `ehuge`.
Uses allstates.dta & scheme vg_s1m

```
twoway scatter ownhome borninstate, msymbol(Oh) msize(*2)
```

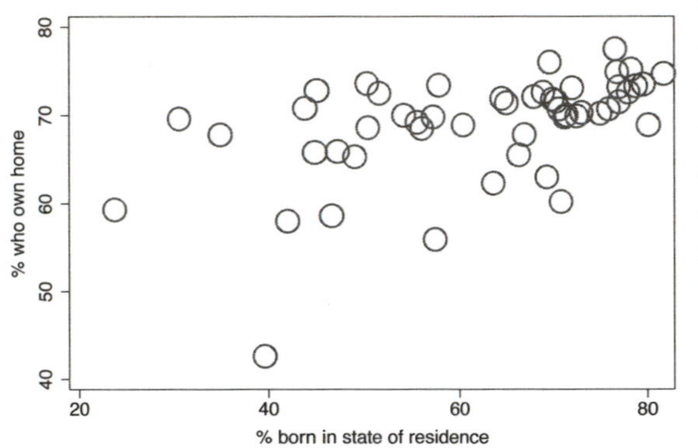

We can specify the sizes as multiples of the original size of the marker. In this example, we make the markers twice their original size by specifying `msize(*2)`. Specifying a value less than one reduces the marker size; e.g., `msize(*.5)`, would make the marker half its normal size. See Styles : Markersize (340) for more details.

Uses allstates.dta & scheme vg_s1m

```
twoway scatter ownhome borninstate [aweight=propval100],
   msymbol(oh)
```

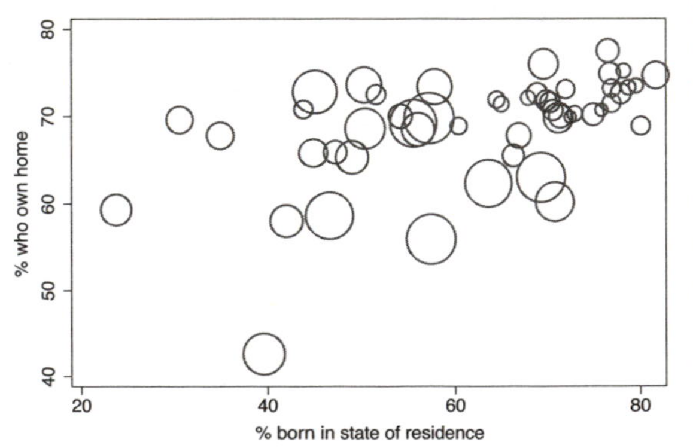

Stata even allows us to size the symbols based on the values of another variable in your data file. This allows us, in a sense, to graph three variables at once. Here, we look at the relationship between `borninstate` and `ownhome` and then size the markers based on `propval100` using `[aweight=propval100]`, weighting the markers by `propval100`.
Uses allstates.dta & scheme vg_s1m

```
twoway scatter ownhome borninstate [aweight=propval100],
    msymbol(oh) msize(small)
```

Even if we weight the size of the markers using **aweight**, we can still control the general size of the markers. Here, we make all markers smaller using the `msize(small)` option. The markers are smaller than they were previously but are still sized according to the value of `propval100`.
Uses allstates.dta & scheme vg_s1m

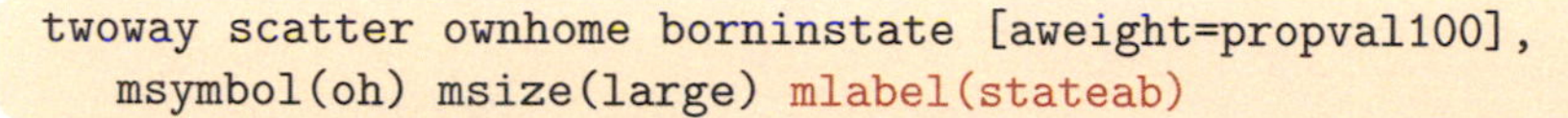

```
twoway scatter ownhome borninstate [aweight=propval100],
    msymbol(oh) msize(large) mlabel(stateab)
```

We might try to even graph a fourth variable in the plot by using the `mlabel()` (marker label) option. Here, we try to use the `mlabel(stateab)` option to label each marker with the abbreviation of the state. However, note that when we add the `mlabel()` option, the weights no longer affect the size of the markers. See the following example for a solution to this.
Uses allstates.dta & scheme vg_s1m

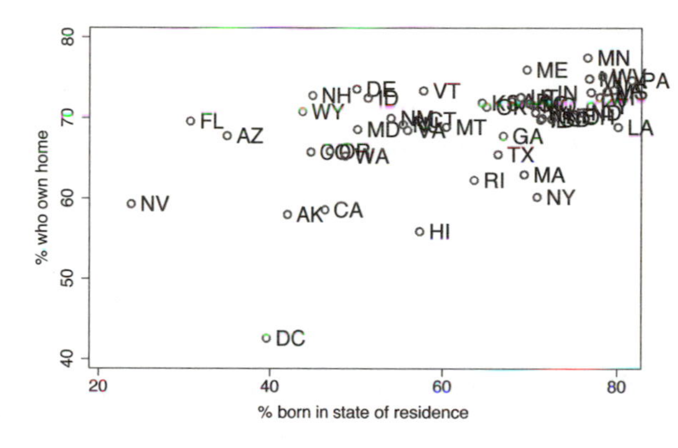


```
twoway (scatter ownhome borninstate [aweight=propval100],
    msymbol(oh) msize(large))
    (scatter ownhome borninstate, mlabel(stateab) msymbol(i) mlabpos(center))
```

We can solve the problem from the previous example by overlaying a scatterplot that has the symbols weighted by `propval100` with a scatterplot that shows just the marker labels. The second scatterplot uses the `mlabel(stateab) msymbol(i) mlabpos(center)` options to label the markers with the state abbreviation. See Options : Marker labels (247) for more details.
Uses allstates.dta & scheme vg_s1m

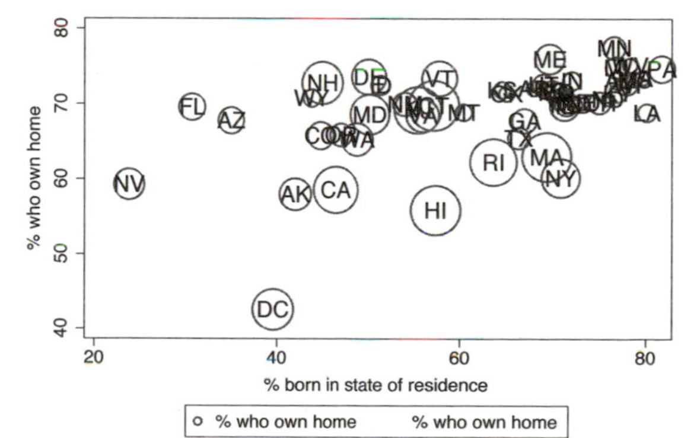

Introduction Twoway Matrix Bar Box Dot Pie Options Standard options Styles Appendix
Markers Marker labels Connecting Axis titles Axis labels Axis scales Axis selection By Legend Adding text Textboxes

Stata also allows us to control the color of the markers. We can control the overall color of the marker, create a solid color, or make the inner part of the marker one color (called a fill color) and the outline of the marker a different color. We can also vary the thickness of the outline of the marker. These next examples will use the `vg_rose` scheme.

```
twoway scatter ownhome borninstate, mcolor(navy)
```

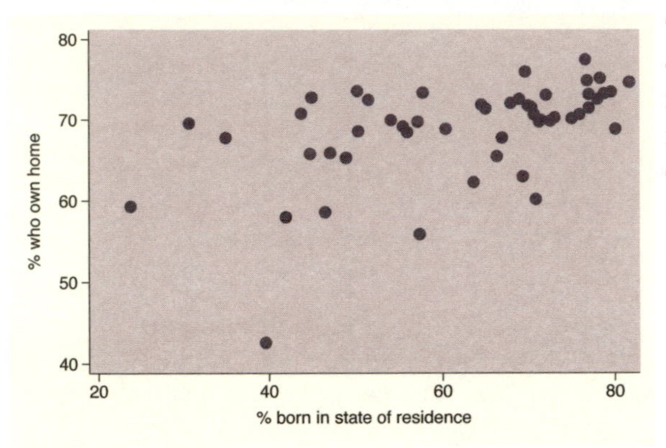

The `mcolor()` (marker color) option can be used to control the color of the markers. Here, we make the markers navy blue using the `mcolor(navy)` option. See Styles : Colors (328) for more information about specifying colors
Uses allstates.dta & scheme vg_rose

```
twoway scatter ownhome borninstate, mfcolor(ltblue) mlcolor(navy)
```

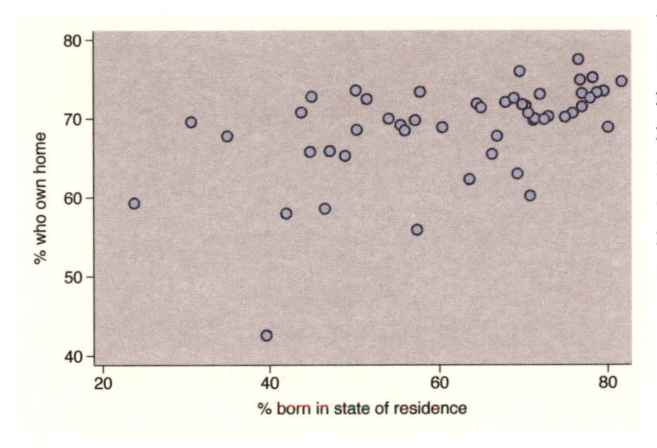

We can separately control the fill color (inside color) and outline color with the `mfcolor()` (marker fill color) and `mlcolor()` (marker line color) options, respectively. Here, we make the fill color light blue by specifying `mfcolor(ltblue)` and the line color navy by specifying `mlcolor(navy)`.
Uses allstates.dta & scheme vg_rose

```
twoway scatter ownhome borninstate, mlcolor(black)
```

We can change the line color surrounding the marker with the `mlcolor()` option. Here, we specify `mlcolor(black)` to make the line surrounding the markers black.
Uses allstates.dta & scheme vg_rose

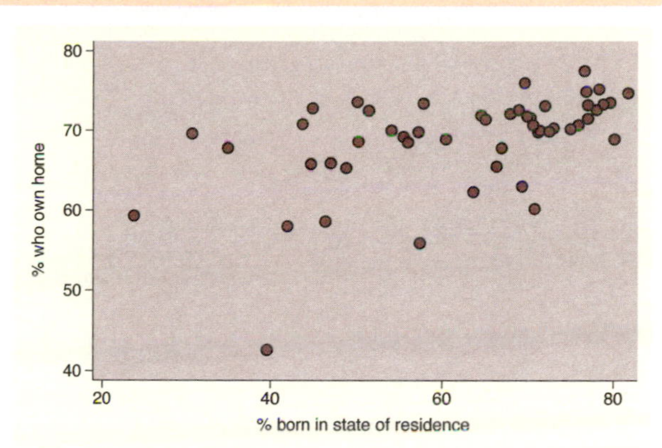


```
twoway scatter ownhome borninstate, mfcolor(ltblue)
```

We can also separately control the fill color using the `mfcolor()` option. If we choose `mfcolor(ltblue)`, the fill color is light blue.
Uses allstates.dta & scheme vg_rose

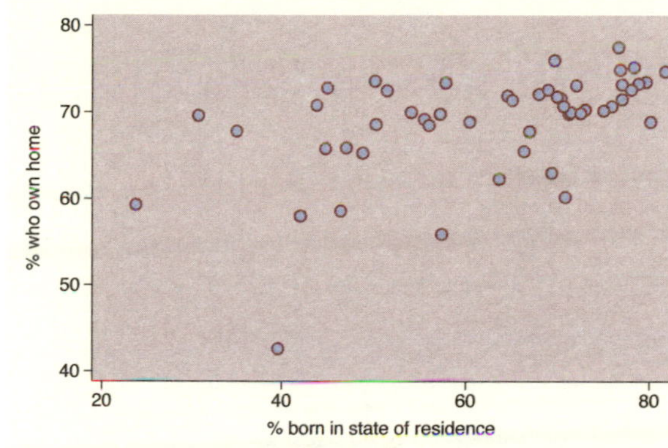


```
twoway scatter ownhome borninstate, mfcolor(eltgreen)
   mlcolor(dkgreen) mlwidth(vthick)
```

We can control the width of the line that surrounds the marker using the `mlwidth()` option. Here, we make the width very thick by specifying the `mlwidth(vthick)` (marker line width) option. We can also indicate the thickness as a multiple of the original thickness; e.g., `mlwidth(*3)` indicates the line should be three times as thick as it would normally be. See Styles : Linewidth (337) for more details.
Uses allstates.dta & scheme vg_rose

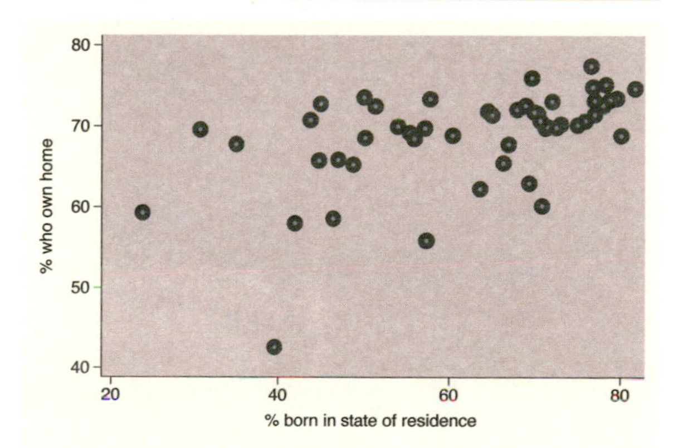

```
twoway scatter ownhome borninstate, mlwidth(medthick)
```

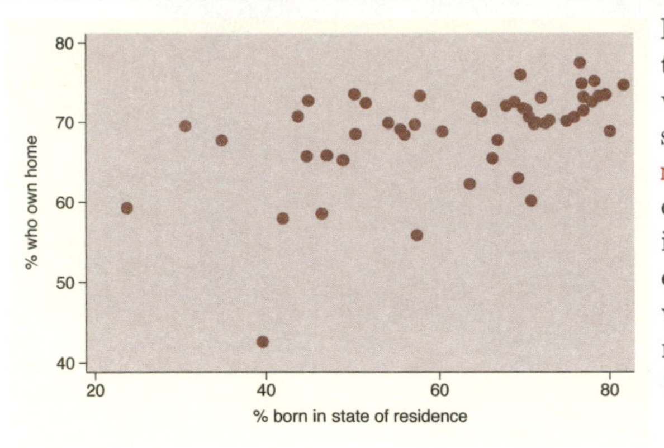

If we do not specify a different color for the line that outlines the marker (e.g., via the `mlcolor()` option), we may not see any effect in specifying the `mlwidth()` option. This is because the color of the line surrounding the marker is the same as the fill color, so we cannot see the effect of modifying the width of the line surrounding the marker, as illustrated here.
Uses allstates.dta & scheme vg_rose

So far, we have focused on controlling the individual elements of markers, the marker symbol, color, size, fill color, line color, and so forth. There is another way to change the appearance of a marker, and that is by specifying a marker style. The marker style controls all these attributes at once, and in some situations, it can be more efficient to use a marker style to control the elements individually, as we will see in the following examples. The next examples will use the `vg_s2m` scheme.

```
twoway scatter ownhome borninstate
```

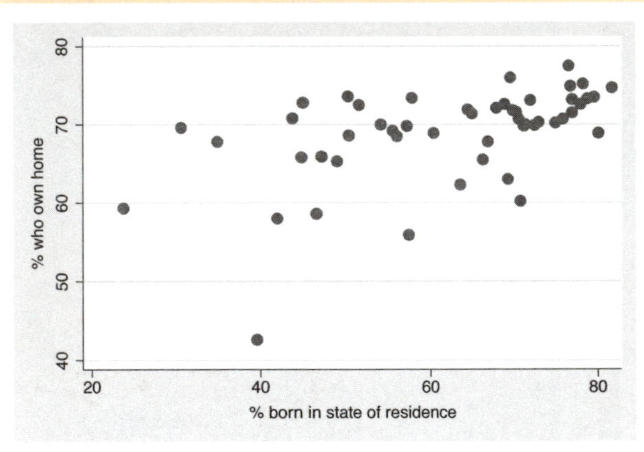

The marker styles are named/numbered `p1` to `p15`. The markers in this example are displayed using the `p1` style because we are plotting only one y-variable and have not specified a marker style.
Uses allstates.dta & scheme vg_s2m

```
twoway scatter ownhome borninstate, mstyle(p1)
```

Here, we explicitly select the default marker style using the `mstyle(p1)` (marker style) option, and the markers look identical to the previous graph.
Uses allstates.dta & scheme vg_s2m

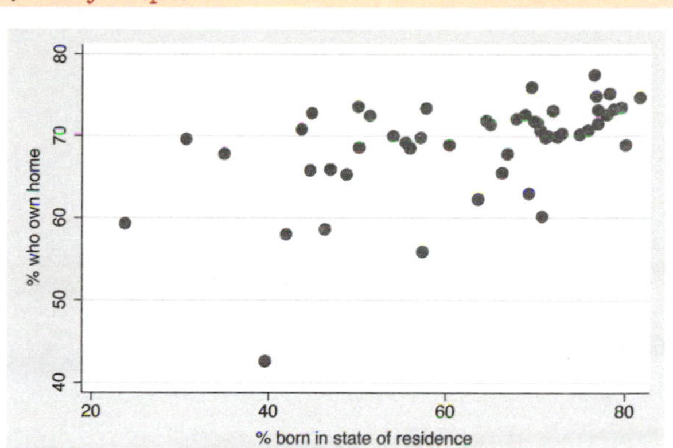


```
twoway scatter ownhome borninstate, mstyle(p2)
```

Here, we use `mstyle(p2)` to explicitly select the `p2` style for displaying the markers, and now the markers are different in size, shape, and color. The markers are now larger diamonds that are a middle-level gray color.
Uses allstates.dta & scheme vg_s2m

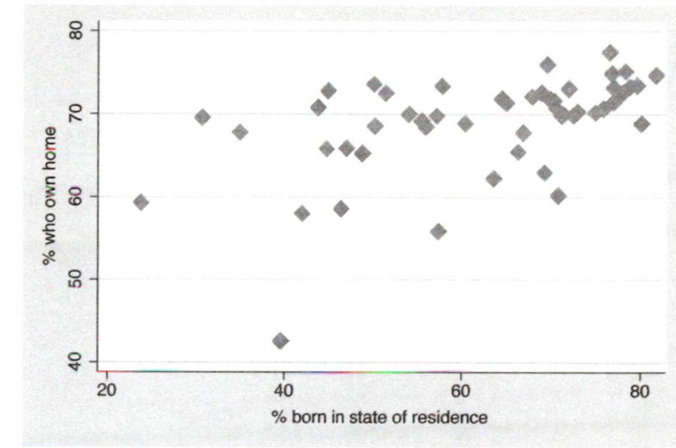


```
twoway scatter ownhome propval100 borninstate
```

Now, if we plot two variables, notice how the first variable is plotted using the `p1` style and the second variable is plotted using the `p2` style. We would have gotten the same result if we had specified the option `mstyle(p1 p2)`.
Uses allstates.dta & scheme vg_s2m

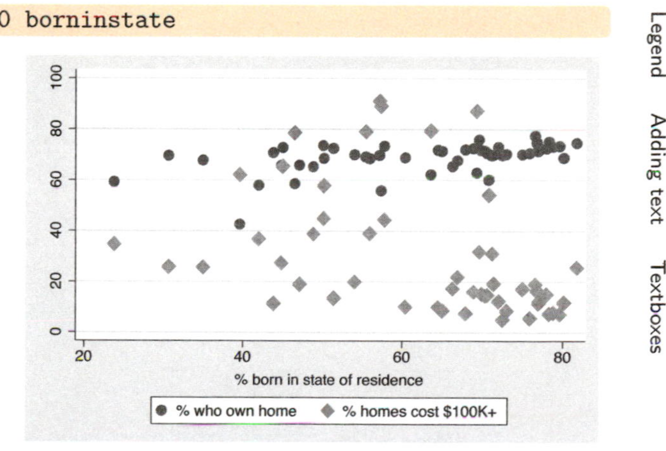

```
twoway scatter ownhome propval100 borninstate, mstyle(p1 p10)
```

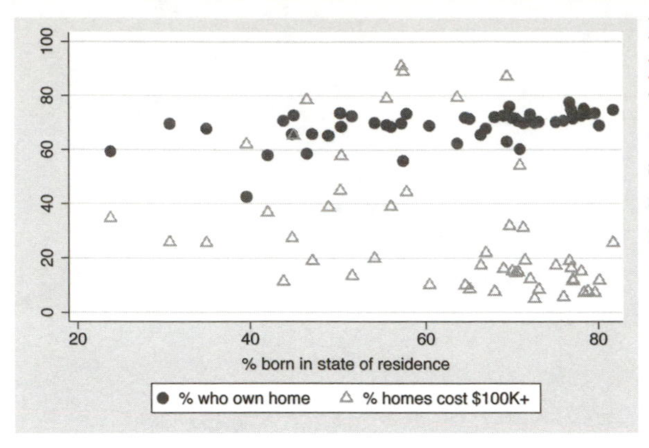

In this graph, we use the `mstyle(p1 p10)` option to request that the first variable be plotted with the `p1` style and the second be plotted with the `p10` style. A style is just a starting point, and we can use additional options to modify the markers to suit our taste.
Uses allstates.dta & scheme vg_s2m

```
twoway scatter ownhome propval100 borninstate, mstyle(p1 p10)
  msize(. medium)
```

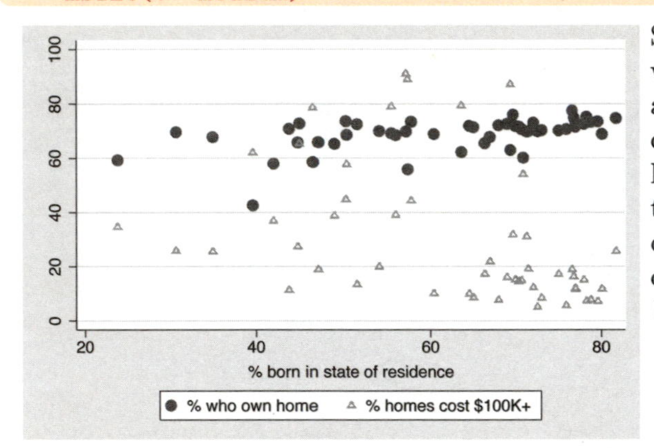

Say that in the previous graph you wanted medium-sized triangles. We can add the `msize(. medium)` option to control the size of the second marker, leaving the first unchanged. So, even though a style chooses a number of characteristics for the markers, we can override them.
Uses allstates.dta & scheme vg_s2m

```
twoway scatter ownhome propval100 borninstate, mstyle(p1 p1)
  mfcolor(. white)
```

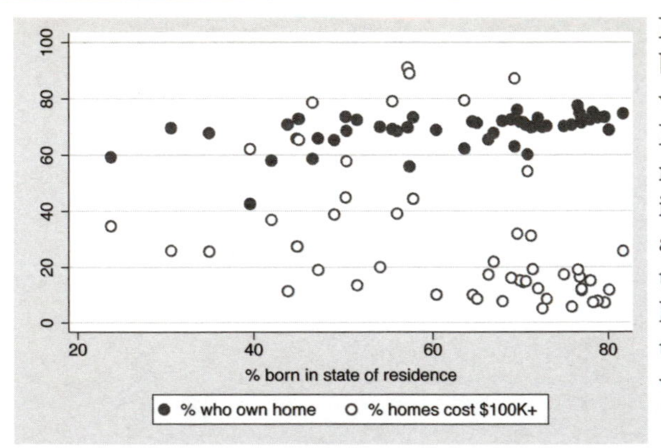

In this example, we use the `p1` style for both the first and second markers, which are small, dark gray, filled circles. If no other options are specified, the markers for the first variable will be identical to those for the second. But adding the `mfcolor(. white)` option, the fill color for the first variable was left alone, and the second was changed to white. This easily gave us solid and white-filled circles for the two markers.
Uses allstates.dta & scheme vg_s2m

```
twoway scatter ownhome propval100 borninstate, msymbol(.  Sh)
```

Another strategy for controlling the marker symbols is choosing or creating a scheme. The `vg_samem` scheme makes all markers the same size, shape, color, etc., allowing you to customize them all from a common base. Here, we use the `vg_samem` scheme, making all markers solid, dark gray circles, but use the `msymbol(.  Sh)` option to make the second symbol hollow squares.
Uses allstates.dta & scheme vg_samem

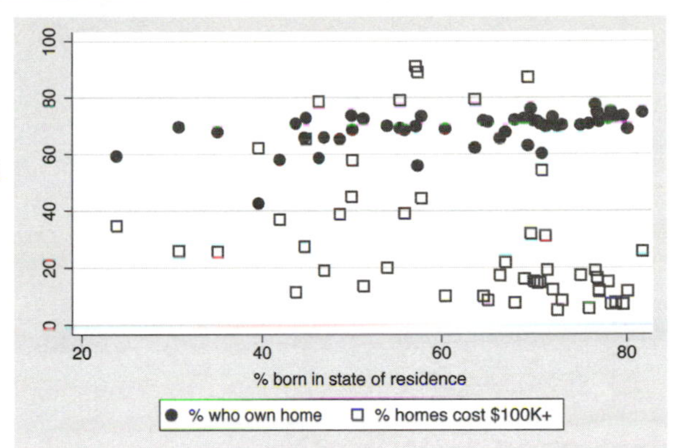


```
twoway scatter ownhome propval100 borninstate
```

Say that you wanted the markers to be displayed as outlines filled with white. Rather than specifying the `mfcolor()` option, you could use the `vg_outm` scheme, as shown here. Even if you overlaid multiple commands, using this scheme would display the markers, by default, as white-filled outlines.
Uses allstates.dta & scheme vg_outm

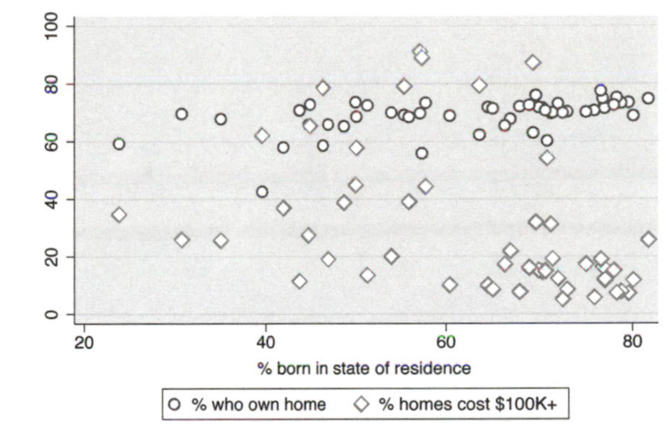

8.2 Creating and controlling marker labels

This section looks at the details of using marker labels. Marker labels can be used to identify the markers with `graph twoway` but also can be used with other types of graphs, such as `graph matrix` and `graph box`, affecting the outside values. You can even use marker labels in lieu of markers. For more information, see [G] ***marker_label_options***. For this section, we will use the `vg_s2c` scheme and the `allstates3` file, which keeps the states that are in the South, i.e., if `region` is equal to 3.

```
twoway scatter ownhome borninstate, mlabel(stateab)
```

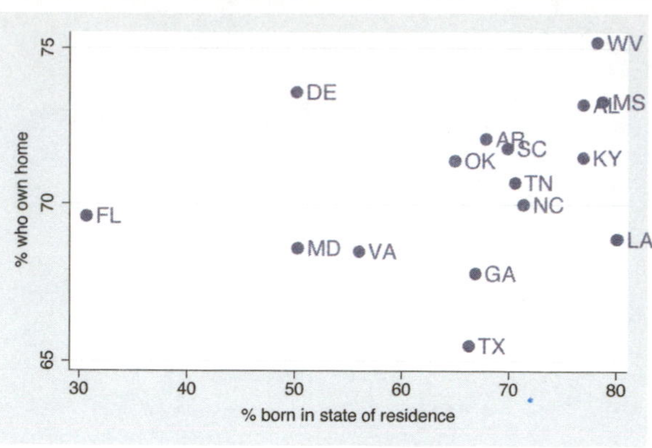

Consider this scatterplot showing the relationship between the percentage of people in a state who own their home and the percentage of people born in their state of residence. We might want to be able to identify some of the observations, and we can use the `mlabel()` (marker label) option to label the markers with the two-letter abbreviation of the state.
Uses allstates3.dta & scheme vg_s2c

```
twoway scatter ownhome borninstate, mlabel(stateab) mlabpos(12)
```

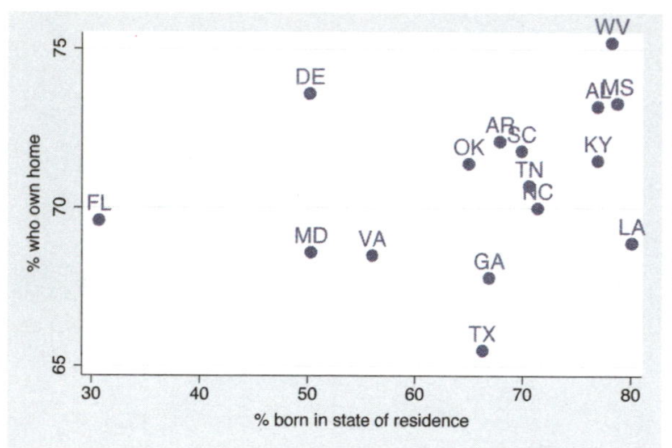

In the previous graph, the marker labels were all at the 3 o'clock position with respect to the markers. We can use the `mlabpos()` (marker label position) option to give the marker labels a different position. In this example, we place the marker labels in the 12 o'clock position above the markers.
Uses allstates3.dta & scheme vg_s2c

```
twoway scatter ownhome borninstate, mlabel(stateab) mlabvpos(pos)
```

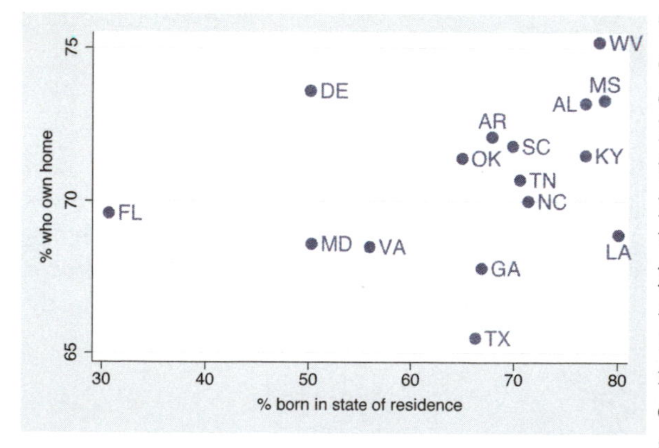

There are a few markers whose corresponding marker labels overlap each other. The `mlabvpos()` (marker label variable position) option allows us to assign a different marker label position for each observation via a variable in the data file. The variable `pos` has a value of 3, except for states AL, MS, AR, and LA, where `pos` is 9, 12, 12, and 6, respectively. Note how the markers are in the 3 o'clock position, except for AL, MS, AR, and LA, which are in the 9, 12, 12, and 6 o'clock positions, respectively.
Uses allstates3.dta & scheme vg_s2c

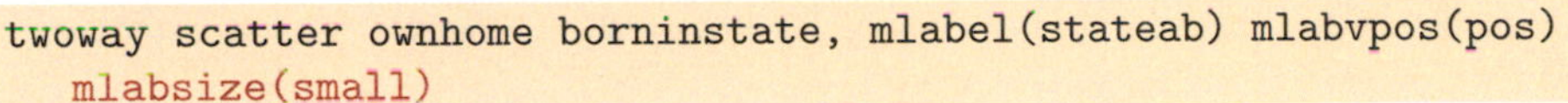

```
twoway scatter ownhome borninstate, mlabel(stateab) mlabvpos(pos)
   mlabsize(small)
```

We can use the `mlabsize()` (marker label size) option to control the size of the markers. In this example, we make the markers small. Some of the sizes you could choose include `small`, `medsmall`, `medium`, `medlarge`, `large`, and `vlarge`; see Styles : Textsize (344) for more options.

Uses allstates3.dta & scheme vg_s2c

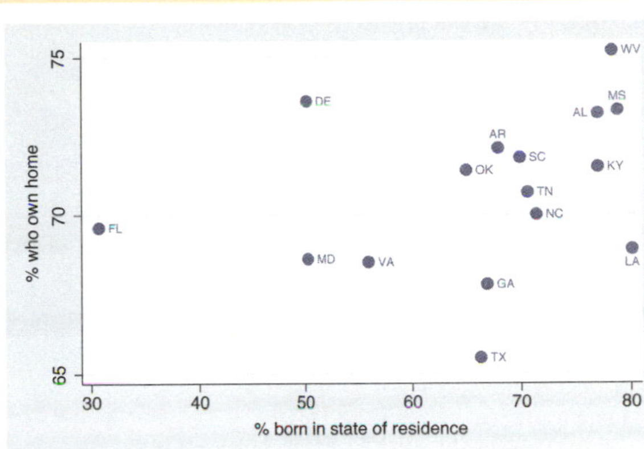


```
twoway scatter ownhome borninstate, mlabel(stateab) mlabvpos(pos)
   mlabsize(*.6)
```

We can also specify the `mlabsize()` as a relative size, a multiple of the original size. In this example, the labels are .6 times their normal size.

Uses allstates3.dta & scheme vg_s2c

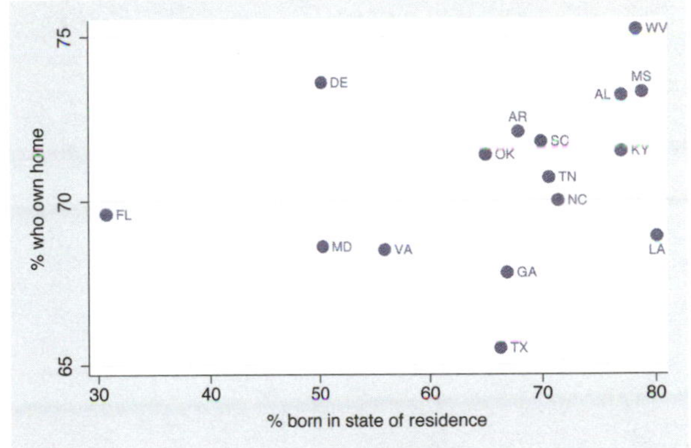


```
twoway scatter ownhome borninstate, mlabel(stateab) mlabangle(45)
```

The `mlabangle()` (marker label angle) option can be used to control the angle of the marker label. 0 degrees indicates horizontal text, 90 degrees vertical text, 180 degrees reverse horizontal text, and 270 degrees reverse vertical text. You can also specify negative degrees (for example, −90 degrees is the same as 270 degrees). See Styles : Angles (327) for more details.

Uses allstates3.dta & scheme vg_s2c

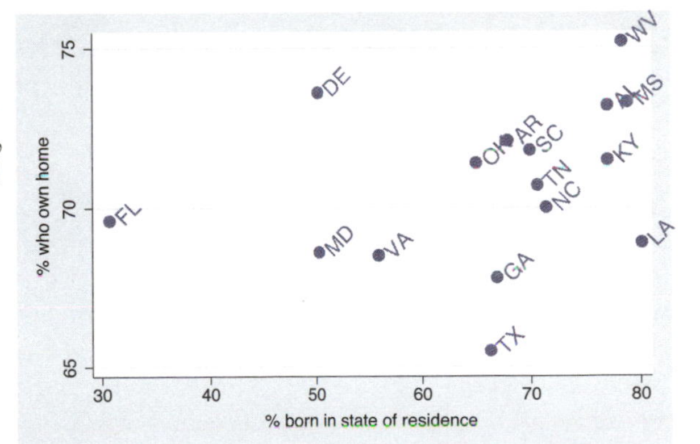

```
twoway scatter ownhome borninstate, mlabel(stateab) mlabpos(7)
   mlabcolor(red)
```

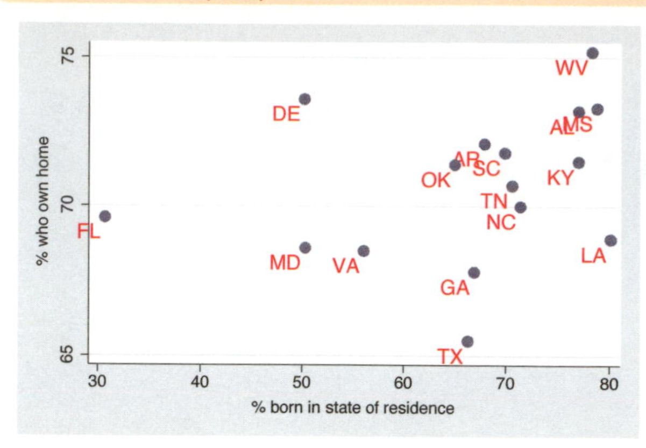

The `mlabcolor()` (marker label color) option controls the color of the marker labels. In this example, we make the marker labels red. See Styles : Colors (328) for more details.
Uses allstates3.dta & scheme vg_s2c

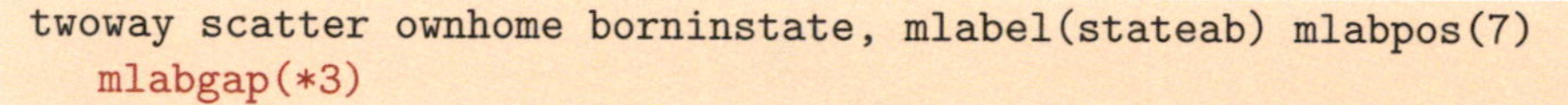

```
twoway scatter ownhome borninstate, mlabel(stateab) mlabpos(7)
   mlabgap(*3)
```

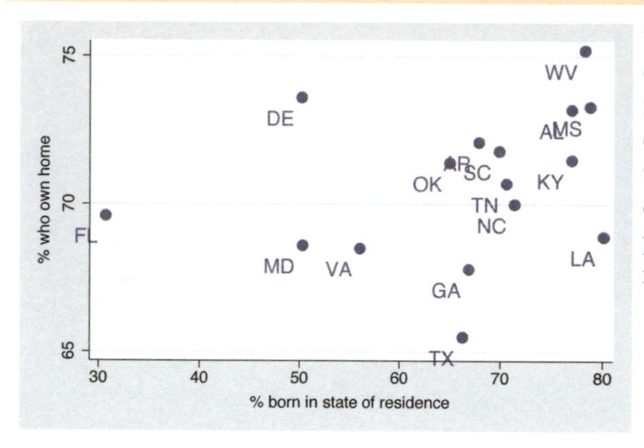

The `mlabgap()` (marker label gap) option controls the gap between the marker and the marker label. In this example, we make the gap three times the size that it would normally. You can also specify a value less than 1 to place the marker label closer to the marker.
Uses allstates3.dta & scheme vg_s2c

8.3 Connecting points and markers

Stata supports a variety of methods for connecting points using different values for the *connectstyle*. These include `l` to connect with a straight line, `L` to connect with a straight line only if the current x-value is greater than the prior x-value, `J` for stairstep, `stepstair` for step then stair, and `i` for invisible connections. For the next few examples, let's switch to using the `spjanfeb2001` data file, keeping just the data for January and February of 2001. These examples of connect styles do not demonstrate how you would normally use these styles but illustrate the different ways you can connect points. See [G] ***connectstyle*** for more information. For this section, we will use the `vg_blue` scheme.

```
twoway scatter close tradeday
```

Consider this graph, which shows the closing price of the S&P 500 index for January and February of 2001 by **tradeday**, the trading day numbered from 1 to 40.
Uses spjanfeb2001.dta & scheme vg_blue

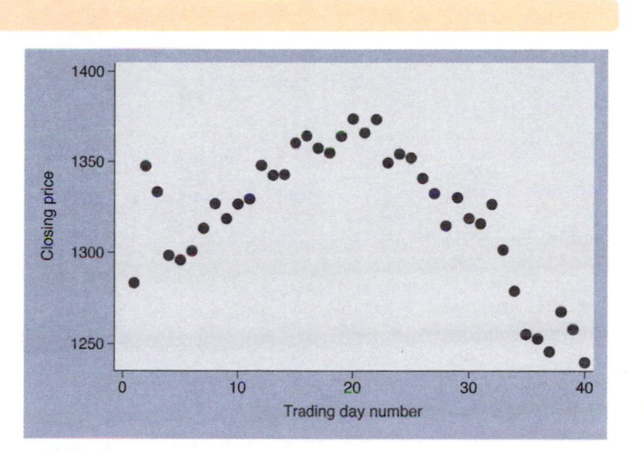


```
twoway scatter close tradeday, connect(l)
```

We use `connect(l)` to connect the points, but this does not lead to the kind of graph we really wanted to create. This is because the observations in the data file are not sorted according to **tradeday**, yet the observations are connected based on the order in which they appear in the data file.
Uses spjanfeb2001.dta & scheme vg_blue

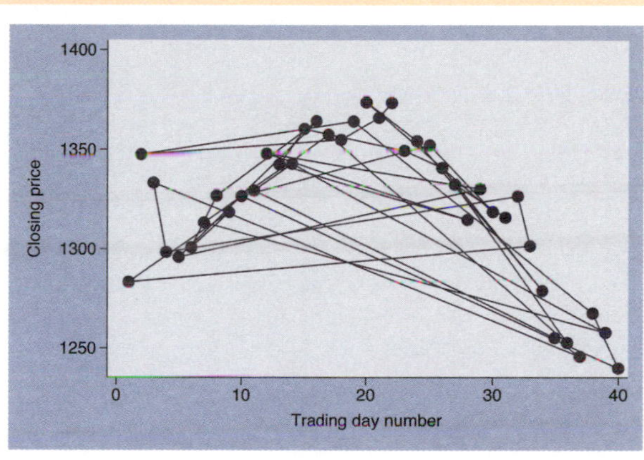


```
twoway scatter close tradeday, connect(l) sort
```

If we add the `sort` option, the observations are connected after sorting them by **tradeday**, which leads to the kind of graph we wanted to create. Alternatively, we could have typed **sort tradeday**, and all ensuing graphs would have been ordered on **tradeday**, even without the **sort** option.
Uses spjanfeb2001.dta & scheme vg_blue

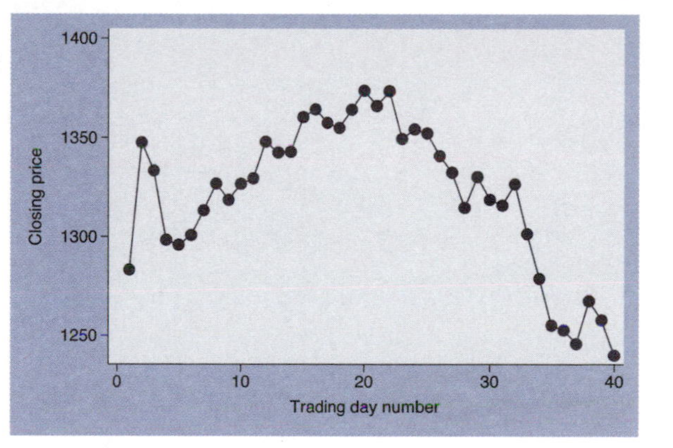

```
twoway scatter close tradeday, connect(J) sort
```

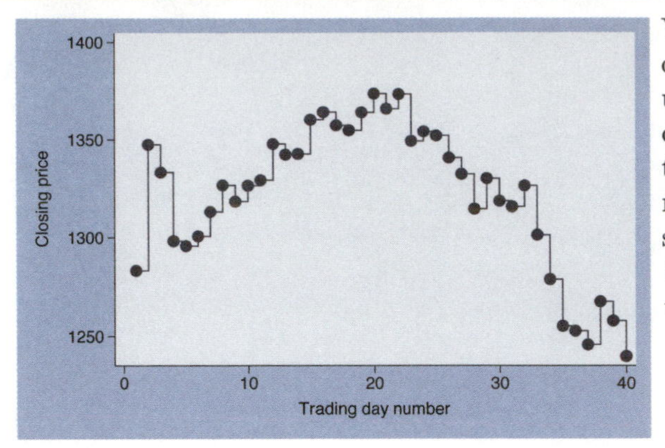

You would not normally connect observations for this kind of graph using a stairstep pattern. This connection method, obtained by using the `connect(J)` option, would more normally be used in a graph showing a survival function over time.
Uses spjanfeb2001.dta & scheme vg_blue

```
twoway scatter close tradeday, connect(stepstair) sort
```

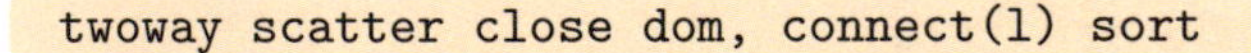

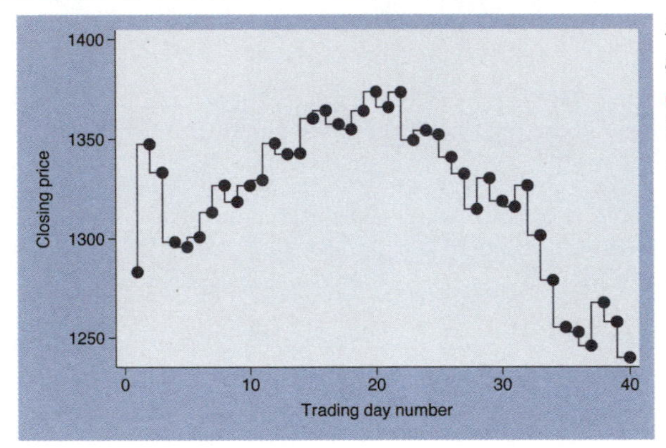

A connection method related to the one above can be obtained using the `connect(stepstair)` option.
Uses spjanfeb2001.dta & scheme vg_blue

```
twoway scatter close dom, connect(l) sort
```

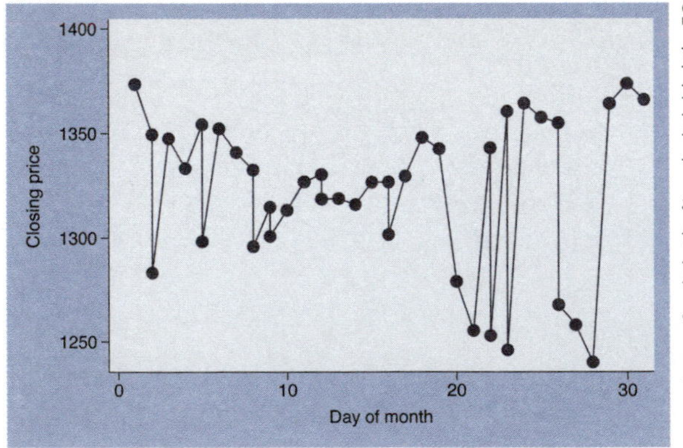

Say that we wanted to show the closing price as a function of the day of the month for the two months for which we have data. In this example, we have the variable `dom` (day of the month) on the x-axis. If we include the `sort` option, the data are shown as one continuous line, as opposed to having one line for January and a second line for February.
Uses spjanfeb2001.dta & scheme vg_blue

```
twoway scatter close dom, connect(l) sort(tradeday)
```

We need to sort the observations by **tradeday**, using the sort(tradeday) option. This graph is almost what we want, but the observation for January 31 is connected to the observation for February 1.
Uses spjanfeb2001.dta & scheme vg_blue

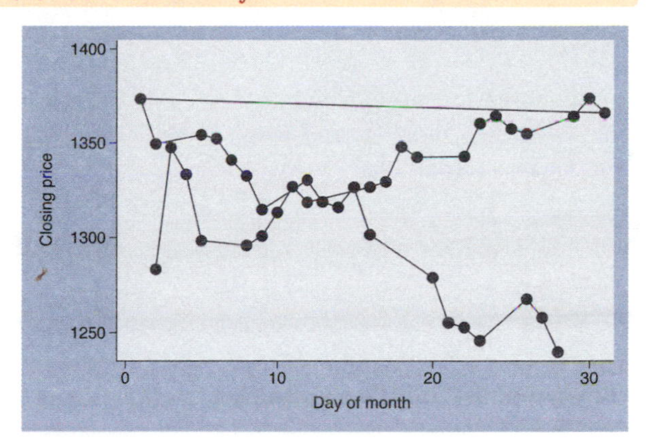


```
twoway scatter close dom, connect(L) sort(tradeday)
```

This graph is what we wanted to create. The connect(L) option avoids the line connecting January 31 and February 1 because it connects points only as long as **dom** is increasing. When **dom** decreases from 31 to 1, the connect(L) option does not connect those two points. See Styles : Connect (332) for more details on connect() options.
Uses spjanfeb2001.dta & scheme vg_blue

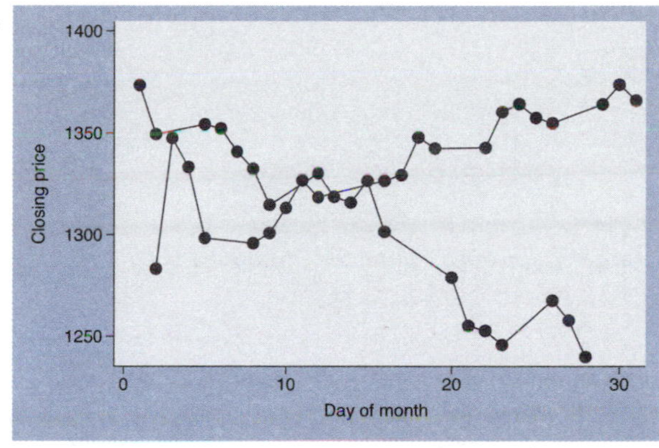


```
twoway scatter close tradeday, connect(l) sort
   clcolor(green) clwidth(thick) clpattern(dash)
```

The connect() option determines how the markers are connected but not the color, width, or pattern of the line. Here, we use the clcolor() (connect line color), clwidth() (connect line width), and clpattern() (connect line pattern) options to make the line green, thick, and dashed. See Styles : Colors (328), Styles : Linewidth (337), and Styles : Linepatterns (336) for more information.
Uses spjanfeb2001.dta & scheme vg_blue

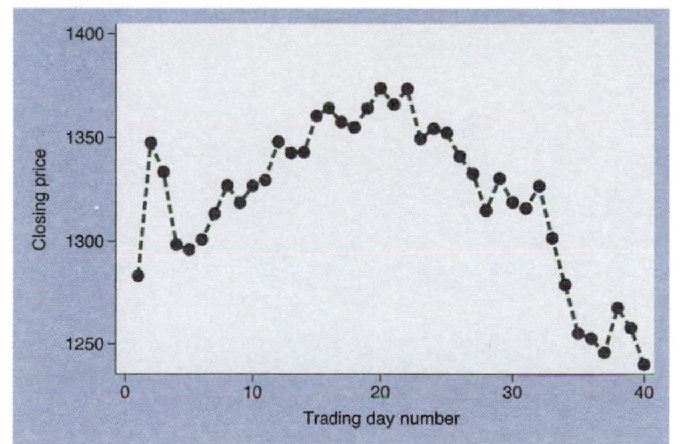

8.4 Setting and controlling axis titles

This section provides more details about the use of axis title options for providing titles for axes. For more information, see [G] ***axis_title_options***. For this section, we will use the vg_past scheme.

```
twoway scatter ownhome propval100
```

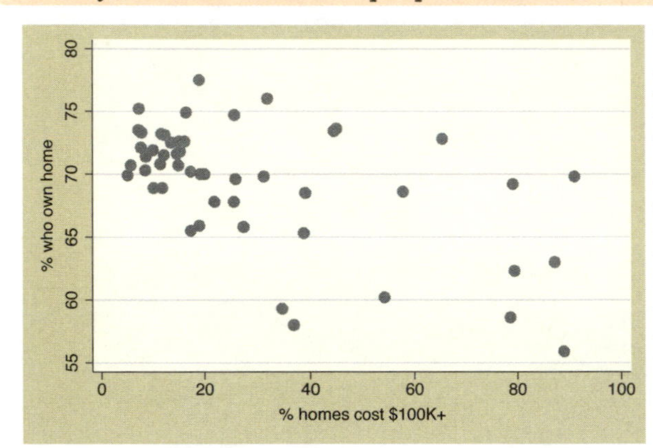

Consider this graph of the percentage of home owners by the percentage of homes that cost over one hundred thousand dollars. The titles of the x- and y-axes are the names of the variables, unless the variables are labeled, in which case the default title is the variable label. In this example, the axes are labeled with the variable labels.
Uses allstatesdc.dta & scheme vg_past

```
twoway scatter ownhome propval100,
   ytitle("Percent of households that own their homes")
   xtitle("Percent of homes that cost over $100,000")
```

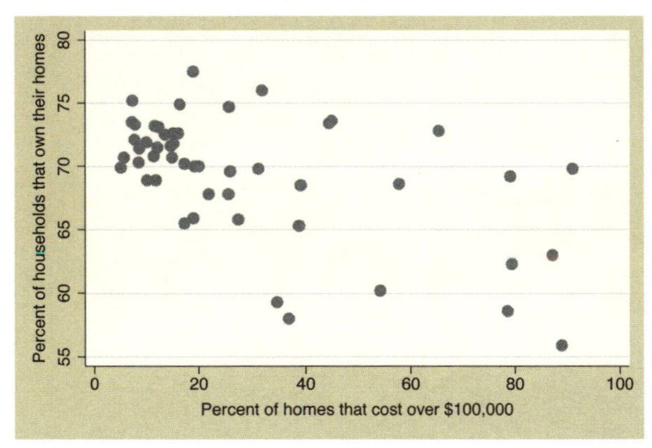

We can use the xtitle() and ytitle() options to supply our own titles.
Uses allstatesdc.dta & scheme vg_past

```
twoway scatter ownhome propval100,
   ytitle("Percent of households that own their homes")
   xtitle("Percent of homes that cost over $100,000", size(small) box)
```

Because an axis title is considered a textbox, you can use textbox options, as illustrated here, to control the look of the axis title. Here, we add the `size()` and `box` options to `xtitle()` to make the x-axis title small with a box around it. See Options : Textboxes (303) for additional examples of how to use textbox options to control the display of text.
Uses allstatesdc.dta & scheme vg_past

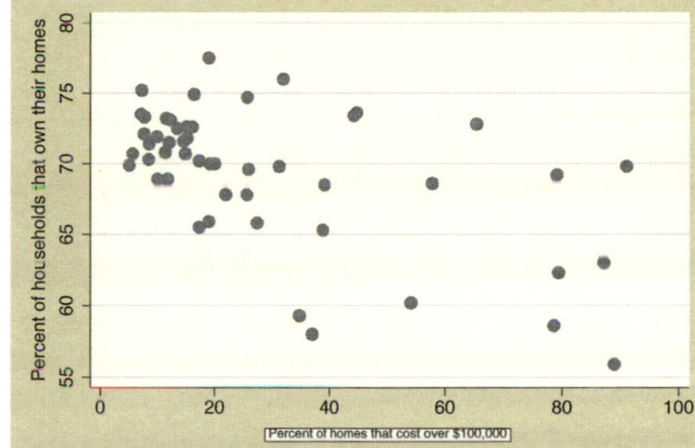


```
twoway scatter ownhome propval100,
   ytitle("Percent of households" "that own their homes")
   xtitle("Percent of homes" "that cost over $100,000")
```

In this example, we supply the same titles but divide them into two separate quoted strings, which then are displayed on separate lines.
Uses allstatesdc.dta & scheme vg_past

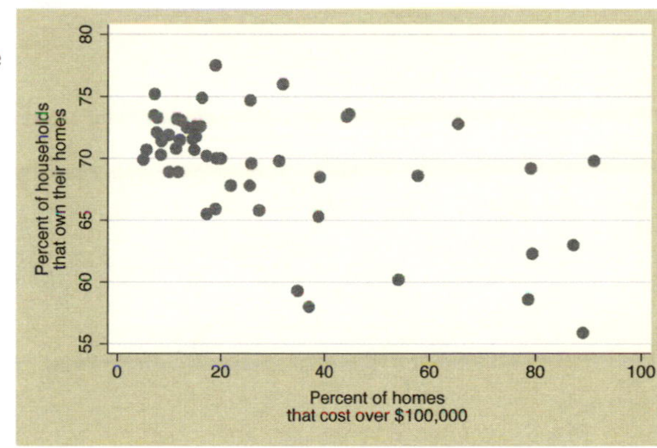


```
twoway scatter ownhome propval100,
   ytitle("1990 Census Data", suffix)
   xtitle("In 1990 dollars", prefix)
```

We can use the `prefix` and `suffix` options to add information before or after the existing title, respectively.
Uses allstatesdc.dta & scheme vg_past

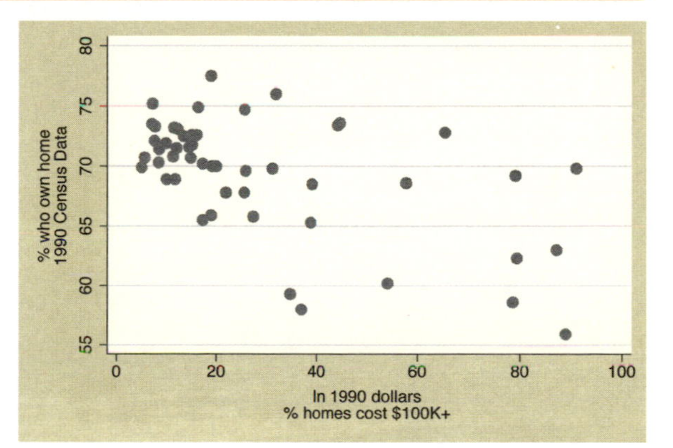

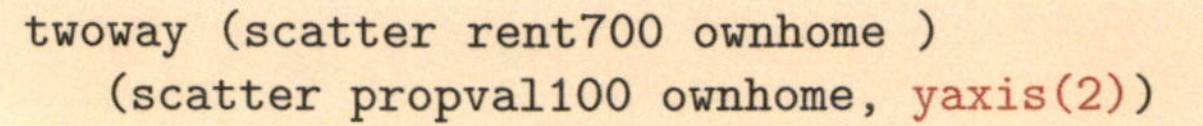

```
twoway (scatter rent700 ownhome )
  (scatter propval100 ownhome, yaxis(2))
```

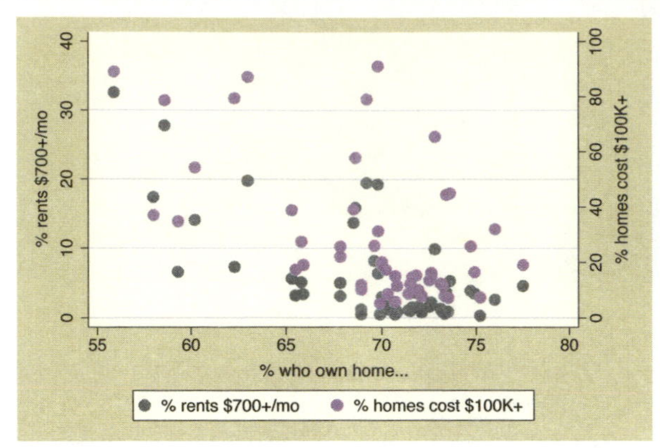

Consider this overlaid twoway graph. The two y-variables are both scaled in percentages, but they have different ranges. We use the `yaxis(2)` option on the second `scatter` command to place that axis on the second y-axis, which is then placed on the right axis.
Uses allstatesdc.dta & scheme vg_past

```
twoway (scatter rent700 ownhome) (scatter propval100 ownhome, yaxis(2)),
  ytitle("Percent rents over $700", axis(1))
  ytitle("Percent homes over $100,000", axis(2))
```

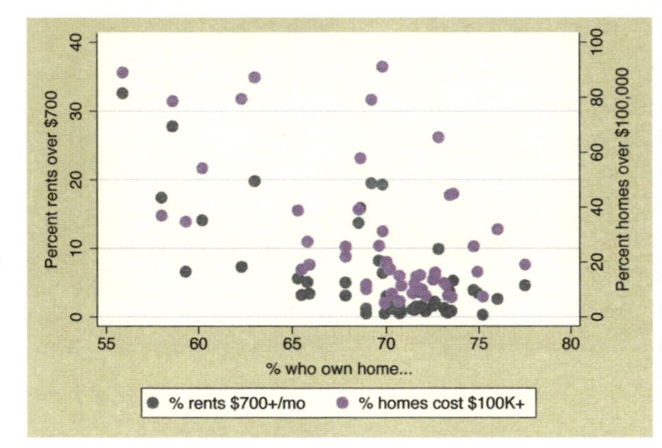

Now that we have two y-axes, the `ytitle()` option would change the y-title for the first y-axis, unless we specify otherwise. In this example, we supply a `ytitle()` option with the `axis(1)` option to indicate that the title belongs to the first y-axis, and a second `ytitle()` option using the `axis(2)` option to indicate that the second title belongs to the second y-axis.
Uses allstatesdc.dta & scheme vg_past

8.5 Setting and controlling axis labels

This section describes more details about axis labels, including major and minor (numeric) labels, major and minor tick marks, and grid lines. This section also shows how to control the appearance of these objects (e.g., size, color, thickness, or angle). For more information, see [G] ***axis_label_options***. For this section, we will use the `vg_s1c` scheme.

```
twoway scatter propval100 faminc
```

Let's start with a basic graph showing the percent of homes costing over $100,000 by the median family income.
Uses allstatesdc.dta & scheme vg_s1c

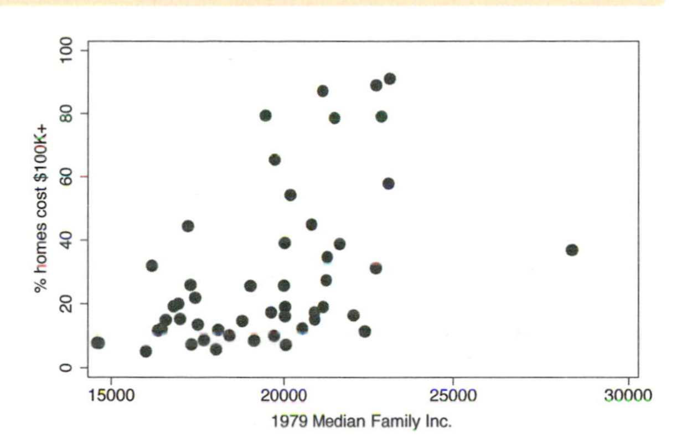


```
twoway scatter propval100 faminc, xlabel(#10) ylabel(#10)
```

Using the `xlabel(#10)` and `ylabel(#10)` options, we ask for about 10 values to be labeled on each axis. Stata chose to use 10 values for the y-axis, labeling it from 0 to 90, incrementing by 10, and 8 values for the x-axis going from 14,000 to 28,000, incrementing by 2,000. As you can see from this example, sometimes Stata follows your suggestion exactly, and sometimes it chooses a different number of values to make more logical labels.
Uses allstatesdc.dta & scheme vg_s1c

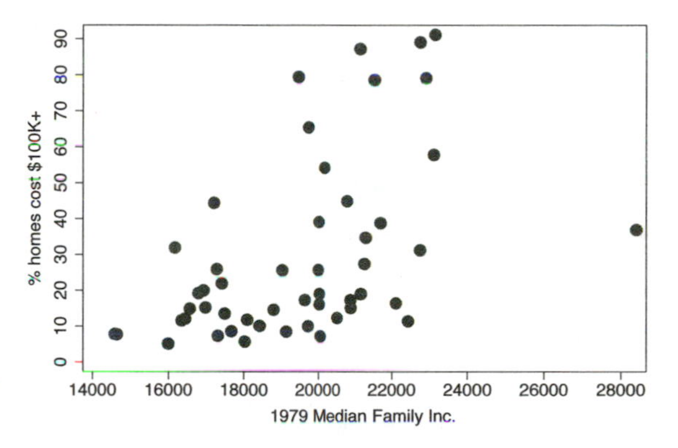


```
twoway scatter propval100 faminc, ylabel(0(10)100)
```

We can change the major labels for the y-variable to range from 0 to 100, incrementing by 10, using the `ylabel(0(10)100)` option.
Uses allstatesdc.dta & scheme vg_s1c

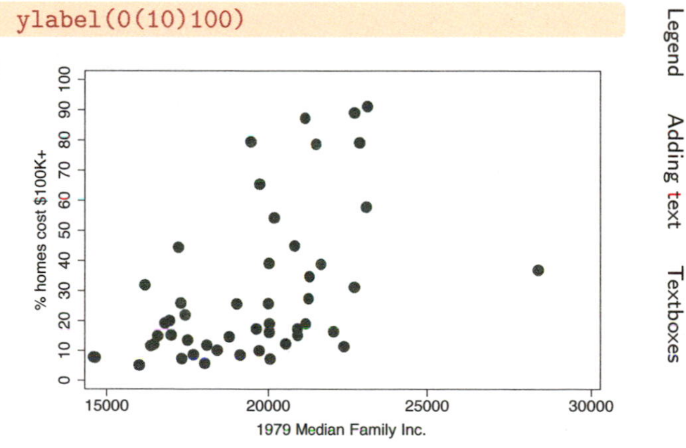

```
twoway scatter propval100 faminc, xlabel(minmax) ylabel(none)
```

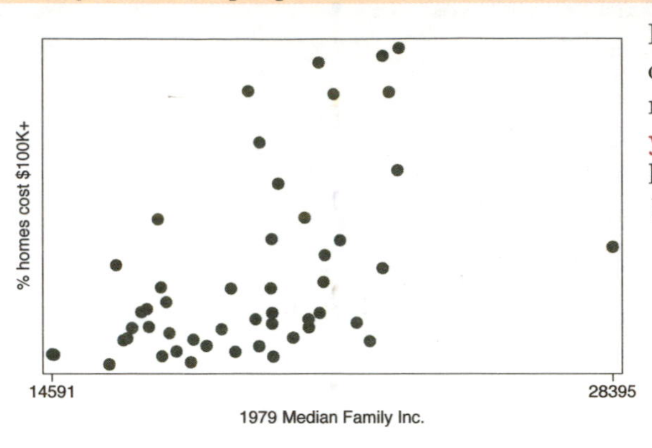

Here, we use the `xlabel(minmax)` option to label the x-axis only with the minimum and maximum and use `ylabel(none)`, so that the y-axis will have no major labels or ticks.
Uses allstatesdc.dta & scheme vg_s1c

```
twoway scatter propval100 faminc, ymlabel(10(20)90)
```

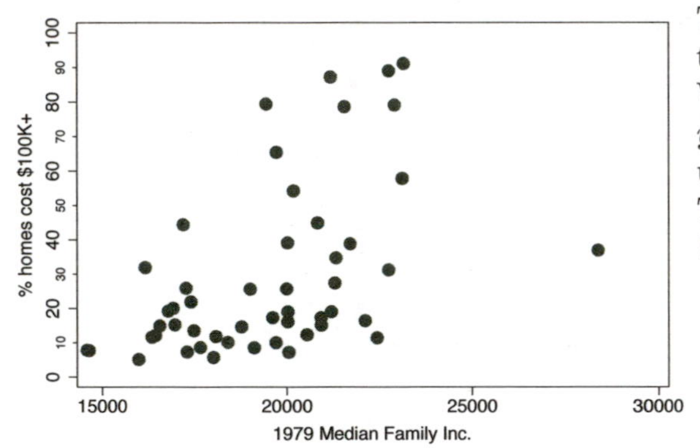

The default graph had major labels for the y-axis at 0, 20, 40, 60, 80, and 100. We could add minor labels for the y-variable at 10, 30, 50, 70, and 90 using the `ymlabel(10(20)90)` option. The `m` in `ymlabel()` stands for minor.
Uses allstatesdc.dta & scheme vg_s1c

```
twoway scatter propval100 faminc, ytick(10(10)90)
```

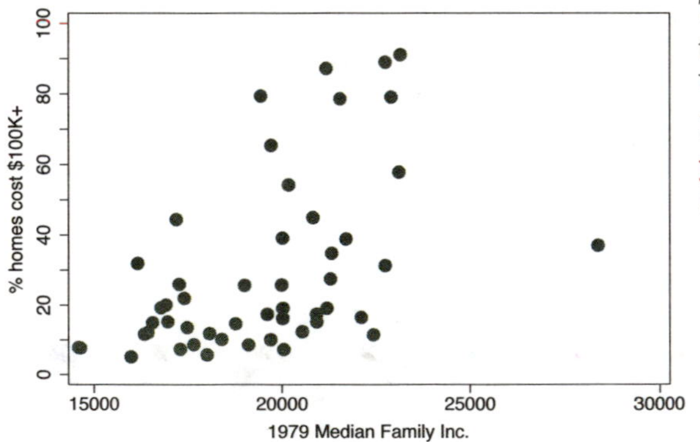

The default graph had major ticks for the y-axis at 0, 20, 40, 60, 80, and 100. We can add major ticks ranging from 10 to 90, incrementing by 10, using the `ytick(10(10)90)` option.
Uses allstatesdc.dta & scheme vg_s1c

```
twoway scatter propval100 faminc, ymtick(10(20)90)
```

We can use the `ymtick()` option to add minor ticks to the graph. For example, here we add minor ticks at 10, 30, 50, 70, and 90. The `m` in `ymtick()` stands for minor.
Uses allstatesdc.dta & scheme vg_s1c

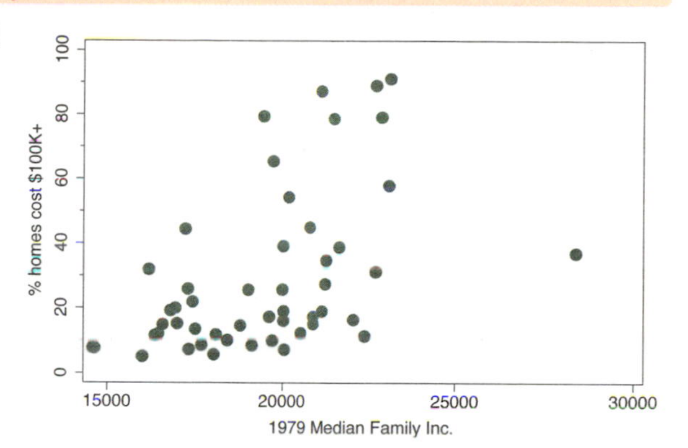


```
twoway scatter propval100 faminc, ymtick(##10)
```

The default graph had major labels for the y-axis at 0, 20, 40, 60, 80, and 100. We can place 9 minor ticks between major ticks with the `ymtick(##10)` option. Note that the value of 10 includes the 9 minor ticks plus the 10th major tick.
Uses allstatesdc.dta & scheme vg_s1c

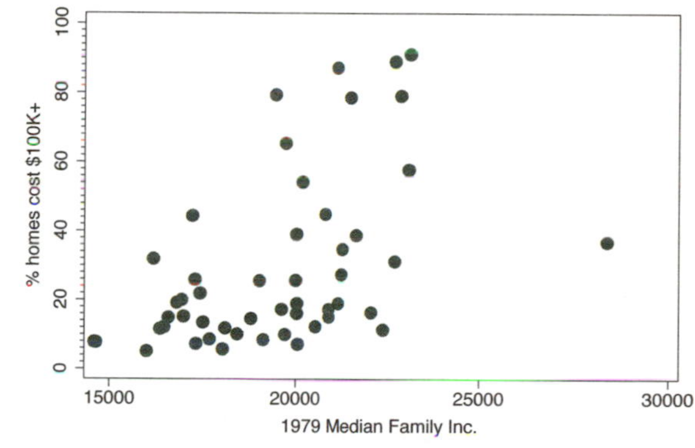


```
twoway scatter propval100 faminc, ylabel(0(10)100, noticks)
```

If we wanted to label the y-axis using values ranging from 0 to 100, incrementing by 10 but suppressing the display of ticks, we could use the `noticks` option.
Uses allstatesdc.dta & scheme vg_s1c

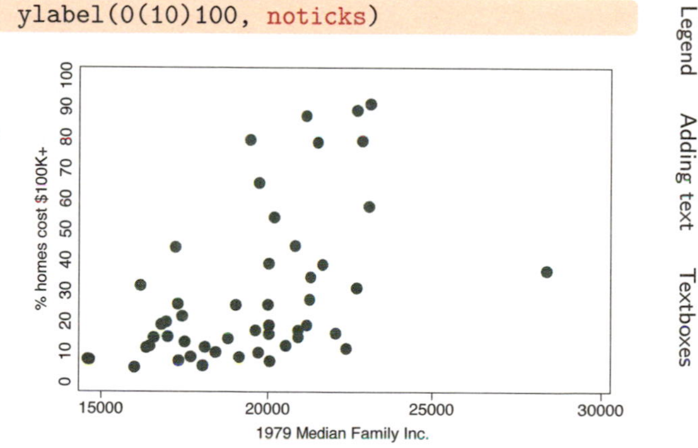

Introduction Twoway Matrix Bar Box Dot Pie Options Standard options Styles Appendix
Markers Marker labels Connecting Axis titles Axis labels Axis scales Axis selection By Legend Adding text Textboxes

```
twoway scatter propval100 faminc, ylabel(, nolabel)
```

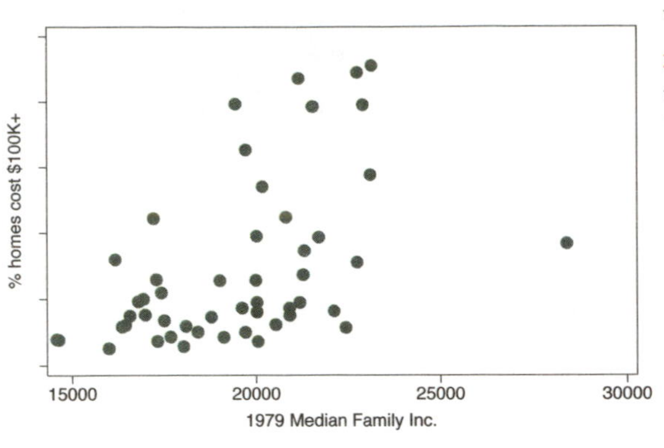

We could suppress the labels using the nolabel option, and only the ticks would be shown.
Uses allstatesdc.dta & scheme vg_s1c

```
twoway scatter propval100 region
```

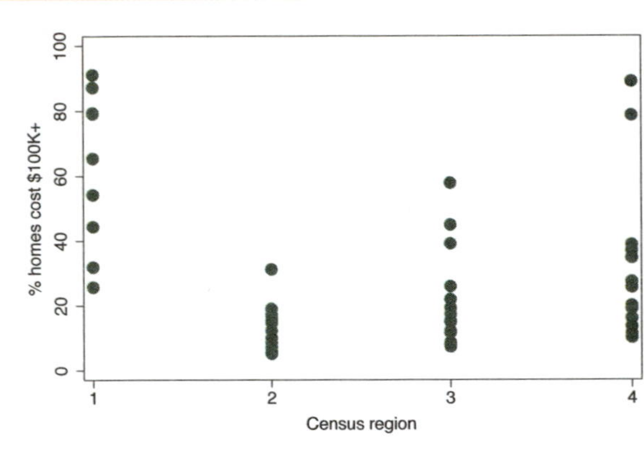

If a variable has meaningful value labels, we can display the value labels in place of the values. For example, we can look at the propval100 broken down by census region, but we do not know which regions correspond to the values 1 to 4.
Uses allstatesdc.dta & scheme vg_s1c

```
twoway scatter propval100 region, xlabel(, valuelabels)
```

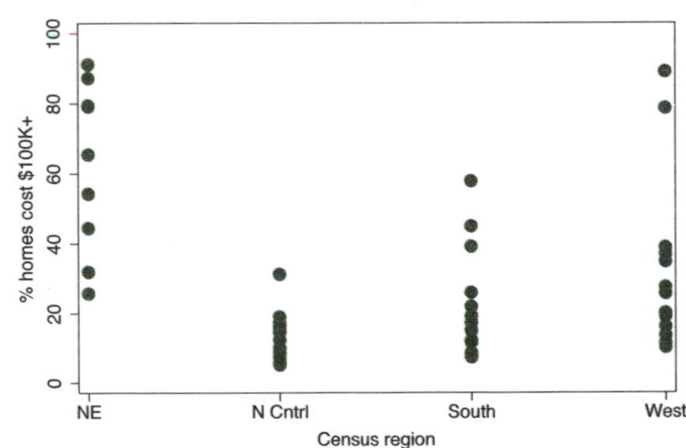

If we include the xlabel(, valuelabels) option, the value labels are displayed instead, making the graph much easier to understand.
Uses allstatesdc.dta & scheme vg_s1c

```
twoway scatter propval100 region,
  xlabel(1 "NorthEast" 2 "NorthCentral" 3 "South" 4 "West")
```

If **region** were not labeled, or if we wanted different labels, we could indicate those labels using the `xlabel()` option, as illustrated here.
Uses allstatesdc.dta & scheme vg_s1c

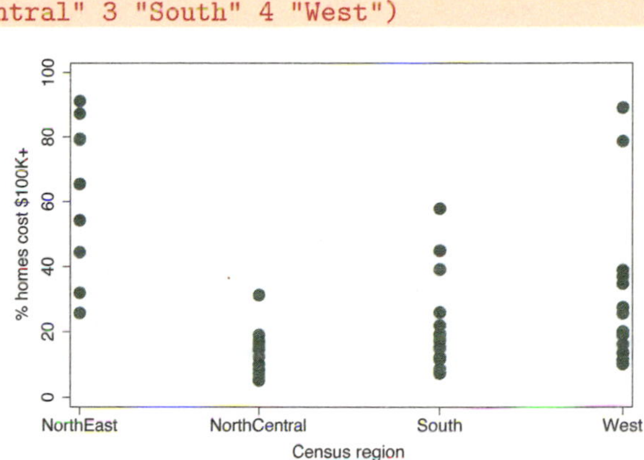


```
twoway scatter propval100 faminc, xlabel(, format(%8.0gc))
```

We can change the formatting of the labels using the `format()` option, just as we would using a **format** statement. In this example, we format income using a comma format to make the larger numbers more readable.
Uses allstatesdc.dta & scheme vg_s1c

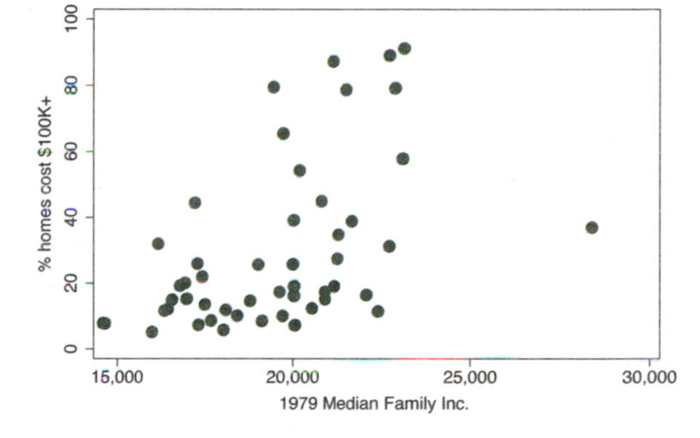


```
twoway scatter propval100 faminc, ylabel(, angle(0))
```

We can change the angles of the labels from their default orientation. By default, the values on the y-axis are shown at a 90-degree angle, but we can use the `ylabel(, angle(0))` to display the labels without rotation.
Uses allstatesdc.dta & scheme vg_s1c

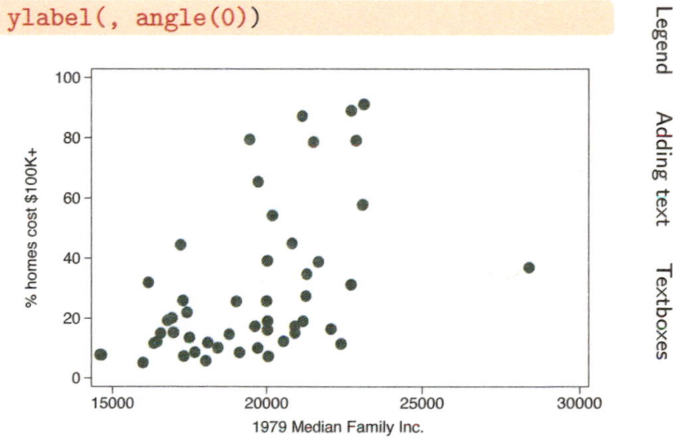

```
twoway scatter propval100 faminc, xlabel(15000(1000)30000, angle(45))
```

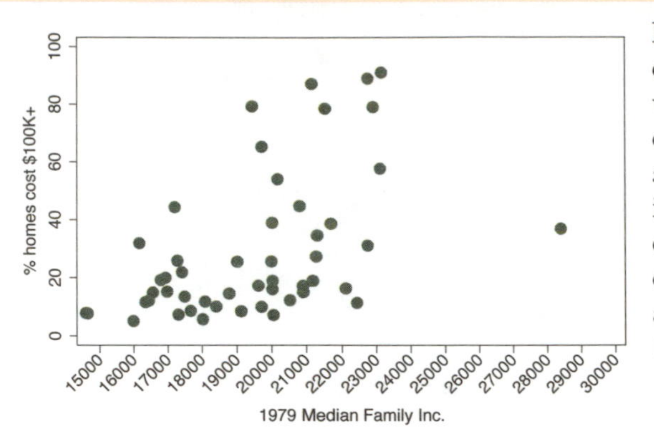

If we label an axis with a large number of values (and especially with wide values), the labels may crowd each other and overlap. Here, we label the x-axis from 15000 to 30000 in increments of 1000. To avoid overlapping, we add the `angle(45)` option to show the labels at a 45-degree angle.
Uses allstatesdc.dta & scheme vg_s1c

```
twoway scatter propval100 faminc, xlabel(15000(1000)30000, alternate)
```

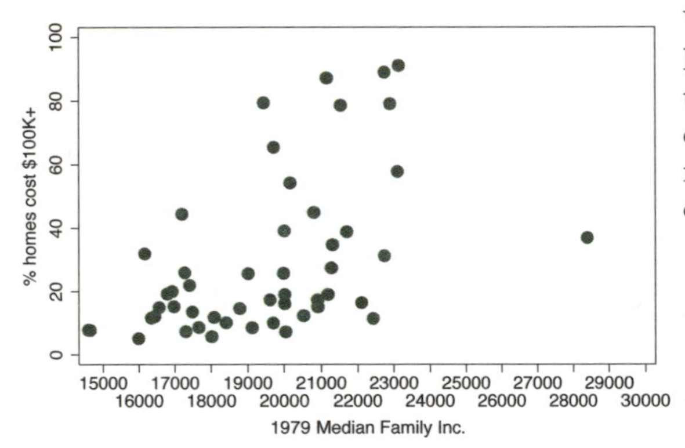

We can also avoid overlapping the axis labels by adding the `alternate` option to `xlabel()`. The labels are now displayed in two rows in alternating rows, so they are not crowded or overlapped.
Uses allstatesdc.dta & scheme vg_s1c

```
twoway scatter propval100 faminc, ylabel(0(5)90, labsize(vsmall))
```

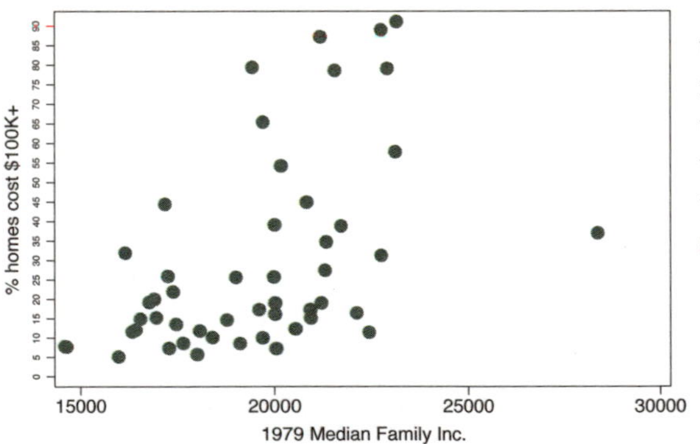

We can control the size of labels with the `labsize()` option. For example, we might want to label our y-axis from 0 to 90, incrementing by 5. The labels would ordinarily overlap, but if we add the `labsize(vsmall)` option, the very small labels no longer overlap.
Uses allstatesdc.dta & scheme vg_s1c

```
twoway scatter propval100 faminc, ylabel(, labgap(*5))
```

We can control the gap between the label and the tick with the `labgap()` option. In this example, we increase the gap between the y-labels and the y-ticks to five times the original size.
Uses allstatesdc.dta & scheme vg_s1c

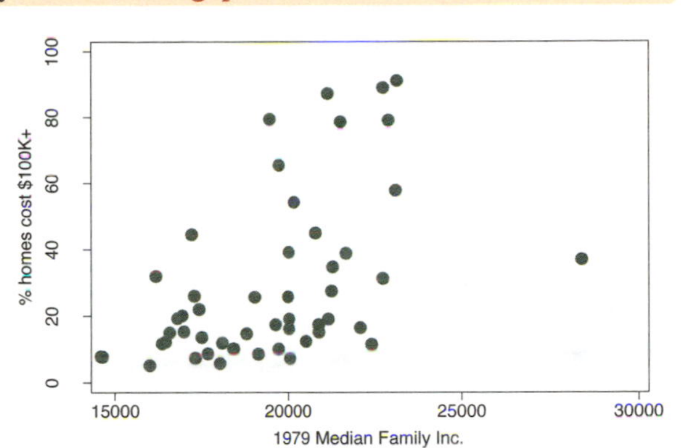


```
twoway scatter propval100 faminc,
   ylabel(, tlength(*1.5) tlwidth(*3) tposition(crossing))
```

You can control the tick length with the `tlength()` option, the tick line width with the `tlwidth()` option, and the tick position with the `tposition()` option. In this example, we make the tick length 1.5 times normal and the width three times normal, with the ticks crossing the y-axis.
Uses allstatesdc.dta & scheme vg_s1c

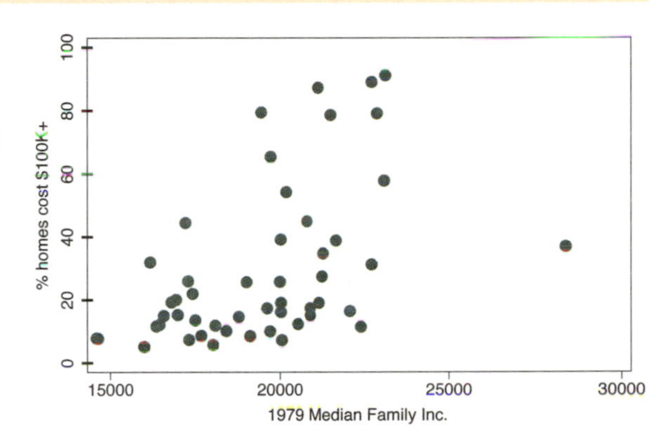


```
twoway scatter propval100 faminc,
   ytick(0(10)100, tposition(outside))
   ymtick(5(10)95, tposition(inside))
```

In this example, we place major ticks from 0 to 100, incrementing by 10, locating the ticks on the outside of the plot, and place minor ticks from 5 to 95, incrementing by 10, placing the ticks on the inside of the plot region.
Uses allstatesdc.dta & scheme vg_s1c

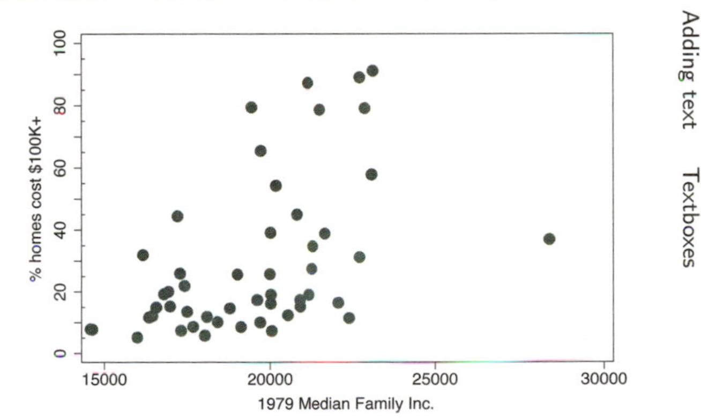

```
twoway scatter propval100 faminc, ylabel(, nogrid)
```

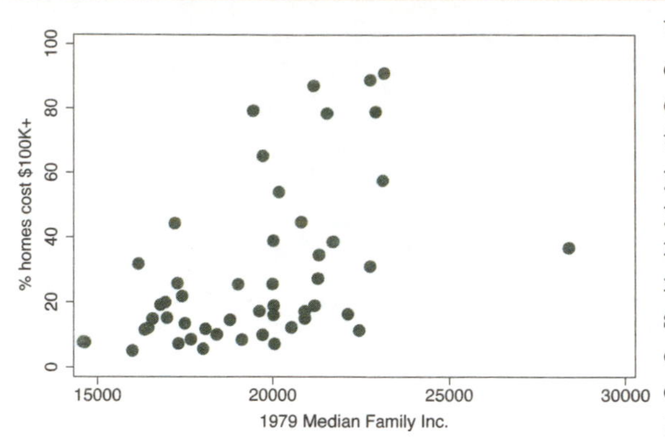

We can use the `grid` and `nogrid` options to display or suppress the display of grid lines corresponding to the labels and ticks associated with the `ylabel()`, `ymlabel()`, `ytick()`, or `ymtick()` options (this also applies to `xlabel()`, `xmlabel()`, `xtick()`, or `xmtick()`). Say that we want to suppress the grid on the y-axis. We can do this with the `ylabel(, nogrid)` option.
Uses allstatesdc.dta & scheme vg_s1c

```
twoway scatter propval100 faminc, ylabel(, grid) xlabel(, grid)
```

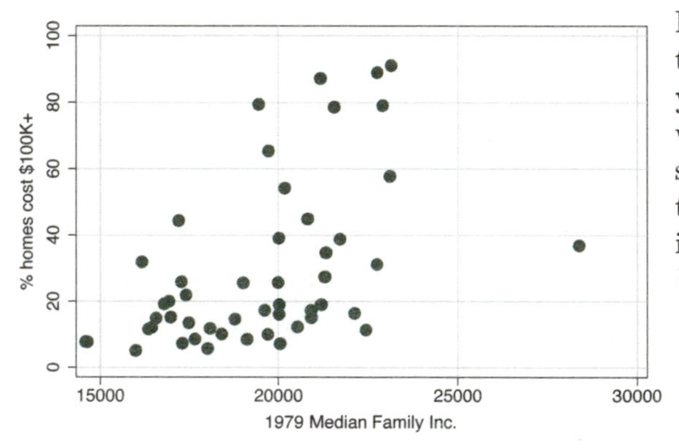

If we want a grid to be displayed for the values that correspond to the `ylabel()` and the `xlabel()` options, we can specify the `grid` option, as shown in this example. Depending on the scheme you choose, grids may be included or omitted by default.
Uses allstatesdc.dta & scheme vg_s1c

```
twoway scatter propval100 faminc,
  ylabel(, grid glwidth(vthin) glcolor(gs10) glpattern(shortdash))
```

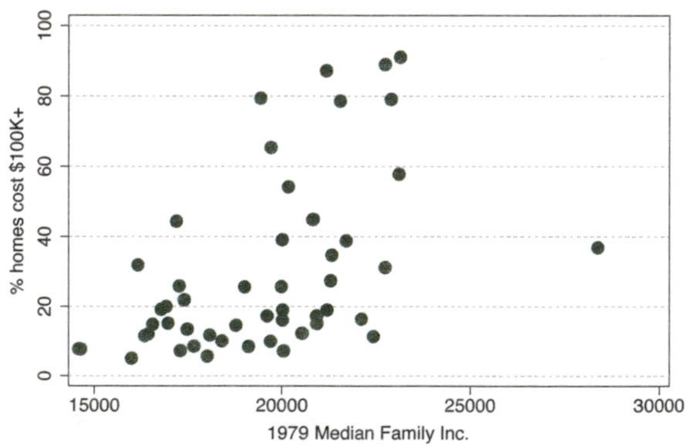

You can control the grid line width, grid line color, and grid line pattern with the `glwidth()`, `glcolor()`, and `glpattern()` options. In this example, we make the grid line very thin, the color gray (`gs10`), and the pattern of the lines short dashes. See **Styles : Linewidth** (337), **Styles : Colors** (328), and **Styles : Linepatterns** (336) for additional details.
Uses allstatesdc.dta & scheme vg_s1c

```
twoway scatter propval100 faminc,
  ylabel(0(20)100, grid glcolor(gs8) glpattern(solid))
  ymlabel(10(20)90, grid glcolor(gs11) glpattern(shortdash))
```

We can use different kinds of grid lines for the major and minor axis labels. In this example, we have a solid, darker gray line for the major axis labels and a lighter gray, short, dashed line for the minor axis labels. We include the grid option to ensure that the grid is displayed.
Uses allstatesdc.dta & scheme vg_s1c

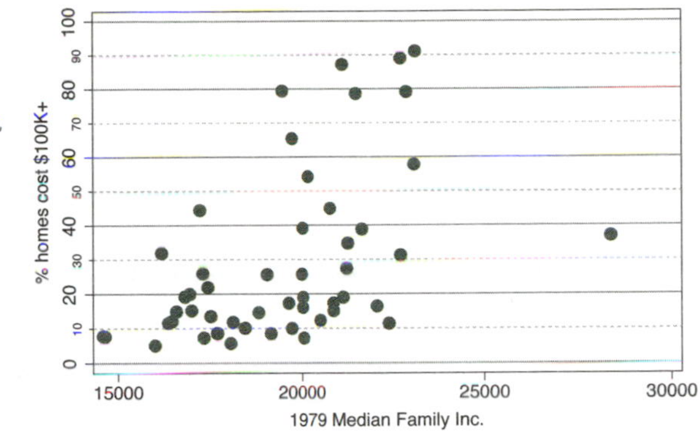

8.6 Controlling axis scales

This section provides more details about axis scale options, which allow us to control whether an axis is displayed, where it is displayed, the direction it is displayed, and the scale of the axis. For more information about these options, see [G] ***axis_scale_options***. This section begins by using data on the S&P 500 from January 2, 2001, to December 31, 2001, stored in the file sp2001. For simplicity, we will use tradeday on the x-axis, representing the trading day of the year. For this section, we will use the vg_s2m scheme.

```
twoway rspike high low tradeday
```

First, consider this rspike graph, which shows the high and low prices across 248 trading days.
Uses sp2001.dta & scheme vg_s2m

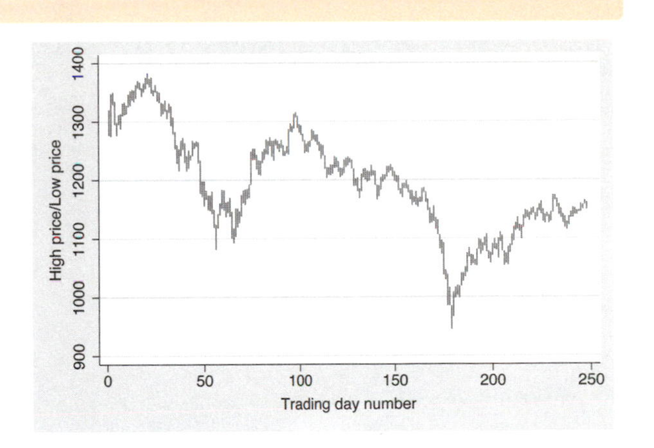

```
twoway rspike high low tradeday, xscale(off)
```

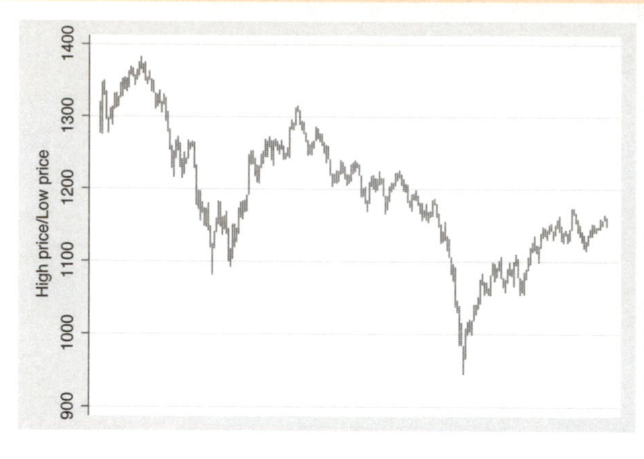

If we wish, we could remove the display of the x-axis entirely with the `xscale(off)` option. Although it is not shown, the same could be done for the y-axis if we were to use the `yscale(off)` option. This is not normally an option we would use, but it can be useful for combining multiple graphs on the same scale without having to show the scale on some of the graphs.
Uses sp2001.dta & scheme vg_s2m

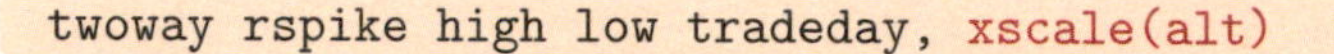

```
twoway rspike high low tradeday, xscale(alt)
```

We could shift the display of the x-axis from the bottom of the graph to the top of the graph with the `xscale(alt)` option. Likewise, we could have chosen to supply the `yscale(alt)` option to shift the y-axis from the left to the right.
Uses sp2001.dta & scheme vg_s2m

```
twoway rspike high low tradeday, xscale(reverse)
```

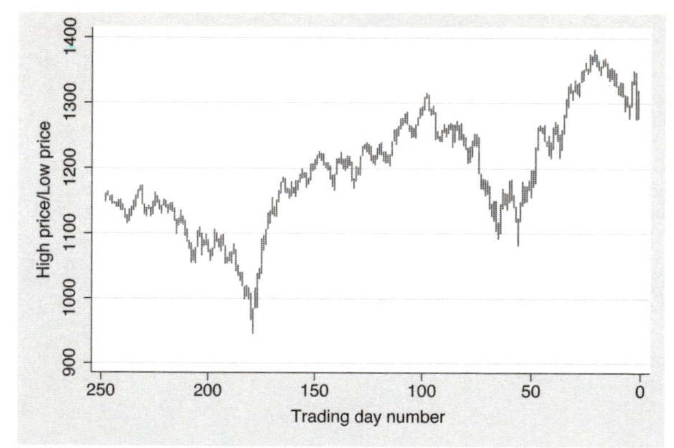

We can reverse the scale of the x-axis by specifying the `xscale(reverse)` option, as illustrated here. We can reverse the y-axis by indicating the `yscale(reverse)` option.
Uses sp2001.dta & scheme vg_s2m

```
twoway scatter educ popden, xscale(log)
```

We briefly return to the `allstates` file to illustrate the `xscale(log)` option. The `xscale(log)` option indicates that the x-axis should be displayed on a log scale. Note that the labels for the x-axis overlap each other.
Uses allstates.dta & scheme vg_s2m

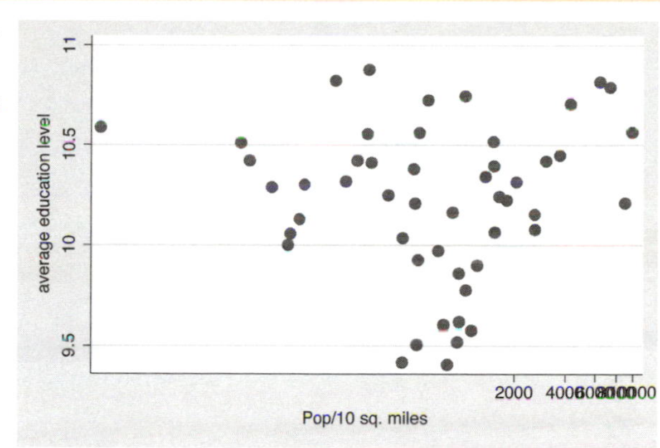


```
twoway scatter educ popden, xscale(log)
   xlabel(1 10 100 1000 10000)
```

Here, we use the `xlabel()` option to change the labels for the x-axis using the values 1, 10, 100, 1000, and 10,000, and you can see how these powers of 10 are more equally spaced, reflecting the log scale of the x-axis.
Uses allstates.dta & scheme vg_s2m

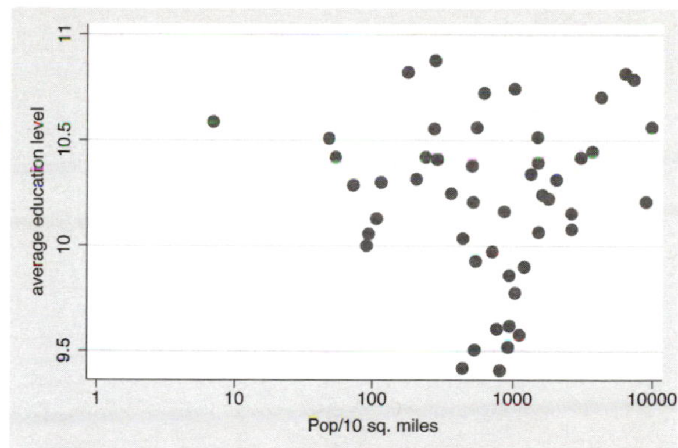


```
twoway rspike high low tradeday, xscale(lwidth(thick))
```

We now return to the `sp2001` data. You can use the `xscale()` and `yscale()` options to control the axis lines. In this example, we make the x-axis line thick by specifying `xscale(lwidth(thick))`.
Uses sp2001.dta & scheme vg_s2m

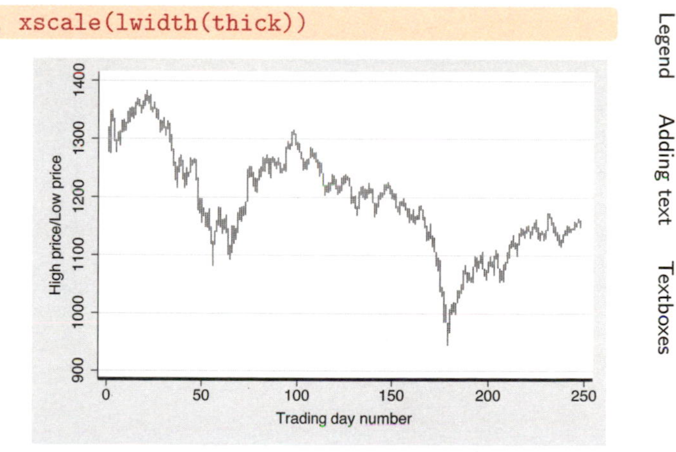

```
twoway rspike high low tradeday, xscale(off noline)
```

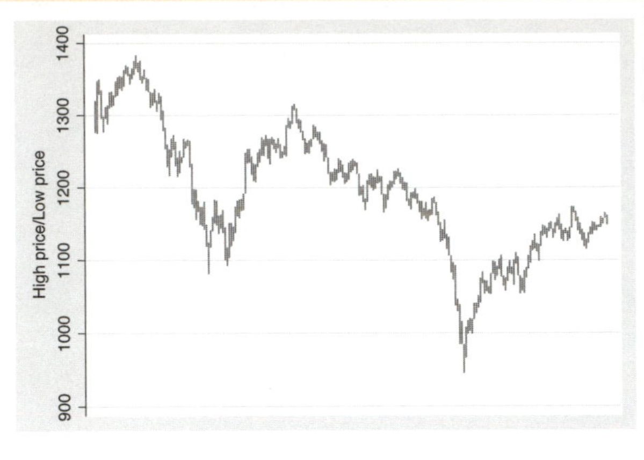

We could suppress the display of the x-axis line completely by using the `xscale(noline)` option.
Uses sp2001.dta & scheme vg_s2m

```
twoway rspike high low tradeday, yscale(range(700 1400))
```

The `yscale(range())` option can be used to expand the scale of the y-axis without needing to expand the labels for the axis (as the `ylabel()` option would). In this example, we have expanded the range of the y-axis from 700 to 1400. However, this example does not show the real utility of this option. Note that `range()` can only be used to expand the scale, not contract it.
Uses sp2001.dta & scheme vg_s2m

```
twoway (rspike high low tradeday)
   (line volmil tradeday, sort yaxis(2))
```

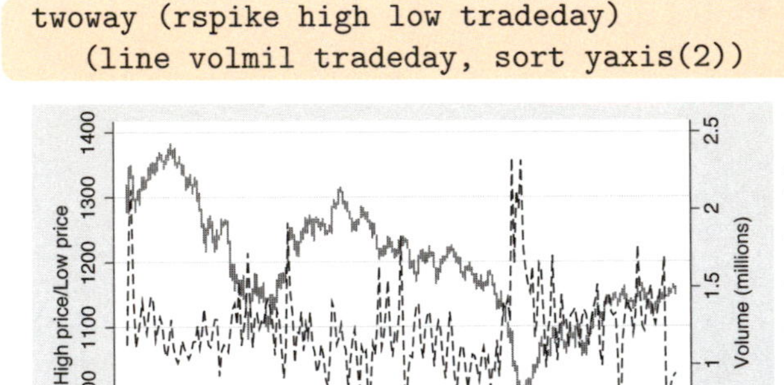

Consider that, in addition to the spike graph that shows the high and low values for a given trading day, we wish to see the volume for a given trading day. We can combine the plots into a single graph, but this is difficult to read because the two plots overlap.
Uses sp2001.dta & scheme vg_s2m

```
twoway (rspike high low tradeday)
   (line volmil tradeday, sort yaxis(2)),
   yscale(range(700 1400) axis(1)) yscale(range(0 10) axis(2))
```

This example shows the utility of the `yscale(range())` option. The `yscale(range(700 1400) axis(1))` option sets the range of price to be from 700 to 1400, shifting that series up to the upper third of the graph. The `yscale(range(0 10) axis(2))` option sets the range of volume to occupy the lower third of the graph.
Uses sp2001.dta & scheme vg_s2m

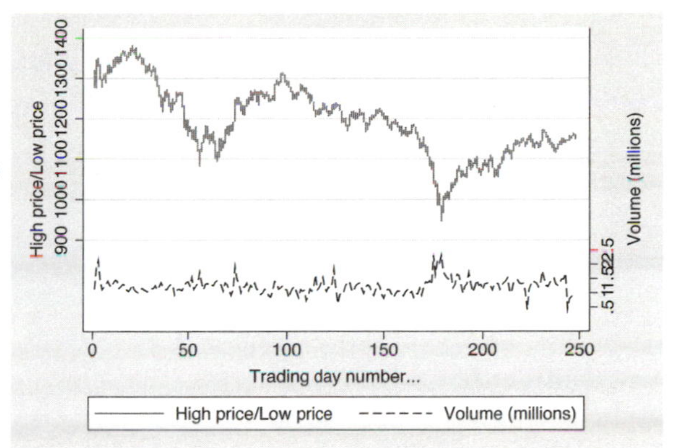


```
twoway (rspike high low tradeday)
   (line volmil tradeday, sort yaxis(2)),
   yscale(range(700 1400) axis(1)) yscale(range(0 10) axis(2))
   ylabel(1000 1200 1400, axis(1)) ylabel(0 1 2, axis(2))
```

Because we manipulated the scale of the y-axes, the labels were pushed together. We can add the `ylabel(1000 1200 1400, axis(1))` and `ylabel(0 1 2, axis(2))` options to the previous example to make the labels for the y-axes more readable.
Uses sp2001.dta & scheme vg_s2m

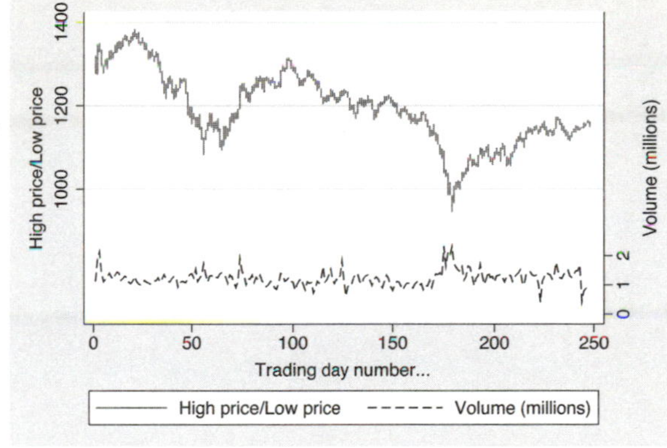

8.7 Selecting an axis

This section provides more details about how to select different axes and modify them. By default, any modifications you make to an axis are applied to the first axis, so you need to take extra action to modify other axes that you may create. For more information about these options, see [G] ***axis_selection_options***. For this section, we will use the `vg_outc` scheme.

Introduction Twoway Matrix Bar Box Dot Pie Options Standard options Styles Appendix
Markers Marker labels Connecting Axis titles Axis labels Axis scales Axis selection By Legend Adding text Textboxes

```
twoway scatter faminc educ
```

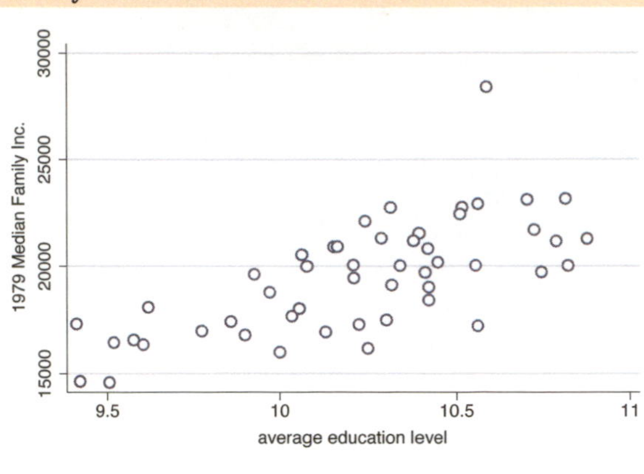

This section focuses on the options that we can use to select axes and shows examples of graphing multiple variables in a single graph. This graph shows the relationship between one x-variable, `educ`, and one y-variable, `faminc`.
Uses allstatesdc.dta & scheme vg_outc

```
twoway (scatter faminc educ, xaxis(1) yaxis(1))
```

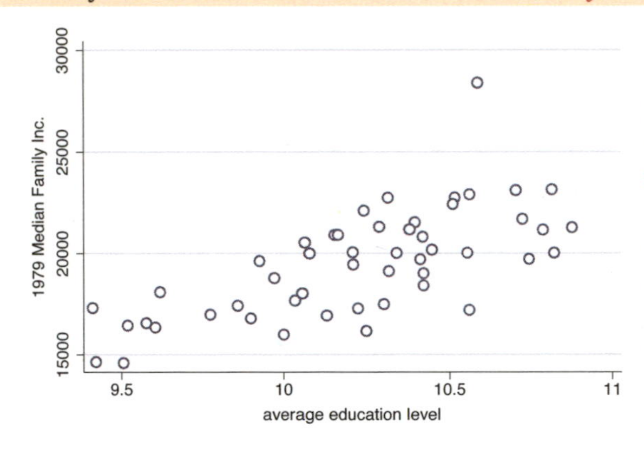

By default, the x-variable is placed on the first x-axis, and the y-variable is placed on the first y-axis. It is as though you had added the options `xaxis(1)` and `yaxis(1)`, as illustrated here. Note that we add parentheses to emphasize that the options `xaxis(1)` and `yaxis(1)` belong to the scatter command and are not general options for the overall graph, which would appear after the parentheses.
Uses allstatesdc.dta & scheme vg_outc

```
twoway (scatter faminc educ)
   (scatter workers2 educ)
```

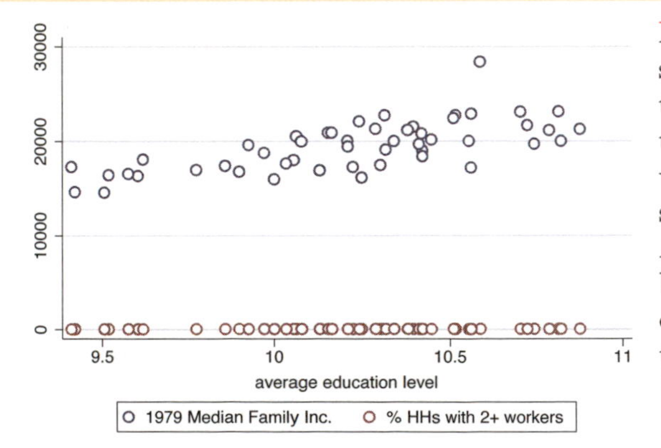

Now let's overlay a second scatterplot showing `workers2` by `educ`, which has the effect of adding a second variable to the y-axis. Stata assumes that all variables are on the first (and thus, the same) axis, unless we specify otherwise. As a result, this graph is hard to read because `faminc` is scaled very differently from `workers2` but scaled on the same axis.
Uses allstatesdc.dta & scheme vg_outc

```
twoway (scatter faminc educ, yaxis(1))
   (scatter workers2 educ, yaxis(2))
```

Stata permits you to have multiple axes for the x-variables and the y-variables. In this example, we use the `yaxis(1)` option to place `faminc` on the first y-axis and the `yaxis(2)` option to place `workers2` on the second y-axis. To make the graph more readable, Stata moved the second y-axis over to the right side. Note that the `yaxis(1)` option was not needed but was included for clarity.
Uses allstatesdc.dta & scheme vg_outc

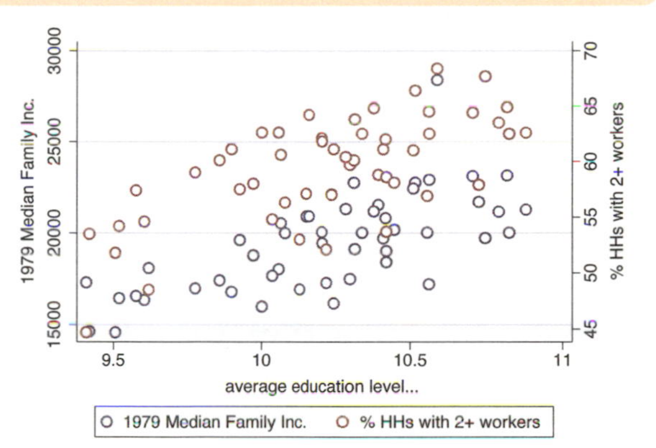


```
twoway (scatter faminc educ)
   (scatter workers2 educ, yaxis(2)), ylabel(40(5)80, axis(2))
```

Say that you wished to label `workers2` starting at 40, incrementing by 5 until 80. Since `workers2` is on the second y-axis, you would specify `ylabel(40(5)80, axis(2))`. Without the `axis(2)` option, Stata would assume that you are referring to the first y-axis and would change the scaling of `faminc`.
Uses allstatesdc.dta & scheme vg_outc

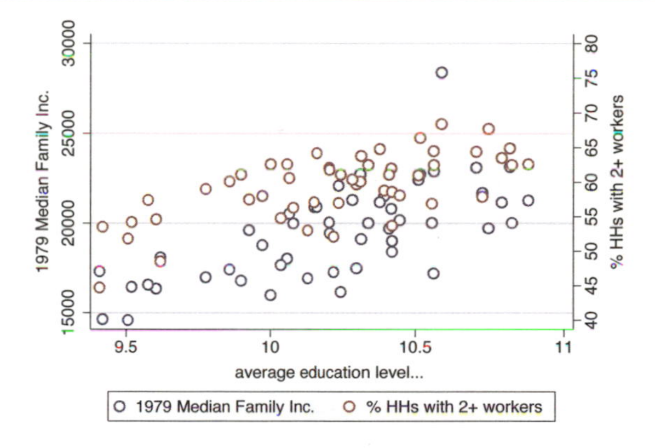


```
twoway (scatter faminc educ)
   (scatter workers2 educ, yaxis(2) ylabel(40(5)80))
```

You might be tempted to enter the `ylabel()` option as an option of the second `scatter` statement and expect the `ylabel()` to modify the scaling of `workers2`. However, we can see in this example that this does not work.
Uses allstatesdc.dta & scheme vg_outc

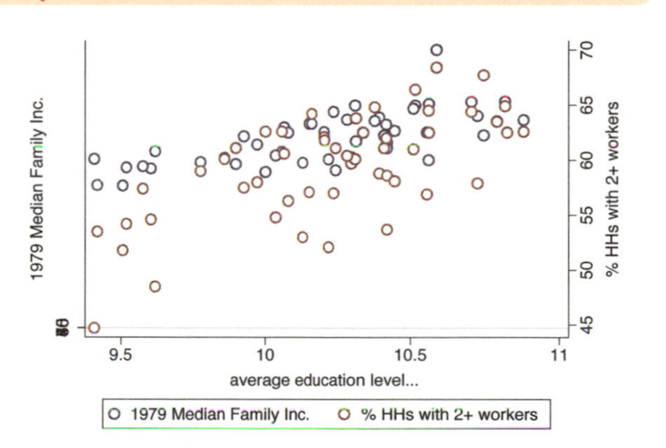

Introduction Twoway Matrix Bar Box Dot Pie Options Standard options Styles Appendix
Markers Marker labels Connecting Axis titles Axis labels Axis scales Axis selection By Legend Adding text Textboxes

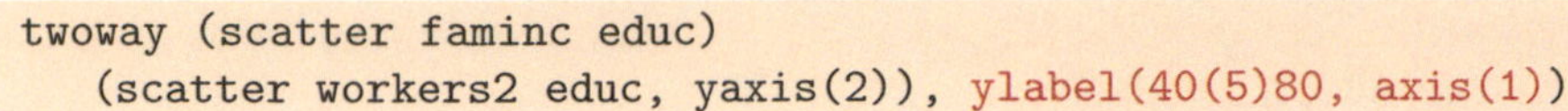

```
twoway (scatter faminc educ)
   (scatter workers2 educ, yaxis(2)), ylabel(40(5)80, axis(1))
```

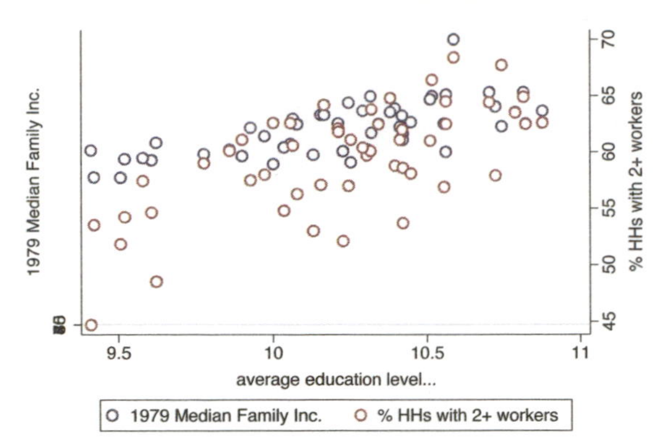

`ylabel()` is really an overall option, but Stata is willing to pretend that you specified this option globally, as though you had typed `ylabel()` as a global option as specified in this example. To make this clearer, we have added the default `axis(1)` to `ylabel()` to illustrate why this usage does not change the second y-axis.
Uses allstatesdc.dta & scheme vg_outc

```
twoway (scatter faminc educ)
   (scatter workers2 educ, yaxis(2)),
   ytitle("Family income", axis(1)) ytitle("Two+ workers", axis(2))
```

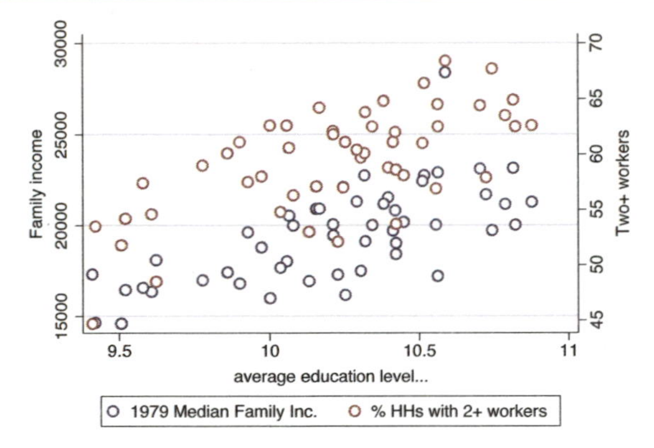

These same rules apply to modifying the axis titles and labeling. In this example, we use the `ytitle()` option to change the titles for the first and second y-axes.
Uses allstatesdc.dta & scheme vg_outc

8.8 Graphing by groups

This section provides more details about repeating graphs using the `by()` option to show separate graphs for each by-group. For more information, see [G] ***by_option***. For this section, we will use the `vg_brite` scheme.

```
twoway scatter ownhome borninstate
```

We start by looking at a scatterplot of `ownhome` and `borninstate`, and we see a general positive relationship such that the higher the percentage of those who were born in the state, the higher the percentage of home owners in the state.
Uses allstatesdc.dta & scheme vg_brite

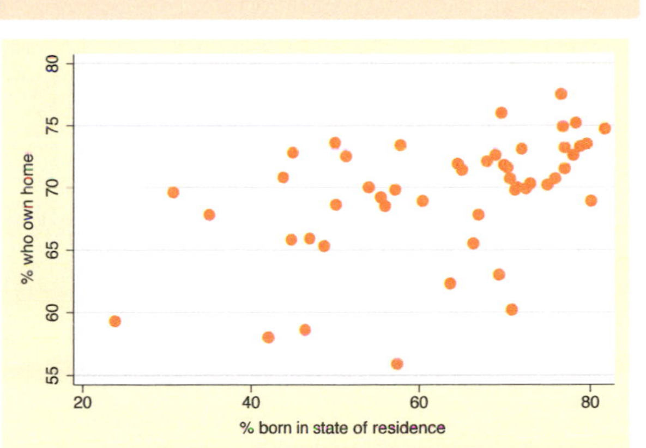


```
twoway scatter ownhome borninstate, by(north)
```

We can use the `by(north)` option to look at this relationship broken down by whether the state is considered to be in the North.
Uses allstatesdc.dta & scheme vg_brite

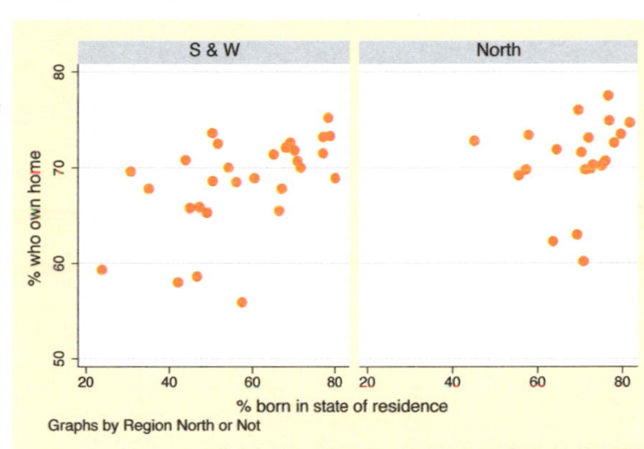


```
twoway scatter ownhome borninstate, by(north, total)
```

We can use the `total` option to see the overall relationship for all 50 states, as well as the two plots separately, by the levels of `north`.
Uses allstatesdc.dta & scheme vg_brite

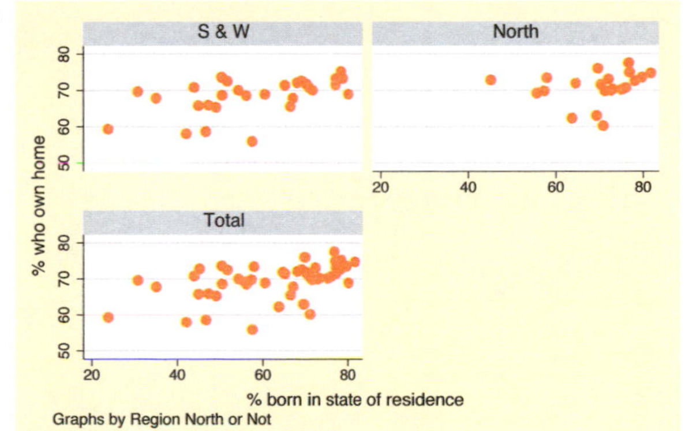

Introduction Twoway Matrix Bar Box Dot Pie Options Standard options Styles Appendix
Markers Marker labels Connecting Axis titles Axis labels Axis scales Axis selection By Legend Adding text Textboxes

```
twoway scatter ownhome borninstate, by(north, total colfirst)
```

We can add the `colfirst` option to show the graphs going down columns first rather than going across rows first, which is the default.
Uses allstatesdc.dta & scheme vg_brite

```
twoway scatter ownhome borninstate, by(north, total holes(2))
```

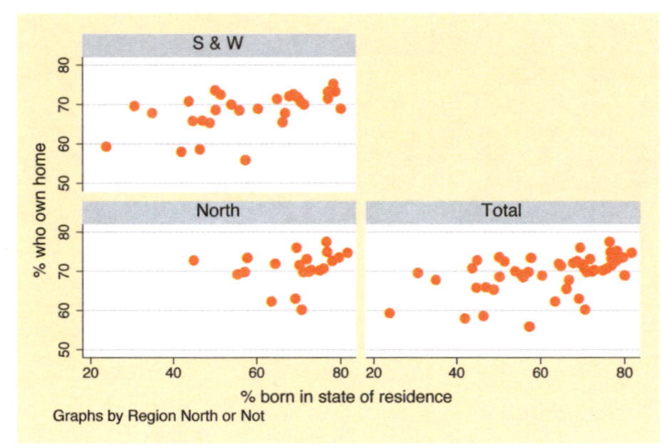

The `holes(2)` option leaves the second position empty. Here, we specify a single position to leave empty, but you can specify multiple positions within the `holes()` option.
Uses allstatesdc.dta & scheme vg_brite

```
twoway scatter ownhome borninstate, by(north, total rows(1))
```

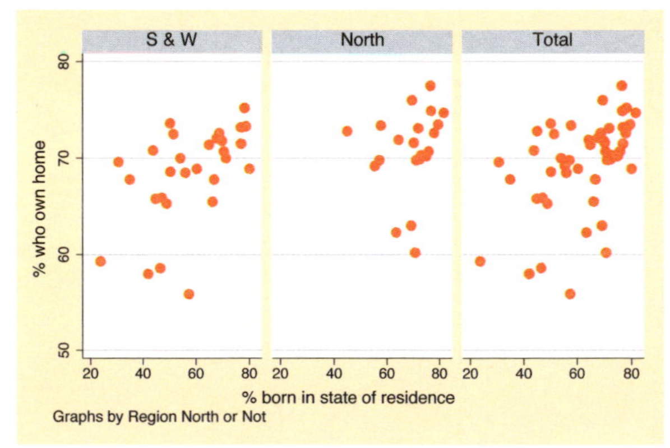

The `rows(1)` option indicates that the graph should be displayed in one row.
Uses allstatesdc.dta & scheme vg_brite

```
twoway scatter ownhome borninstate, by(north, total cols(1))
```

The `cols(1)` option shows the graph in a single column.
Uses allstatesdc.dta & scheme vg_brite

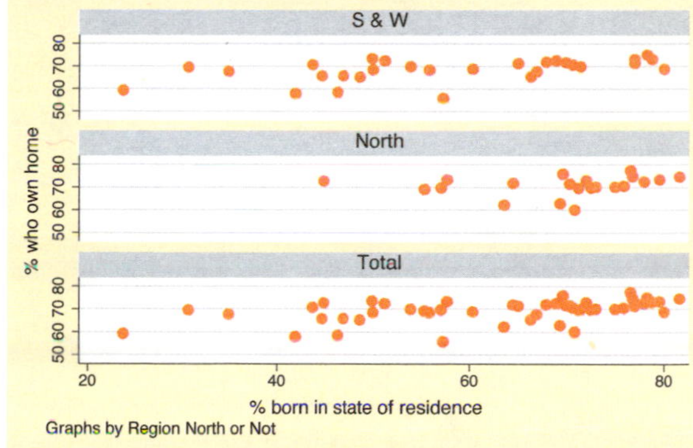


```
twoway scatter ownhome borninstate, by(north, total iscale(*1.5))
```

Sometimes when you use the `by()` option, the graph can become small, making the text and symbols difficult to see. You can use the `iscale()` option to magnify the size of these elements. In this example, we increase the size of these elements by a factor of 1.5.
Uses allstatesdc.dta & scheme vg_brite

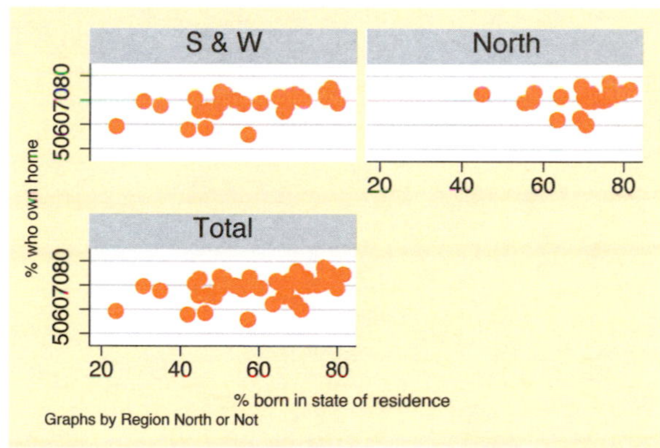


```
twoway scatter ownhome borninstate, by(north, total compact)
```

The `compact` option displays the graph using a compact style, pushing the graphs tightly together. This is almost the same as specifying `style(compact)`.
Uses allstatesdc.dta & scheme vg_brite

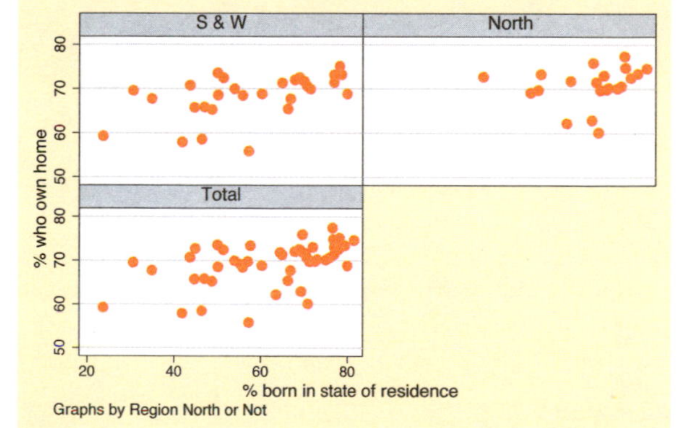

```
twoway scatter ownhome borninstate, by(north, total noedgelabel)
```

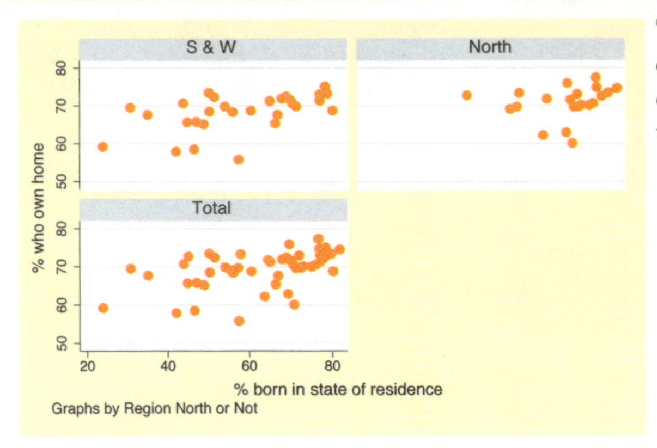

The `noedgelabel` option suppresses the display of the x-axis for the graphs that do not appear on the bottom row, in this case the graph for the North.
Uses allstatesdc.dta & scheme vg_brite

```
twoway scatter ownhome borninstate, by(north, yrescale)
```

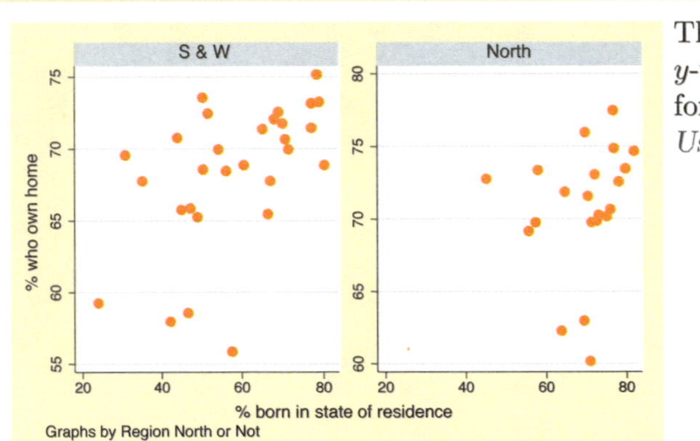

The `yrescale` option allows the y-variables to be scaled independently for each by-group.
Uses allstatesdc.dta & scheme vg_brite

```
twoway scatter ownhome borninstate, by(north, xrescale)
```

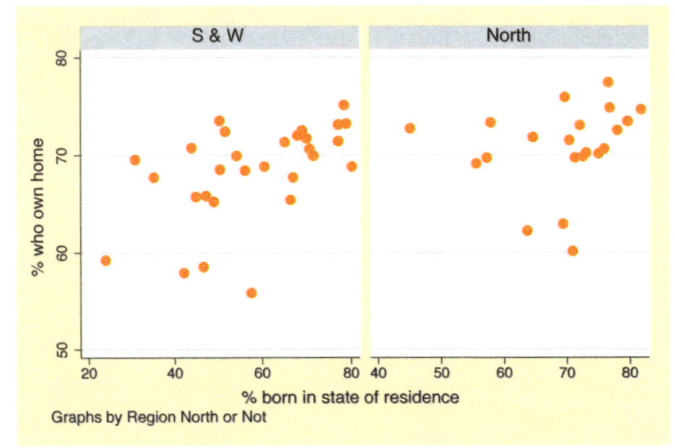

Likewise, the `xrescale` option allows the x-variable to be scaled differently across all the by-groups.
Uses allstatesdc.dta & scheme vg_brite

```
twoway scatter ownhome borninstate, by(north, rescale)
```

If you want both the x-variable and y-variable to be scaled differently across the by-groups, you can use the rescale option, and both axes are separately rescaled.
Uses allstatesdc.dta & scheme vg_brite

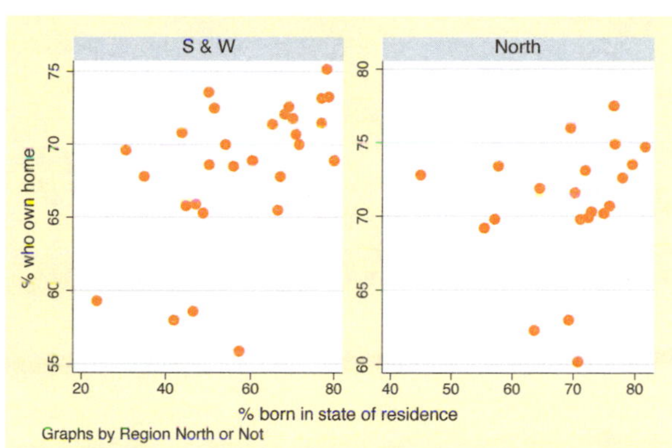


```
twoway scatter ownhome borninstate, by(north, iyaxes)
```

You can use the iyaxes option so the y-axes for each individual graph will be displayed.
Uses allstatesdc.dta & scheme vg_brite

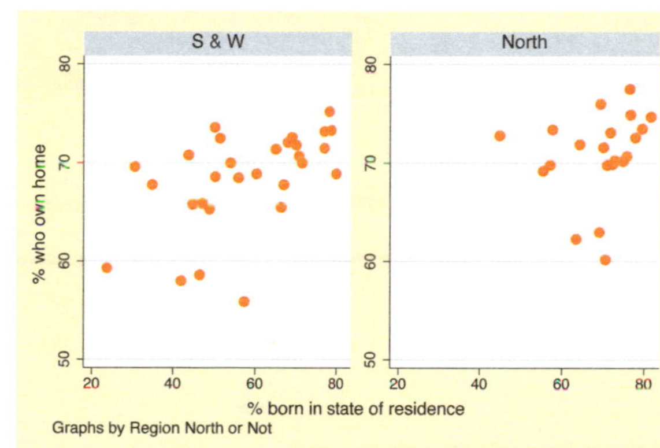


```
twoway scatter ownhome borninstate, by(north, cols(1))
```

Likewise, the **ixaxes** option will display the x-axis for all graphs. In this graph, we omit this option. If we display two graphs in a single column, Stata displays the top graph, omitting the x-axis.
Uses allstatesdc.dta & scheme vg_brite

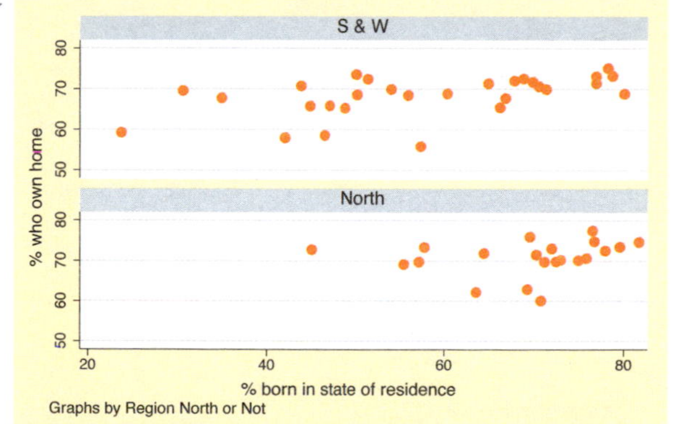

```
twoway scatter ownhome borninstate, by(north, ixaxes cols(1))
```

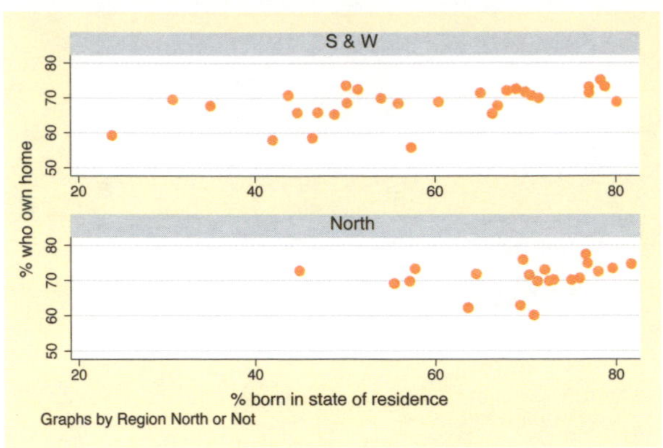

We now include the `ixaxes` option and see that the x-axis is now displayed on the top graph.
Uses allstatesdc.dta & scheme vg_brite

```
twoway scatter ownhome borninstate, by(north, total iytitle)
```

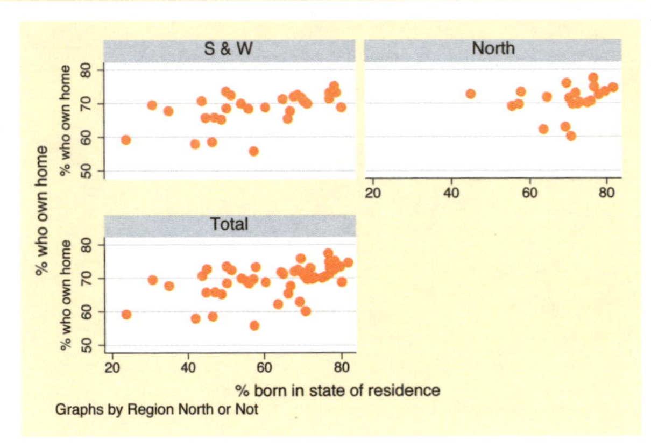

We can display the title for each y-axis using the `iytitle` option.
Uses allstatesdc.dta & scheme vg_brite

```
twoway scatter ownhome borninstate, by(north, total iyaxes iytitle)
```

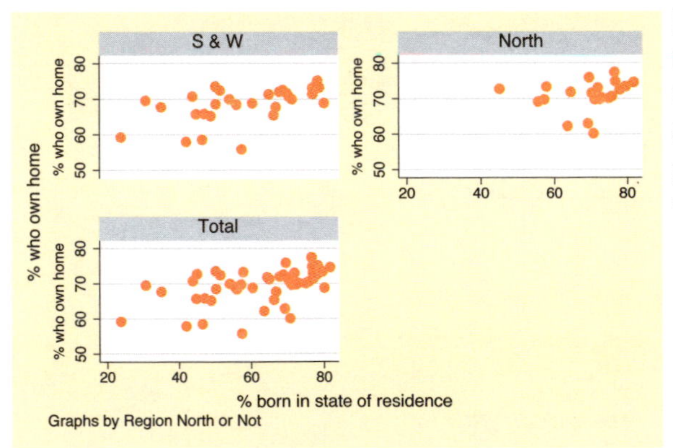

Note that the y-title is not displayed for the North since the y-axis is omitted for that graph. If we include the `iyaxes` and `iytitle` options, the y-axis and y-title are displayed for that graph as well.
Uses allstatesdc.dta & scheme vg_brite

```
twoway scatter ownhome borninstate, by(north, total ixaxes ixtitle)
```

Likewise, we can display the x-title on each graph using the `ixaxes` and `ixtitle` options.
Uses allstatesdc.dta & scheme vg_brite

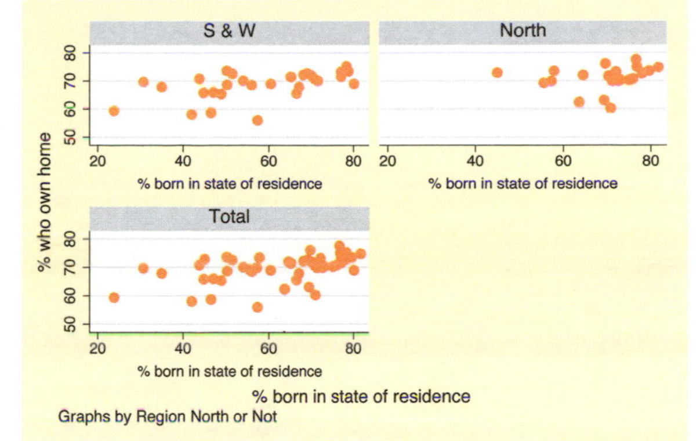


```
twoway scatter ownhome borninstate, by(north) title("My title")
```

If we include a `title()` option with `by()`, Stata creates each graph separately using the title we specify.
Uses allstatesdc.dta & scheme vg_brite

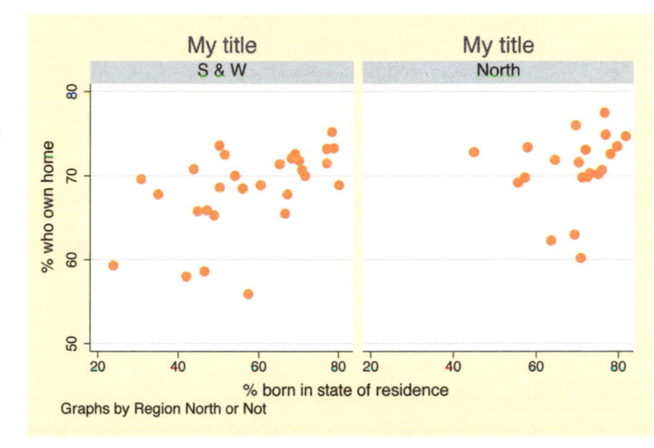


```
twoway scatter ownhome borninstate, by(north, title("My title"))
```

If we make the `title()` an option within the `by()` option, Stata will make this an overall title for the graph.
Uses allstatesdc.dta & scheme vg_brite

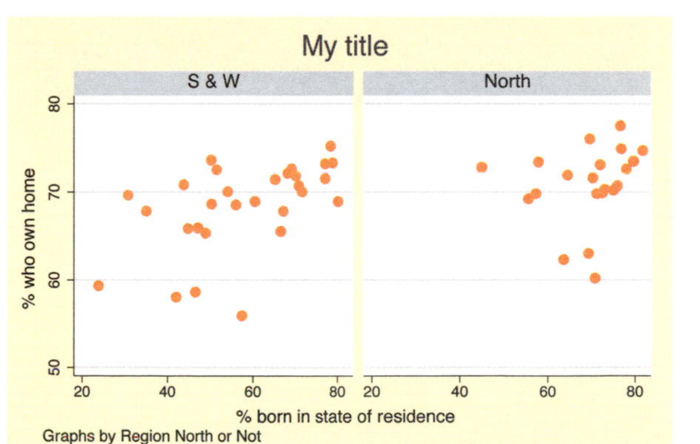

```
twoway scatter ownhome borninstate, by(north, title("By title"))
   title("Regular title")
```

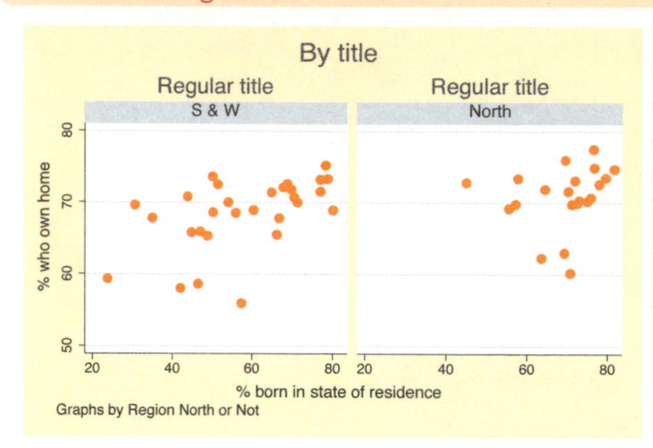

This example should help you to understand how these two types of titles work. When the `title()` is used overall, it applies to all graphs that are created because it is repeated via the `by()` option. The `by(title())` is applied after all smaller graphs are created, providing an overall title for the graph.
Uses allstatesdc.dta & scheme vg_brite

```
twoway scatter ownhome borninstate, by(north)
   caption("Regular caption")
```

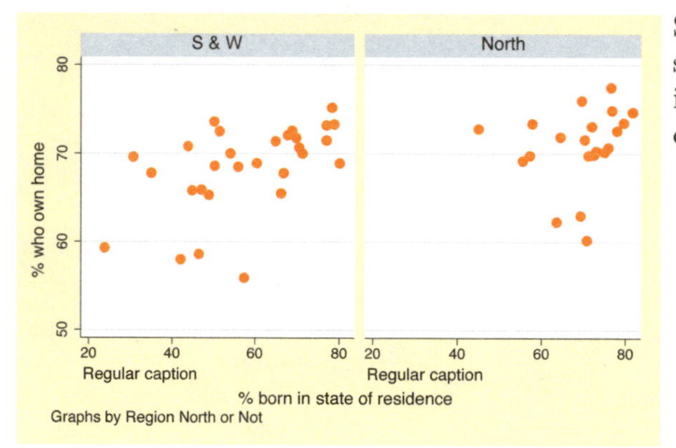

Stata treats the `caption()` option the same way that it treats titles. Here, we include an overall caption, which is displayed with each graph.
Uses allstatesdc.dta & scheme vg_brite

```
twoway scatter ownhome borninstate, by(north, caption("By caption"))
```

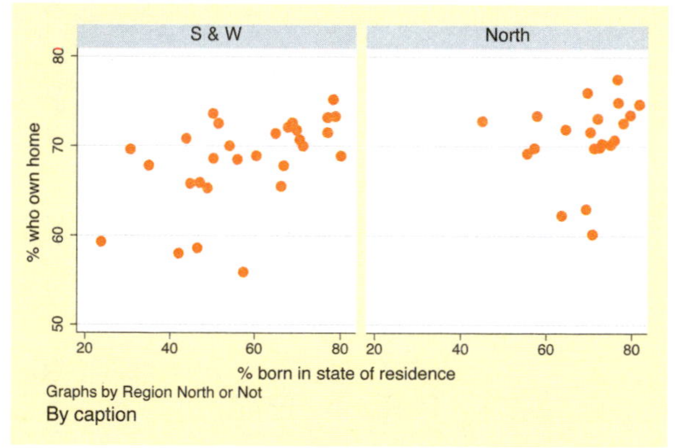

When we include the `caption()` inside the `by()` option, it is displayed as a caption for the full graph.
Uses allstatesdc.dta & scheme vg_brite

```
twoway scatter ownhome borninstate, by(north)
   subtitle("This is a subtitle")
```

Stata treats the subtitle() option differently than the title() and caption() options. Here, we include a subtitle() option, and we see that it has replaced the title above each graph that represented the names of the by-group.
Uses allstatesdc.dta & scheme vg_brite

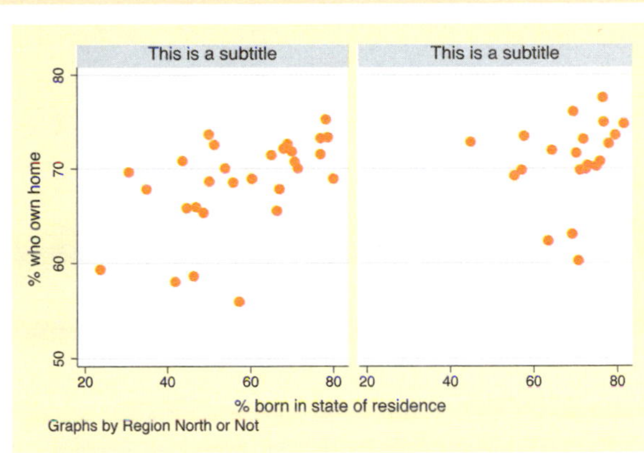


```
twoway scatter ownhome borninstate, by(north)
   subtitle("Region of state", prefix)
```

We can use the subtitle() option to add more labeling to the by-group names. Here, we use the prefix option to insert text that appears in the subtitle before the name of the by-group.
Uses allstatesdc.dta & scheme vg_brite

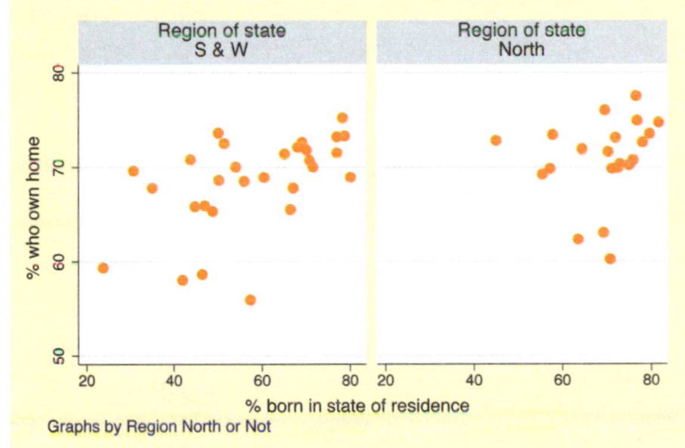


```
twoway scatter ownhome borninstate, by(north)
   subtitle("Region of state", suffix)
```

We can use the **suffix** option to insert text that appears in the subtitle after the name of the by-group.
Uses allstatesdc.dta & scheme vg_brite

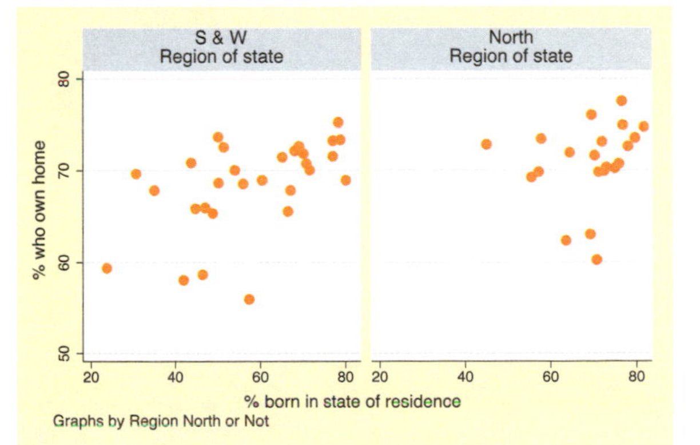

```
twoway scatter ownhome borninstate, by(north)
  subtitle("State's location", prefix)
  subtitle("Based on Region", suffix)
```

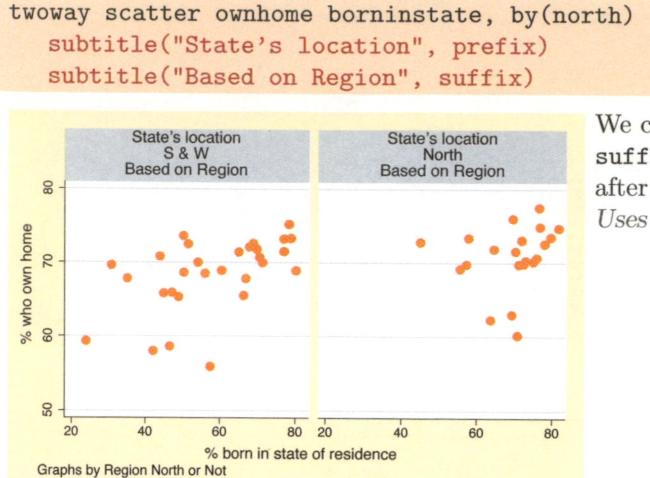

We can even combine the **prefix** and **suffix** option to insert text before and after the label of the by-group.
Uses allstatesdc.dta & scheme vg_brite

```
twoway scatter ownhome borninstate,
  by(north, subtitle("This is a subtitle"))
```

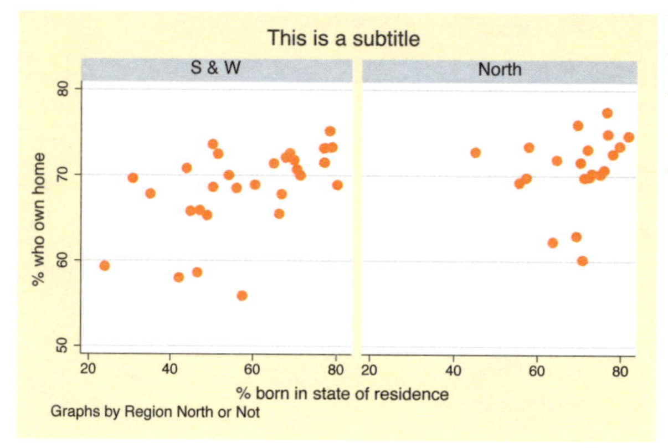

When used as an option within the **by()** option, the **subtitle()** option works just like the **title()** and **caption()** options, placing a subtitle on the overall graph.
Uses allstatesdc.dta & scheme vg_brite

```
twoway scatter ownhome borninstate, by(north) note("Regular note")
```

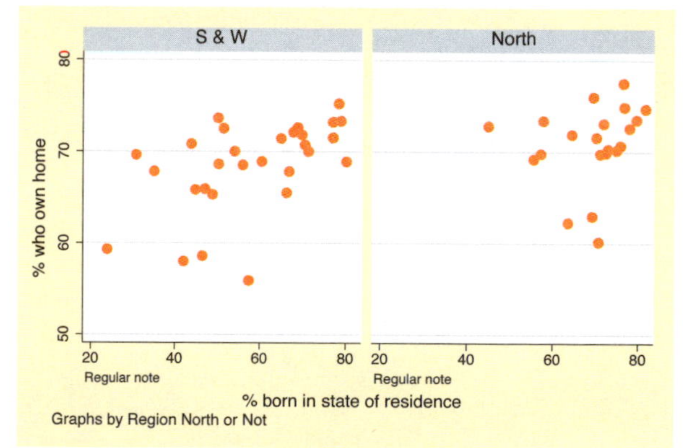

Stata treats the **note()** option much as it does the **title()**, **caption()**, and **subtitle()** options. Here, we include a **note()** option and see that it is shown beneath both graphs.
Uses allstatesdc.dta & scheme vg_brite

```
twoway scatter ownhome borninstate, by(north, note("By note"))
```

If we include the note() option within the by() option, we see that our note overrides the note that Stata provided to indicate that the graphs were separated by the variable **north**.
Uses allstatesdc.dta & scheme vg_brite

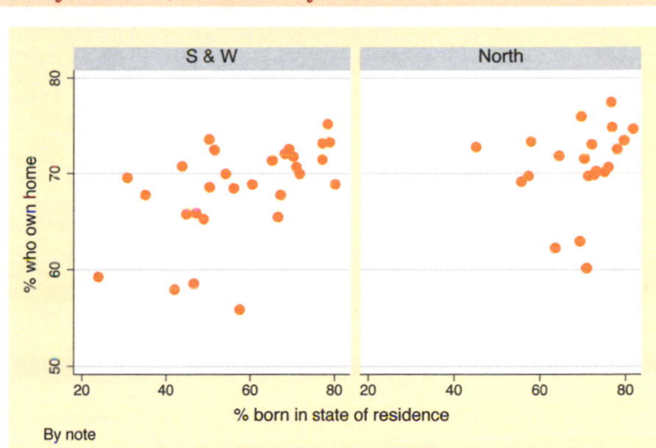


```
twoway scatter ownhome borninstate,
  by(north, note("North N=21, Not North N=29", suffix))
```

As with the subtitle() option, we can use the prefix or suffix option to add our own text before or after the existing note.
Uses allstatesdc.dta & scheme vg_brite

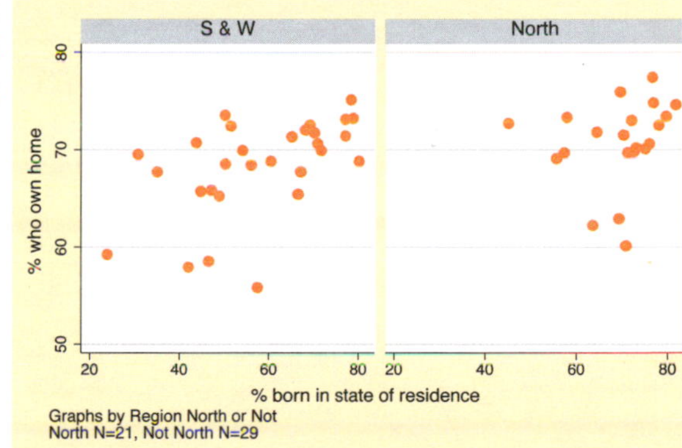


```
twoway scatter ownhome borninstate,
  by(north, total) subtitle(, position(11))
```

Previously, we saw that the subtitle() option could be used to modify the by-group names above each graph. We can also use the subtitle(, position()) option to modify the placement of this text. Here, we move the text to appear in the 11 o'clock position.
Uses allstatesdc.dta & scheme vg_brite

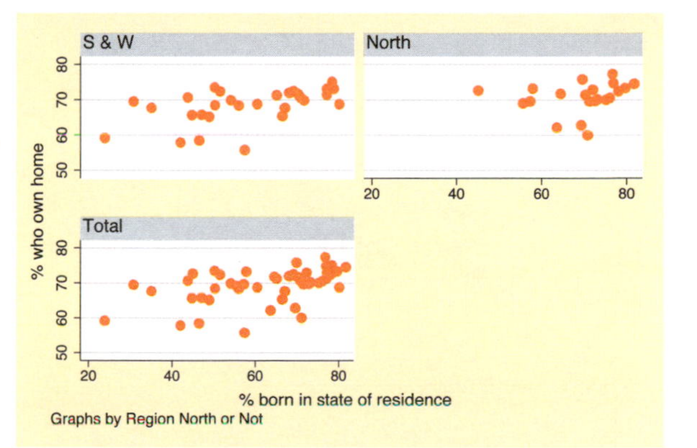

```
twoway scatter ownhome borninstate,
  by(north, total) subtitle(, pos(5) ring(0) nobexpand)
```

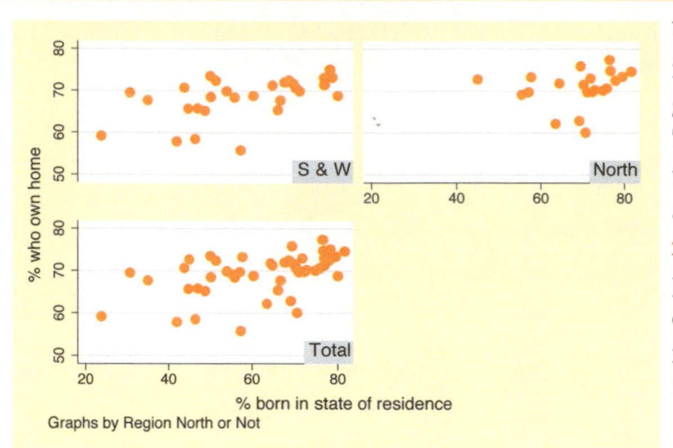

We can place the name of the by-group in the bottom right corner of each graph using the `subtitle()` option. The options `pos(5)` and `ring(0)` move the subtitle to the 5 o'clock position and inside the plot region. The `nobexpand` (no box expand) option prevents the by-group name from expanding to consume the entire plot region.
Uses allstatesdc.dta & scheme vg_brite

```
twoway scatter ownhome borninstate,
  by(north, total title("My title", ring(0) position(5)))
```

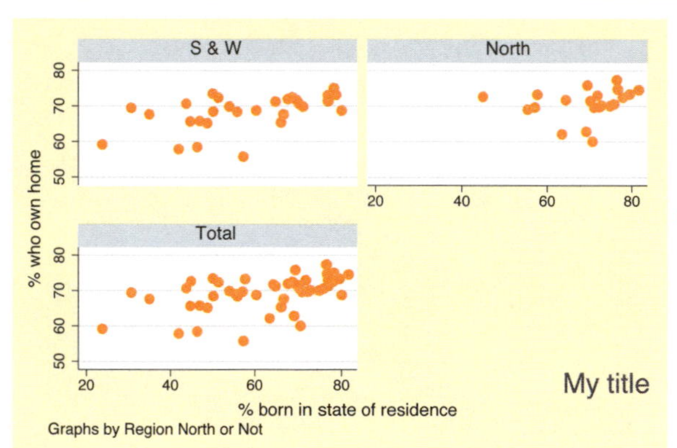

We can also use the `ring()` and `pos()` options with `title()`, `note()`, and `caption()` to alter their placement. Here, we use `position(5)` to put the title in the bottom right corner and `ring(0)` to locate it inside the plot region.
Uses allstatesdc.dta & scheme vg_brite

```
twoway scatter ownhome borninstate,
  by(north, total title("My title", position(5)))
```

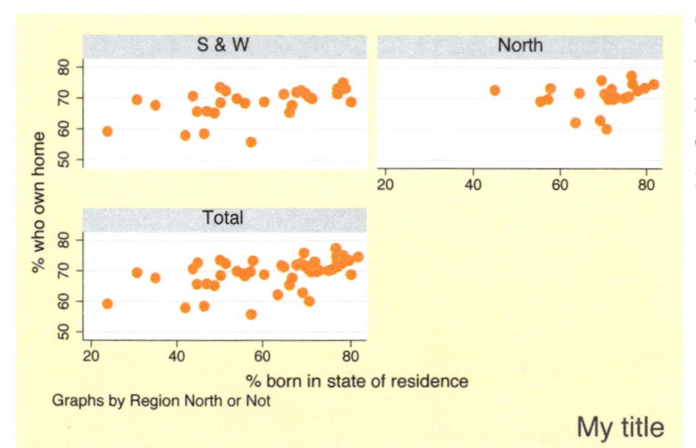

The previous graph is repeated with the `position(5)` option but not the `ring(0)` option to illustrate the impact of `ring(0)`. Without `ring(0)`, the title is placed outside the plot region.
Uses allstatesdc.dta & scheme vg_brite

```
twoway scatter ownhome borninstate,
   by(north, total) l1title("left title") b1title("bottom title")
```

Including the `l1title()` option adds a title to the left (on the y-axis) of each of the graphs. Likewise, the `b1title()` option adds a title to the bottom (on the x-axis) of each of the graphs.
Uses allstatesdc.dta & scheme vg_brite

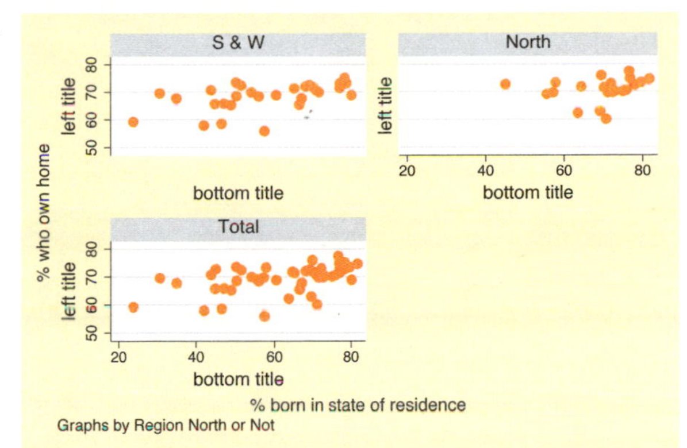


```
twoway scatter (borninstate propval100 ownhome), by(nsw)
   legend(label(1 "Born in state") label(2 "% > 100K"))
```

Here, we use the `legend()` option to change the labels associated with the first two keys. These options modify the contents of the legend, so they should appear outside of the `by()` option.
Uses allstatesdc.dta & scheme vg_brite

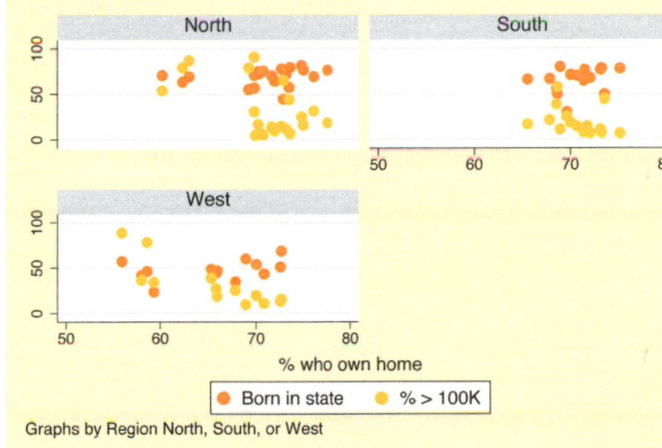


```
twoway scatter (borninstate propval100 ownhome),
   by(nsw, legend(position(12)))
```

In this graph, we use the `position()` option to modify the position of the legend. Such options that modify the position of the legend must be placed as an option within the `by()` option.
Uses allstatesdc.dta & scheme vg_brite

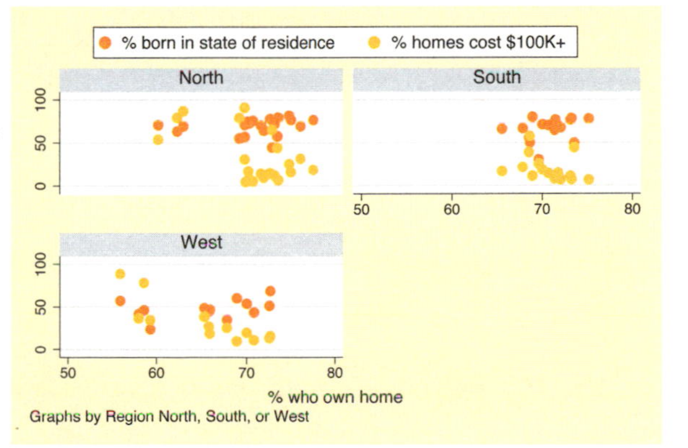

Introduction Twoway Matrix Bar Box Dot Pie Options Standard options Styles Appendix
Markers Marker labels Connecting Axis titles Axis labels Axis scales Axis selection By Legend Adding text Textboxes

```
twoway scatter (borninstate propval100 ownhome),
  by(nsw, legend(pos(12)))
  legend(label(1 "Born in state") label(2 "% > 100K"))
```

Here, we use both of the options from the previous two graphs, and the `legend()` option is used twice: inside the `by()` option to modify the position and outside the `by()` option to modify its contents. The use of `legend()` with the `by()` option is covered more thoroughly in Options : Legend (287).
Uses allstatesdc.dta & scheme vg_brite

```
twoway scatter ownhome borninstate,
  by(north, title("% own home" "by % born in state"))
  title("Region of state")
```

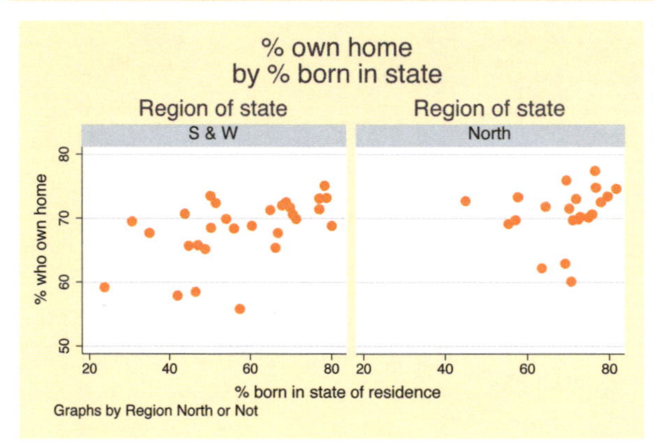

We can use the `title()` option on its own to make a title that is displayed with each graph, and the `title()` option within the `by()` option to make an overall title.
Uses allstatesdc.dta & scheme vg_brite

```
twoway scatter ownhome borninstate,
  by(north, total rescale ixtitle iytitle b1title("") l1title(""))
```

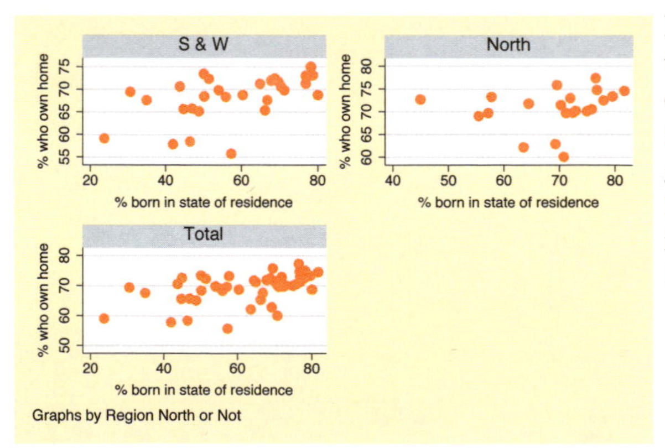

Here, we obtain separate graphs for the three groups, using `rescale` to obtain different x- and y-axis labels and scales, `ixtitle` and `iytitle` to title the graphs separately, and `b1title()` and `l1title()` to suppress the overall titles for the x- and y-axes.
Uses allstatesdc.dta & scheme vg_brite

8.9 Controlling legends

This section describes more details about using legends. Legends can be useful in a number of situations, and this section shows how to customize them. For more information about legend options, see [G] ***legend_option***. Also, for controlling the text and textbox of the legend, see Options : Textboxes (303) and Options : Adding text (299). We will use the vg_s2c scheme.

```
twoway scatter ownhome propval100 urban
```

Legends can be created in a variety of ways. For example, here we have two y-variables, `ownhome` and `propval100`, on the same plot, and Stata creates a legend labeling the different points. The default legend, in this case, is quite useful.
Uses allstatesdc.dta & scheme vg_s2c

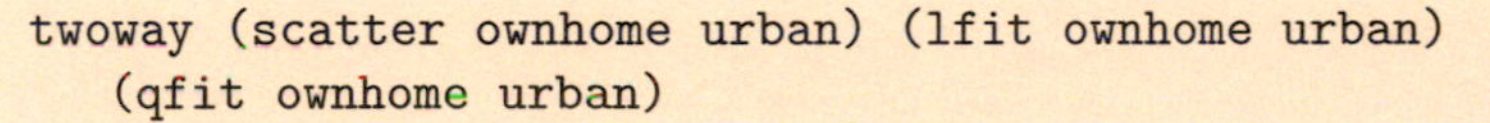

```
twoway (scatter ownhome urban) (lfit ownhome urban)
   (qfit ownhome urban)
```

Legends are also created when you overlay plots. Here, Stata adds a legend entry for each of the overlaid plots. The default legend, in this case, is less useful since it does not help us differentiate between the kinds of fit values.
Uses allstatesdc.dta & scheme vg_s2c

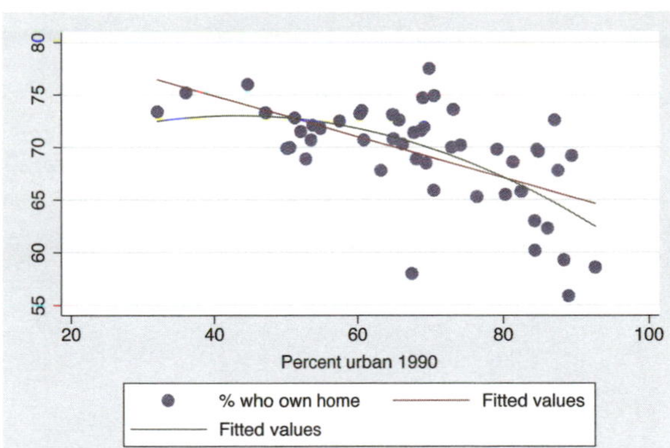

Introduction Twoway Matrix Bar Box Dot Pie Options Standard options Styles Appendix
Markers Marker labels Connecting Axis titles Axis labels Axis scales Axis selection By Legend Adding text Textboxes

```
twoway (scatter ownhome urban if north==0)
   (scatter ownhome urban if north==1)
```

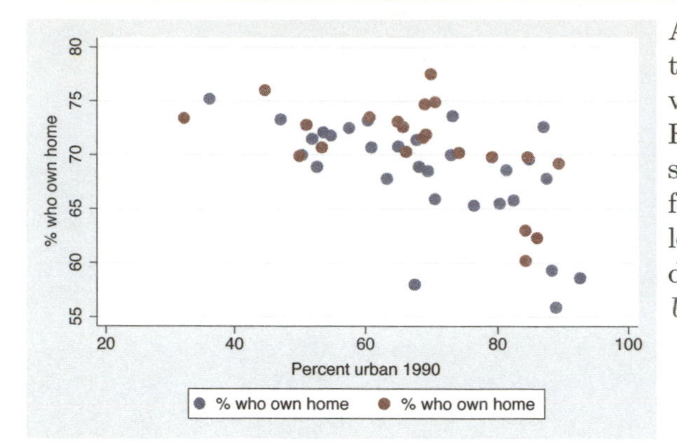

A third example is when you overlay two plots using `if` to display the same variables but for different observations. Here, we show the same scatterplot separately for states in the North and for those not in the North. Here, the legend does not help us at all to differentiate the kinds of values.
Uses allstatesdc.dta & scheme vg_s2c

```
twoway (scatter ownhome urban) (lfit ownhome urban)
   (qfit ownhome urban)
```

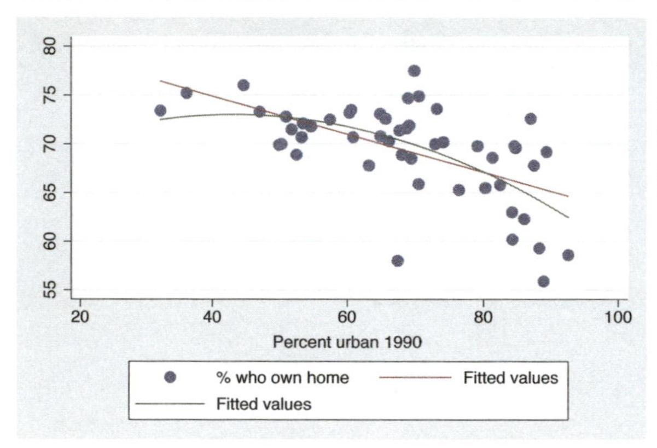

Regardless of the graph command(s) that generated the legend, it can be customized the same way. For many of the examples, we will use this graph for customizing the legend.
Uses allstatesdc.dta & scheme vg_s2c

```
twoway (scatter ownhome urban) (lfit ownhome urban)
   (qfit ownhome urban),
   legend(label(1 "% Own home") label(2 "Lin.  Fit") label(3 "Quad.  Fit"))
```

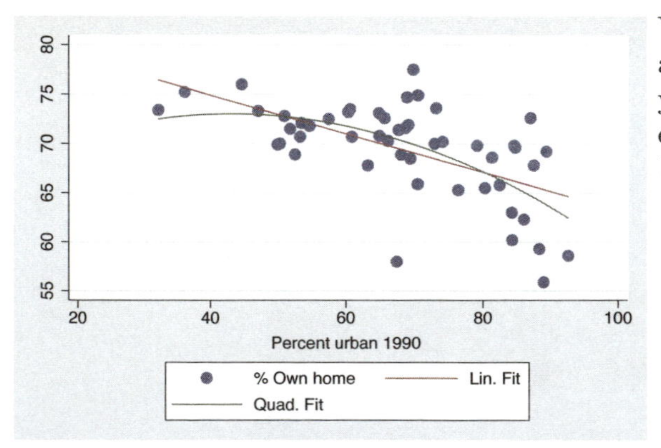

You can use the `label()` option to assign labels for the keys. Note that you use a separate `label()` option for each key that you wish to modify.
Uses allstatesdc.dta & scheme vg_s2c

```
twoway (scatter ownhome urban) (lfit ownhome urban)
   (qfit ownhome urban),
   legend(label(2 "Lin.  Fit") label(3 "Quad.  Fit"))
```

You can use the label() option to modify just some of the keys; for example, here we just modify the second and third key.
Uses allstatesdc.dta & scheme vg_s2c

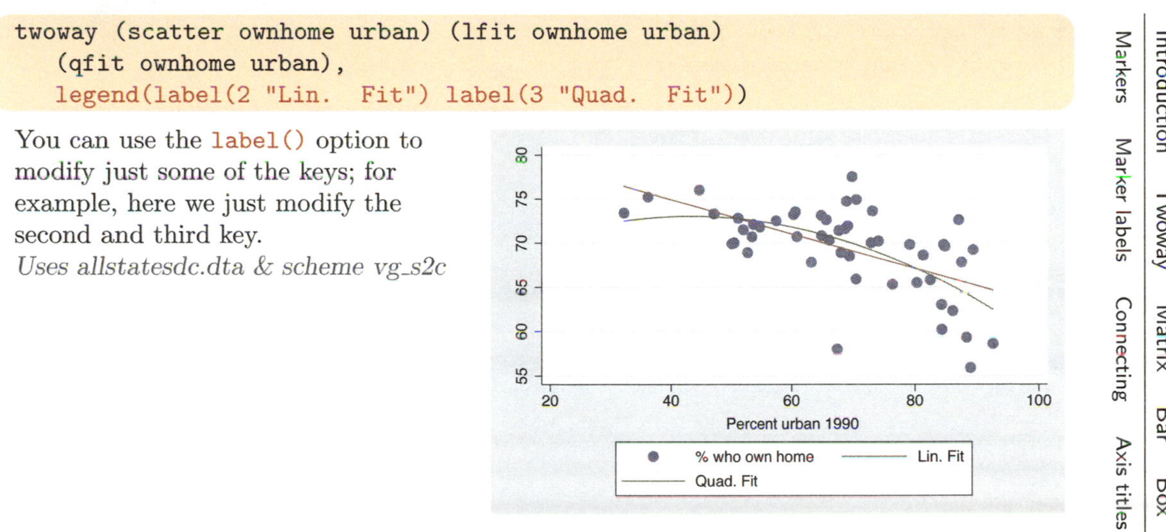

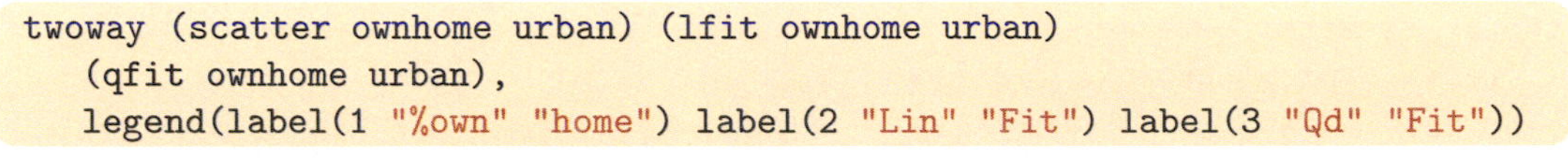

```
twoway (scatter ownhome urban) (lfit ownhome urban)
   (qfit ownhome urban),
   legend(label(1 "%own" "home") label(2 "Lin" "Fit") label(3 "Qd" "Fit"))
```

You can put the label on multiple lines by including multiple quoted strings.
Uses allstatesdc.dta & scheme vg_s2c

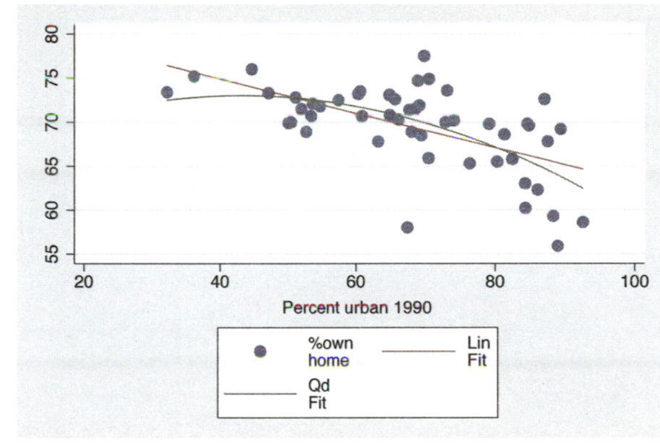


```
twoway (scatter ownhome urban) (lfit ownhome urban)
   (qfit ownhome urban), legend(order(2 3 1))
```

You can use the order() option to change the order of the keys in the legend.
Uses allstatesdc.dta & scheme vg_s2c

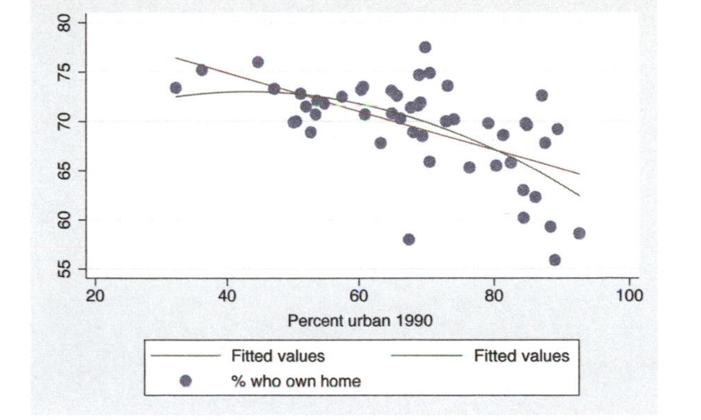

```
twoway (scatter ownhome urban) (lfit ownhome urban)
   (qfit ownhome urban), legend(order(2 3))
```

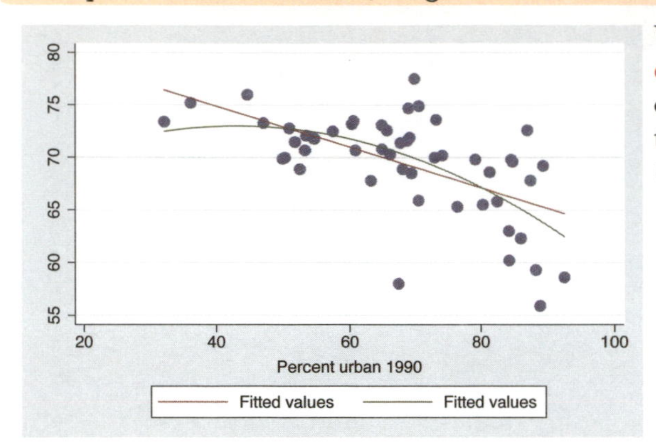

We can also omit keys from the `order()` option to suppress their display in the legend. Here, we suppress the display of the first key.
Uses allstatesdc.dta & scheme vg_s2c

```
twoway (scatter ownhome urban) (lfit ownhome urban)
   (qfit ownhome urban), legend(order(2 "Lin.  fit" 3 "Quad.  fit" 1))
```

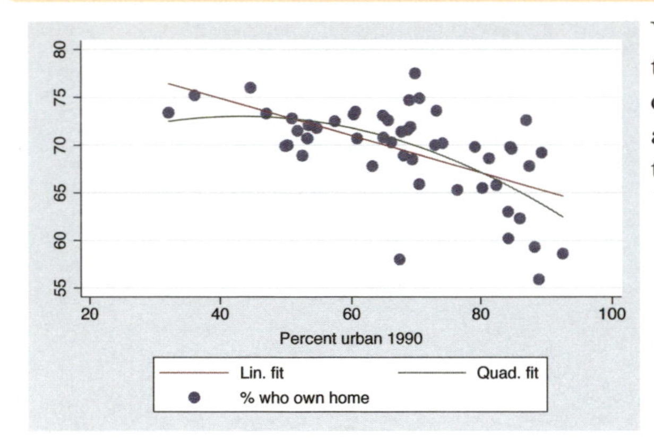

You can also insert and replace text for the keys when using the `order()` option. Here, we order the keys 2, 3, and 1, and at the same time, replace the text for keys 2 and 3.
Uses allstatesdc.dta & scheme vg_s2c

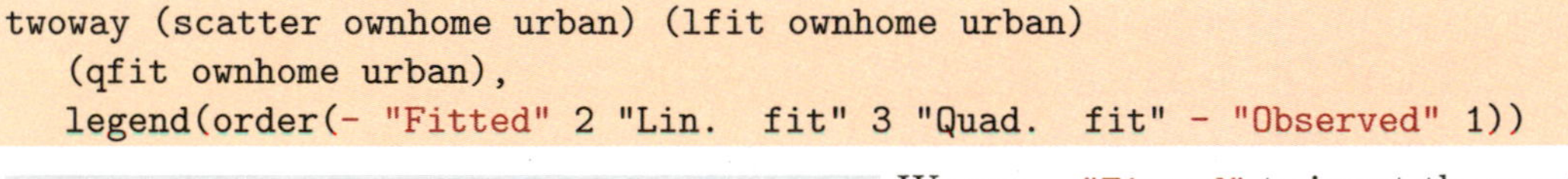

```
twoway (scatter ownhome urban) (lfit ownhome urban)
   (qfit ownhome urban),
   legend(order(- "Fitted" 2 "Lin.  fit" 3 "Quad.  fit" - "Observed" 1))
```

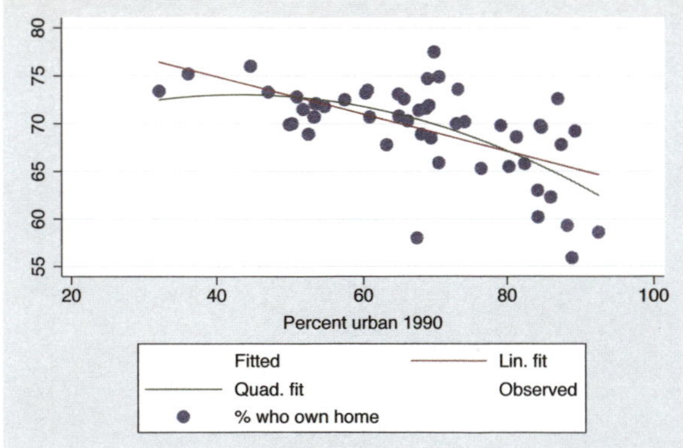

We use `- "Fitted"` to insert the word Fitted and `- "Observed"` to insert the word Observed. Due to the organization of the keys in the legend, this is hard to follow.
Uses allstatesdc.dta & scheme vg_s2c

```
twoway (scatter ownhome urban) (lfit ownhome urban) (qfit ownhome urban),
   legend(order(- "Fitted" 2 "Lin.  fit" 3 "Quad.  fit" - "Observed" 1)
   cols(1))
```

We can use the cols() option to display the legend in a single column. Here, the added text makes more sense, but the legend uses quite a bit of space.
Uses allstatesdc.dta & scheme vg_s2c

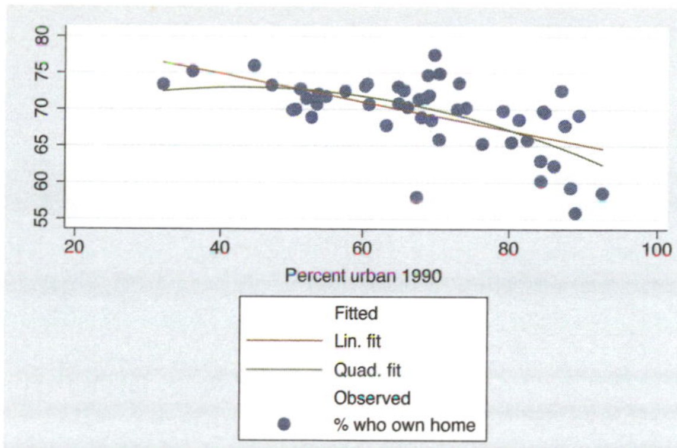


```
twoway (scatter ownhome urban) (lfit ownhome urban) (qfit ownhome urban),
   legend(order(- "Fitted" 2 "Lin.  fit" 3 "Quad.  fit" - "Observed" 1)
   rows(3))
```

We can use the rows() option to display the legend in three rows. If we want to display the fitted keys in the left column and the observed keys in the right column, we can order the keys according to columns instead of according to rows. See the next example.
Uses allstatesdc.dta & scheme vg_s2c

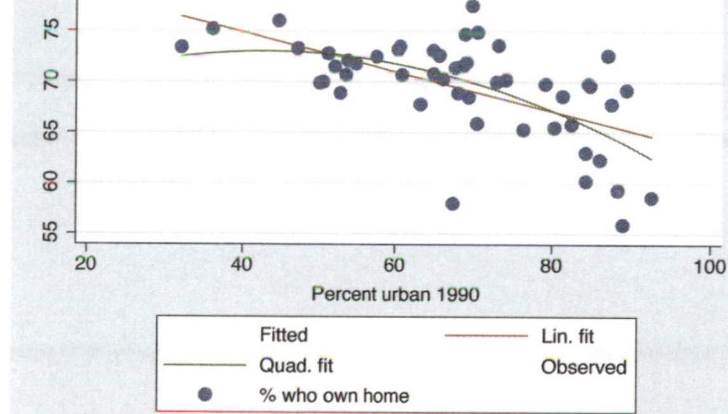


```
twoway (scatter ownhome urban) (lfit ownhome urban) (qfit ownhome urban),
   legend(order(- "Fitted" 2 "Lin.  fit" 3 "Quad.  fit" - "Observed" 1)
   rows(3) colfirst)
```

Adding the colfirst option displays the keys in column order instead of row order, with the Fitted keys in the left column and the Observed keys in the right column.
Uses allstatesdc.dta & scheme vg_s2c

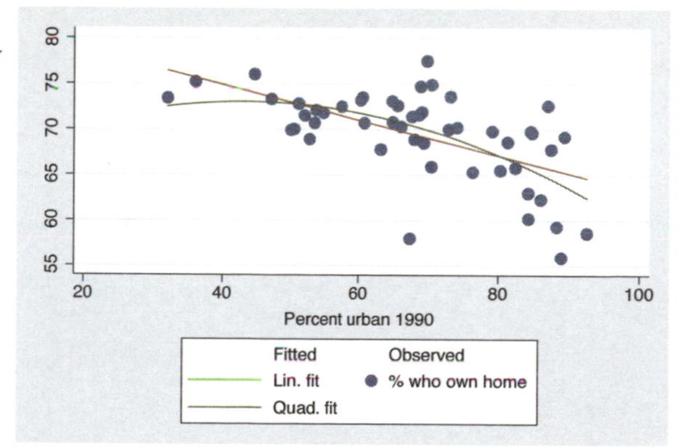

```
twoway (scatter ownhome urban) (lfit ownhome urban) (qfit ownhome urban),
   legend(order(- "Observed" 1 - "Fitted" 2 "Lin.  fit" 3 "Quad.  fit")
   rows(3) holes(3) colfirst)
```

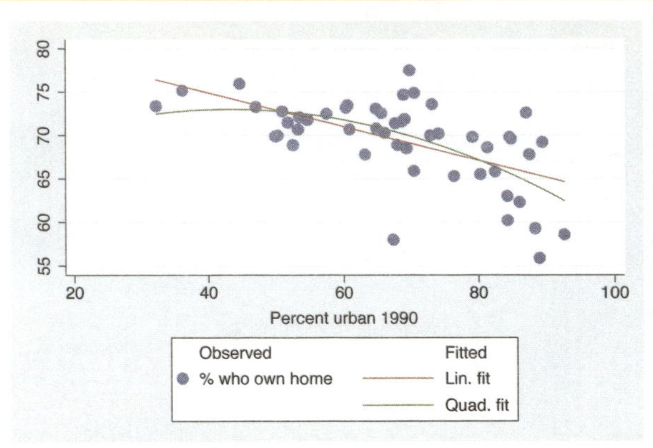

This legend is the same as the one in the previous example but places the Observed keys in the left column and the Fitted keys in the right column. To do this, we changed the order of the keys but also added the `holes(3)` option so that Fitted would be in the fourth position at the top of the second column.
Uses allstatesdc.dta & scheme vg_s2c

```
twoway (scatter ownhome urban) (lfit ownhome urban) (qfit ownhome urban),
   legend(order(- "Observed" 1 - " " - "Fitted" 2 "Lin fit" 3 "Qd fit")
   rows(3) colfirst)
```

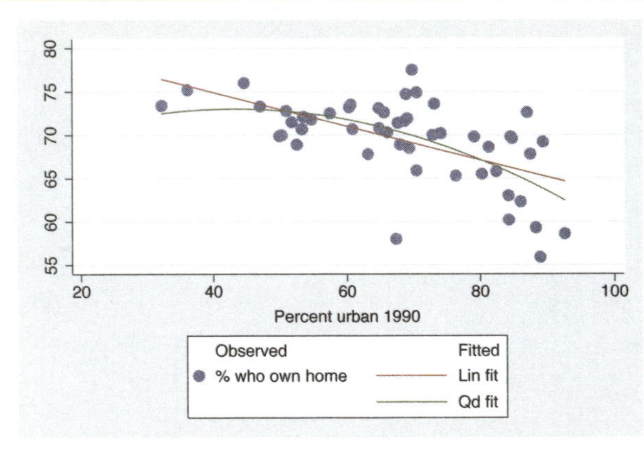

Referring to the last graph, instead of using `holes()`, we can put in a blank key, `- " "`, in the `order()` option, which pushes the word Fitted over to the next column.
Uses allstatesdc.dta & scheme vg_s2c

```
twoway (scatter ownhome urban) (lfit ownhome urban) (qfit ownhome urban),
   legend(order(- "Observed" 1 - " " - "Fitted" 2 "Lin fit" 3 "Qd fit")
   rows(3) colfirst textfirst)
```

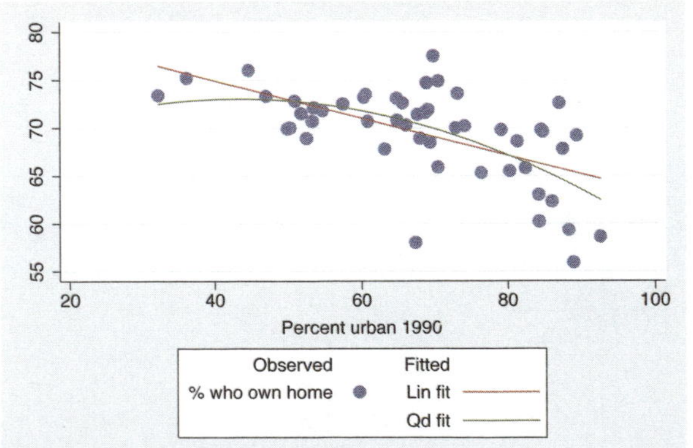

Using the `textfirst` option, we can make the text for the key appear first, followed by the symbol.
Uses allstatesdc.dta & scheme vg_s2c

```
twoway (scatter ownhome urban) (lfit ownhome urban) (qfit ownhome urban),
   legend(order(2 "Linear" "Fit" 3 "Quadratic" "Fit")
   stack cols(1))
```

Using the stack option, we can stack the symbols above the labels. We use this here to make a tall, narrow legend.
Uses allstatesdc.dta & scheme vg_s2c

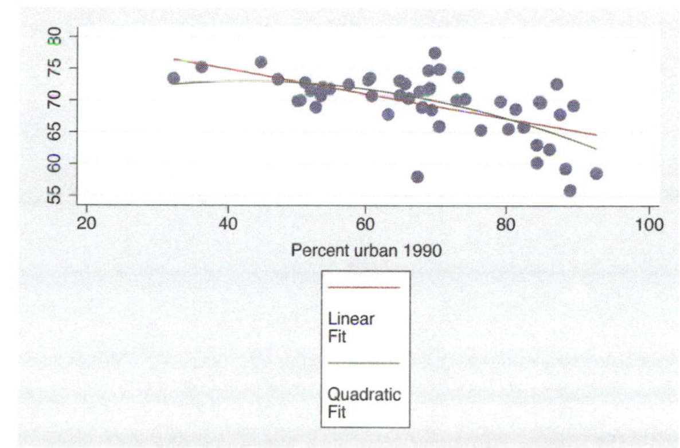


```
twoway (scatter ownhome urban) (lfit ownhome urban) (qfit ownhome urban),
   legend(order(2 "Linear" "Fit" 3 "Quadratic" "Fit")
   stack cols(1) position(3))
```

We can use the position() option to change where the legend is displayed. Here, we take the narrow legend from the previous graph and put it to the right of the graph, making good use of space.
Uses allstatesdc.dta & scheme vg_s2c

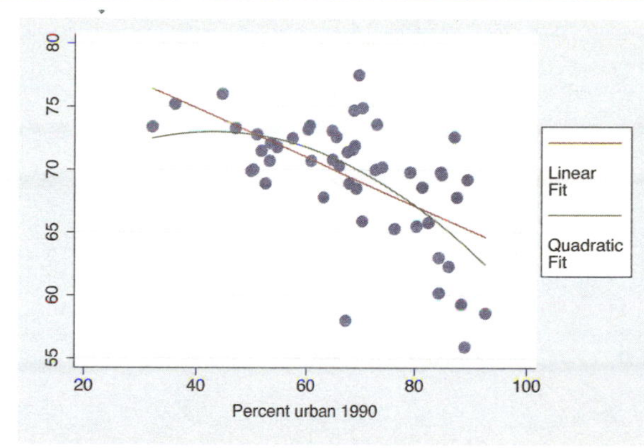


```
twoway (scatter ownhome urban) (lfit ownhome urban) (qfit ownhome urban),
   legend(order(2 "Linear" "Fit" 3 "Quadratic" "Fit")
   stack cols(1) ring(0) position(7))
```

We can use the ring(0) option to place the legend inside the plot area and use position(7) to put it in the bottom left corner, using the empty space in the plot for the legend.
Uses allstatesdc.dta & scheme vg_s2c

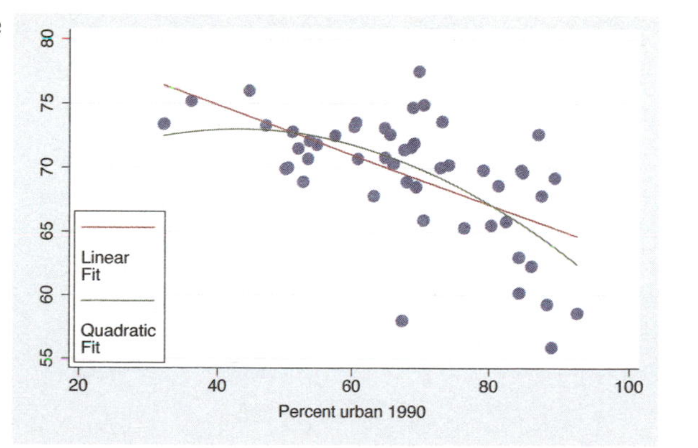

```
twoway (scatter ownhome urban) (lfit ownhome urban) (qfit ownhome urban),
   legend(order(1 "% Own Home" 2 "Linear" 3 "Quad")
   rows(1) position(12))
```

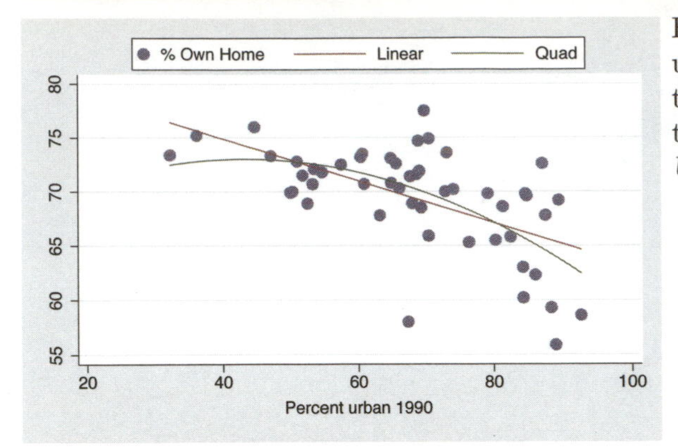

Here, we make the legend a thin row using the rows(1) option and then use the position(12) option to put it at the top of the graph.
Uses allstatesdc.dta & scheme vg_s2c

```
twoway (scatter ownhome urban) (lfit ownhome urban) (qfit ownhome urban),
   legend(order(1 "% Own Home" 2 "Linear" 3 "Quad")
   rows(1) position(12) bexpand)
```

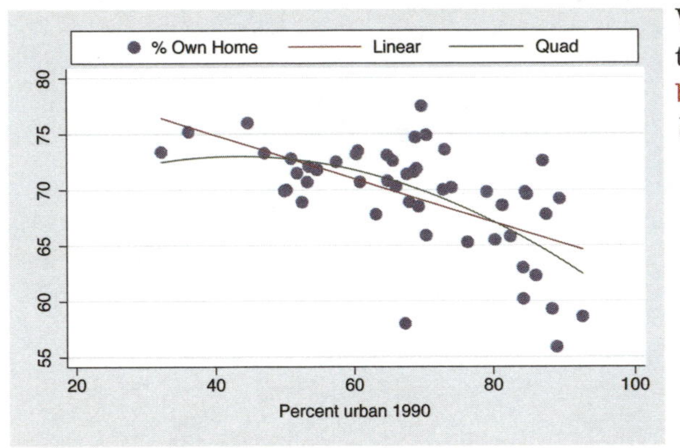

We can expand the width of the legend to the width of the plot area using the bexpand (box expand) option.
Uses allstatesdc.dta & scheme vg_s2c

```
twoway (scatter ownhome urban) (lfit ownhome urban) (qfit ownhome urban),
   legend(order(2 "Linear Fit" 3 "Quadratic Fit")
   rows(1) position(12) bexpand span)
```

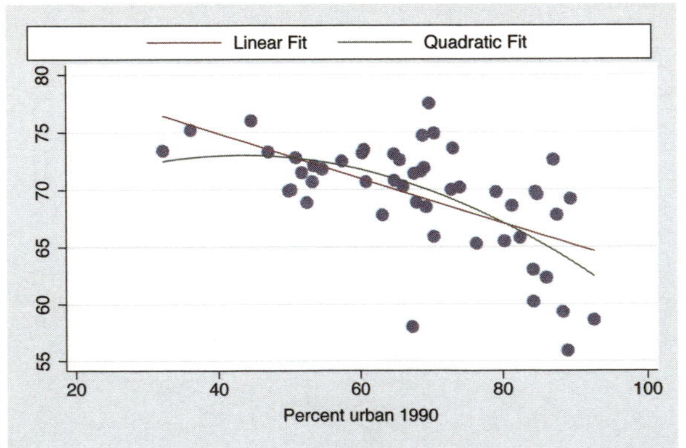

If we wanted to expand the legend to the entire width of the graph area (not just the plot area), we would add the span option.
Uses allstatesdc.dta & scheme vg_s2c

```
twoway (scatter ownhome urban) (lfit ownhome urban) (qfit ownhome urban),
   legend(order(2 "Linear Fit" 3 "Quadratic Fit")
   rows(1) pos(5) title("Legend", position(11)))
```

We can add a title, subtitle, note, or caption to the legend using all the features described in Standard options : Titles (313). Here, we add a title() and use the position() option to position it in the top left corner. A simple way to get a smaller title is to use the subtitle() option instead.
Uses allstatesdc.dta & scheme vg_s2c

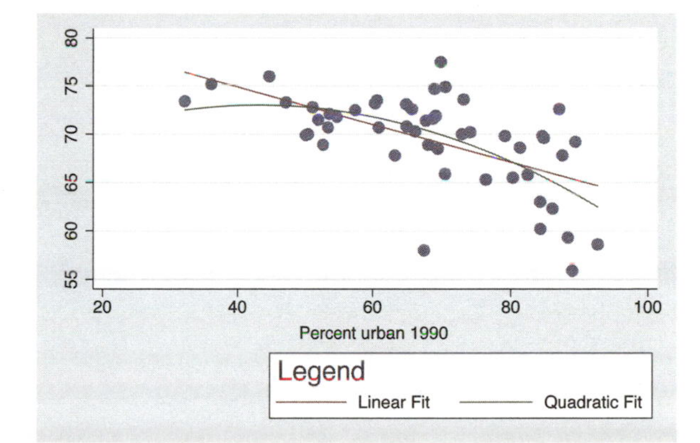


```
twoway (scatter ownhome urban) (lfit ownhome urban) (qfit ownhome urban),
   legend(order(2 "Linear Fit" 3 "Quadratic Fit")
   rows(1) pos(5) subtitle("Legend", box bexpand))
```

To emphasize all the control we have, we could put the subtitle for the legend in a box and use bexpand to make it expand to the width of the legend.
Uses allstatesdc.dta & scheme vg_s2c

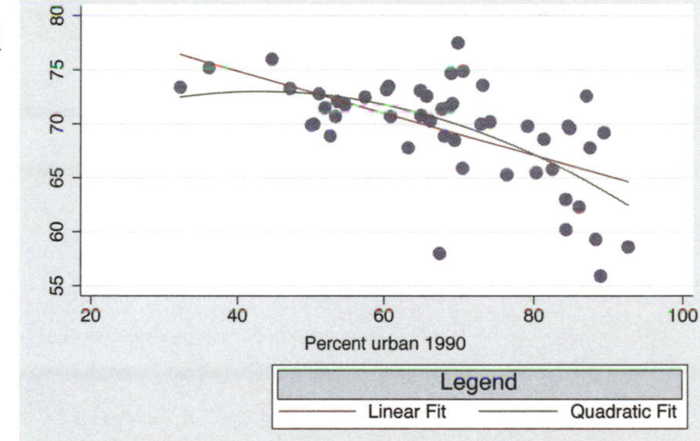


```
twoway (scatter ownhome urban) (lfit ownhome urban) (qfit ownhome urban),
   legend(order(2 "Linear Fit" 3 "Quadratic Fit")
   rows(1) pos(5) note("Fit obtained with lfit and qfit"))
```

Here, we use the note() option, showing that we can even add a note to the legend.
Uses allstatesdc.dta & scheme vg_s2c

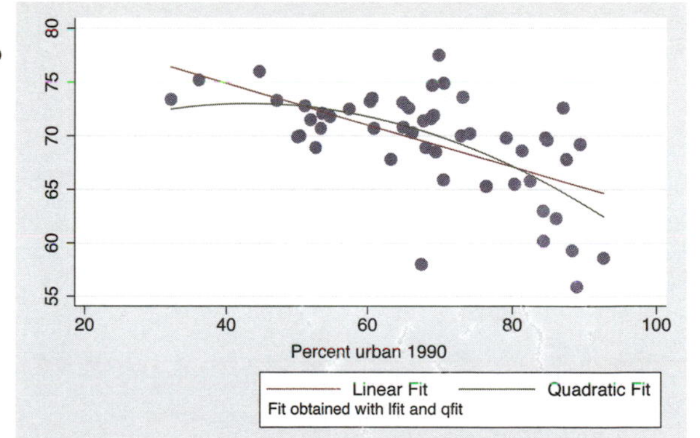

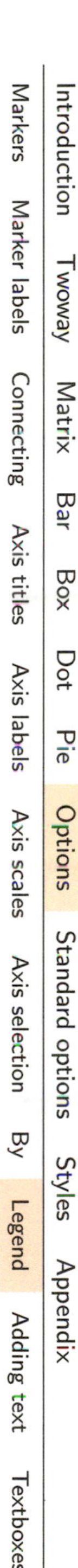

```
twoway (scatter ownhome urban) (lfit ownhome urban) (qfit ownhome urban),
   legend(size(medium) color(maroon) bfcolor(eggshell) box)
```

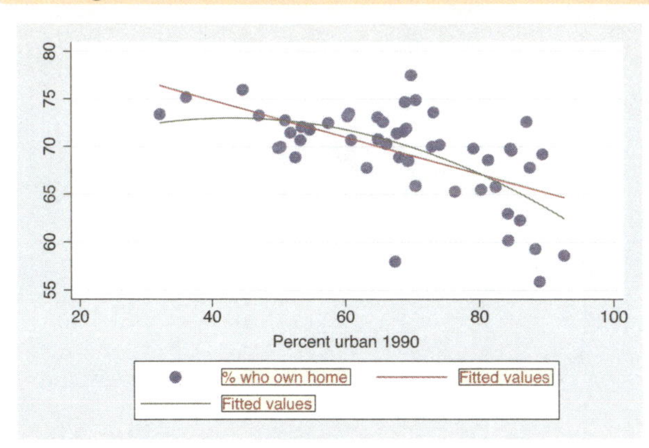

The `legend()` option permits us to supply options that control the display of the labels for the keys. Here, we request that those labels be maroon, medium in size, displayed with an eggshell background, and surrounded by a box.
Uses allstatesdc.dta & scheme vg_s2c

```
twoway (scatter ownhome urban) (lfit ownhome urban) (qfit ownhome urban),
   legend(region(fcolor(dimgray) lcolor(gs8) lwidth(thick) margin(medium)))
```

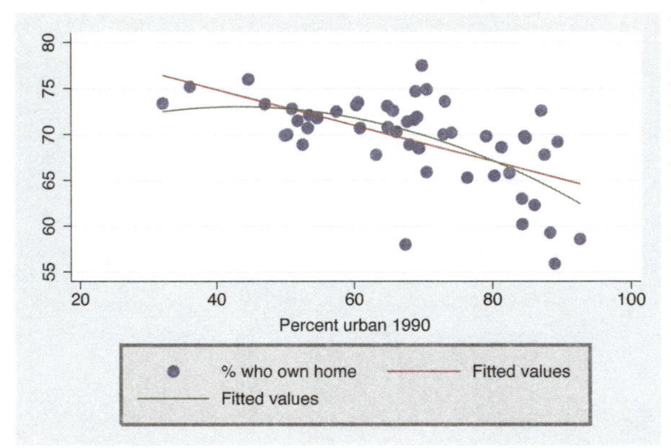

The `region()` option can be used to control the overall box in which the legend is placed. Here, we specify the fill color to be a dim gray, the line color to be a medium gray (`gs8` = gray scale 8), the line to be thick, and the margin between the text and the box to be medium.
Uses allstatesdc.dta & scheme vg_s2c

```
twoway (scatter ownhome urban) (lfit ownhome urban) (qfit ownhome urban),
   legend(rows(1) bmargin(t=10))
```

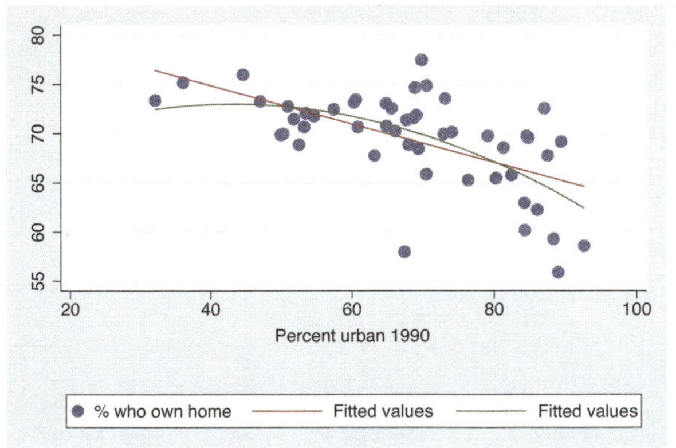

We can adjust the margin around the box of the legend with the `bmargin()` option. Here, we use `t=10` to make the margin 10 at the top, increasing the gap between the legend and the title of the x-axis.
Uses allstatesdc.dta & scheme vg_s2c

```
twoway (scatter ownhome urban) (lfit ownhome urban) (qfit ownhome urban),
   legend(symxsize(30) symysize(20))
```

We can control the width allocated to symbols with the `symxsize()` option and the height with the `symysize()` option.
Uses allstatesdc.dta & scheme vg_s2c

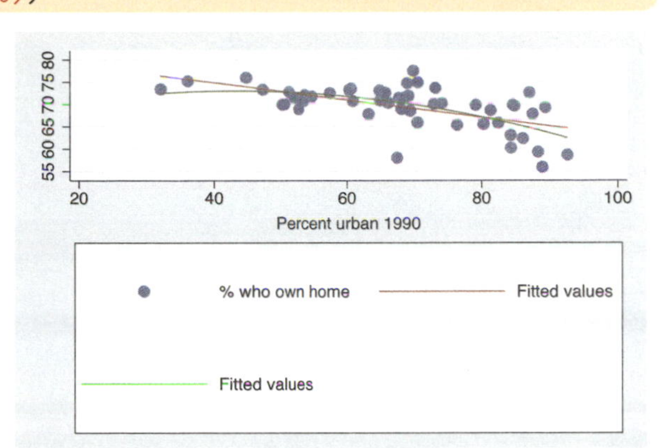


```
twoway (scatter ownhome urban) (lfit ownhome urban) (qfit ownhome urban),
   legend(colgap(20) rowgap(20))
```

We can control the space between columns of the legend with the `colgap()` option and the space between the rows with the `rowgap()` option. Note that the `rowgap()` option does not affect the border between the top row and the box or the border between the bottom row and the box.

Uses allstatesdc.dta & scheme vg_s2c

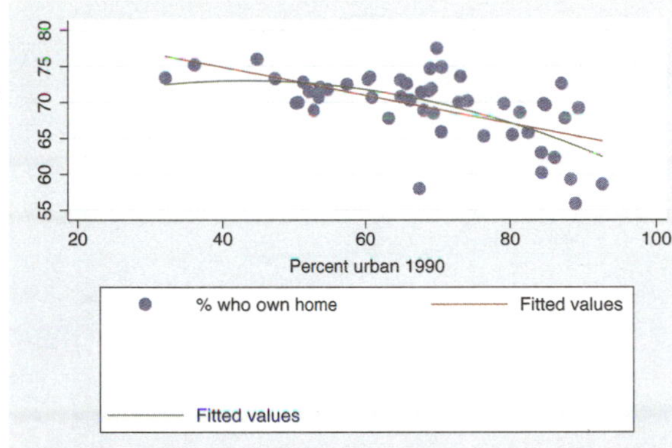


```
twoway (scatter ownhome urban) (qfit ownhome urban),
   by(nsw)
```

Consider this graph, which shows two overlaid scatterplots shown separately by the location of the state. We will explore how to modify the legend for this kind of graph.
Uses allstatesdc.dta & scheme vg_s2c

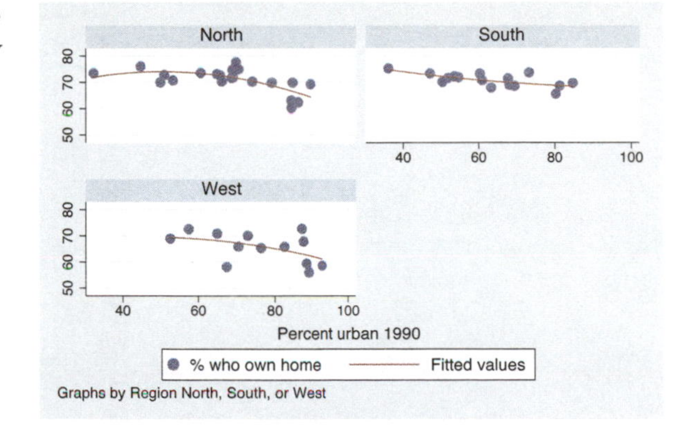

```
twoway (scatter ownhome urban) (qfit ownhome urban),
   by(nsw) legend(position(12) label(2 "Quadratic Fit"))
```

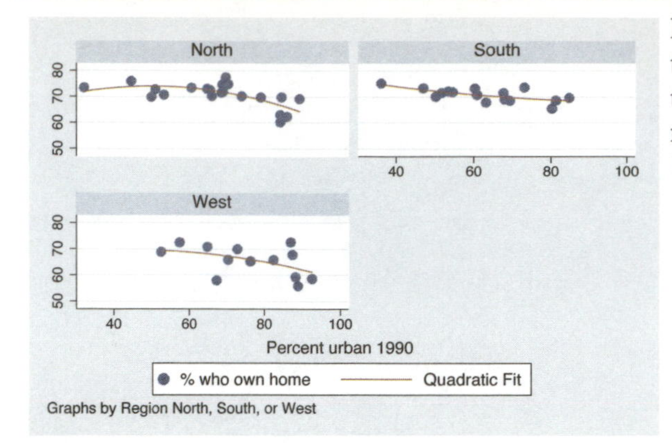

Here, we add a `legend()` option, but the `position()` option does not seem to have any effect since it does not move the position of the legend.
Uses allstatesdc.dta & scheme vg_s2c

```
twoway (scatter ownhome urban) (qfit ownhome urban),
   by(nsw, legend(position(12))) legend(label(2 "Quadratic Fit"))
```

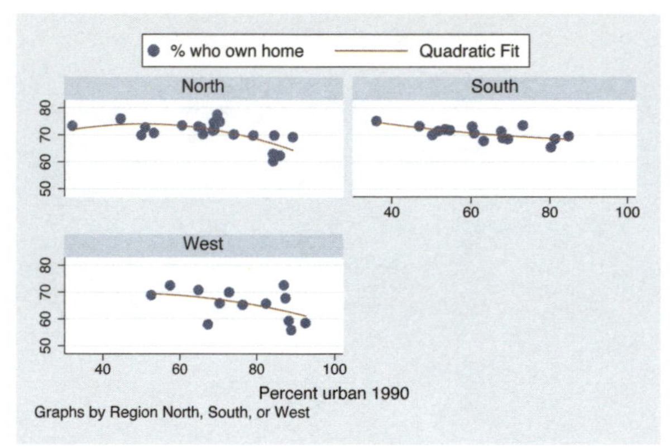

The graph command from the last example did not change the position of the legend because options for positioning the legend must be placed within the `by()` option. Here, we place the `legend(position())` option within the `by()` option, and the legend is now placed above the graph.
Uses allstatesdc.dta & scheme vg_s2c

```
twoway (scatter ownhome urban) (qfit ownhome urban), by(nsw, legend(off))
```

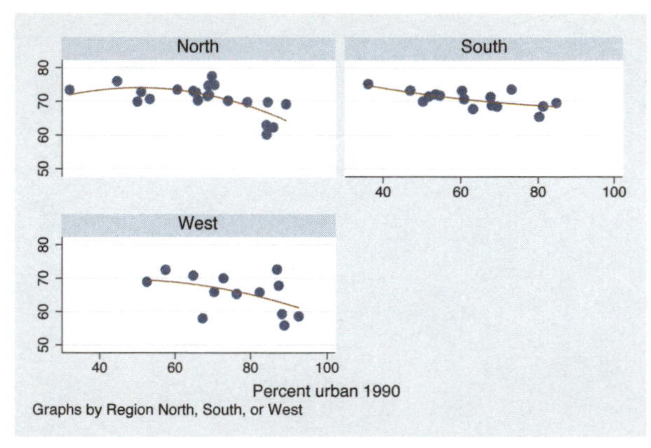

Likewise, if we wish to turn the legend off, we must place `legend(off)` within the `by()` option.
Uses allstatesdc.dta & scheme vg_s2c

```
twoway (scatter ownhome urban) (qfit ownhome urban),
  by(nsw, legend(at(4))) legend(cols(1))
```

To place the legend in one of the holes, we can use the at() option within the by() option. Here, the legend is placed inside the fourth position. To display the legend in one column, we use the legend(cols(1)) option outside of the by() option since this does not control the position of the legend.
Uses allstatesdc.dta & scheme vg_s2c

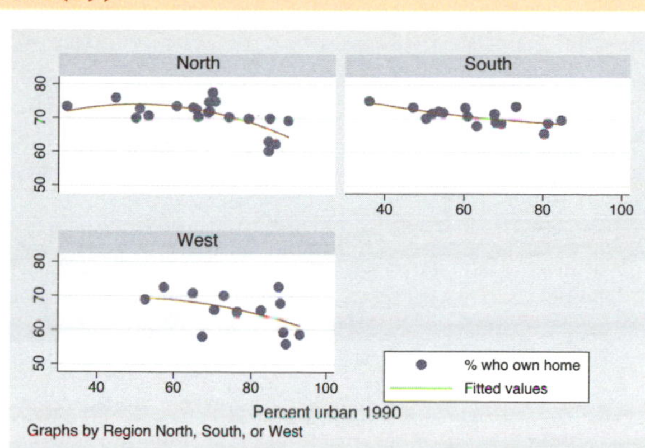


```
twoway (scatter ownhome urban) (qfit ownhome urban),
  by(nsw, legend(position(center) at(4))) legend(cols(1))
```

To position the legend, we can add the position(center) option within the by() option to make the legend appear in the center of the fourth position.
Uses allstatesdc.dta & scheme vg_s2c

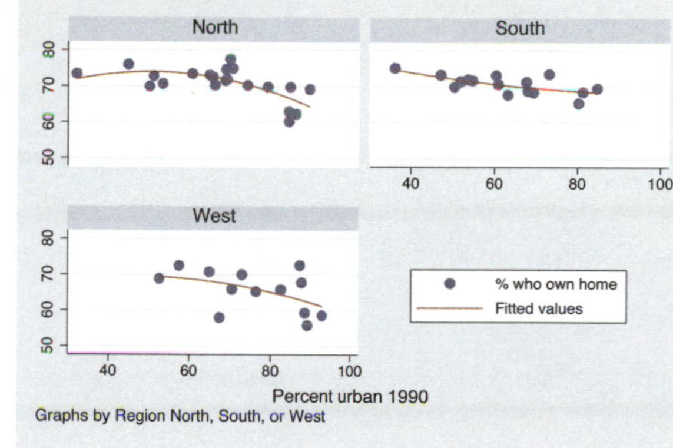

8.10 Adding text to markers and positions

This section provides more details about the text() option for adding text to a graph. Although added text can be used in a wide variety of situations, we will focus on how it can be used to label points and lines and to add descriptive text to your graph. For more information about this option, see [G] ***added_text_option***. To learn more about how the text can be customized, see Options : Textboxes (303). For this section, we will use the vg_teal scheme.

Introduction Twoway Matrix Bar Box Dot Pie Options Standard options Styles Appendix
Markers Marker labels Connecting Axis titles Axis labels Axis scales Axis selection By Legend Adding text Textboxes

```
twoway scatter ownhome borninstate
```

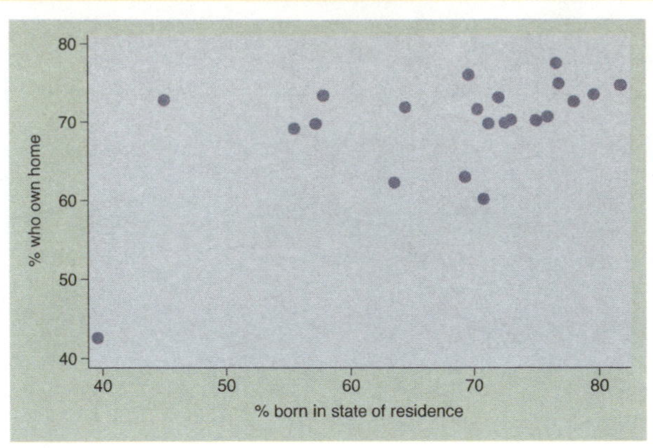

In this scatterplot, one point appears to be an outlier from the rest, but since it is not labeled, we cannot tell from which state it originates.
Uses allstatesn.dta & scheme vg_teal

```
scatter ownhome borninstate, mlabel(stateab)
```

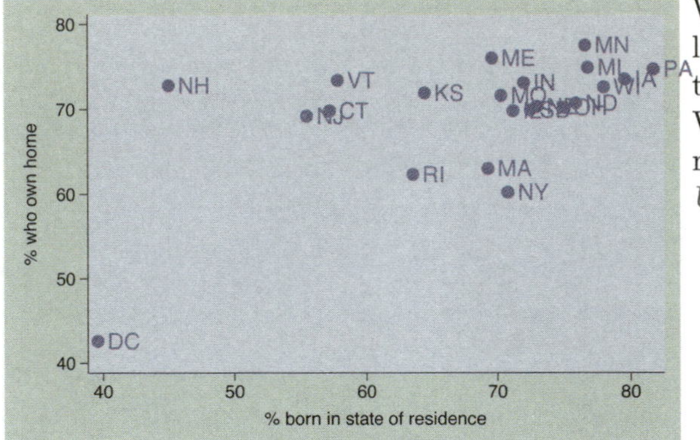

We can use the `mlabel(stateab)` to label all points, which helps us see that the outlying point comes from Washington, DC. However, this plot is rather cluttered by all the labels.
Uses allstatesn.dta & scheme vg_teal

```
twoway (scatter ownhome borninstate)
   (scatter ownhome borninstate if stateab == "DC", mlabel(stateab))
```

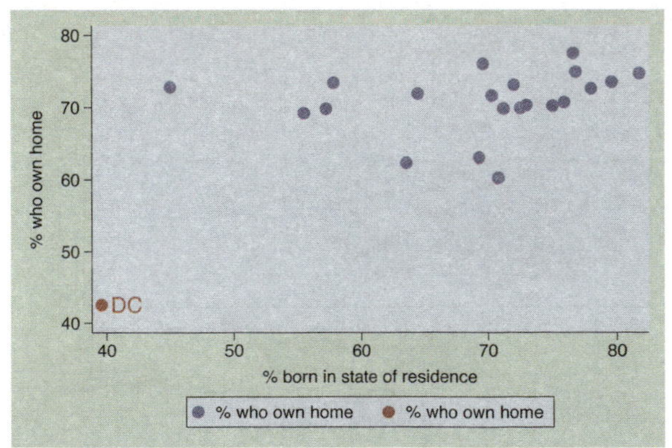

We could repeat a second scatterplot just to label DC, but this is a bit cumbersome.
Uses allstatesn.dta & scheme vg_teal

```
twoway scatter ownhome borninstate, text(43 40 "DC")
```

Instead, we can use the `text()` option to add text to our graph. Looking at the values of `ownhome` and `borninstate` for DC, we see that their values are about 43 and 40, respectively. We use these as coordinates to label the point, but the `text()` option places the label at the center of the specified y x coordinate, sitting right over the point.
Uses allstatesn.dta & scheme vg_teal

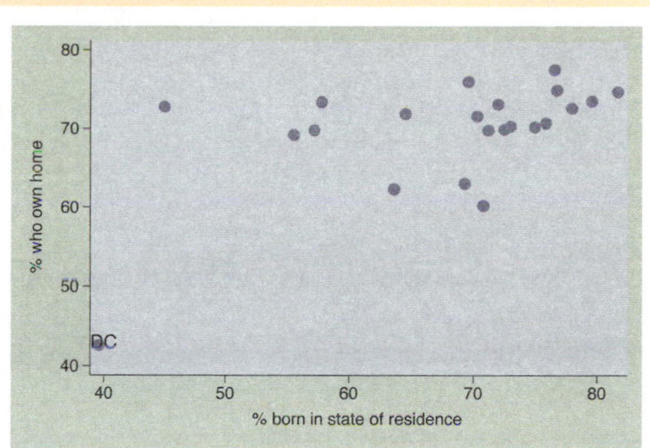


```
twoway scatter ownhome borninstate, text(43 40 "DC", placement(ne))
```

Adding the `placement(ne)` option places the label above and to the right (northeast) of the point. Other options you could choose include `n`, `ne`, `e`, `se`, `s`, `sw`, `w`, `nw`, and `c` (center); see Styles : Compassdir (331) for more details.
Uses allstatesn.dta & scheme vg_teal

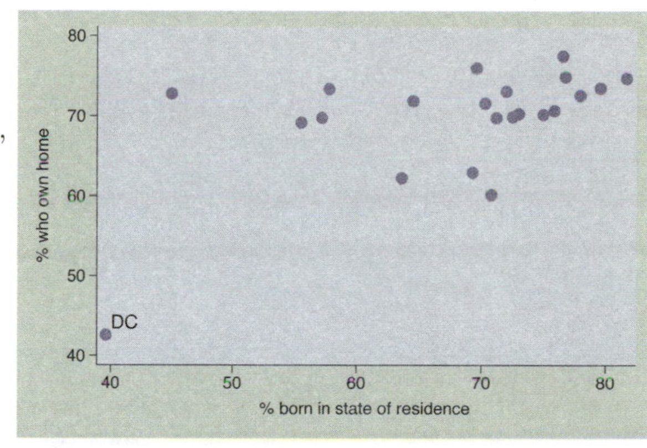


```
twoway (scatter ownhome borninstate, text(43 40 "DC", placement(e)))
   (lfit ownhome borninstate) (lfit ownhome borninstate if stateab !="DC")
```

Consider this scatterplot showing a linear fit between the two variables: one including Washington, DC, and one omitting Washington, DC. See the next graph, which uses the `text()` option to label the graph instead of the legend.
Uses allstatesn.dta & scheme vg_teal

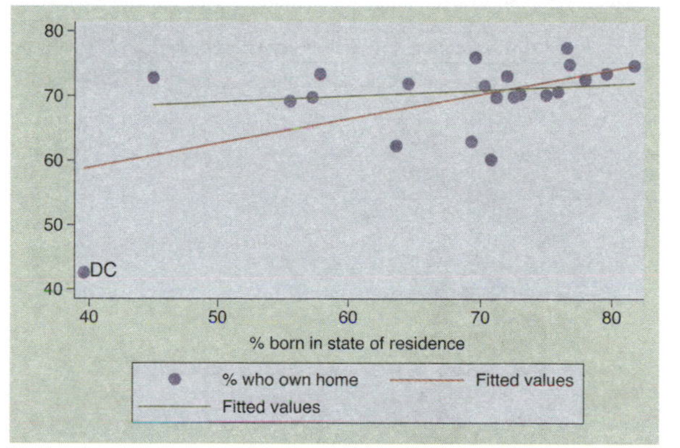

```
twoway (scatter ownhome borninstate, text(43 40 "DC", placement(ne)))
   (lfit ownhome borninstate) (lfit ownhome borninstate if stateab !="DC",
   text(72 50 "Without DC") text(60 50 "With DC")), legend(off)
```

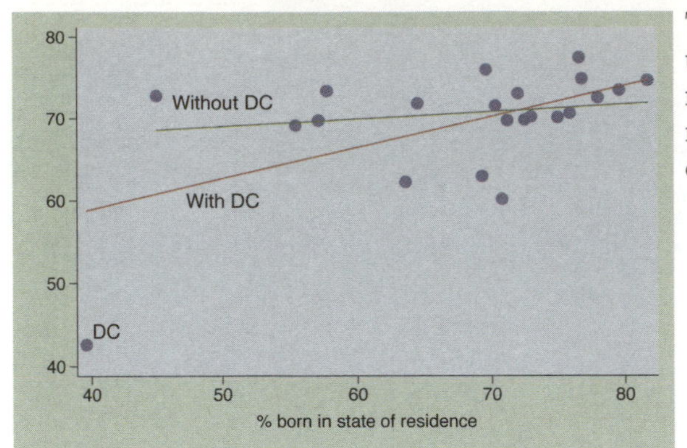

This graph turns the legend off and uses the `text()` option to label each regression line to indicate which regression line includes DC and which excludes DC.
Uses allstatesn.dta & scheme vg_teal

```
twoway (scatter ownhome borninstate, text(43 40 "DC", placement(ne)))
   (lfit ownhome borninstate) (lfit ownhome borninstate if stateab !="DC",
   text(71 50 "Without DC") text(60 50 "With DC")
   text(50 70 "Coef with DC .16" "Coef without DC .44")), legend(off)
```

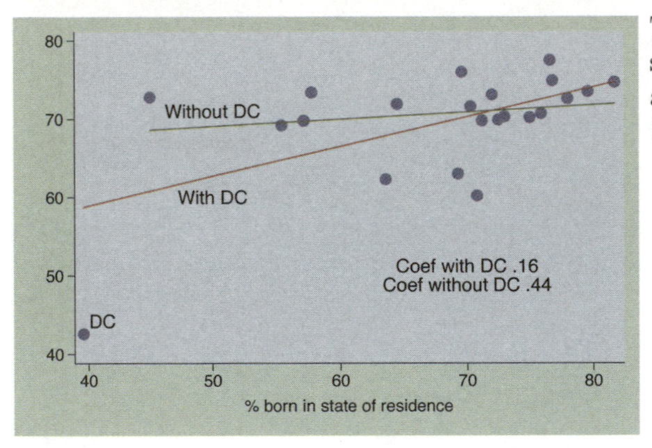

This graph adds explanatory text showing the regression coefficient with and without DC.
Uses allstatesn.dta & scheme vg_teal

```
twoway (scatter ownhome propval100, xaxis(1) mlabel(stateab))
   (scatter ownhome borninstate, xaxis(2) mlabel(stateab))
```

Consider this graph in which we overlay two scatterplots. We place `propval100` on the first x-axis and `borninstate` on the second x-axis.
Uses allstatesn.dta & scheme vg_teal

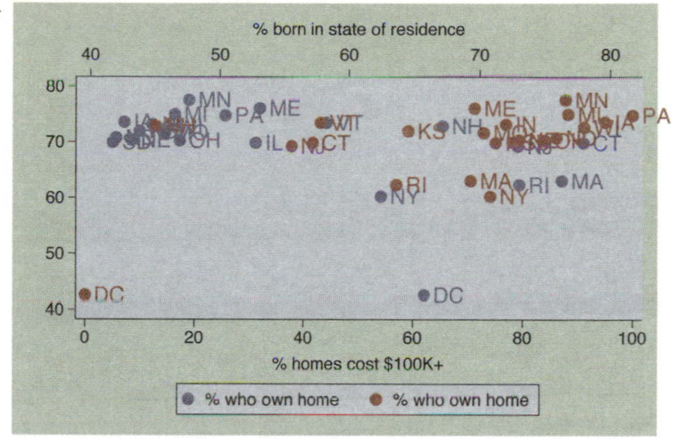


```
twoway (scatter ownhome propval100, xaxis(1))
   (scatter ownhome borninstate, xaxis(2)),
   text(43 66 "DC") text(43 42 "DC", xaxis(2))
```

Rather than labeling all the points, we can label just the point for DC. We have to be very careful because we have two different x-axes. The first `text()` option uses the first x-axis, so no special option is required. The second `text()` option uses the second x-axis, so we must specify the `xaxis(2)` option.
Uses allstatesn.dta & scheme vg_teal

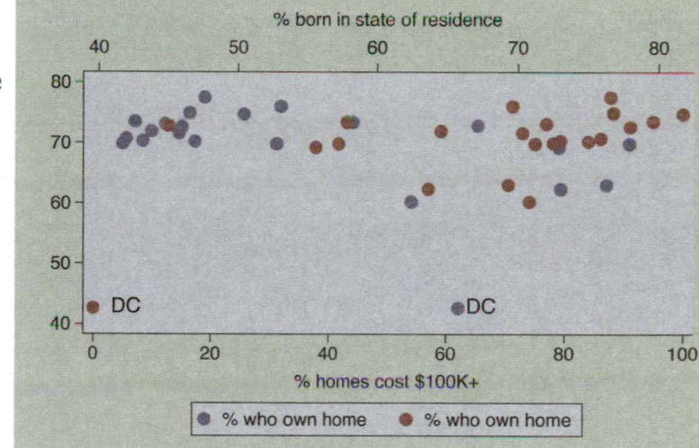

8.11 More options for text and textboxes

This section describes more options for modifying textbox elements: titles, captions, notes, added text, and legends. Technically, all text in a graph is displayed within a textbox. We can modify the box's attributes, such as its size and color, the margin around the box, and the outline; and we can modify the attributes of the text within the box, such as its size, color, justification, and margin. We sometimes use the `box` option to see how both the textbox and its text are being displayed. This helps us to see if we should modify the attributes of the box containing the text or the text within the box. For more information, see [G] ***textbox_options*** and Options : Adding text (299). In this section, we will begin by showing examples illustrating how to control the placement, size, color, and orientation of text. We will begin this section using the `vg_s1m` scheme.

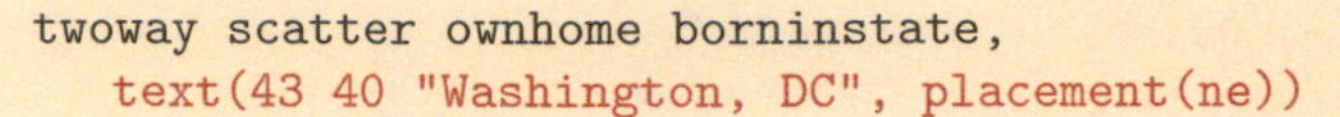

```
twoway scatter ownhome borninstate,
   text(43 40 "Washington, DC", placement(ne))
```

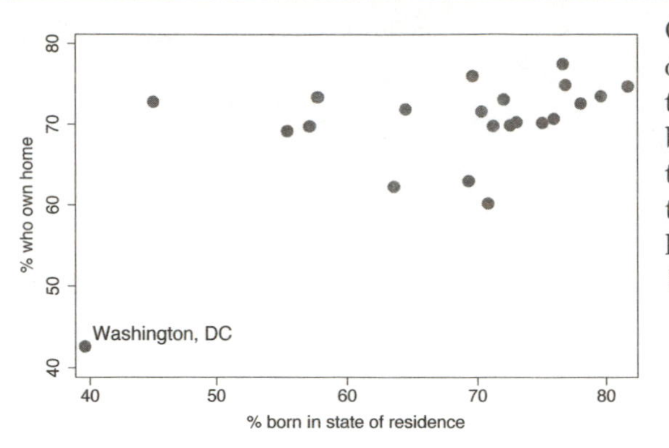

Consider this scatterplot, which has a dramatic outlying point. We have used the `text()` option to label that point, but, perhaps, we might want to control the size of the text for this label. See the next example for an illustration of how to do this.
Uses allstatesn.dta & scheme vg_s1m

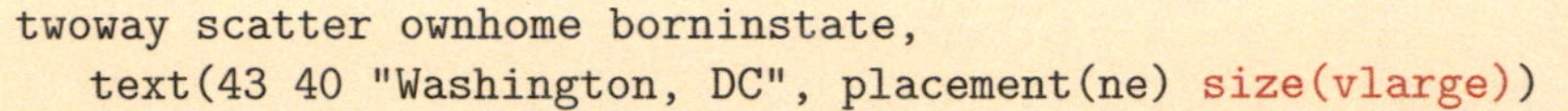

```
twoway scatter ownhome borninstate,
   text(43 40 "Washington, DC", placement(ne) size(vlarge))
```

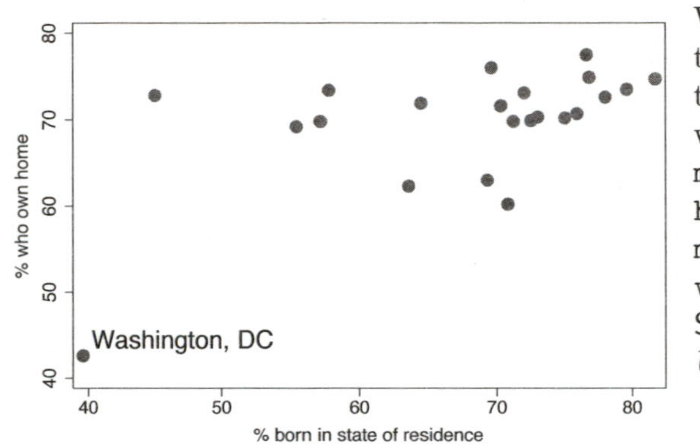

We can alter the size of the text using the `size()` option. Here, we make the text large. Other values we could use with the `size()` option include `zero`, `miniscule`, `quarter_tiny`, `third_tiny`, `half_tiny`, `tiny`, `vsmall`, `small`, `medsmall`, `medium`, `medlarge`, `large`, `vlarge`, `huge`, and `vhuge`; see Styles : Textsize (344) for more details.
Uses allstatesn.dta & scheme vg_s1m

```
twoway scatter ownhome borninstate,
   text(43 40 "Washington, DC", placement(ne) color(gs9))
```

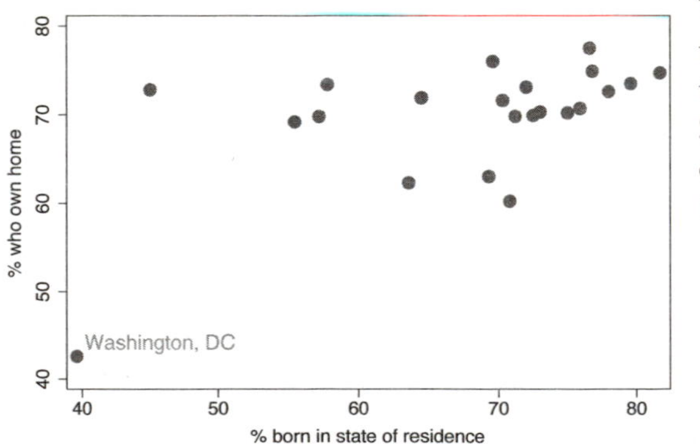

We can alter the color of the text using the `color()` option. Here, we make the text a middle-level gray. See Styles : Colors (328) for other colors you could select.
Uses allstatesn.dta & scheme vg_s1m

```
twoway scatter ownhome borninstate,
   text(43 40 "Washington, DC", placement(ne) orientation(vertical))
```

We can use the orientation() option to change the direction of the text. Other values you can choose are horizontal for 0 degrees, vertical for 90 degrees, rhorizontal for 180 degrees, and rvertical for 270 degrees, see Styles : Orientation (341) for more details.
Uses allstatesn.dta & scheme vg_s1m

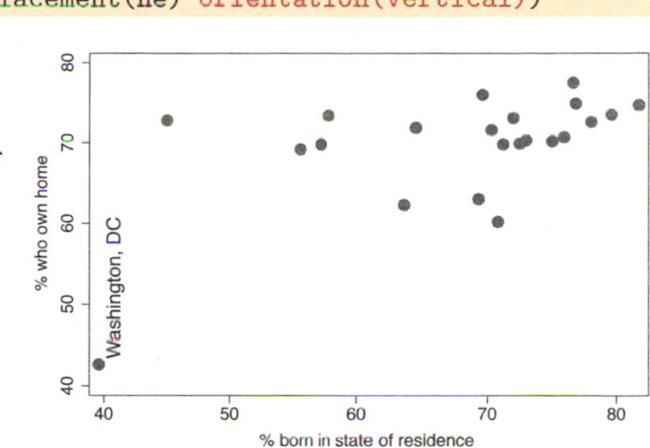

This next set of examples considers options for justifying text within a box, sizing the box, and creating margins around the box. This is followed by options that control margins within the textbox. This next set of graphs use the vg_rose scheme

```
twoway (scatter ownhome borninstate),
   title("% who own home by" "% that reside in state of birth", box)
```

Consider this example where we place a title on our graph. To help show how the options work, we will put a box around the title.
Uses allstatesn.dta & scheme vg_rose

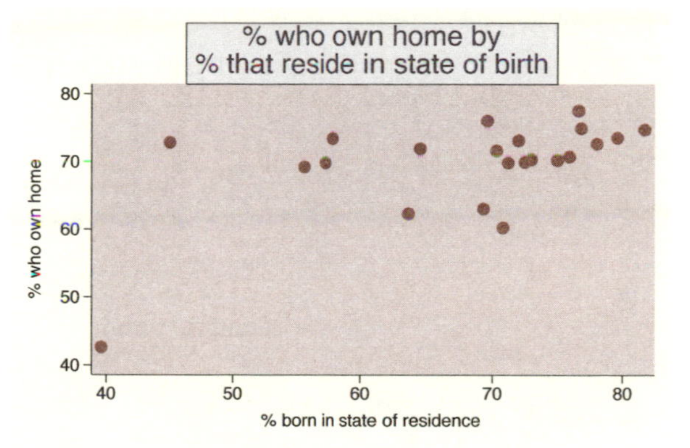

```
twoway (scatter ownhome borninstate),
   title("% who own home by" "% that reside in state of birth", box
   justification(left))
```

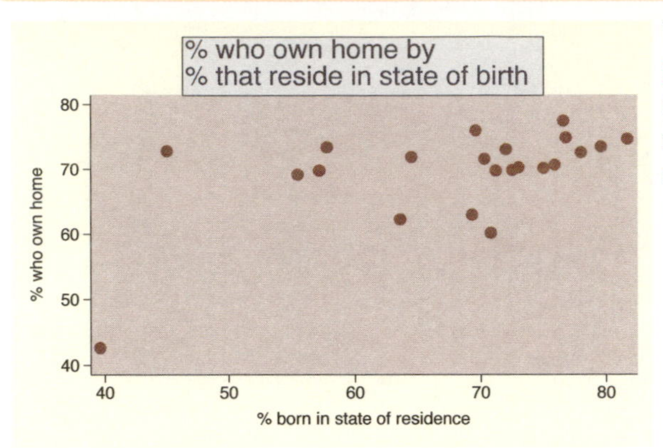

We can left-justify the text using the `justification()` option. Note that the title is justified within the textbox, not with respect to the entire graph area.
Uses allstatesn.dta & scheme vg_rose

```
twoway (scatter ownhome borninstate),
   title("% who own home by" "% that reside in state of birth", box
   bexpand)
```

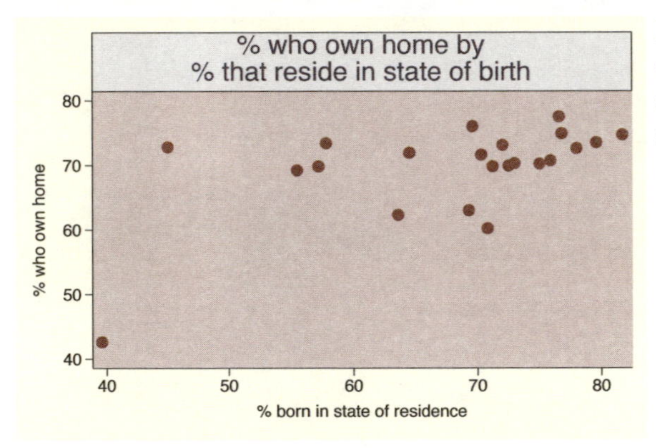

If we use the `bexpand` (box expand) option, the textbox containing the title expands to fill the width of the plot area.
Uses allstatesn.dta & scheme vg_rose

```
twoway (scatter ownhome borninstate),
   title("% who own home by" "% that reside in state of birth", box
   bexpand justification(left))
```

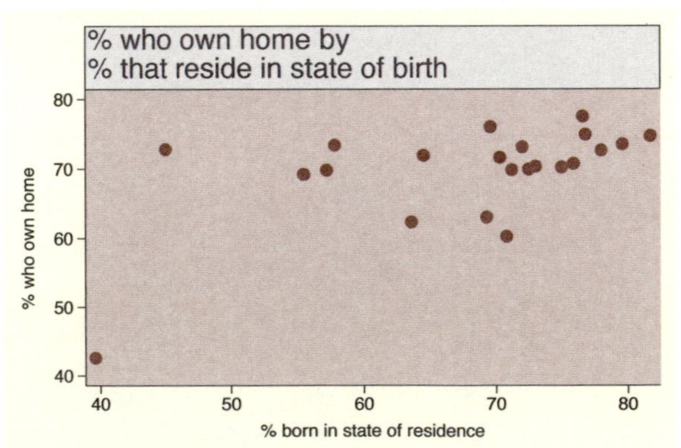

With the box expanded, the `justification(left)` option now makes the title flush left with the plot area.
Uses allstatesn.dta & scheme vg_rose

```
twoway (scatter ownhome borninstate),
   title("% who own home by" "% that reside in state of birth",
   box bexpand justification(left) bmargin(medium))
```

We can change the size of the margin around the outside of the box using the `bmargin(medium)` (box margin) option, making the margin a medium size at all four edges: left, right, top, and bottom.
Uses allstatesn.dta & scheme vg_rose

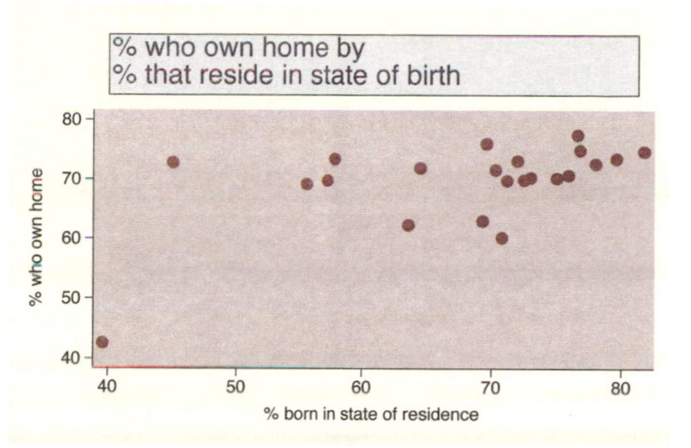


```
twoway (scatter ownhome borninstate),
   title("% who own home by" "% that reside in state of birth",
   box bexpand justification(left) bmargin(0 0 3 3))
```

If we wanted the margin for the left and right to be 0 and for the top and bottom to be 3, we could use the `bmargin(0 0 3 3)` option. The order of the margins is `bmargin(`$\#_{\text{left}}$ $\#_{\text{right}}$ $\#_{\text{top}}$ $\#_{\text{bottom}}$`)`.
Uses allstatesn.dta & scheme vg_rose

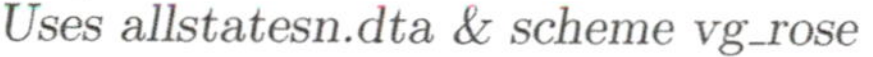

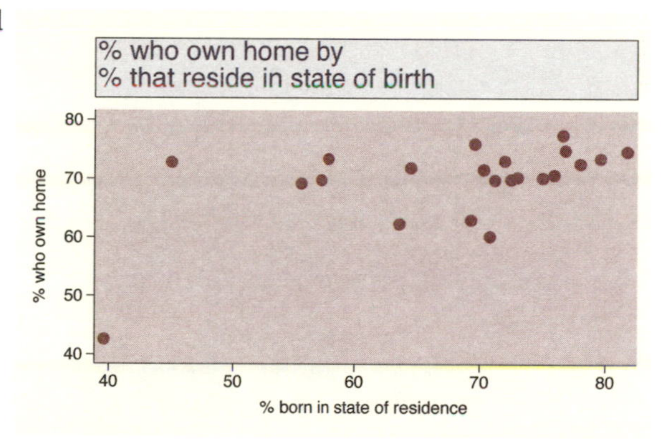


```
twoway (scatter ownhome borninstate),
   title("% who own home by" "% that reside in state of birth",
   box bexpand justification(left) bmargin(b=3))
```

To make only the bottom margin 3, we could specify `bmargin(b=3)`, where `b=3` means to change the bottom margin to 3. The top, left, bottom, and top margins can be changed individually using `t=`, `l=`, `b=`, and `t=`, respectively.
Uses allstatesn.dta & scheme vg_rose

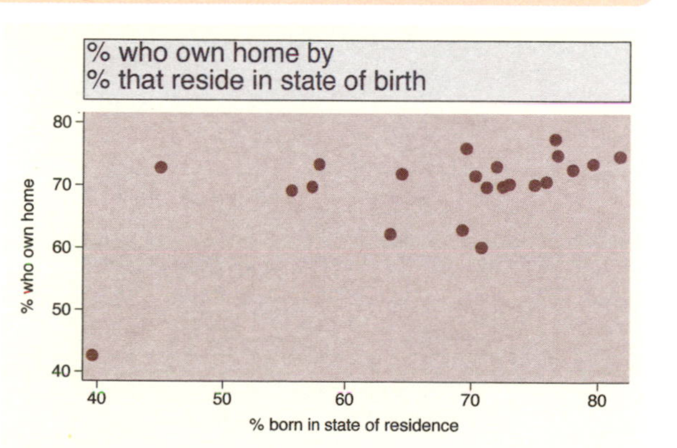

```
twoway (scatter ownhome borninstate) (lfit ownhome borninstate)
   (lfit ownhome borninstate if stateab !="DC",
   text(45 70 "Coef with DC .16" "Coef without DC .44", box))
```

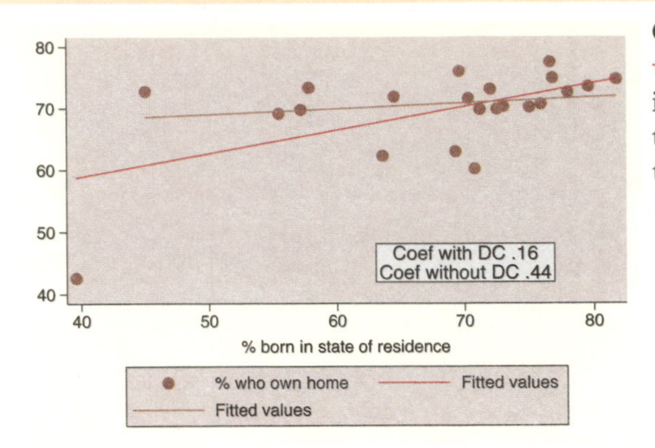

Consider this graph, which uses the text() option to place an annotation in the middle of the plot region. The text might look better if we increased the margin around the text.
Uses allstatesn.dta & scheme vg_rose

```
twoway (scatter ownhome borninstate) (lfit ownhome borninstate)
   (lfit ownhome borninstate if stateab !="DC",
   text(45 70 "Coef with DC .16" "Coef without DC .44", box margin(medium)))
```

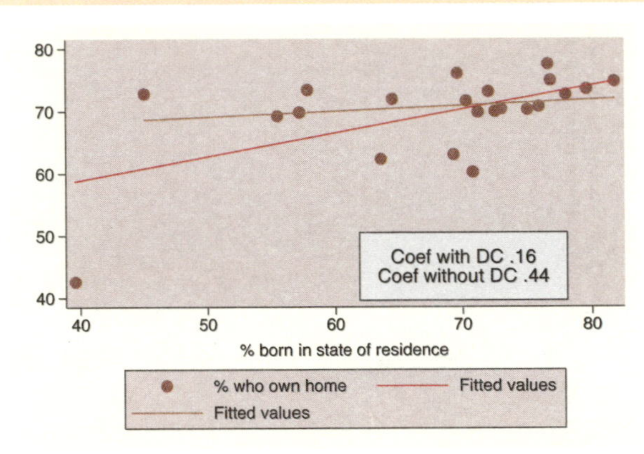

We can expand the margin between the text and the box with the margin() option. Note the difference between this and the **bmargin()** option (illustrated previously), which increased the margin around the outside of the box.
Uses allstatesn.dta & scheme vg_rose

```
twoway (scatter ownhome borninstate) (lfit ownhome borninstate)
   (lfit ownhome borninstate if stateab !="DC",
   text(45 70 "Coef with DC .16" "Coef w/out DC .44", box margin(5 5 2 2)))
```

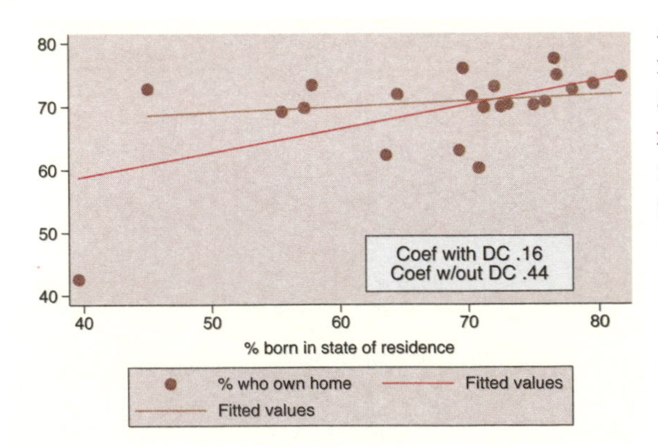

As with the **bmargin()** option, we can more precisely modify the margin around the text. Here, we use the margin() option to make the size of the margin 5, 5, 2, and 2 for the left, right, top, and bottom, respectively.
Uses allstatesn.dta & scheme vg_rose

```
twoway (scatter ownhome borninstate) (lfit ownhome borninstate)
   (lfit ownhome borninstate if stateab !="DC",
   text(45 70 "Coef with DC .16" "Coef without DC .44", box linegap(4)))
```

We can change the gap between the lines with the `linegap()` option. Here, we make the gap larger than it normally would be. See Styles : Margins (338) for more details.
Uses allstatesn.dta & scheme vg_rose

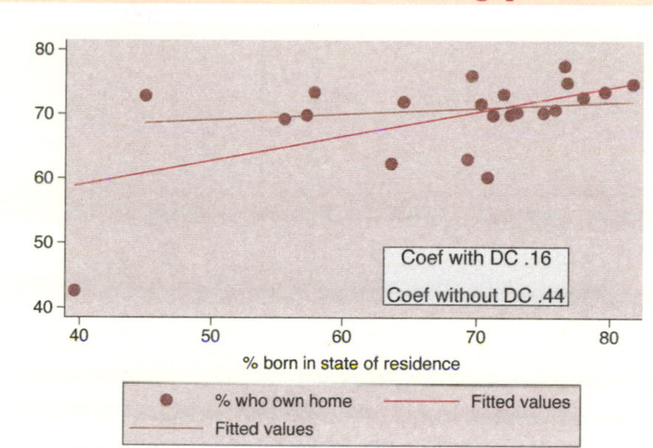

Let's now consider options that control the color of the textbox and the characteristics of the outline of the box (including the color, thickness, and pattern). This next set of graphs uses the `vg_past` scheme.

```
twoway (scatter ownhome borninstate),
   title("% own home by % reside in state")
```

Consider this graph with a title at the top.
Uses allstatesn.dta & scheme vg_past

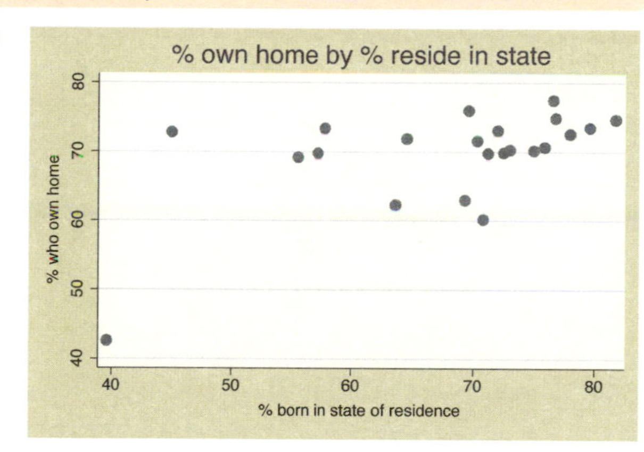

```
twoway (scatter ownhome borninstate),
  title("% own home by % reside in state", box)
```

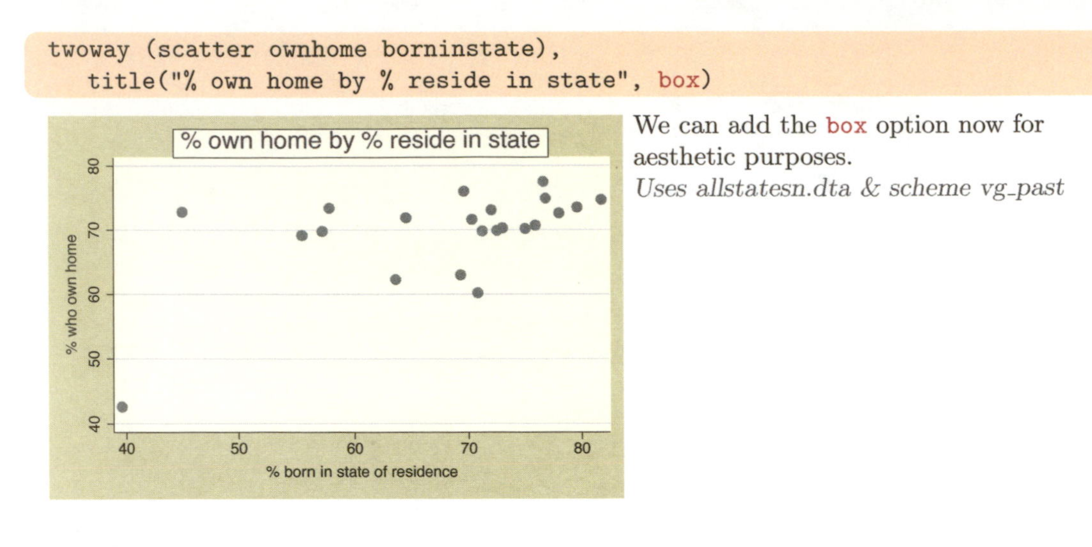

We can add the box option now for aesthetic purposes.
Uses allstatesn.dta & scheme vg_past

```
twoway (scatter ownhome borninstate),
  title("% own home by % reside in state",
  box bfcolor(ltblue) blcolor(gray) blwidth(thick))
```

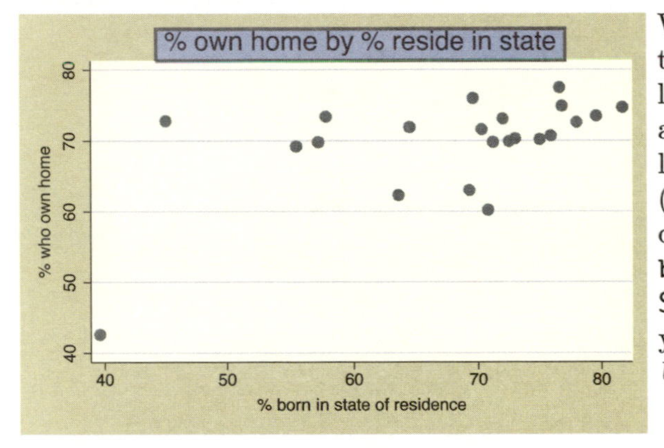

We can change the box fill color with the bfcolor() option, the color of the line around the box with blcolor(), and the width of the surrounding box line with blwidth(). See Styles : Colors (328) for other possible values you could use with the bfcolor() and blcolor() options and Styles : Linewidth (337) for other values you could choose for blwidth().
Uses allstatesn.dta & scheme vg_past

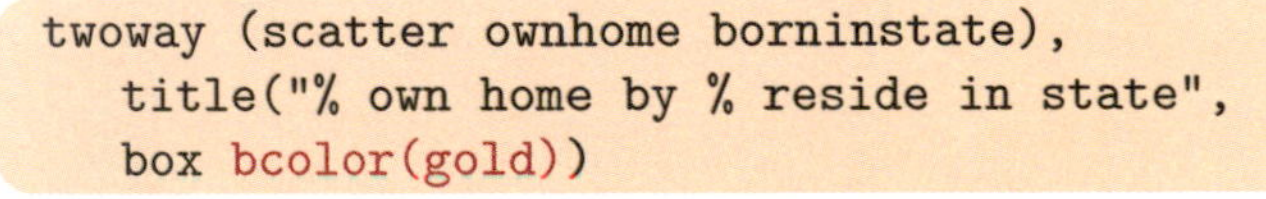

```
twoway (scatter ownhome borninstate),
  title("% own home by % reside in state",
  box bcolor(gold))
```

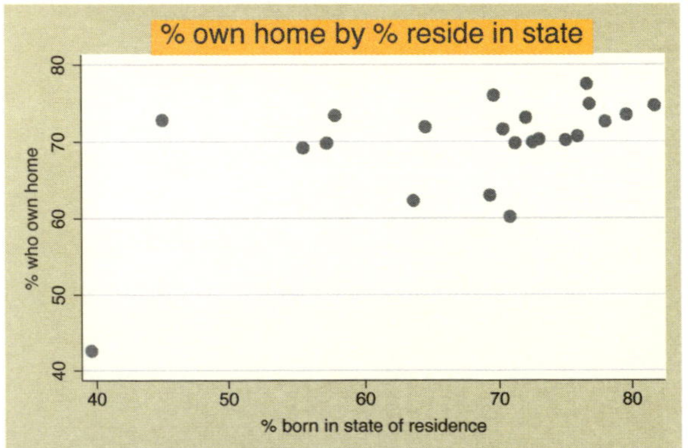

We can change the box color with the bcolor() option. Here, we make the fill and outline color of the title box gold.
Uses allstatesn.dta & scheme vg_past

Let's now use the allstates file and consider some examples in which we use the by() option to display multiple graphs broken down by the location of the state. We will look at options for placing and aligning text in graphs that use the by() option. This next set of graphs uses the vg_s2c scheme.

```
scatter ownhome borninstate,
   by(nsw, title("% own home" "by % born in state"))
```

Consider this graph in which we use the by() option to show this scatterplot separately for states in the North, South, and West.
Uses allstates.dta & scheme vg_s2c

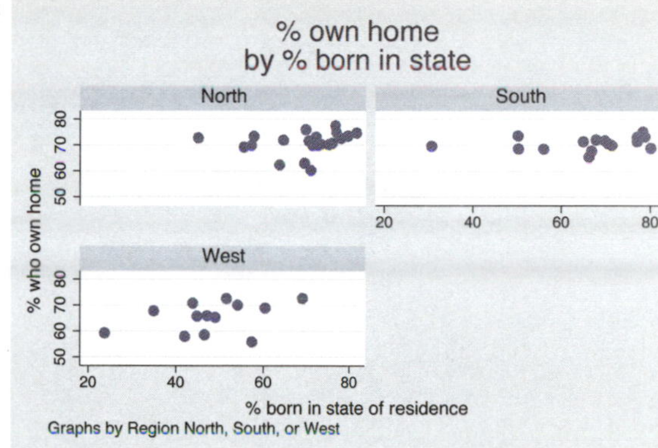


```
scatter ownhome borninstate,
   by(nsw, title("% own home" "by % born in state",
   ring(0) position(5) box))
```

Let's put the title in the open hole in the right corner of the graph using the ring(0) and position(5) options. We include the box option only to show the outline of the textbox, not for aesthetics.
Uses allstates.dta & scheme vg_s2c

Introduction Twoway Matrix Bar Box Dot Pie Options Standard options Styles Appendix
Markers Marker labels Connecting Axis titles Axis labels Axis scales Axis selection By Legend Adding text Textboxes

```
scatter ownhome borninstate,
   by(nsw, title("% own home" "by % born in state",
   ring(0) position(5) box width(65) height(40)))
```

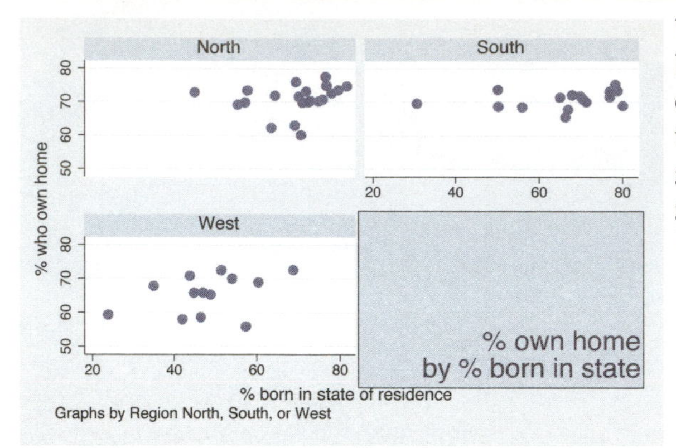

We can make the area for the textbox bigger using the `width()` and `height()` options. We change the value to make the box approximately as tall as the graph for the West and as wide as the graph for the South.
Uses allstates.dta & scheme vg_s2c

```
scatter ownhome borninstate,
   by(nsw, title("% own home" "by % born in state", ring(0) position(5)
   box width(65) height(40) justification(left) alignment(top)))
```

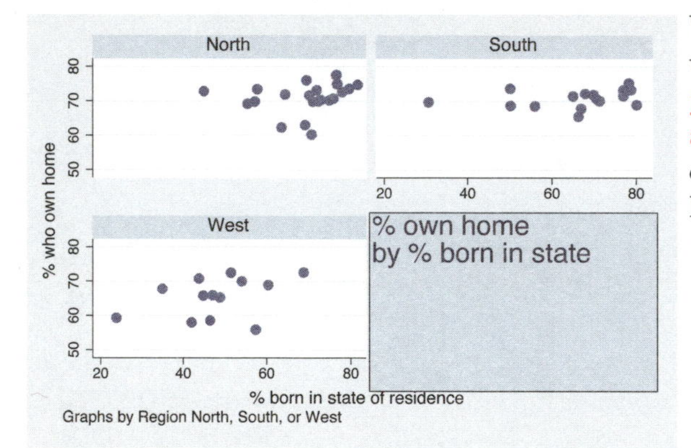

We can left-justify the text and align it with the top using the `justification(left)` and `alignment(top)` options. These options make the title appear in the top left corner of the empty hole.
Uses allstates.dta & scheme vg_s2c

```
scatter ownhome borninstate,
   by(nsw, title("% own home" "by % born in state", ring(0) position(5)
   width(65) height(40) justification(left) alignment(top)))
```

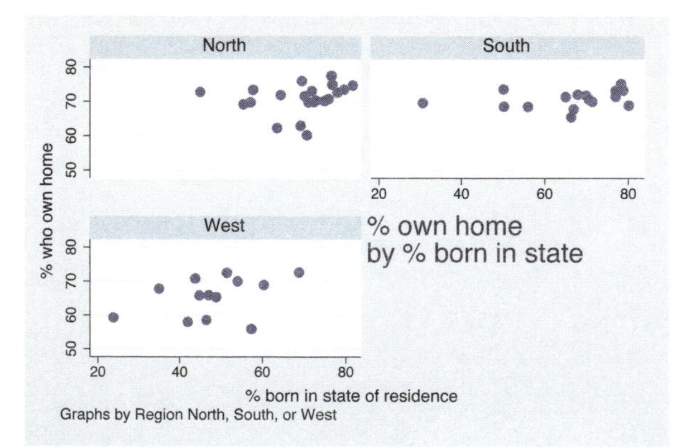

Now that we have aligned the text as we would like, we can take away the box by omitting the `box` option.
Uses allstates.dta & scheme vg_s2c

9 Standard options available for all graphs

This chapter discusses a class of options Stata refers to as *standard options*, because these options can be used in most graphs. This chapter will begin by discussing options that allow you to add or change the titles in the graph and then showing you how to use schemes to control the overall look and style of your graph. Next, we demonstrate options for controlling the size of the graph and the scale of items within graphs. The chapter will conclude by illustrating options that allow you to control the colors of the plot region, the graph region, and the borders that surround these regions. For further details, see [G] ***std_options***.

9.1 Creating and controlling titles

Titles are useful for providing additional information that explains the contents of a graph. Stata includes four standard options for adding explanatory text to graphs: `title()`, `subtitle()`, `note()`, and `caption()`. This section will illustrate how to use these titles and how to customize their content and their placement. For further information about customizing the appearance of such titles (e.g., color, size, orientation, etc.), see Options : Textboxes (303). For more information about titles, see [G] ***title_options***. This section uses the `vg_s1m` scheme.

```
scatter propval100 ownhome, title("My title")
```

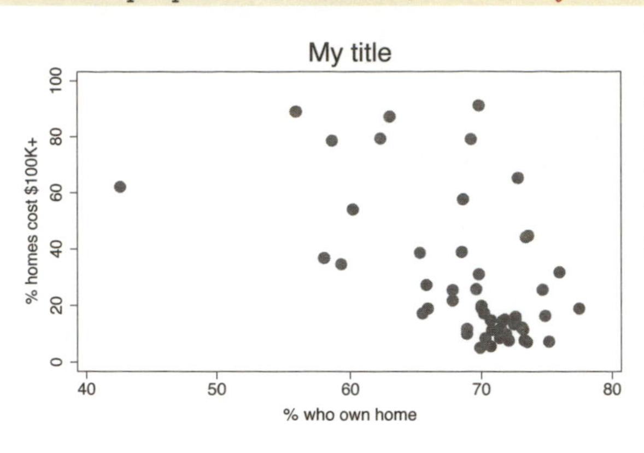

The `title()` option adds a title to a graph. Here, we add a simple title to the graph. Although the title includes quotes, we could have omitted them in this case. Later, we will see examples where the quotes become very important.
Uses allstates.dta & scheme vg_s1m

```
scatter propval100 ownhome, title("My title") subtitle("My subtitle")
```

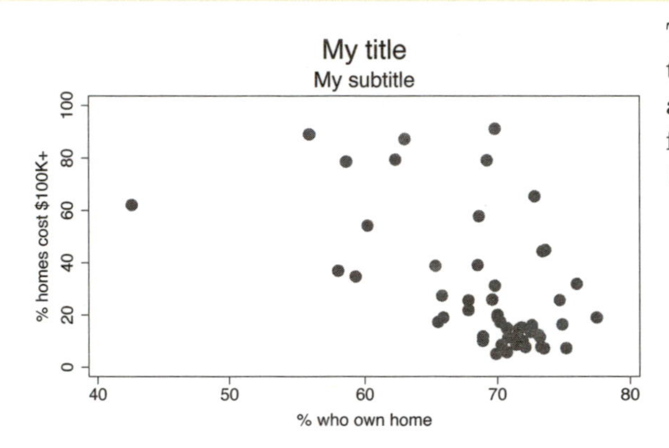

The `subtitle()` option adds a subtitle to a graph. The subtitle, by default, appears below the title in a smaller font.
Uses allstates.dta & scheme vg_s1m

```
scatter propval100 ownhome, subtitle("My smaller title")
```

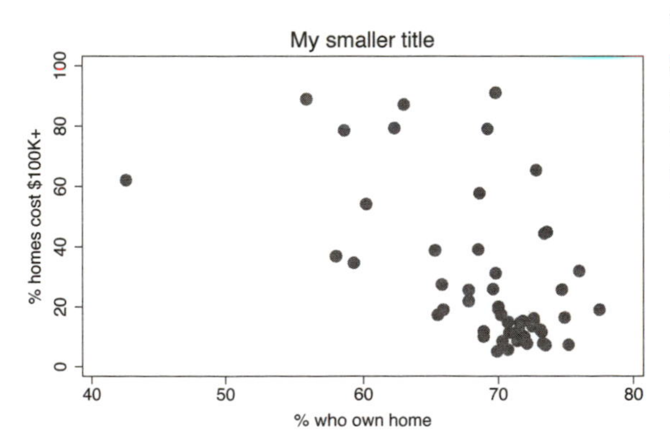

We do not have to specify a `title()` to specify a `subtitle()`. For example, we might want a title that is smaller in size than a regular title, so we could specify a subtitle alone.
Uses allstates.dta & scheme vg_s1m

```
scatter propval100 ownhome, caption("My caption") note("My note")
```

In this example, the `caption()` option adds a small-sized caption in the lower corner, and the `note()` option places a smaller-sized note in the bottom left corner. If both options are specified, the note appears above the caption. We do not need to include both of these options in the same graph.
Uses allstates.dta & scheme vg_s1m

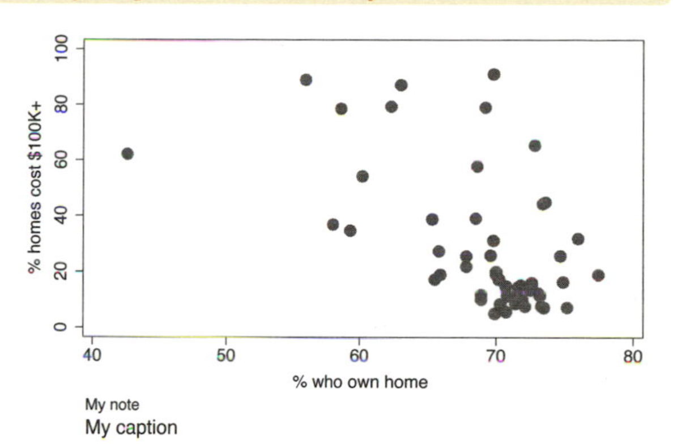

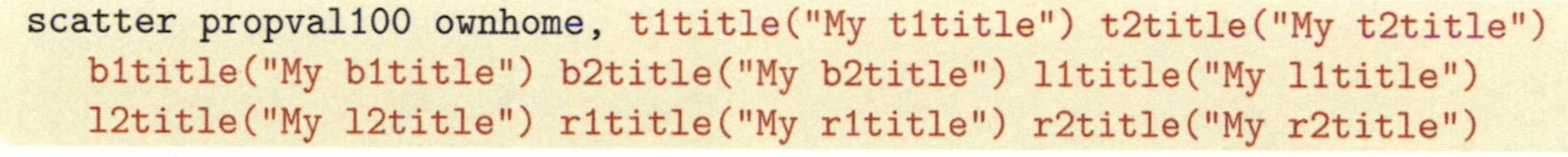

```
scatter propval100 ownhome, t1title("My t1title") t2title("My t2title")
   b1title("My b1title") b2title("My b2title") l1title("My l1title")
   l2title("My l2title") r1title("My r1title") r2title("My r2title")
```

Although these are not as commonly used, Stata offers a number of additional title options for titling the top of the graph (`t1title()` and `t2title()`), the bottom of the graph (`b1title()` and `b2title()`), the left side of the graph (`l1title()` and `l2title()`), and the right side of the graph (`r1title()` and `r2title()`).
Uses allstates.dta & scheme vg_s1m

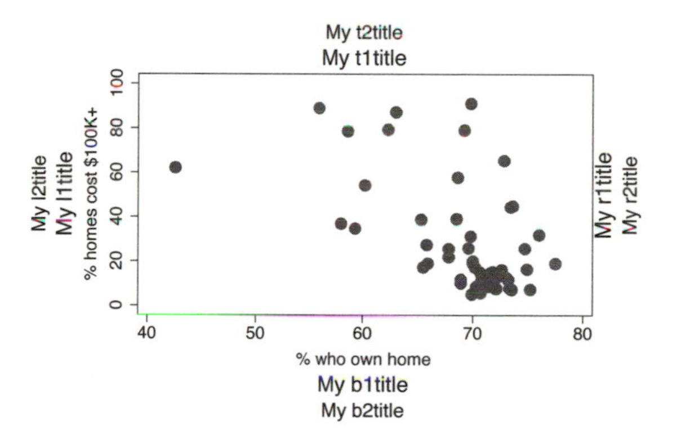

Stata gives you considerable flexibility in the placement of these titles, notes, and captions, as well as controlling the size, color, and orientation of the text. This is illustrated below using the `title()` option, but the same options apply equally to the `subtitle()`, `note()`, and `caption()` options.

```
scatter propval100 ownhome, title("My" "title")
```

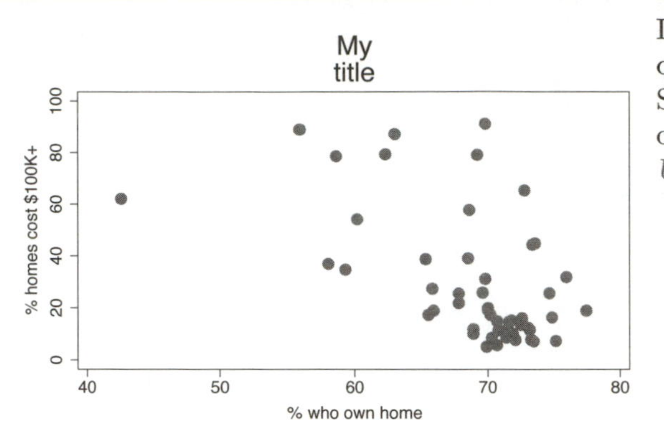

In this example, we use multiple sets of quotes in the `title()` option to tell Stata that we want the title to appear on two separate lines.
Uses allstates.dta & scheme vg_s1m

```
scatter propval100 ownhome, title(`"A "title" with quotes"')
```

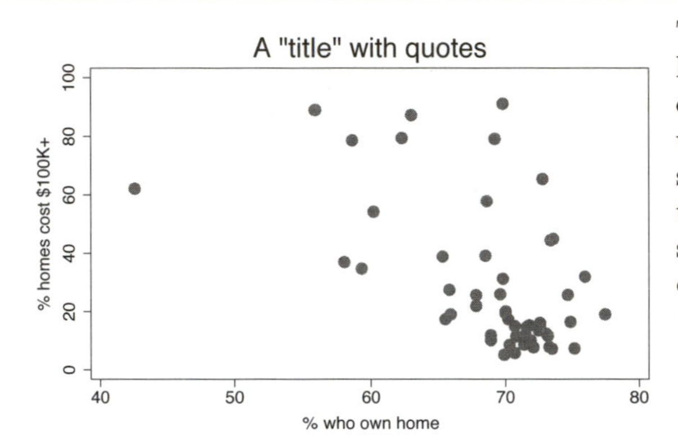

This example illustrates that we can have quotation marks in the `title()` option, as long as we open the title with `` `" `` and close it with `"'`. (The open single quote is often located below the tilde on your keyboard, and the close single quote is often located below the double quote on your keyboard.)
Uses allstates.dta & scheme vg_s1m

```
scatter propval100 ownhome, title("My title", position(7))
```

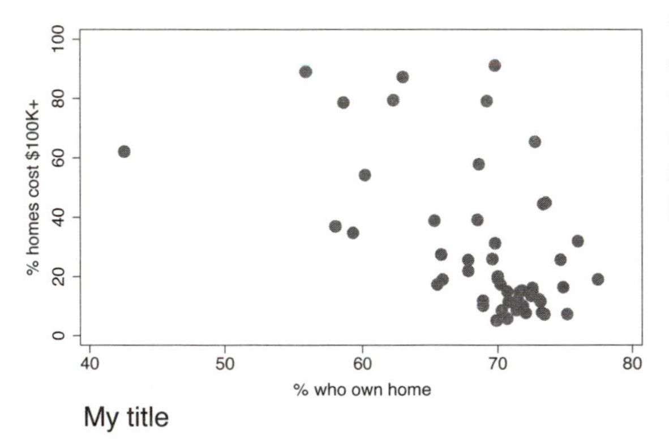

The `position()` option can be used to change the position of the title. Here, we place the title in the bottom left corner of the graph by indicating that it should be at the 7 o'clock position. See Styles : Clockpos (330) for more details.
Uses allstates.dta & scheme vg_s1m

```
scatter propval100 ownhome, title("My title", position(1) ring(0))
```

As we saw in the last example, we can use the `position()` option to control the placement of the title, but this option does not control the distance between the title and center of the plot region. That is controlled by the `ring()` option. `ring(0)` means that the item is inside the plot region, and higher values for `ring()` place the item farther away from the plot region. Imagine concentric rings around the plot area with higher values corresponding to the rings that are farther from the center.
Uses allstates.dta & scheme vg_s1m

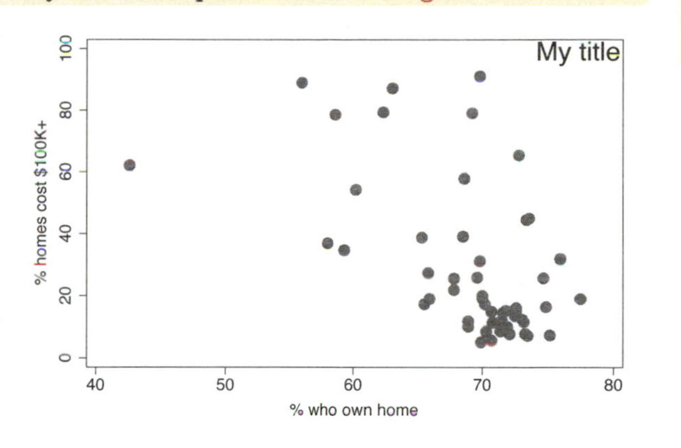


```
scatter propval100 ownhome, title("This is my" "title",
   position(11) box)
```

Because titles, subtitles, notes, and captions are considered textboxes, you can use the options associated with textboxes to customize their display. Here, we place a box around the title using the `box` option. We also use the `position(11)` option to place the title in the 11 o'clock position.
Uses allstates.dta & scheme vg_s1m

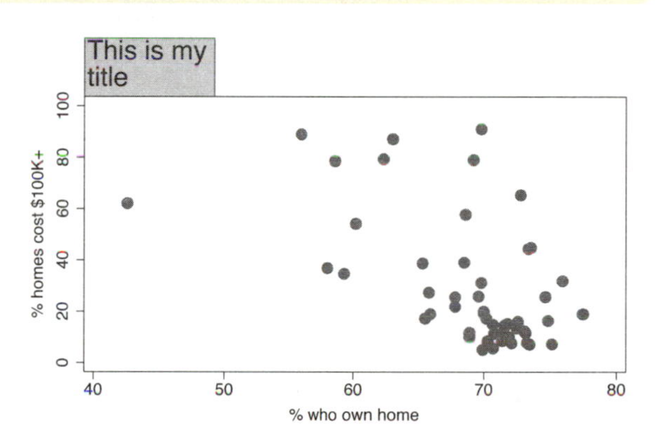


```
scatter propval100 ownhome, title("This is my" "title",
   position(11) box span)
```

Here, we add the `span` option, so the title spans the width of the graph, positioning the title flush left at the 11 o'clock position. Note that now the title partly obscures the 100 labeling the y-axis.
Uses allstates.dta & scheme vg_s1m

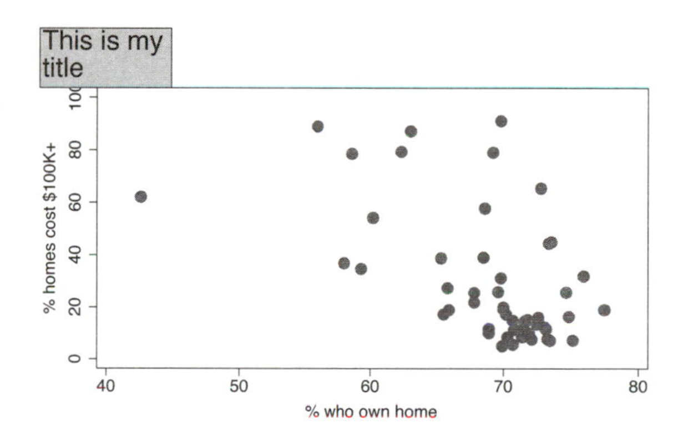

```
scatter propval100 ownhome, title("This is my" "title", box
   justification(right))
```

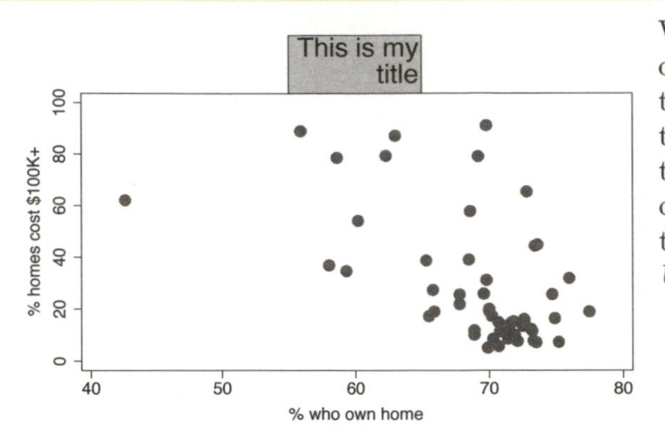

We can use the `justification(right)` option to right-justify the text inside the box. Note the difference between the `position()` option, which positions the textbox, and the `justification()` option, which justifies the text within the textbox.
Uses allstates.dta & scheme vg_s1m

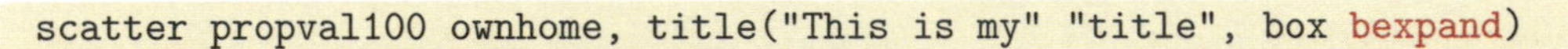

```
scatter propval100 ownhome, title("This is my" "title", box bexpand)
```

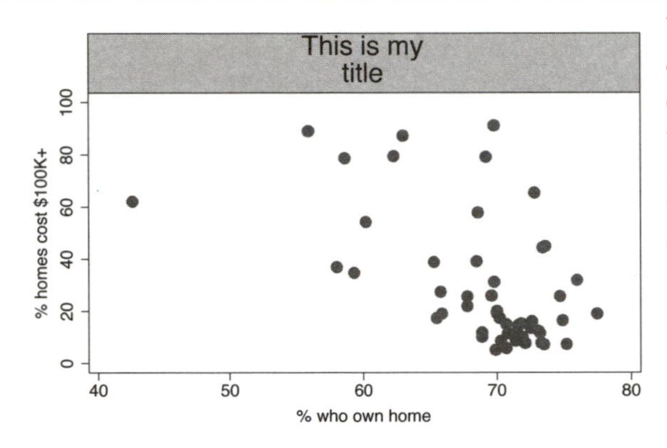

We can expand the box to fill the width of the plot region using the `bexpand` option. If we wanted the box to span the entire width of the graph, we could add the `span` option (not shown). There are numerous other textbox options than can be used with titles; see Options : Textboxes (303) and [G] ***textbox_options*** for more details.
Uses allstates.dta & scheme vg_s1m

9.2 Using schemes to control the look of graphs

Schemes control the overall look of Stata graphs by providing default values for numerous graph options. You can accept these defaults or override them using graph options. This section first examines the kinds of schemes available in Stata, discuss different methods for selecting schemes, and then show how to obtain additional schemes. For more information about schemes, see [G] **schemes**. Stata has two basic families of schemes, the `s2` family and the `s1` family, each sharing similar characteristics. There are also other specialized schemes, including the `sj` scheme for making graphs like those in the *Stata Journal* and the `economist` scheme for making graphs like those that appear in *The Economist*. We will look at these schemes below.

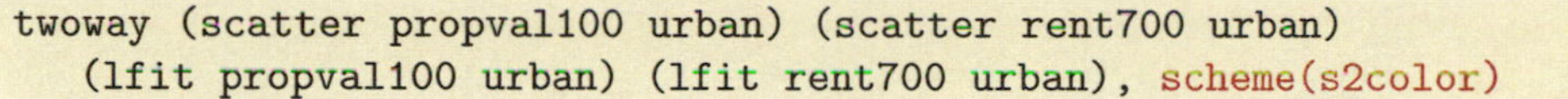

```
twoway (scatter propval100 urban) (scatter rent700 urban)
   (lfit propval100 urban) (lfit rent700 urban), scheme(s2color)
```

This example uses the `scheme(s2color)` option to create a graph using the `s2color` scheme. Using the `scheme()` option, we can manually select which scheme to use for displaying the graph we wish to create. The `s2color` scheme is the default scheme for Stata graphs.
Uses allstates.dta & scheme s2color

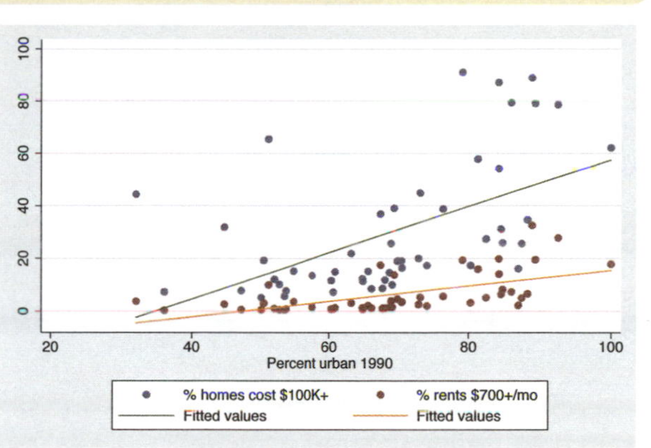


```
twoway (scatter propval100 urban) (scatter rent700 urban)
   (lfit propval100 urban) (lfit rent700 urban), scheme(s2mono)
```

The `s2mono` scheme is a black-and-white version of the `s2color` scheme. In this example, the symbols differ in gray scale and size, and the lines differ in their patterns.
Uses allstates.dta & scheme s2mono

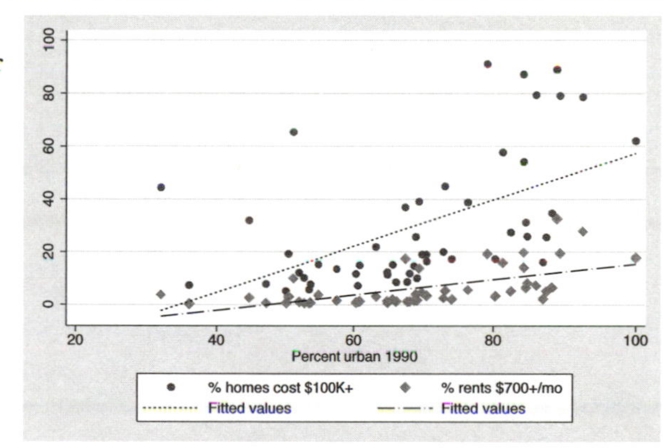


```
twoway (scatter propval100 urban) (scatter rent700 urban)
   (lfit propval100 urban) (lfit rent700 urban), scheme(s2manual)
```

Here is an example using the `s2manual` scheme, which is very similar to the `s2mono` scheme. One difference is that the lines of the fit values are the same pattern (solid) in this graph, but they have different patterns when we use `s2mono`.
Uses allstates.dta & scheme s2manual

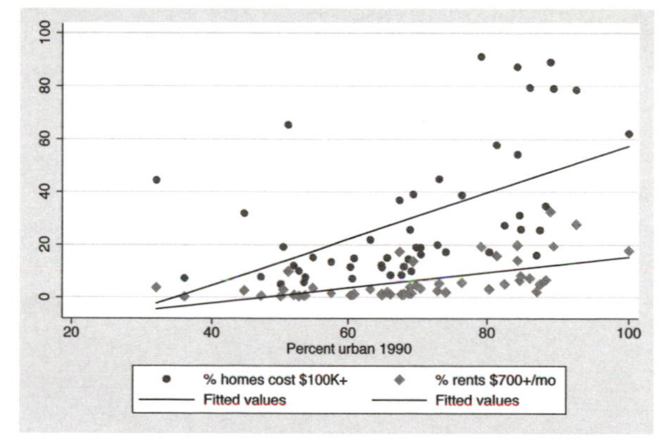

```
twoway (scatter propval100 urban) (scatter rent700 urban)
   (lfit propval100 urban) (lfit rent700 urban), scheme(s1color)
```

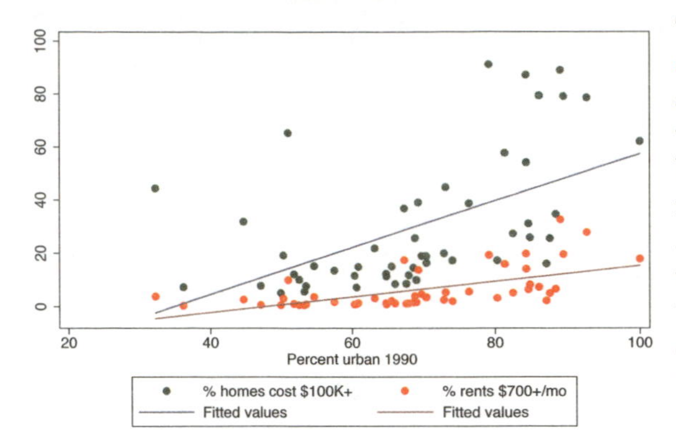

This is an example of a graph using the s1color scheme. Note how the lines and markers are only differentiated by their color. Both the plot area and the border around the plot are white. Also, note the absence of grid lines. (Stata also has an s1rcolor scheme, in which the plot area and border area are black. This is not shown since it would be difficult to read in print.)
Uses allstates.dta & scheme s1color

```
twoway (scatter propval100 urban) (scatter rent700 urban)
   (lfit propval100 urban) (lfit rent700 urban), scheme(s1mono)
```

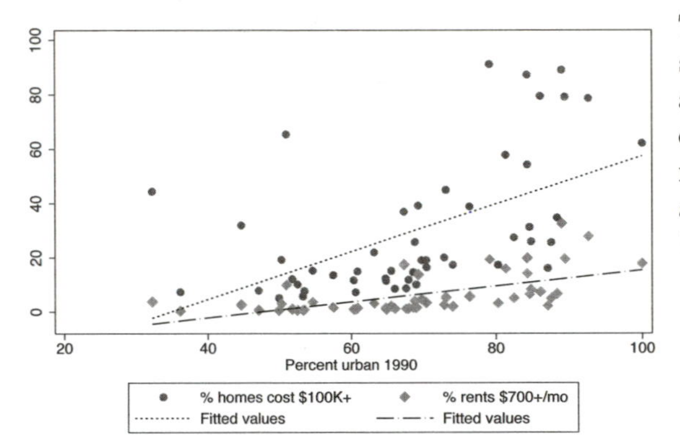

The s1mono scheme is similar to the s1color scheme in that the plot area and border are white and the grid is omitted. In a mono scheme, the markers differ in gray scale and size, and the lines differ in their pattern.
Uses allstates.dta & scheme s1mono

```
twoway (scatter propval100 urban) (scatter rent700 urban)
   (lfit propval100 urban) (lfit rent700 urban), scheme(s1manual)
```

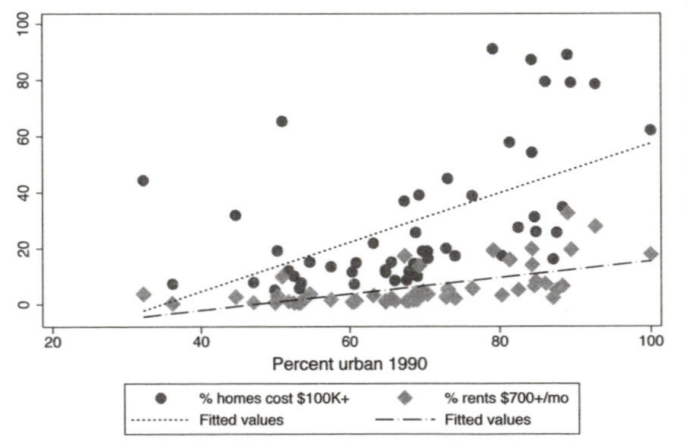

The s1manual is similar to s1mono, but the sizes of the markers and text are increased. This is useful if you are making a small graph and want these small elements to be magnified to be more easily seen.
Uses allstates.dta & scheme s1manual

```
twoway (scatter propval100 urban) (scatter rent700 urban)
   (lfit propval100 urban) (lfit rent700 urban), scheme(sj)
```

The sj scheme is very similar to the s2mono scheme. In fact, a comparison of this graph with an earlier graph that used the s2mono scheme shows no visible differences. The sj scheme is based on the s2mono scheme and only alters xsize() and ysize(). See Appendix : Customizing schemes (379) for more information about how to inspect (and alter) the contents of graph schemes.
Uses allstates.dta & scheme sj

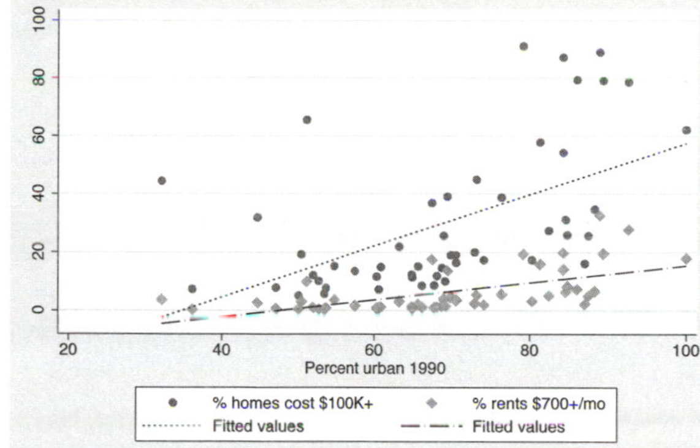


```
twoway (scatter propval100 urban) (scatter rent700 urban)
   (lfit propval100 urban) (lfit rent700 urban), scheme(economist)
```

The economist scheme is quite different from all the other schemes and is a very good example of how much can be controlled with a scheme. Using this scheme modifies the colors of the plot area, border, markers, lines, the position of the y-axis, and the legend. It also removes the line on the y-axis and changes the angle of the labels on the y-axis.
Uses allstates.dta & scheme economist

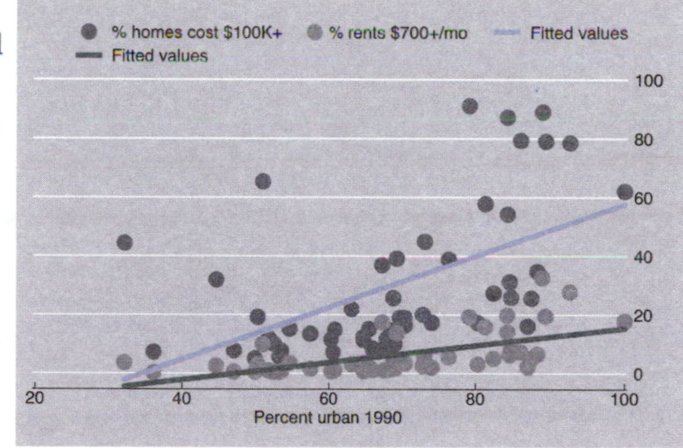

As these examples have shown, we can change the scheme of a graph by supplying the scheme() option on a graph command. If we want to use the same scheme over and over, we can use the set scheme command to set the default scheme. For example, if we typed

```
. set scheme economist
```

the default scheme would become economist until we quit Stata. Or, we could type

```
. set scheme economist, permanently
```

The **economist** scheme would be our default scheme, even after we quit and start Stata again. If we will be creating a series of graphs that we want to have a common look, then schemes are a very powerful tool for accomplishing this. Even though Stata has a variety of built-in schemes, we may want to obtain other schemes. The **findit** command can be used to search for information about schemes and to download schemes that others have developed. To search for schemes, type

```
. findit scheme
```

and Stata will list web pages and packages associated with the word *scheme*.

See Intro : Schemes (14) for an overview of the schemes used in this book and Appendix : Online supplements (382) for instructions for obtaining the schemes for this book.

Seeing how powerful and flexible schemes are, we might be interested in creating our own schemes. Stata gives us complete control over creating schemes. The section Appendix : Customizing schemes (379) provides tips for getting started.

9.3 Sizing graphs and their elements

This section illustrates how to use the **xsize()** and **ysize()** options to control the size and aspect ratio of graphs. It also illustrates the use of the **scale()** option for controlling the size of the text and markers. This section uses the **vg_s1c** scheme.

```
scatter propval100 ownhome
```

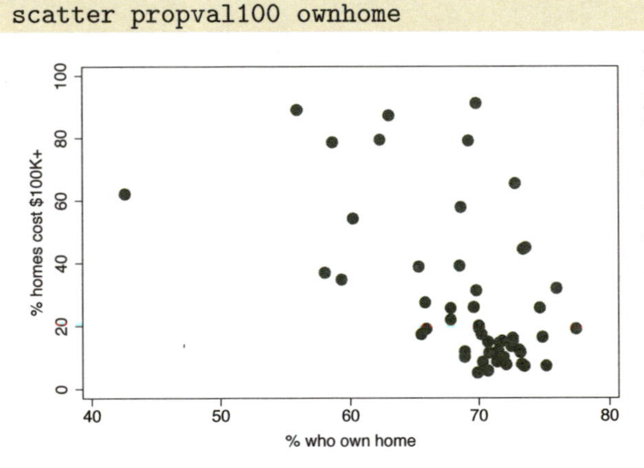

Let's first consider this graph. The graphs in this book have been sized to be 3 inches wide by 2 inches tall. Although we do not see it, some graphs are sized via an **xsize()** and **ysize()** option, and some are sized via schemes. *Uses allstates.dta & scheme vg_s1c*

```
scatter propval100 ownhome, xsize(3) ysize(1)
```

Here, we make a graph to illustrate how to use `xsize()` and `ysize()` to control the aspect ratio of the graph, as well as the size. Note that when we do this, the size of the graph will not change on the screen but the aspect ratio will. Although we can size the graph on the screen, when we export the graph, it will have both the size and aspect ratio we chose using `xsize()` and `ysize()`.
Uses allstates.dta & scheme vg_s1c

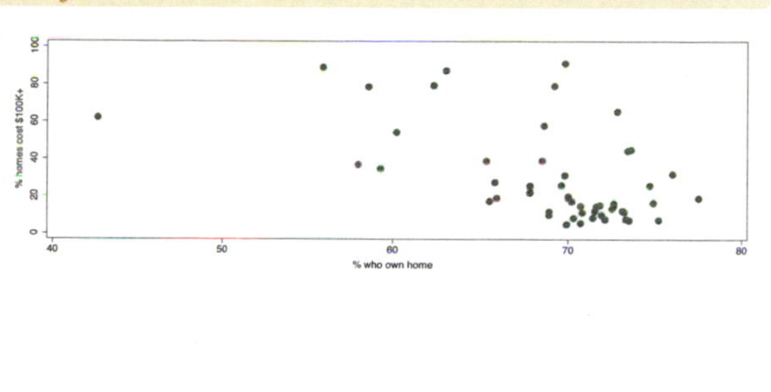

```
scatter propval100 ownhome, xsize(2) ysize(2)
```

Here, we make just one more graph to illustrate that we can use `xsize()` and `ysize()` to control the aspect ratio of the graph, as well as the size. Here, we make the graph square by making the graph 2 inches high by 2 inches tall.
Uses allstates.dta & scheme vg_s1c

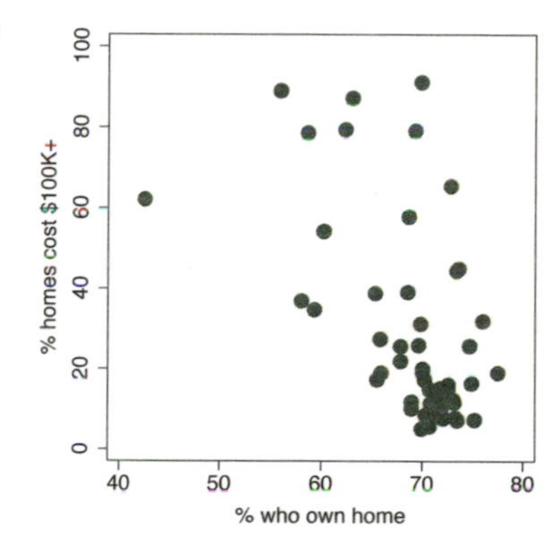


```
scatter propval100 ownhome, scale(1.7)
```

In this example, we add the `scale(1.7)` option to magnify the sizes of the text and markers in the graph, making them 1.7 times their normal sizes. This can be useful when we make small graphs and want to increase the sizes of the text and markers to make them easier to see.
Uses allstates.dta & scheme vg_s1c

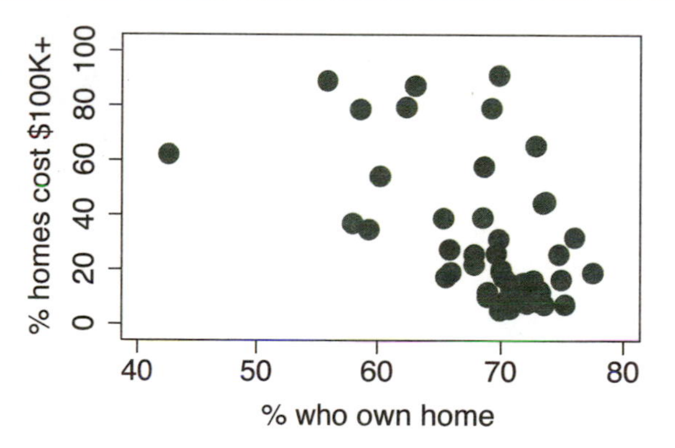

```
scatter propval100 ownhome, scale(.5)
```

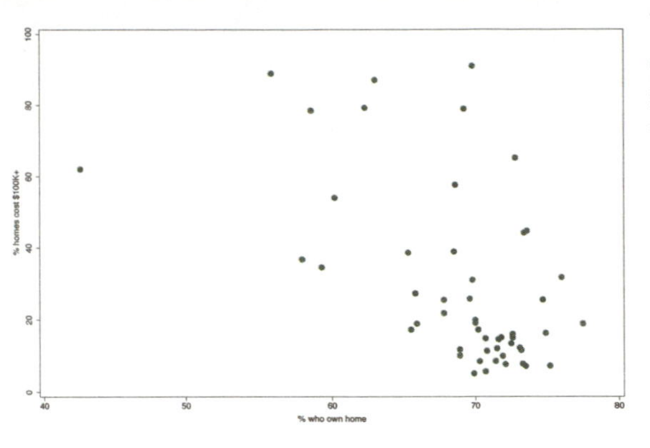

We can also use the `scale()` option to decrease the size of the text and markers. Here, we make the size of these elements half their normal size. *Uses allstates.dta & scheme vg_s1c*

9.4 Changing the look of graph regions

This section discusses the region options that can be controlled via the `plotregion()` and `graphregion()` options. These allow we to control the color of the plot region and graph region, as well as the lines that border these regions. For more information, see [G] ***region_options***. This section uses the `vg_s2c` scheme.

```
scatter propval100 ownhome, title("My title")
```

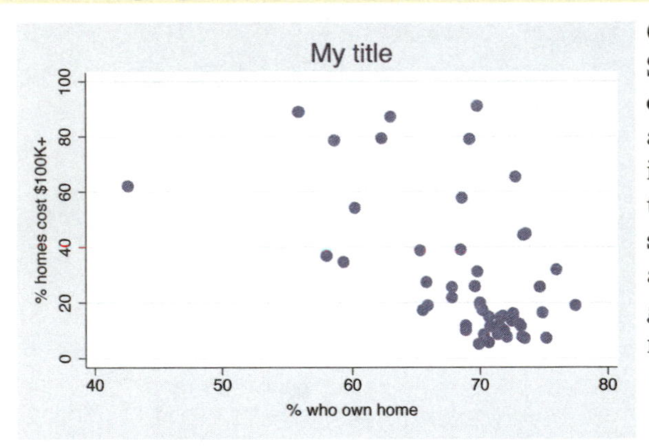

Consider this scatterplot. In general, Stata sees this graph as having two overall regions. The area inside the x- and y-axes where the data are plotted is called the *plot region*. In this graph, the plot region is white. The area surrounding the plot region, where the axes and titles are placed, is called the *graph region*. In this graph, the graph region is shaded light blue. *Uses allstates.dta & scheme vg_s2c*

```
scatter propval100 ownhome, title("My title") plotregion(color(stone))
```

Here, we use `plotregion(color(stone))` to make the color of the plot region stone. The `color()` option controls the color of the plot region.
Uses allstates.dta & scheme vg_s2c

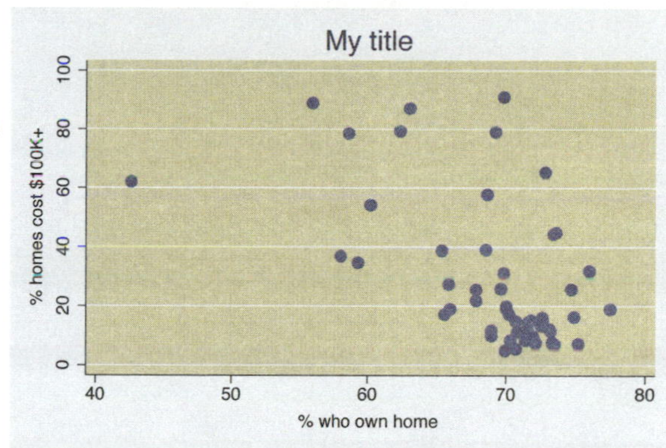


```
scatter propval100 ownhome, title("My title") plotregion(lcolor(navy)
    lwidth(thick) )
```

In this graph, we put a thick, navy blue line around the plot region using the `lcolor()` and `lwidth()` options. This puts a bit of a frame around the plot region.
Uses allstates.dta & scheme vg_s2c

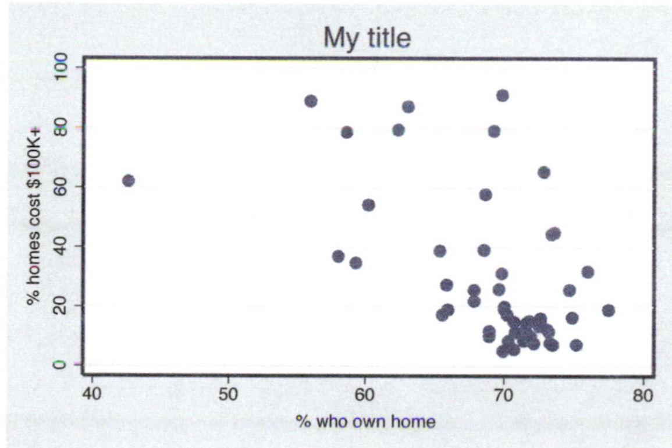


```
scatter propval100 ownhome, title("My title") graphregion(color(erose))
```

Here, we use the `graphregion(color(erose))` option to modify the color of the graph region to be erose, a light rose color. The graph region is the area outside of the plot region where the titles and axes are displayed.
Uses allstates.dta & scheme vg_s2c

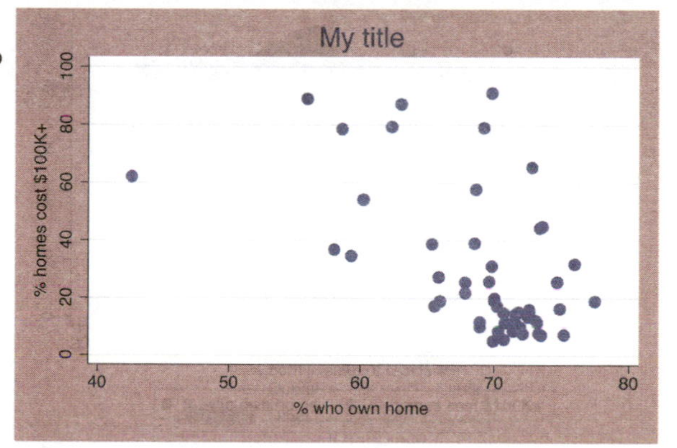

```
scatter propval100 ownhome, title("My title")
   graphregion(ifcolor(erose) fcolor(maroon))
```

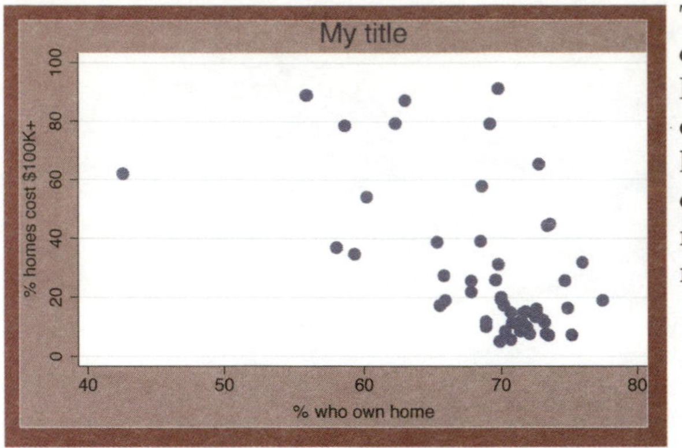

The graph region is actually composed of an inner part and an outer part. Here, we use the `ifcolor(erose)` option to make the inner graph region light rose and the `fcolor(maroon)` option to make the outer graph region maroon. This has the effect of putting a maroon frame around the entire graph. *Uses allstates.dta & scheme vg_s2c*

```
scatter propval100 ownhome, title("My title")
   graphregion(lcolor(navy) lwidth(vthick))
```

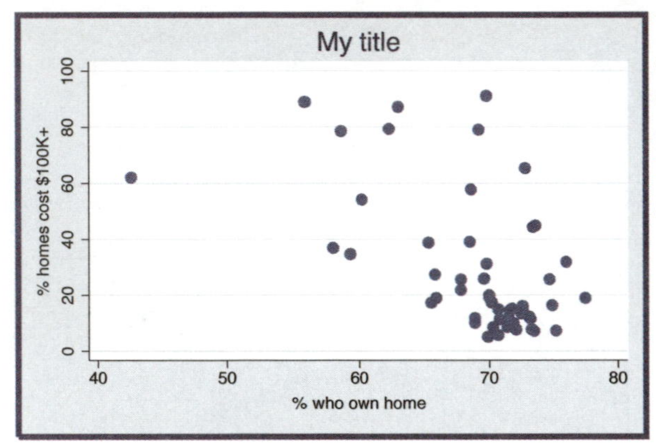

We can put a somewhat different frame around the graph by altering the size and color of the line that surrounds the graph region. Using the `lcolor(navy) lwidth(vthick)` options gives this graph a very thick, navy blue border. *Uses allstates.dta & scheme vg_s2c*

This section omitted numerous options that we could use to control the plot region and graph region, including further control of the inner and outer regions and further control of the lines that surround these regions. Stata gives us more control than we generally need, so rather than covering these options here, I refer you to [G] ***region_options***.

10 Styles for changing the look of graphs

This section focuses on frequently used styles that arise in making graphs, such as *linepatternstyle*, *linewidthstyle*, or *markerstyle*. The styles are covered in alphabetical order, providing more details about the values you can choose. Each section refers to the appropriate section of [G] **graph** to provide complete details on each style. We begin by using the `allstates` file and omitting Washington, DC.

10.1 Angles

An *anglestyle* specifies the angle for displaying an item (or group of items) in the graph. Common examples include specifying the angle for marker labels with `mlabangle()` or the angle of the labels on the y-axis with `ylabel(, angle())`. We can specify an *anglestyle* as a number of degrees of rotation (negative values are permitted, so for example, −90 can be used instead of 270). We can also use the keywords `horizontal` for 0 degrees, `vertical` for 90 degrees, `rhorizontal` for 180 degrees, and `rvertical` for 270 degrees. See [G] ***anglestyle*** for more information.

```
scatter workers2 faminc, mlabel(stateab) mlabangle(45)
```

Here, we use the `mlabangle(45)` (marker label angle) to change the angle of the marker labels to 45 degrees.
Uses allstatesdc.dta & scheme vg_s2c

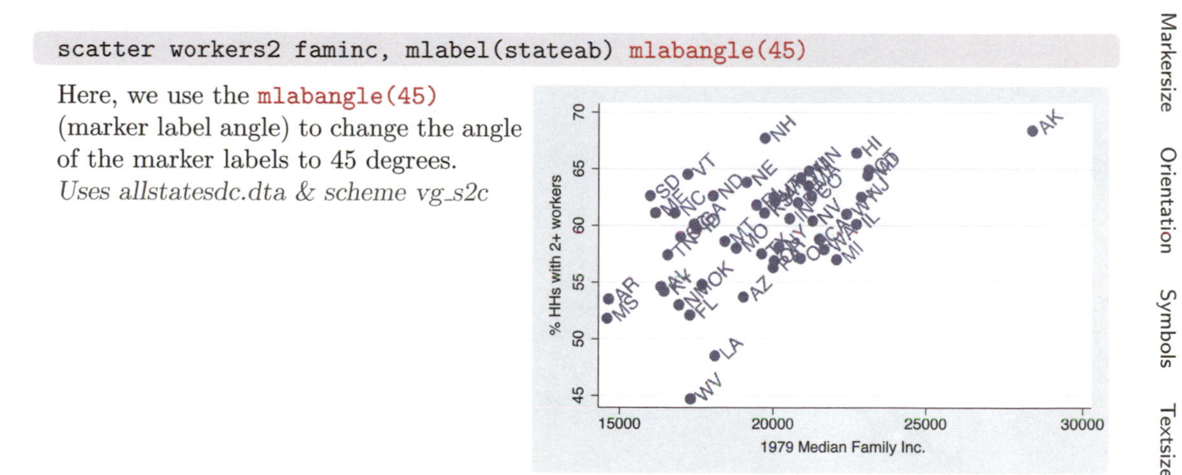

```
scatter workers2 faminc, ylabel(, angle(0))
```

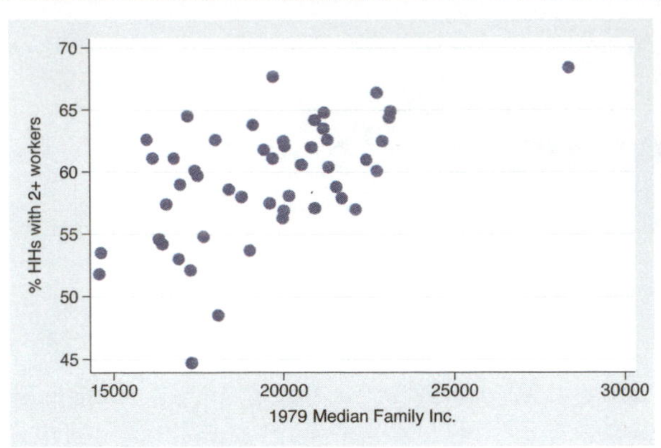

Here, we change the angle of the labels of the y-axis so that they read horizontally by using the `angle(0)` option. We could also have used `horizontal` to obtain the same effect. *Uses allstatesdc.dta & scheme vg_s2c*

```
scatter workers2 faminc, xlabel(15000(1000)30000, angle(45))
```

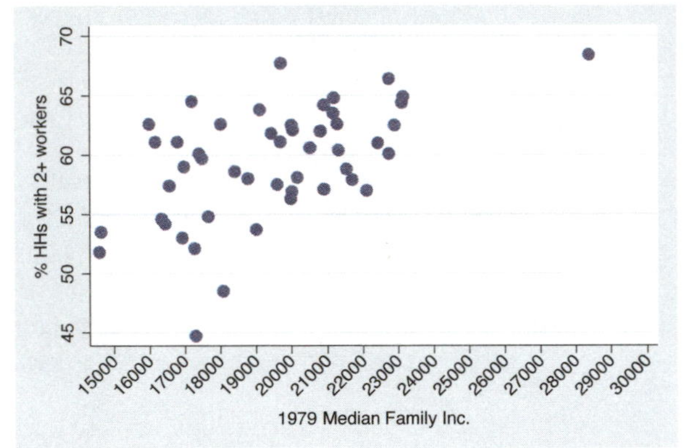

In this example, we label the x-axis from 15,000 to 30,000 incremented by 1,000. When we have so many labels, we can use the `angle(45)` option to display the labels at a 45-degree angle. *Uses allstatesdc.dta & scheme vg_s2c*

10.2 Colors

A *colorstyle* allows us to modify the color of an object, be it a title, a marker, a marker label, a line around a box, a fill color of a box, or practically any other object in graphs. The two main ways to specify a color are either by giving a name of color (e.g., **red**, **pink**, **teal**) or by supplying an RGB value giving the amount of red, green, and blue to be mixed to form a custom color. See [G] ***colorstyle*** for more information.

```
scatter workers2 faminc, mcolor(gs8)
```

The `mcolor()` (marker color) option is used here to make the marker a middle gray. Stata provides 17 levels of gray named `gs0` to `gs16`. The darkest is `gs0` (a synonym for `black`), and the lightest is `gs16` (a synonym for `white`).
Uses allstatesdc.dta & scheme vg_s2c

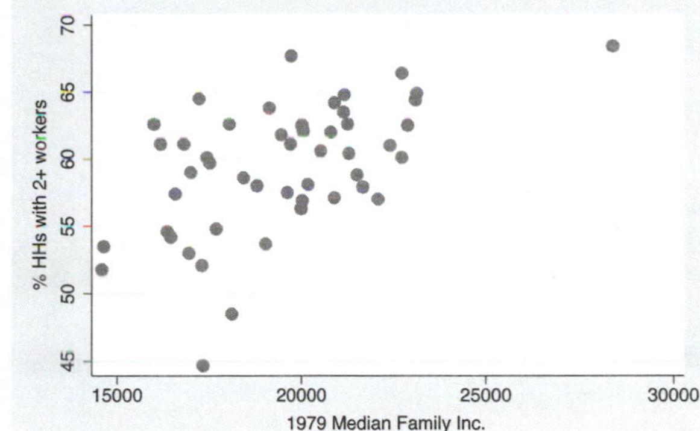


```
scatter workers2 faminc, mcolor(lavender)
```

Here, we use the `mcolor(lavender)` option to make the markers lavender, one of the predefined colors created by Stata. The next example illustrates more of the colors from which you can choose.
Uses allstatesdc.dta & scheme vg_s2c

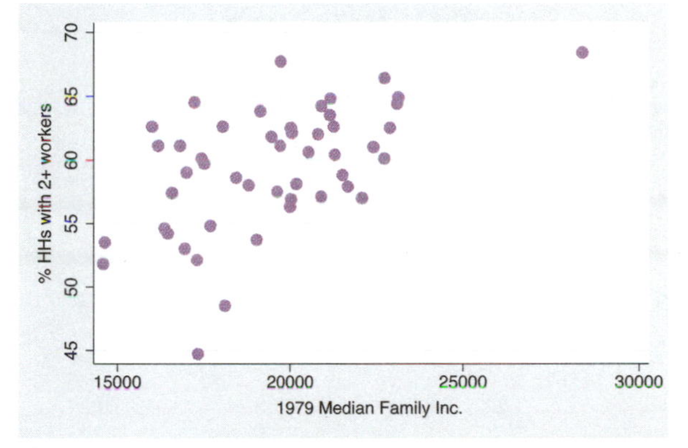


```
vgcolormap, quietly
```

This `vgcolormap` command is a command that I wrote to show the different standard colors available in Stata all at once. We simply issue the command `vgcolormap`, and it creates a scatterplot that shows the colors we can choose from and their names. See the list of colors available in [G] ***colorstyle***, and see how to get `vgcolormap` in Appendix : Online supplements (382).
Uses allstatesdc.dta & scheme s2color

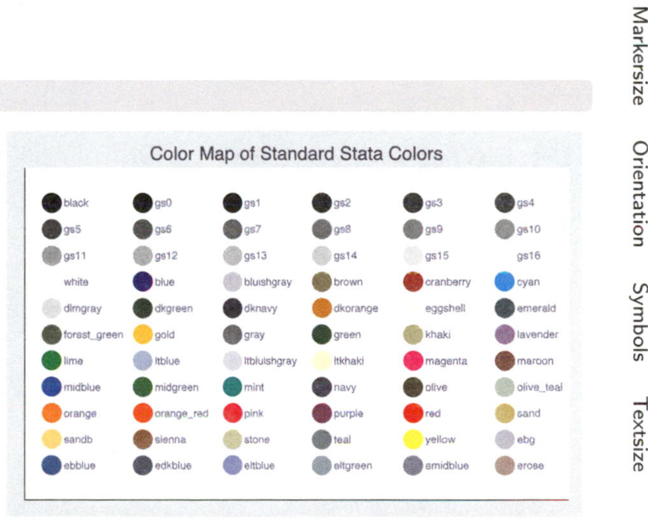

```
scatter workers2 faminc, mcolor("255 255 0")
```

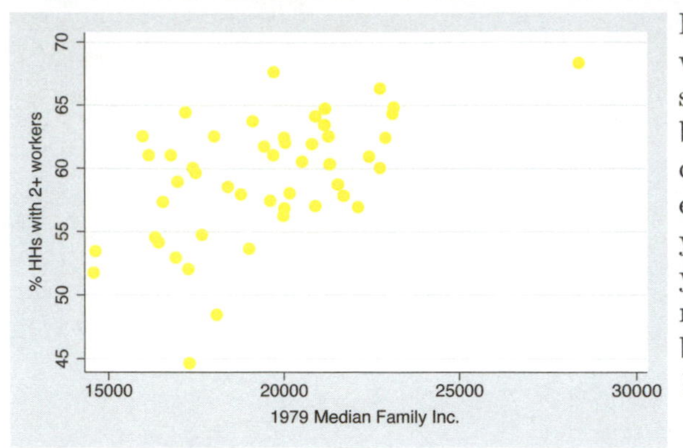

Despite all the standard color choices, we may want to mix our own colors by specifying how much red, green, and blue that we want mixed together. We can mix between 0 and 255 units of each color. Mixing 0 units of each yields black, and 255 units of each yields white. Here, we mix 255 units of red, 255 units of green, and 0 units of blue to get a shade of yellow.
Uses allstatesdc.dta & scheme vg_s2c

```
scatter workers2 faminc, mcolor("255 150 100")
```

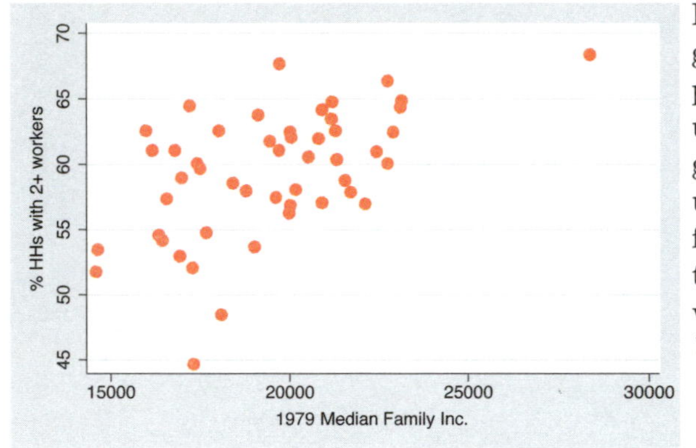

By mixing 255 parts red, 150 parts green, and 100 parts blue, we get a peach color. Since colors for web pages use this same principle of mixing red, green, and blue, we can do a web search using terms like *color mixing html* and find numerous web pages to help us find the right mixture for the colors that we want to make.
Uses allstatesdc.dta & scheme vg_s2c

10.3 Clock position

A clock position refers to a location using the numbers on an analog clock to indicate the location, with 12 o'clock being above the center, 3 o'clock to the right, 6 o'clock below the center, and 9 o'clock to the left. A value of 0 refers to the center but may not always be valid. See [G] ***clockpos*** for more information.

```
scatter workers2 faminc, mlabel(stateab) mlabposition(5)
```

In this example, we add marker labels to a scatterplot and use the `mlabposition(5)` (marker label position) option to place the marker labels in the 5 o'clock position with respect to the markers.
Uses allstatesdc.dta & scheme vg_s2c

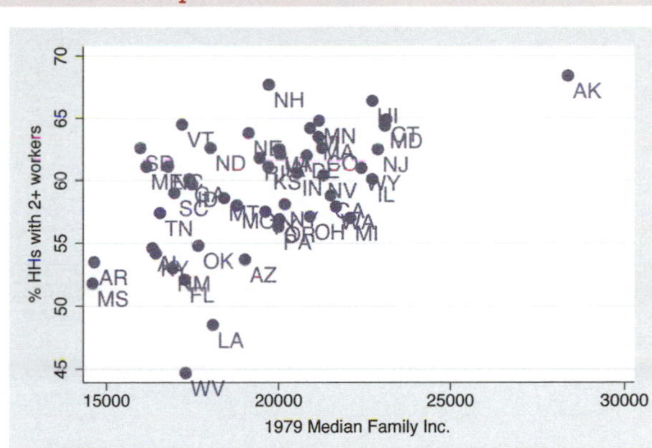


```
scatter workers2 faminc, mlabel(stateab) mlabposition(0) msymbol(i)
```

In this example, we place the markers in the center position using the `mlabposition(0)` option. We also make the symbols invisible using the `msymbol(i)` option. Otherwise, the markers and marker labels would be atop each other.
Uses allstatesdc.dta & scheme vg_s2c

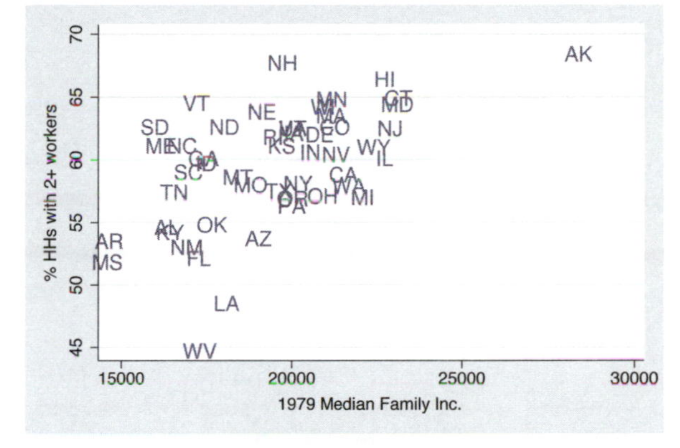

10.4 Compass direction

A *compassdirstyle* is much like *clockpos*, but where a *clockpos* has 12 possible outer positions, like a clock, the *compassdirstyle* has only 9 possible outer positions, like the major labels on a compass: `north`, `neast`, `east`, `seast`, `south`, `swest`, `west`, `nwest`, and `center`. These can be abbreviated as `n`, `ne`, `e`, `se`, `s`, `sw`, `w`, `nw`, and `c`. Stata permits us to use a *clockpos* even when a *compassdirstyle* is called for and makes intuitive translations; for example, `12` is translated to `north`, or `2` is translated to `neast`. See [G] ***compassdirstyle*** for more information.

```
scatter workers2 faminc, title("Work Status and Income",
   ring(0) placement(se))
```

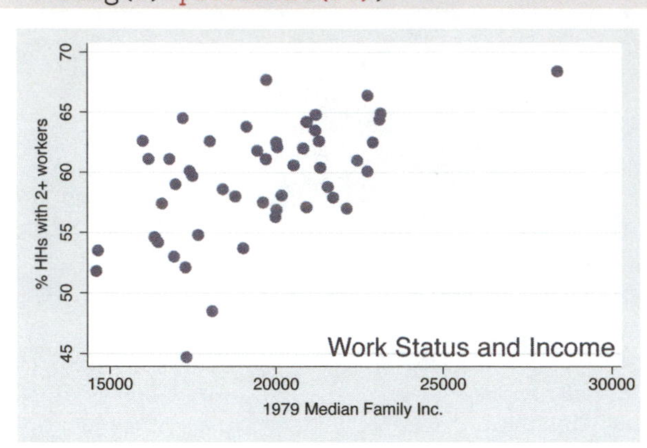

In this example, we use `placement()` to position the title in the southeast (bottom right corner) of the plot region. The `ring(0)` option moves the title inside the plot region.
Uses allstatesdc.dta & scheme vg_s2c

```
scatter workers2 faminc, title("Work Status and Income",
   ring(0) placement(4))
```

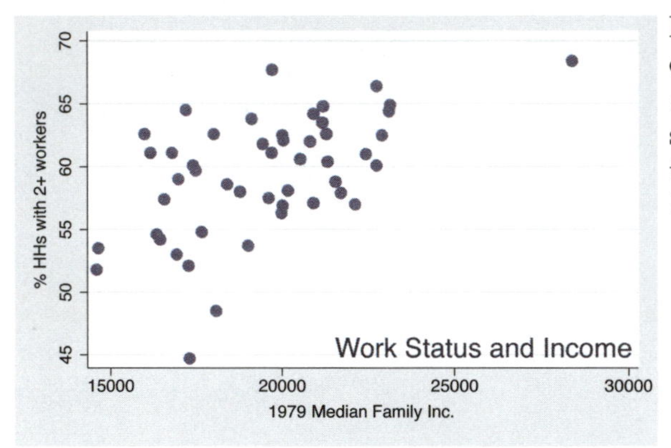

If we instead specify the `placement(4)` option (using a *clockpos* instead of *compassdir*), Stata makes a suitable substitution, and the title is placed in the bottom right corner.
Uses allstatesdc.dta & scheme vg_s2c

10.5 Connecting points

Stata supports a variety of methods for connecting points using different values for the *connectstyle*. These include `l` (lowercase L, as in line) to connect with a straight line, `L` to connect with a straight line only if the current x-value is greater than the prior x-value, `J` for stairstep, `stepstair` for step then stair, and `i` for invisible connections. For the next few examples, let's switch to using the `spjanfeb2001` data file, keeping only the data for January and February of 2001. See [G] ***connectstyle*** for more information.

```
scatter close tradeday
```

Here, we make a scatterplot showing the closing price on the y-axis and the trading day (numbered 1 to 40) on the x-axis. Normally, we would connect these points.
Uses spjanfeb2001.dta & scheme vg_s2c

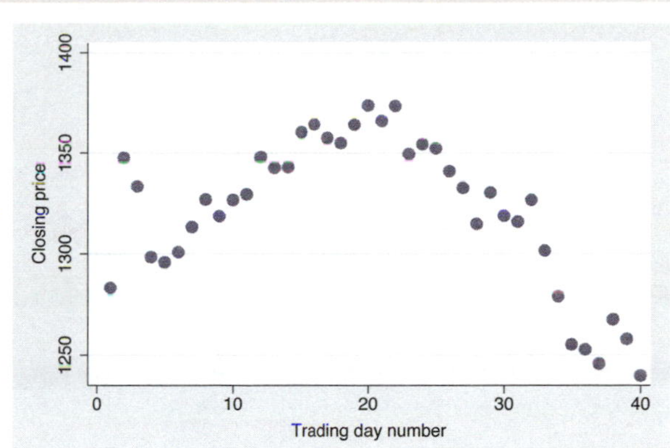


```
scatter close tradeday, connect(l)
```

Here, we add the connect(l) option, but this is probably not the kind of graph we wanted to create. The problem is that the observations are in a random order, but the observations are connected in the same order as they appear in the data. We really want the points to be connected based on the order of `tradeday`.
Uses spjanfeb2001.dta & scheme vg_s2c

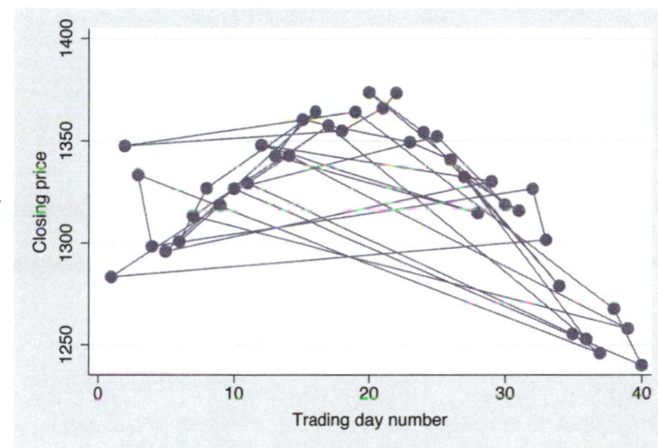


```
scatter close tradeday, connect(l) sort
```

To fix the previous graph, we can either first use the `sort` command to sort the data on `tradeday` or, as we do here, use the `sort` option to tell Stata to sort the data on `tradeday` before connecting the points. We also could have specified `sort(tradeday)`, and it would have had the same effect.
Uses spjanfeb2001.dta & scheme vg_s2c

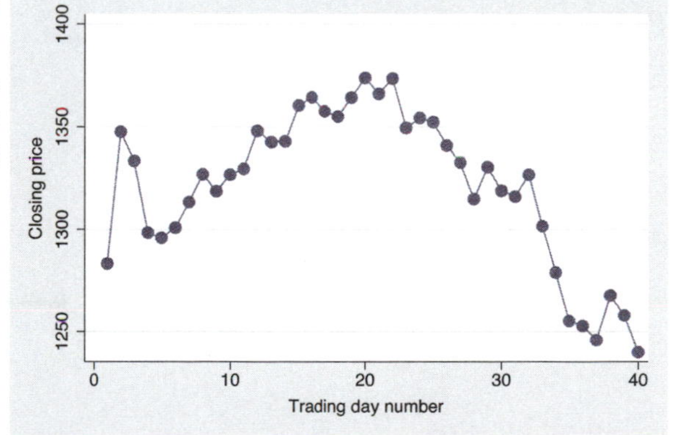

```
scatter close predclose tradeday
```

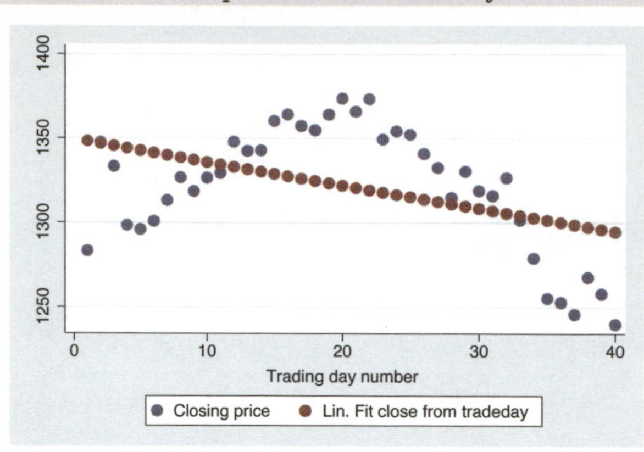

Say that we used the **regress** command to predict **close** from **tradeday** and generated a predicted value called **predclose**. Here, we plot the actual closing prices and the predicted closing prices.
Uses spjanfeb2001.dta & scheme vg_s2c

```
scatter close predclose tradeday, connect(i l) sort
  msymbol(.  i)
```

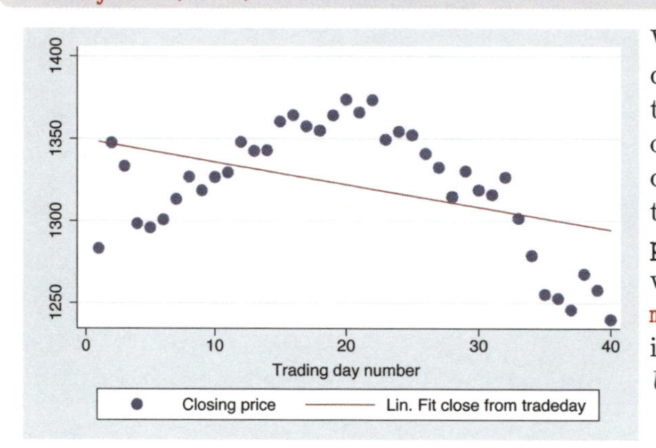

We use the `connect(i l)` option to connect the predicted values and leave the observed values unconnected. The `i` option with `close` indicates that the closing values are not connected, and the `l` (letter l) option indicates that the `predclose` values should be connected with a straight line. We also add `msymbol(.  i)` to make the symbols invisible for the fit values.
Uses spjanfeb2001.dta & scheme vg_s2c

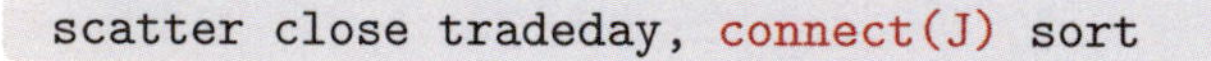

```
scatter close tradeday, connect(J) sort
```

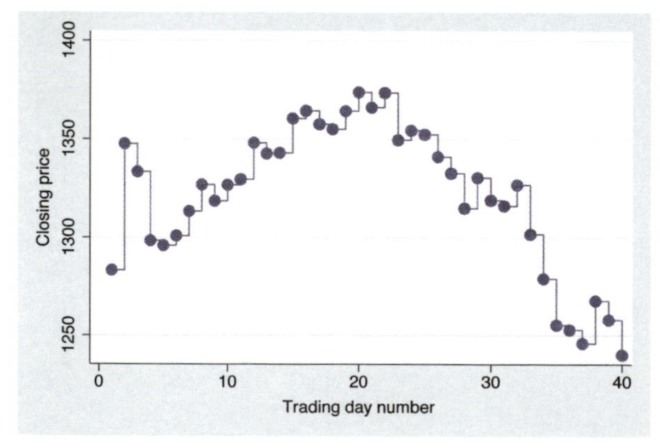

In other contexts (such as survival analysis), we might want to connect points using a stairstep pattern. Here, we connect the observed closing prices with the `J` option (which can also be specified as **stairstep**) to get a stairstep effect.
Uses spjanfeb2001.dta & scheme vg_s2c

```
scatter close tradeday, connect(stepstair) sort
```

In other contexts, we might want to connect points using a stepstair pattern. Here, we connect the observed closing prices with the stepstair option to get a stepstair effect.
Uses spjanfeb2001.dta & scheme vg_s2c

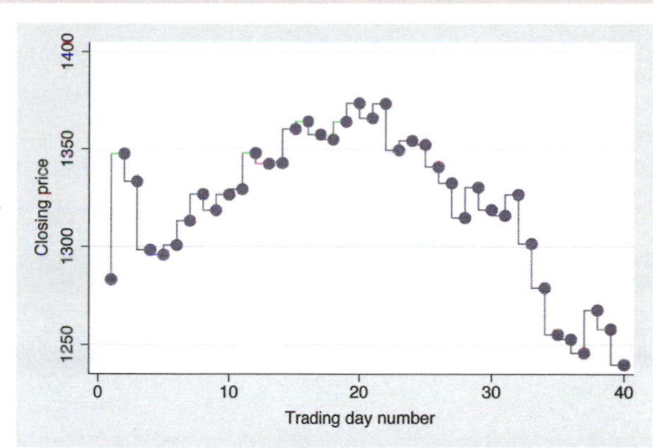


```
scatter close dom, connect(l) sort(date)
```

Say that we created a variable called dom that represented the day of month and wanted to graph the closing prices for January and February against the day of the month. Using the sort(date) option combined with connect(l), we almost get what we want, but we get a line that swoops back connecting January 31 to Feb 1.
Uses spjanfeb2001.dta & scheme vg_s2c

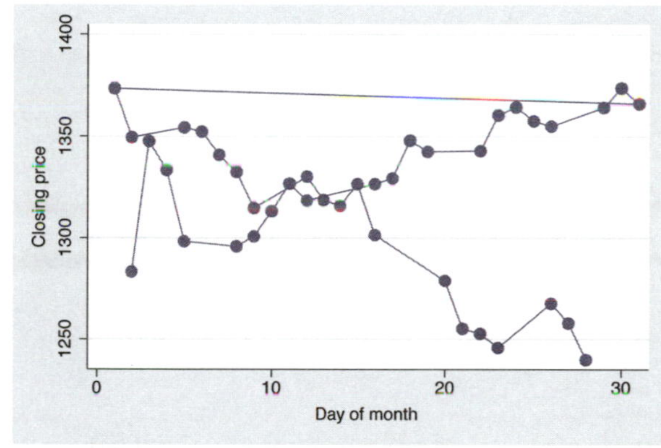


```
scatter close dom, connect(L) sort(date)
```

This kind of example calls for the connect(L) option, which avoids the line that swoops back by connecting points with a straight line, except when the x-value (dom) decreases (e.g., goes from 31 to 1).
Uses spjanfeb2001.dta & scheme vg_s2c

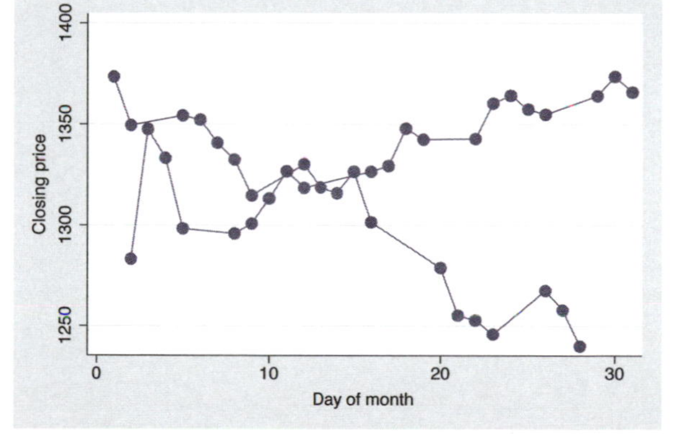

10.6 Line patterns

We can specify the pattern we want for a line in three ways. We can specify a word that selects among a set of predefined styles, including `solid` (solid line), `dash` (a dashed line), `dot` (a dotted line), `shortdash` (short dashes), `longdash` (long dashes), and `blank` (invisible). There are also combination styles `dash_dot`, `shortdash_dot`, and `longdash_dot`. We can also use a formula that combines the following five elements in any way that we wish: `l` (letter l, solid line), `_` (underscore, long dash), `-` (hyphen, medium dash), `.` (period, short dash), and `#` (small amount of space). We could specify `longdash_dot` or `"_."`, and they would be equivalent. See [G] ***linepatternstyle*** for more information.

```
twoway (line close tradeday, clpattern(solid) sort)
   (lfit close tradeday, clpattern(dash))
   (lowess close tradeday, clpattern(shortdash_dot))
```

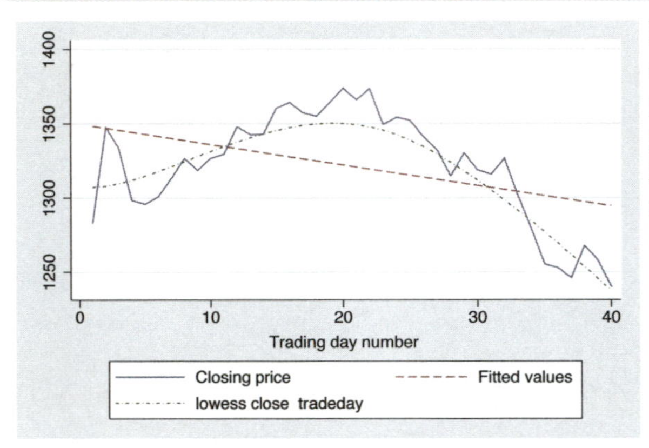

In this example, we make a line plot and use the `clpattern()` (connect line pattern) option to obtain a solid pattern for the observed data, a dash for the linear fit line, and a short dash and dot line for a lowess fit.
Uses spjanfeb2001.dta & scheme vg_s2c

```
twoway (line close tradeday, clpattern("l") sort)
   (lfit close tradeday, clpattern("._"))
   (lowess close tradeday, clpattern("-###"))
```

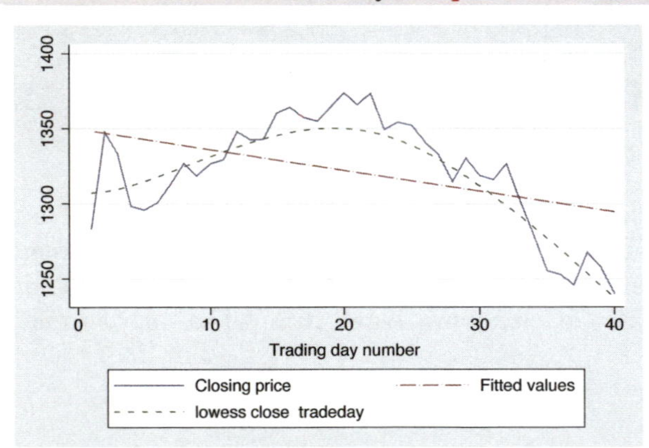

We can use the `clpattern()` option specifying a formula to indicate the pattern for the lines. Here, we specify a solid line for the `line` plot, a dot and dash for the `lfit` plot, and a dash and three spaces for the `lowess` fit.
Uses spjanfeb2001.dta & scheme vg_s2c

```
twoway (line close tradeday, clpattern("l") sort)
   (lfit close tradeday, clpattern("__##"))
   (lowess close tradeday, clpattern("-.#"))
```

This example shows other formulas we could create, including "__##", which yields long dashes with short and then long breaks in the middle, and "-.#", which yields a short dash, a dot, and a space. Using these formulas, we can create a wide variety of line patterns for those instances where we need to differentiate multiple lines.
Uses spjanfeb2001.dta & scheme vg_s2c

```
palette linepalette
```

We can use the built-in Stata command `palette linepalette` to view a variety of line patterns that are available within Stata to help us choose a pattern to our liking.
Uses spjanfeb2001.dta & scheme vg_s2c

10.7 Line width

We can indicate the width of a line in two ways. We can indicate a *linewidthstyle*, which allows us to use a word to specify the width of a line, including `none` (no width, invisible), `vvthin`, `vthin`, `thin`, `medthin`, `medium`, `medthick`, `thick`, `vthick`, `vvthick`, and even `vvvthick`. We can also specify a *relativesize*, which is a multiple of the line's normal thickness (e.g., *2 is twice as thick, or *.7 is .7 times as thick). See [G] ***linewidthstyle*** for more information.

```
twoway (line close tradeday, clwidth(vthick) sort)
   (lfit close tradeday, clwidth(thick))
   (lowess close tradeday, bwidth(.5) clwidth(thin))
```

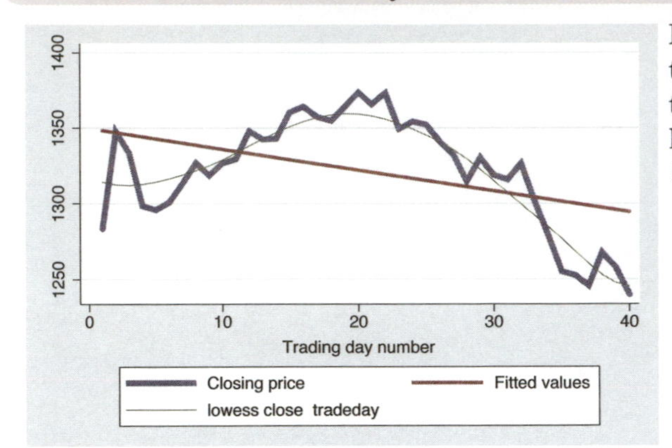

Now, we plot the same three lines but this time differentiate them by line thickness using the `clwidth()` (connect line width) option.
Uses spjanfeb2001.dta & scheme vg_s2c

```
twoway (line close tradeday, clwidth(*4) sort)
   (lfit close tradeday, clwidth(*2))
   (lowess close tradeday, bwidth(.5) clwidth(*.5))
```

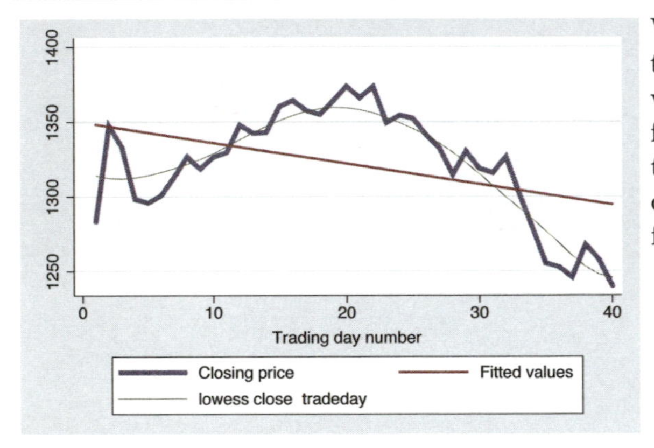

We could create a similar graph using the `clwidth()` option and specify the widths as relative sizes, making the line four times as wide for the `line` plot, two times as wide for the `lfit` command, and half as wide for the line for the `lowess` command.
Uses spjanfeb2001.dta & scheme vg_s2c

10.8 Margins

We can specify the size of a margin in three different ways. We can use a word that represents a predefined margin. These include `zero`, `vtiny`, `tiny`, `vsmall`, `small`, `medsmall`, `medium`, `medlarge`, `large`, and `vlarge`. They also include `top_bottom` to indicate a medium margin at the top and bottom, and `sides` to indicate a medium margin at the left and right. A second method is to give four numbers giving the margins at the left, right, top, and bottom. A third method is to use expressions such as `b=5` to modify one or more of the margins. These are illustrated below. See [G] ***marginstyle*** for more information.

```
scatter workers2 faminc, title("Overall title", margin(large) box)
```

We illustrate the control of margins by adding a title to this scatterplot and putting a box around it. We can then see the effect of the `margin()` option: the gap between the title and the box changes. Here, we specify a `large` margin, and the margin on all four sides is now large.
Uses allstatesdc.dta & scheme vg_s2c

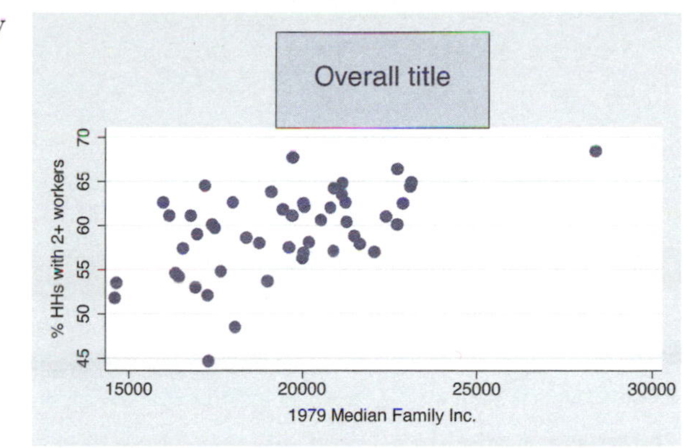


```
scatter workers2 faminc, title("Overall title", margin(top_bottom) box)
```

Using `margin(top_bottom)`, we obtain a margin that is medium on the top and bottom but zero on the left and right.
Uses allstatesdc.dta & scheme vg_s2c

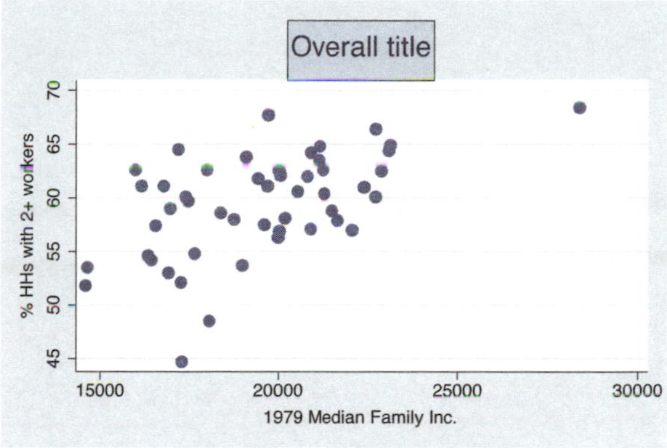

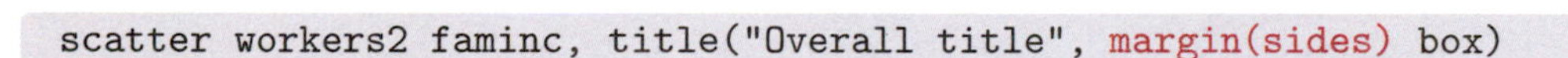

```
scatter workers2 faminc, title("Overall title", margin(sides) box)
```

Using the `margin(sides)` option, we obtain a margin that is medium on the left and right but zero on the top and bottom.
Uses allstatesdc.dta & scheme vg_s2c

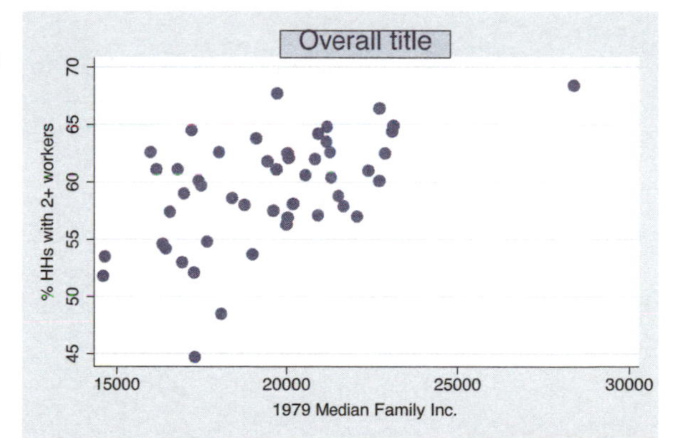

```
scatter workers2 faminc, title("Overall title", margin(9 6 3 0) box)
```

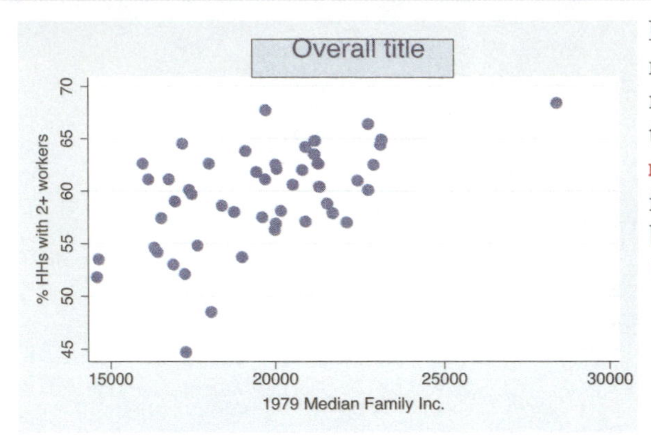

In addition to the words describing margins, we can manually specify the margin for the left, right, bottom, and top. In this example, we specify `margin(9 6 3 0)` and make the margin for the left 9, for the right 6, for the bottom 3, and for the top 0.
Uses allstatesdc.dta & scheme vg_s2c

```
scatter workers2 faminc, title("Overall title", margin(l=9 r=9) box)
```

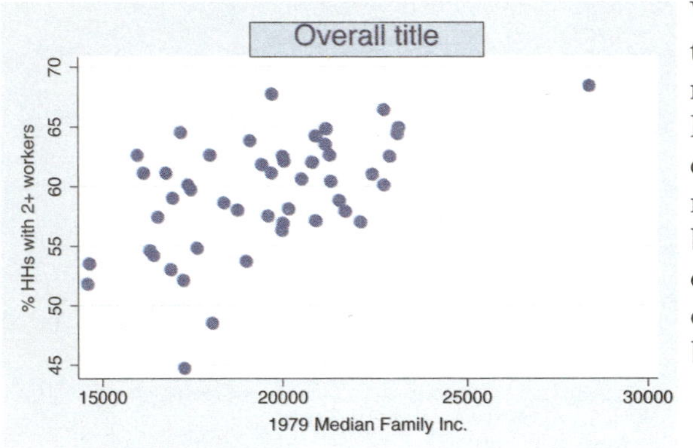

We can also manually change some of the margins without specifying all four margins (as in the previous example). By specifying `margin(l=9 r=9)`, we can make the margin at the left and right 9 units, leaving the top and bottom unchanged. We can specify one or more of the expressions `l=`, `r=`, `t=`, or `b=` to modify the left, right, top, or bottom margins, respectively.
Uses allstatesdc.dta & scheme vg_s2c

10.9 Marker size

We can control the size of the markers by specifying a *markersizestyle* or a *relativesize*. The *markersizestyle* is a word that describes the size of a marker, including `vtiny`, `tiny`, `vsmall`, `small`, `medsmall`, `medium`, `medlarge`, `large`, `vlarge`, `huge`, `vhuge`, and `ehuge`. We could also specify the sizes as a *relativesize*, which is either an absolute size or a multiple of the original size of the marker (e.g., *2 is twice as large, or *.7 is .7 times as large). See [G] ***markersizestyle*** for more information.

```
twoway (scatter propval100 rent700 ownhome urban,
   msize(vsmall medium large))
```

Here, we have an overlaid scatterplot where we graph three variables on the y-axis (`propval100`, `rent700`, and `ownhome`) and use the `msize(vsmall medium large)` option to make the sizes of these markers very small, medium, and large, respectively.
Uses allstatesdc.dta & scheme vg_s2c

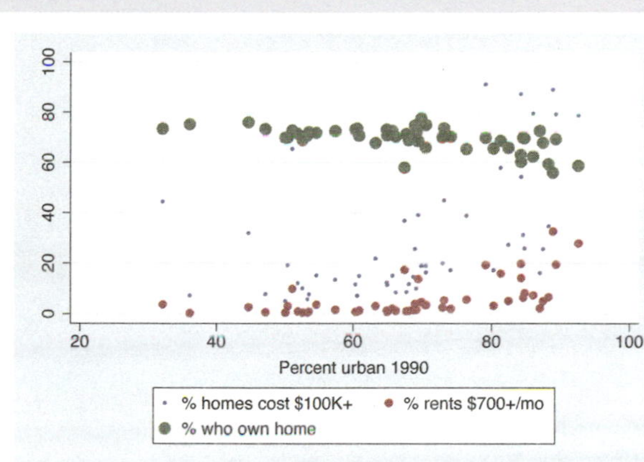


```
twoway (scatter propval100 rent700 ownhome urban, msize(*.5 *1 *1.5))
```

We can repeat the previous graph but use relative sizes within the `msize()` option to control the sizes of the markers, making them, respectively, half the normal size, regular size, and half again the normal size.
Uses allstatesdc.dta & scheme vg_s2c

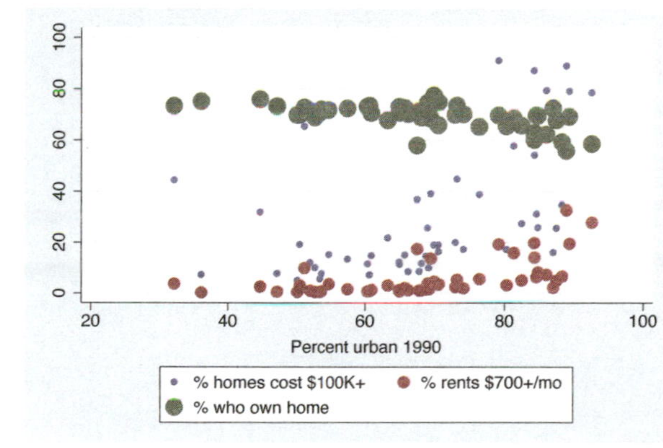

10.10 Orientation

An *orientationstyle* is used to change the orientation of text, such as a y-axis title, an x-axis title, or added text. An *orientationstyle* is similar to an *anglestyle* (see Styles : Angles (327)). We can only specify four different orientations using the keywords `horizontal` for 0 degrees, `vertical` for 90 degrees, `rhorizontal` for 180 degrees, and `rvertical` for 270 degrees. See [G] ***orientationstyle*** for more information.

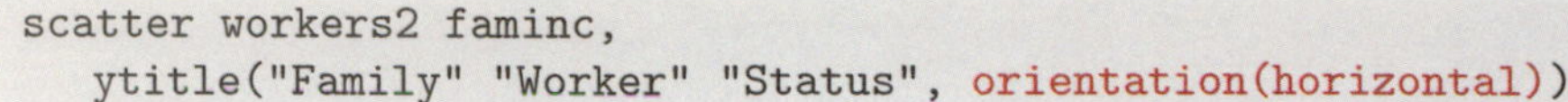

```
scatter workers2 faminc,
  ytitle("Family" "Worker" "Status", orientation(horizontal))
```

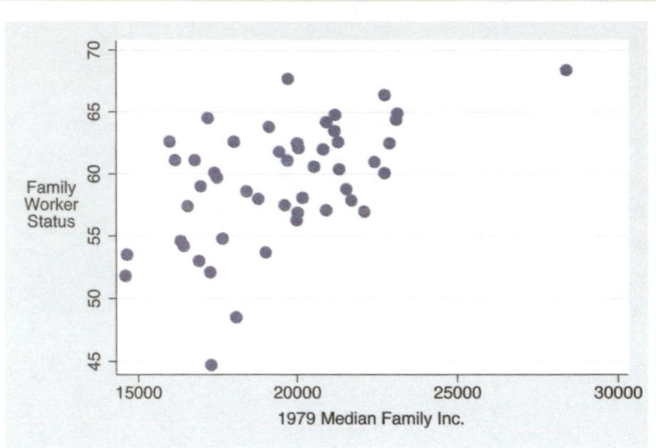

This example shows how we can rotate the title for the y-axis using the `orientation(horizontal)` option to make the title horizontal.
Uses allstatesdc.dta & scheme vg_s2c

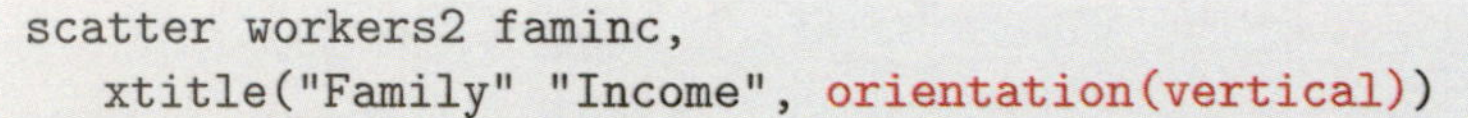

```
scatter workers2 faminc,
  xtitle("Family" "Income", orientation(vertical))
```

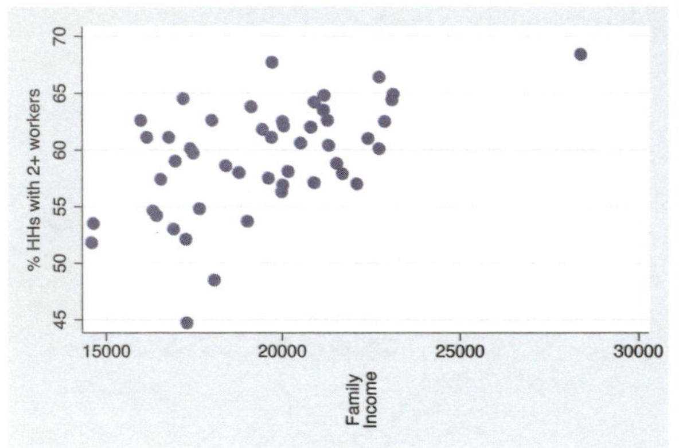

This example shows how we can rotate the title for the x-axis to be vertical using the `orientation(vertical)` option.
Uses allstatesdc.dta & scheme vg_s2c

10.11 Marker symbols

Stata allows a wide variety of marker symbols. We can specify `O` (circle), `D` (diamond), `T` (triangle), `S` (square), `+` (plus sign), `X` (x), `p` (a tiny point), and `i` (invisible). We can also use lowercase letters `o`, `d`, `t`, `s`, and `x` to indicate smaller symbols. For circles, diamonds, triangles, and squares, we can append an `h` to indicate that the symbol should be displayed as hollow (e.g., `Oh` is a hollow circle). See [G] ***symbolstyle*** for more information.

```
twoway (scatter propval100 rent700 ownhome urban, msymbol(S T O))
```

In this example, we use the `msymbol(S T O)` (marker symbol) option to plot the three symbols in this graph using squares, triangles, and circles.
Uses allstatesdc.dta & scheme vg_s2c

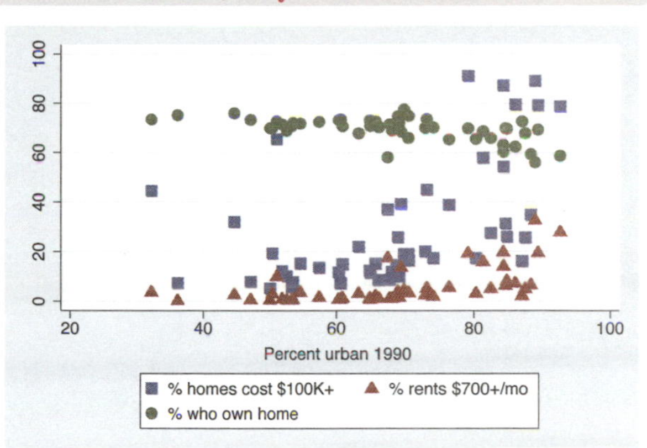


```
twoway (scatter propval100 rent700 ownhome urban, msymbol(Sh Th Oh))
```

We append an `h` to each marker symbol option to indicate that the symbol should be displayed as hollow.
Uses allstatesdc.dta & scheme vg_s2c

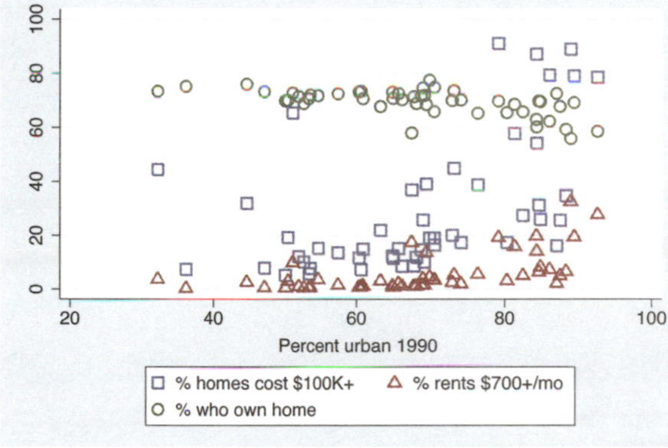


```
twoway (scatter propval100 rent700 ownhome urban, msymbol(s t o))
```

In this example, we use the `msymbol(s t o)` option to specify small squares, small triangles, and small circles.
Uses allstatesdc.dta & scheme vg_s2c

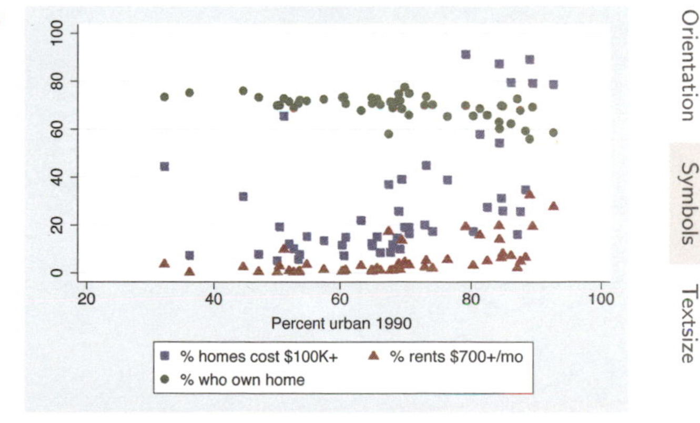

10.12 Text size

The *textsizestyle* is used to control the size of text, either by specifying a keyword that corresponds to a particular size or by specifying a number representing a relative size. The predefined keywords include `zero`, `miniscule`, `quarter_tiny`, `third_tiny`, `half_tiny`, `tiny`, `vsmall`, `small`, `medsmall`, `medium`, `medlarge`, `large`, `vlarge`, `huge`, and `vhuge`. We could also specify the sizes as a relative size, which is a multiple of the original size of the text. See [G] ***textsizestyle*** for more information.

```
scatter workers2 faminc, mlabel(stateab) mlabsize(small)
```

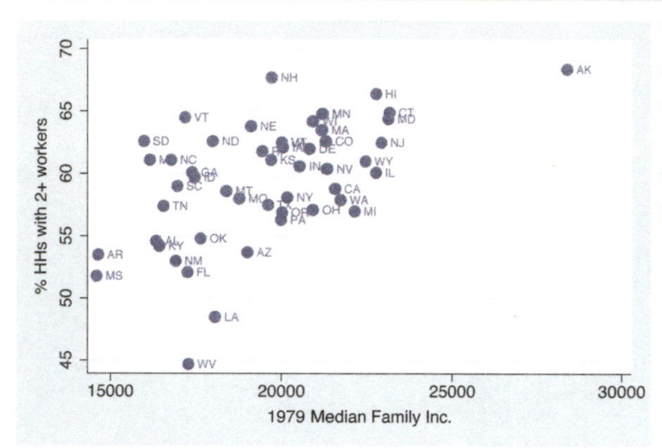

This example uses `mlabel(stateab)` to add marker labels with the state abbreviation labeling each point. We use the `mlabsize(small)` (marker label size) option to modify the size of the marker labels to make the labels small.
Uses allstatesdc.dta & scheme vg_s2c

```
scatter workers2 faminc, mlabel(stateab) mlabsize(*1.5)
```

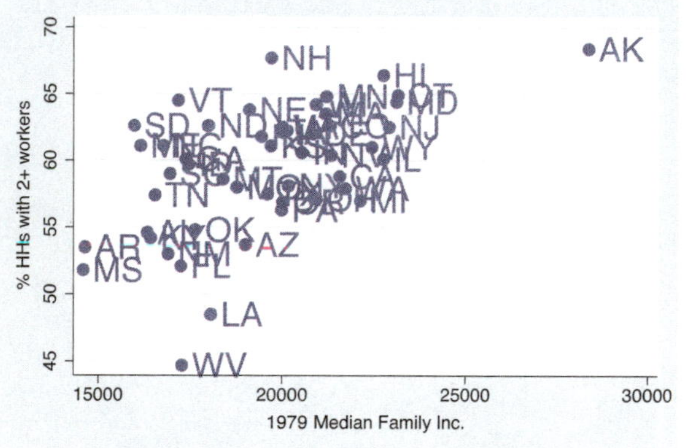

In addition to using the keywords, we can specify a relative size that is a multiple of the normal size. Here, we use the `mlabsize(*1.5)` option to make the marker labels 1.5 times as large as they would normally be.
Uses allstatesdc.dta & scheme vg_s2c

11 Appendix

The appendix contains a hodgepodge of material that did not fit well in any of the previous chapters. We begin by illustrating some of the other kinds of graphs Stata can produce that were not covered in this book and how to use the options illustrated in this book to make them. Next, we look at how to save graphs, redisplay graphs, and combine multiple graphs into a single graph. This is followed by a section with more realistic examples that require a combination of multiple options or data manipulation to create the graph. We review some common mistakes in writing graph commands and showing how to fix them, followed by a brief look at creating custom schemes. This chapter and the book conclude by describing the online supplements to the book and how to get them.

11.1 Overview of statistical graph commands, stat graphs

This section illustrates some of the Stata commands for producing specialized statistical graphs. Unlike other sections of this book, this section merely illustrates these kinds of graphs but does not further explain the syntax of the commands used to create them. The graphs are illustrated on the following six pages, with multiple graphs on each page. The title of each graph is the name of the Stata command that produced the graph. We can use the `help` command to find out more about that command or look up more information in the appropriate Stata manual. The figures are described below.

- Figure 11.1 illustrates a number of graphs used to examine the univariate distribution of variables.
- Figure 11.2 illustrates the `gladder` and `qladder` commands, which show the distribution of a variable according to the *ladder of powers* to help visually identify transformations for achieving normality.
- Figure 11.3 shows a number of graphs that can be used to assess how your data meets the assumptions of linear regression.
- Figure 11.4 shows some plots that help to illustrate the results of a survival analysis.
- Figure 11.5 shows a number of different plots used to understand the nature of time-series data and to select among different time-series models.
- Figure 11.6 shows plots associated with Receiver Operating Characteristic (ROC) analyses, which can also be used with logistic regression analysis.

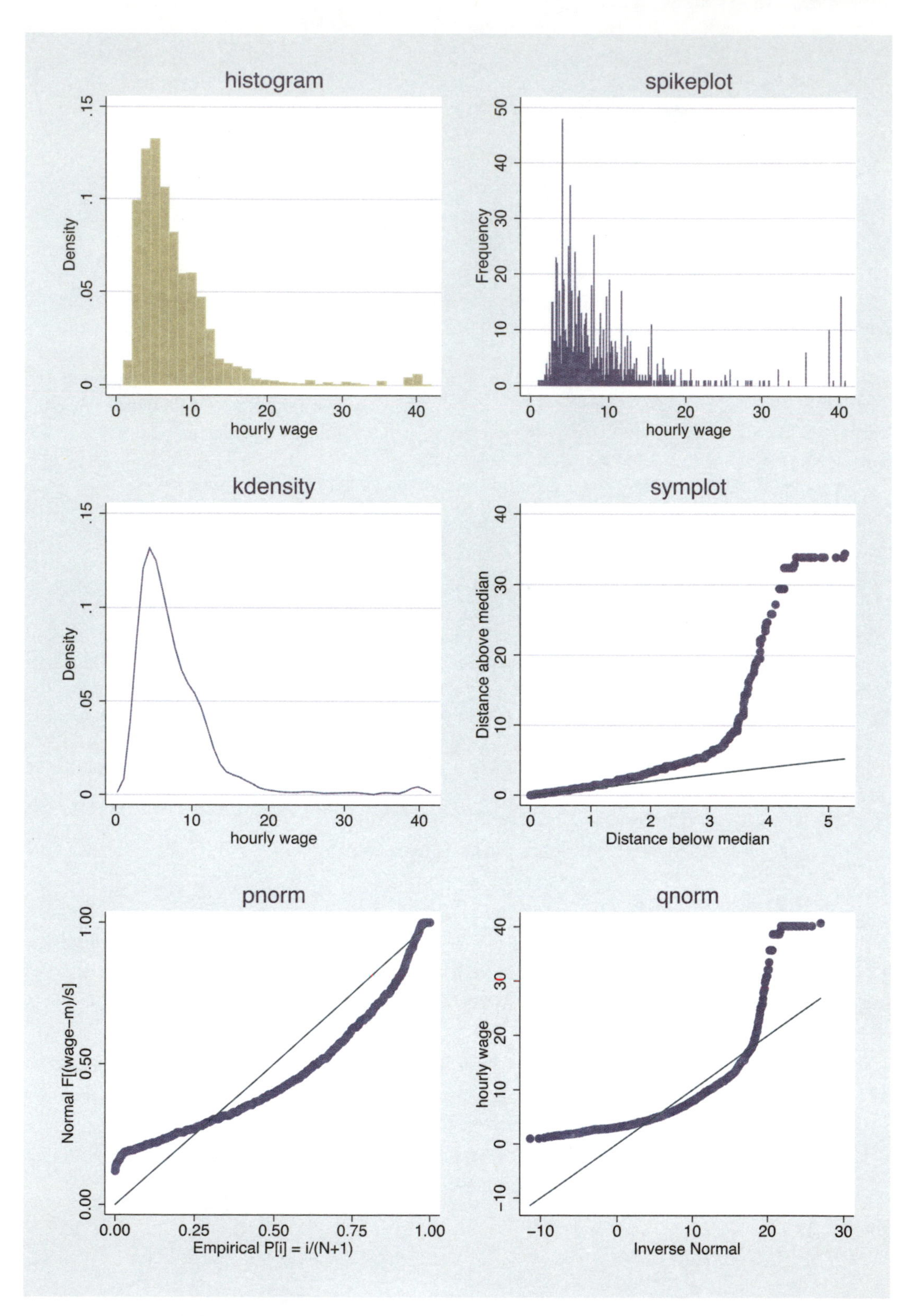

Figure 11.1: Distribution graphs

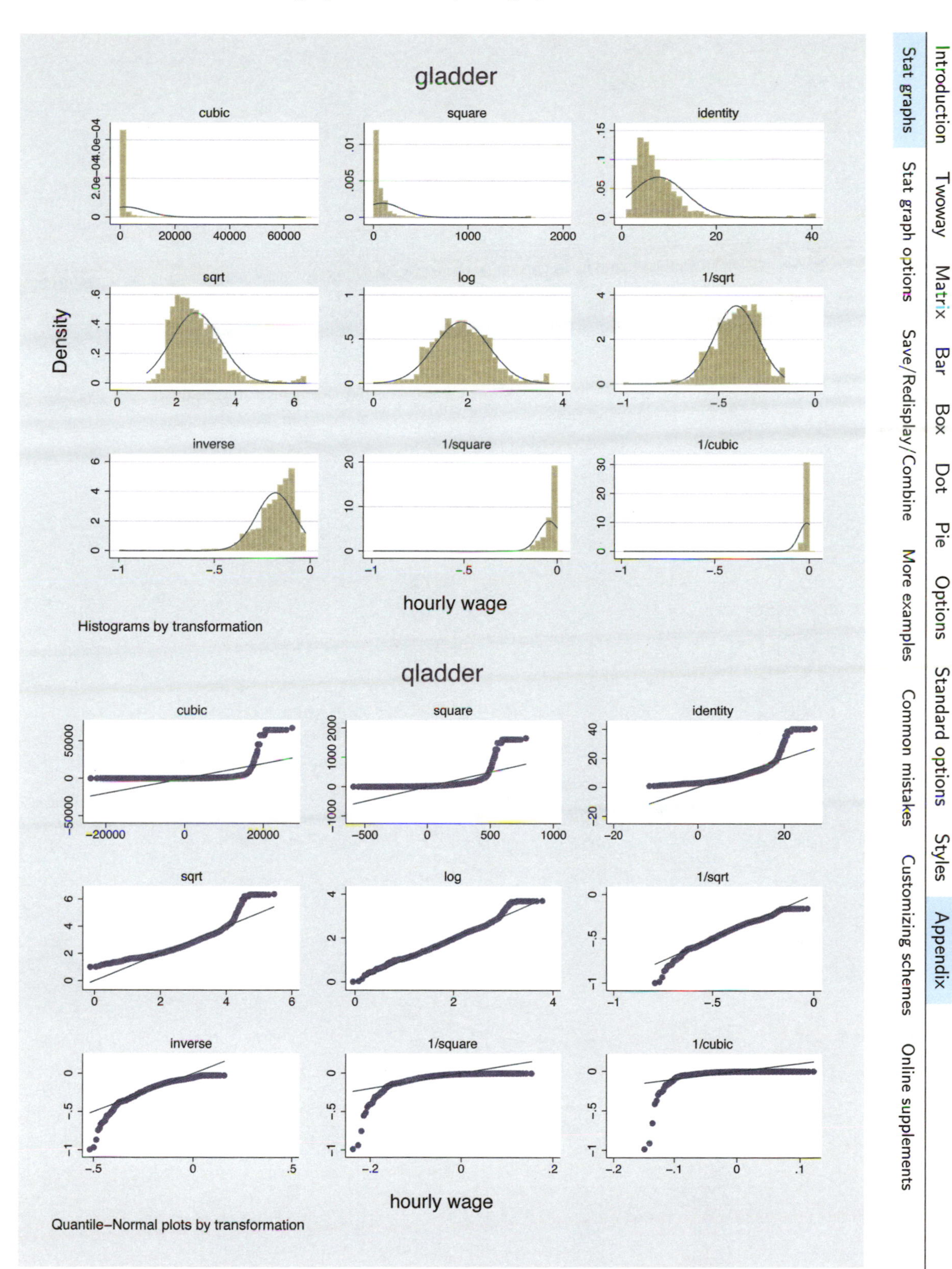

Figure 11.2: Ladder of power graphs

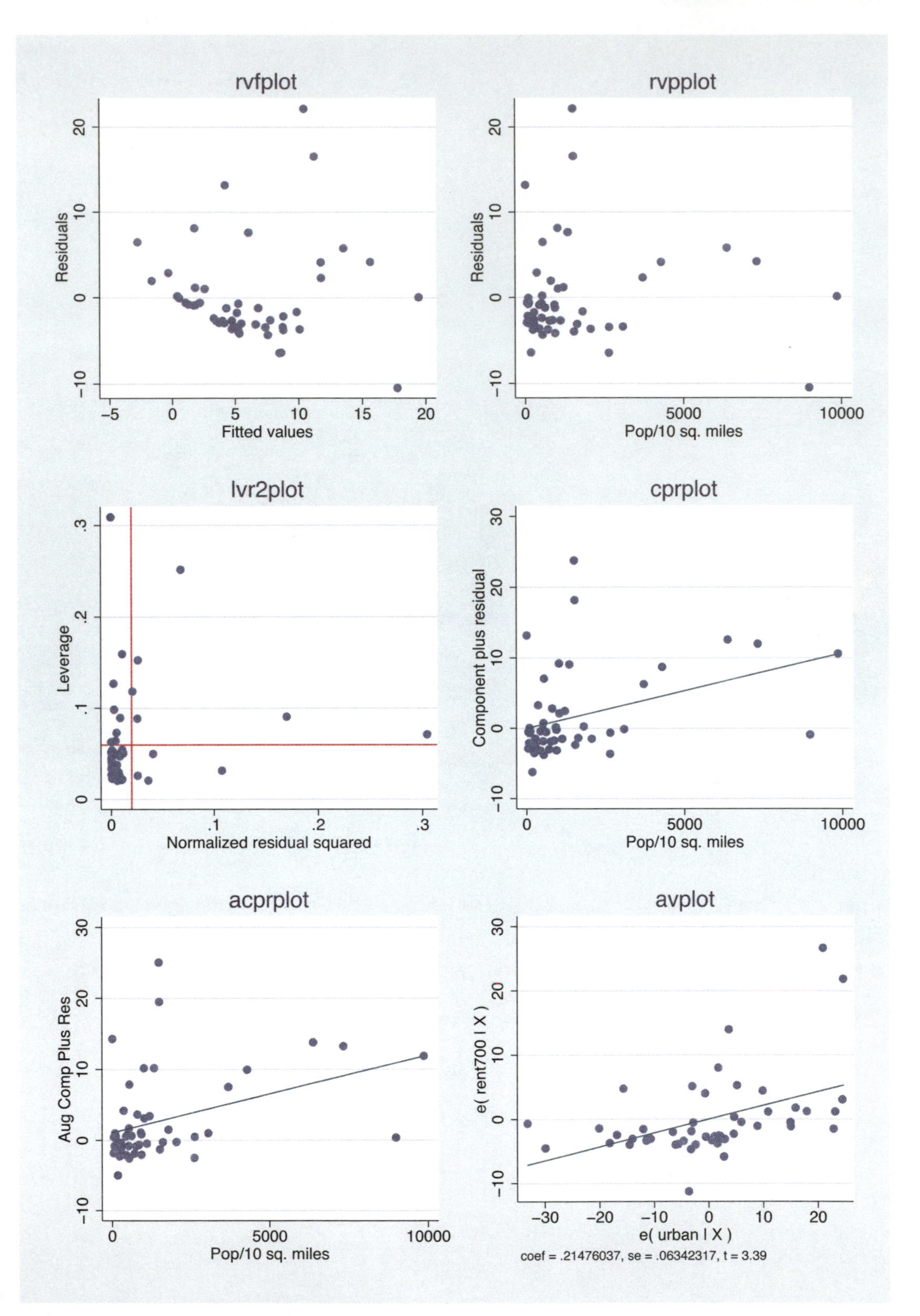

Figure 11.3: Regression diagnostics graphs

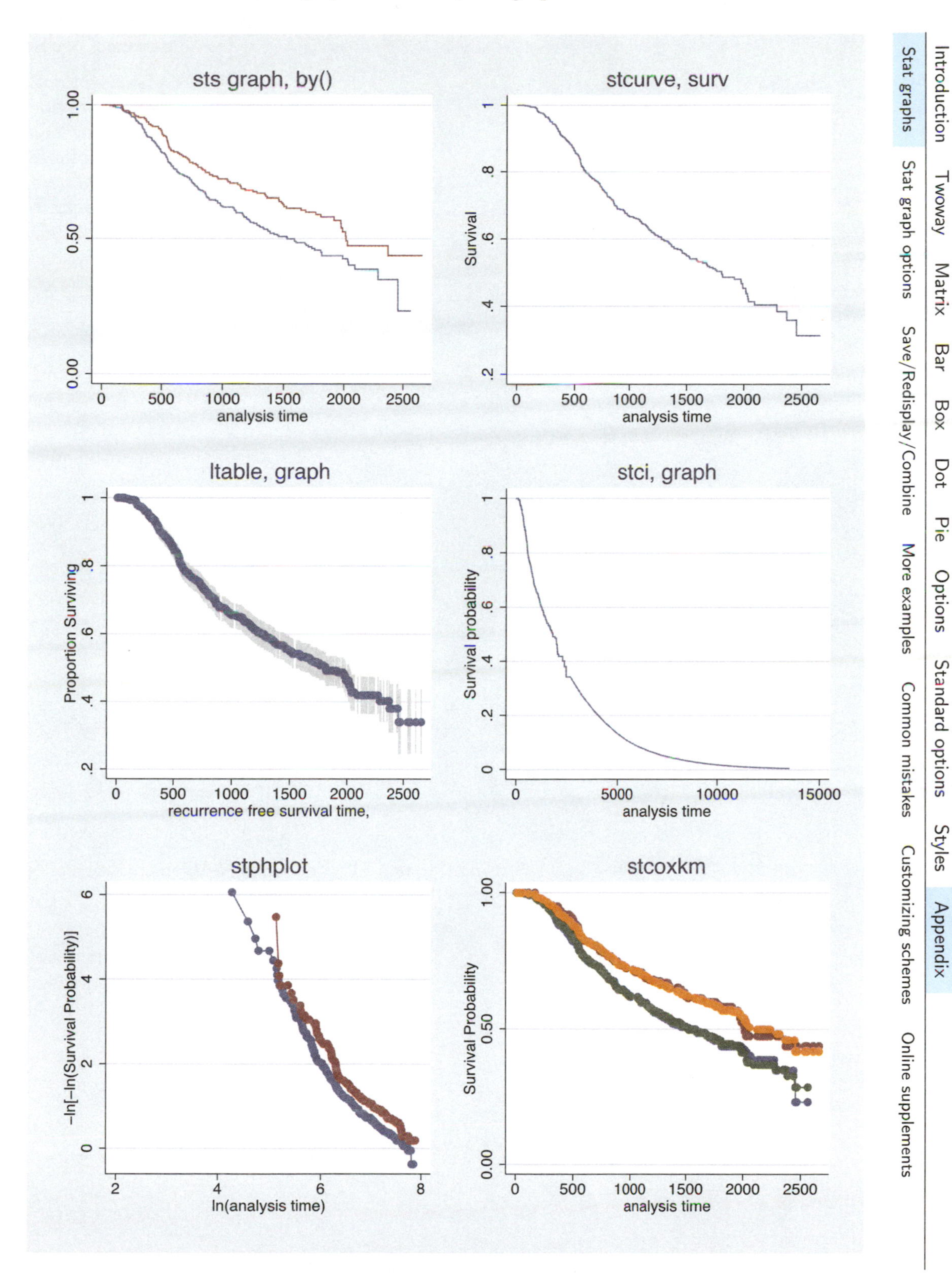

Figure 11.4: Survival graphs

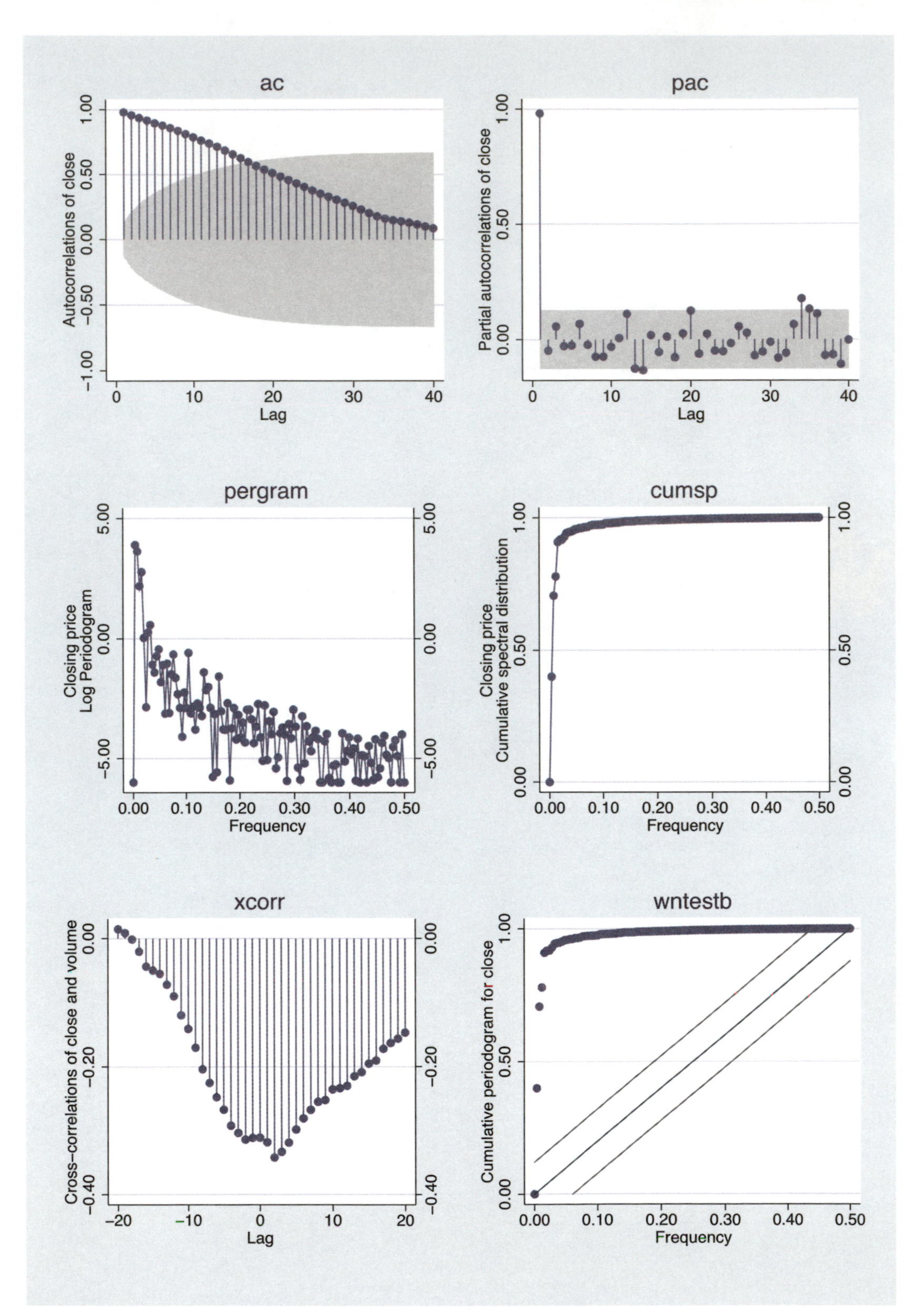

Figure 11.5: Time-series graphs

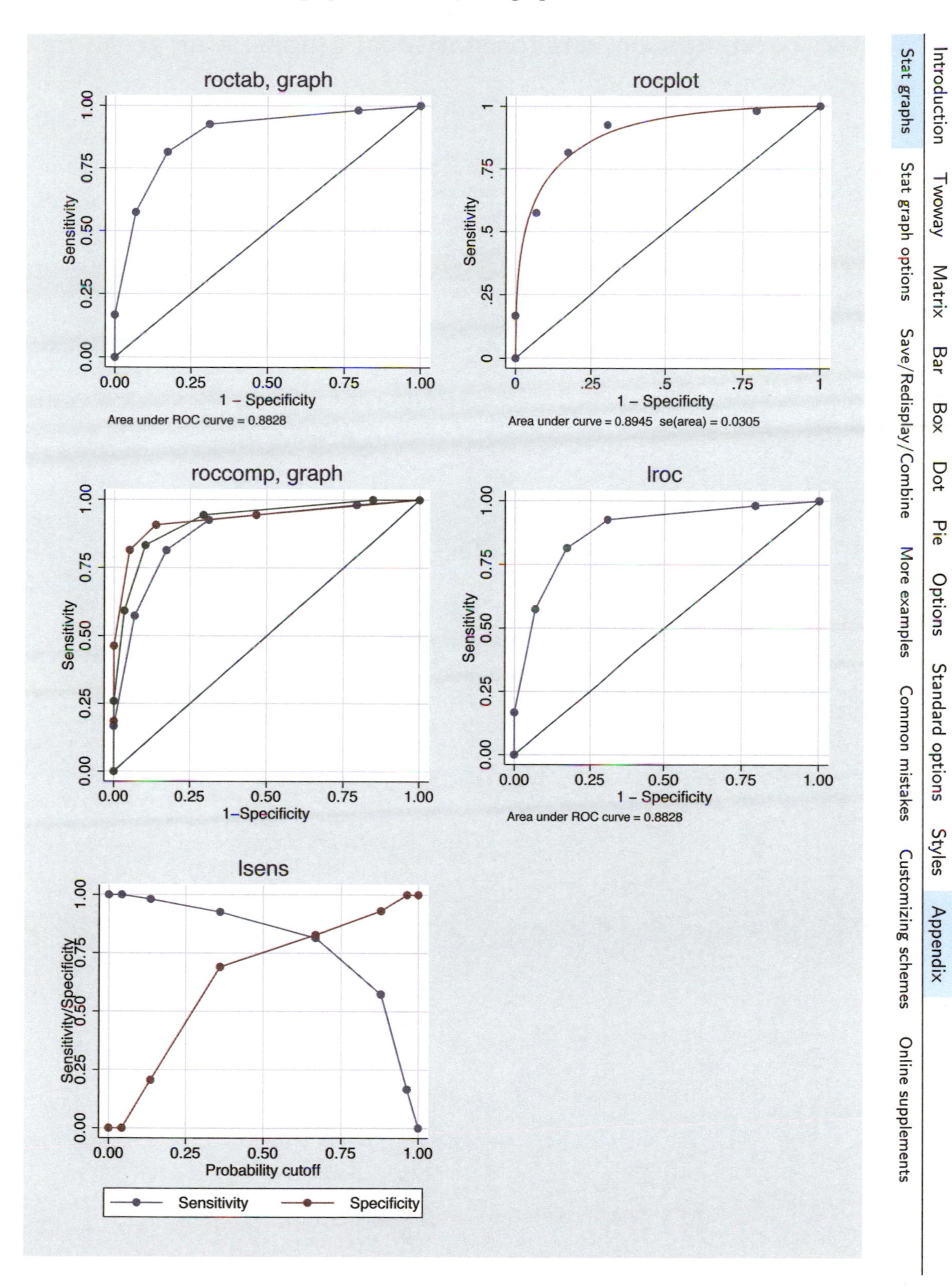

Figure 11.6: ROC graphs

11.2 Common options for statistical graphs, stat graph options

This section illustrates how to use Stata graph options with specialized statistical graph commands. Many of the examples will assume that we have run the command

```
. regress propval100 popden pcturban
```

and will illustrate subsequent commands with options to customize those specialized statistics graphs.

`lvr2plot`

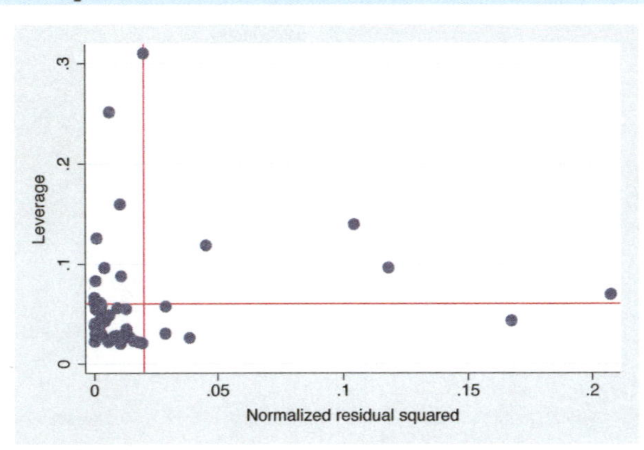

Consider this regression analysis, which predicts `propval100` from two variables, `popden` and `pcturban`. We can use the `lvr2plot` command to produce a leverage-versus-residual squared plot.
Uses allstates.dta & scheme vg_s2c
Before running the graph command, type
`reg propval100 popden pcturban`

`lvr2plot, msymbol(Oh) msize(vlarge)`

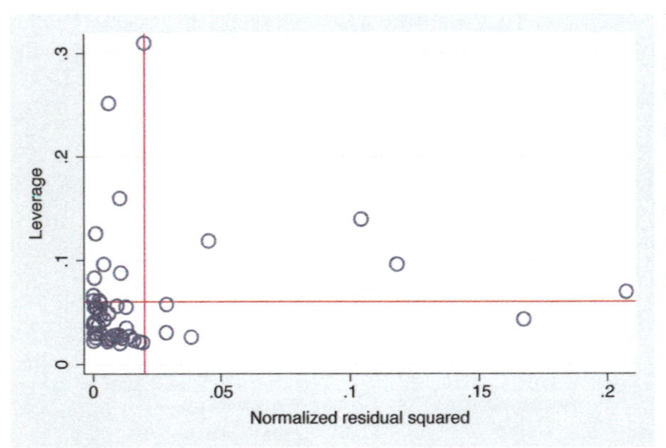

We can add options such as `msymbol()` and `msize()` to control the display of the markers in the graph. See Options : Markers (235) for more details.
Uses allstates.dta & scheme vg_s2c
Before running the graph command, type
`reg propval100 popden pcturban`

```
lvr2plot, mlabel(stateab)
```

We can add the mlabel() option to add marker labels to the graph. We could also add further options to control the size, color, and position of the marker labels; see Options : Marker labels (247) for more details.

Uses allstates.dta & scheme vg_s2c

Before running the graph command, type

```
reg propval100 popden pcturban
```

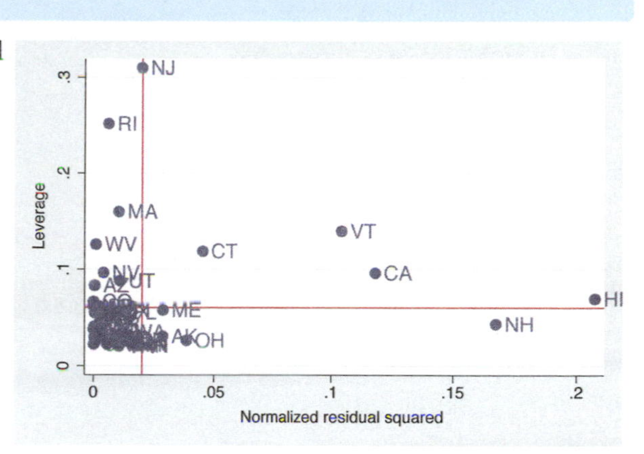


```
kdensity propval100
```

Consider this kernel-density plot for the variable propval100. We could add options to control the display of the line. See the following example.

Uses allstates.dta & scheme vg_s2c

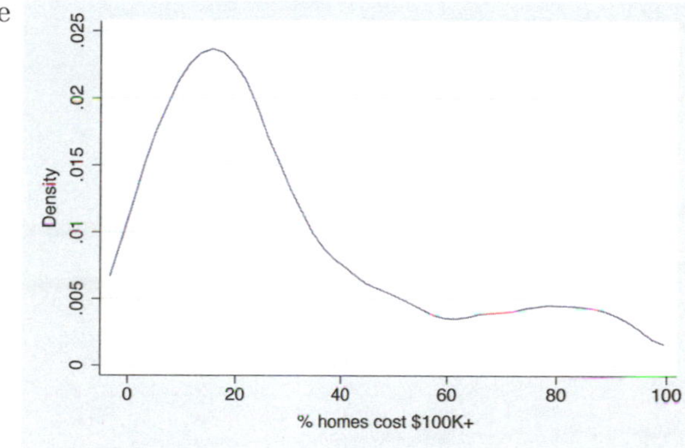


```
kdensity propval100, clwidth(thick) clpattern(dash)
```

The section Options : Connecting (250) shows a number of options we could add to control the display of the line. Here, we add the clwidth() and clpattern() options to make the line thick and dashed.

Uses allstates.dta & scheme vg_s2c

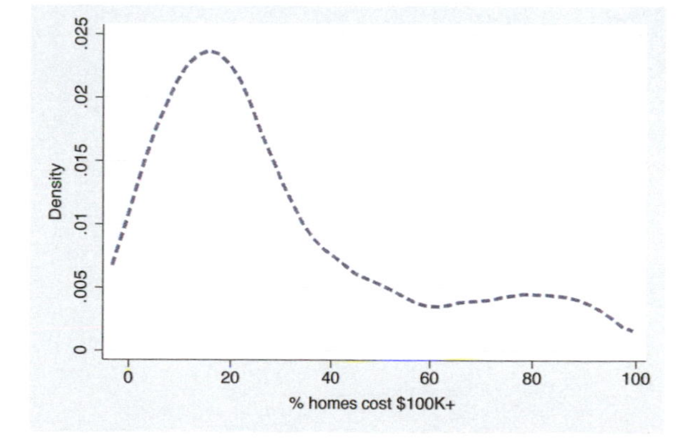

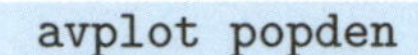

```
avplot popden
```

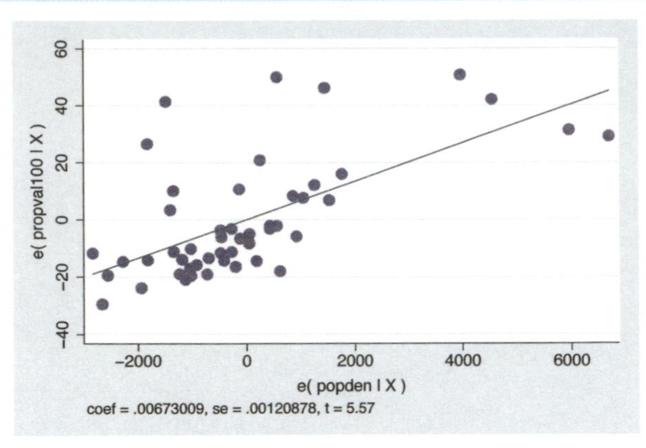

Consider this added-variable plot. We can modify the axis titles as illustrated in the following examples.
Uses allstates.dta & scheme vg_s2c
Before running the graph command, type
`reg propval100 popden pcturban`

```
avplot popden, xtitle("popden adjusted for percent urban")
   ytitle("property value adjusted for percent urban")
```

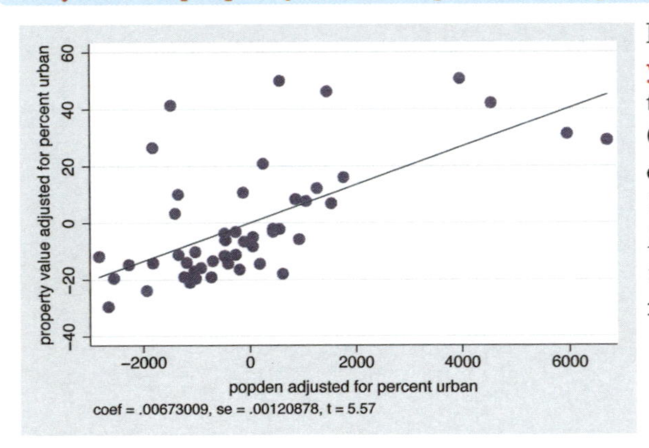

Here, we use the `xtitle()` and `ytitle()` options to change the titles of the x- and y-axes. See Options : Axis titles (254) for more details.
Uses allstates.dta & scheme vg_s2c
Before running the graph command, type
`reg propval100 popden pcturban`

```
avplot popden, note("Regression statistics for popden", prefix)
```

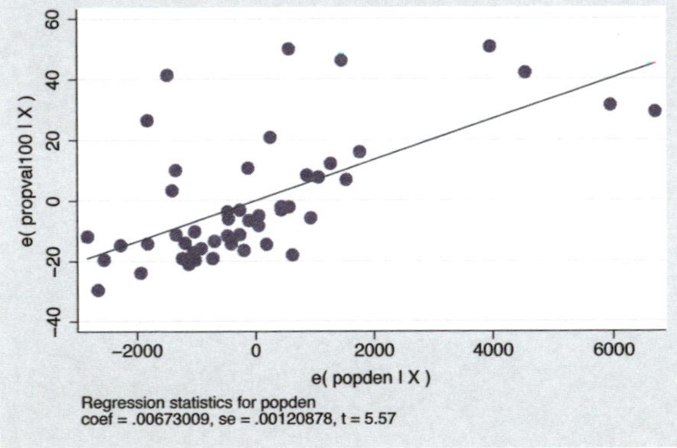

The `prefix` option can be used with the different title options to add a prefix to an existing title. In the `note()` option, for example, we add text before the existing note. In this way, we add additional descriptive information to an existing title, subtitle, note, or caption. We could also use the `suffix` option to add information after an existing title.
Uses allstates.dta & scheme vg_s2c
Before running the graph command, type
`reg propval100 popden pcturban`

```
avplot popden, xtitle(, size(huge))
```

We can modify the look of the existing title without changing the text. For example, we add the `size(huge)` option to make the existing title huge in size. See Options : Axis titles (254) and Options : Textboxes (303) for more details.
Uses allstates.dta & scheme vg_s2c
Before running the graph command, type
`reg propval100 popden pcturban`

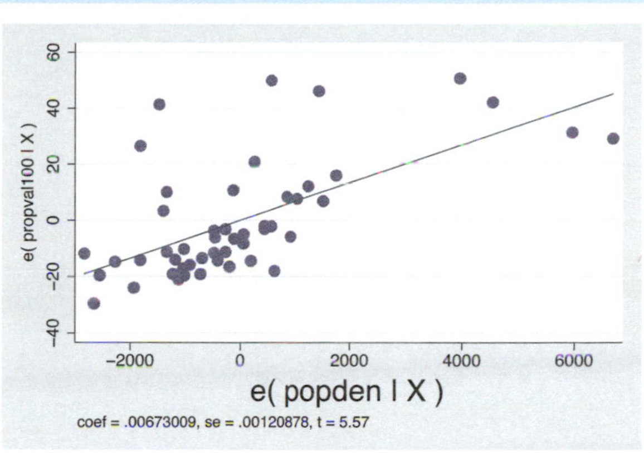


```
rvfplot
```

Consider this residual-versus-fit plot. We often hope to see an even distribution of points around zero on the y-axis. To help evaluate this distribution, we might want to label the y-axis identically for the values above 0 and below 0.
Uses allstates.dta & scheme vg_s2c
Before running the graph command, type
`reg propval100 popden pcturban`

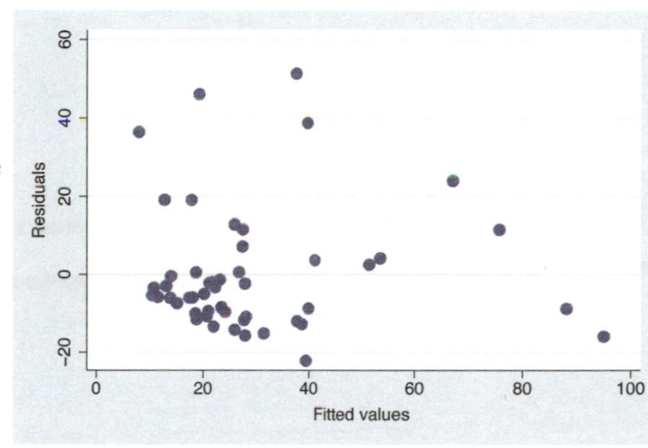


```
rvfplot, ylabel(-60(20)60, nogrid) yline(-20 20)
```

Here, we add the `ylabel()` option to label the y-axis from -60 to 60, incrementing by 20, and suppress the grid. Further, we use the `yline()` option to add a y-line at 20 and -20. For more information about labeling and scaling axes, see Options : Axis labels (256) and Options : Axis scales (265).
Uses allstates.dta & scheme vg_s2c
Before running the graph command, type
`reg propval100 popden pcturban`

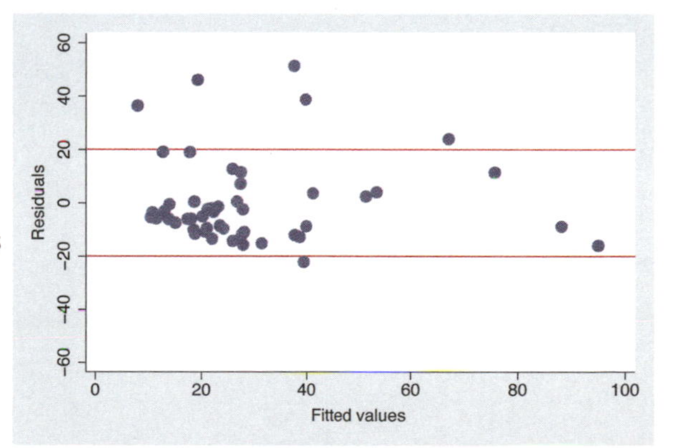

Introduction | Twoway | Matrix | Bar | Box | Dot | Pie | Options | Standard options | Styles | Appendix
Stat graphs | Stat graph options | Save/Redisplay/Combine | More examples | Common mistakes | Customizing schemes | Online supplements

```
sts graph, by(hormon)
```

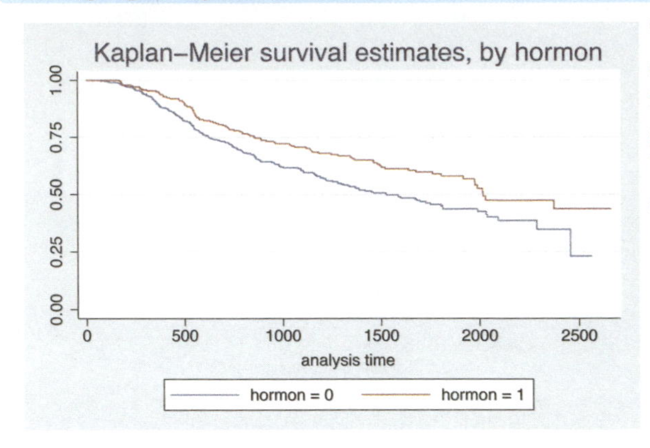

This graph shows survival-time estimates broken down by whether one is in the treatment group or the control group. The legend specifies the groups, but we might want to modify the labels as shown in the next example.
Uses hormone.dta & scheme vg_s2c

```
sts graph, by(hormon) legend(label(1 Control) label(2 Treatment))
```

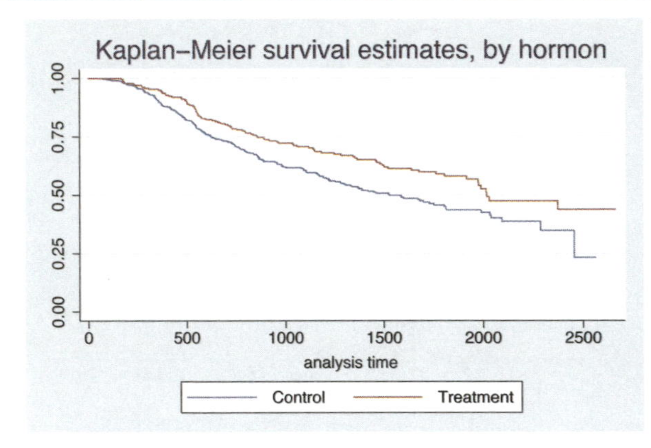

We can use the `legend()` option to use different labels within the legend. See Options : Legend (287) for more details.
Uses hormone.dta & scheme vg_s2c

```
sts graph, by(hormon) legend(off)
   text(.5 800 "Control", box) text(.8 1500 "Treatment", box)
```

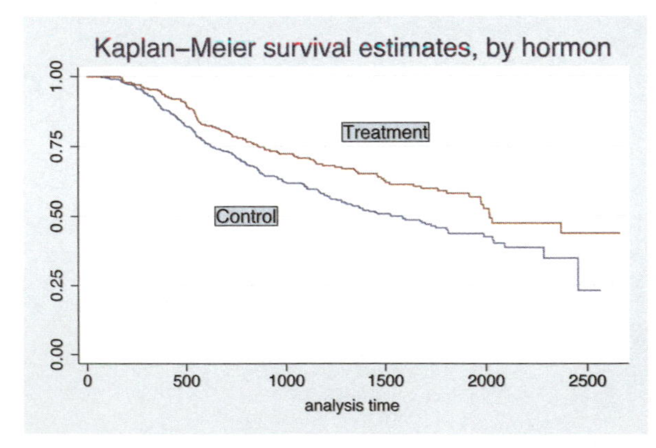

To suppress the display of the legend, we can use the `legend(off)` option. Instead, we can use the `text()` option to add text directly to the graph to label the two lines; see Options : Adding text (299) for more information. We also use the `box` option to surround the text with a box; see Options : Textboxes (303) for more details.
Uses hormone.dta & scheme vg_s2c

```
avplot popden, title("Added variable plot")
```

We return to the regression analysis predicting `propval100` from two variables, `popden` and `pcturban`. Here, we show an added-variable plot with the `title()` option to add a title. We could also add a `subtitle()`, `caption()`, or `note()` to the graph, as well; see Standard options : Titles (313) for more details.

Uses allstates.dta & scheme vg_s2c

Before running the graph command, type

`reg propval100 popden pcturban`

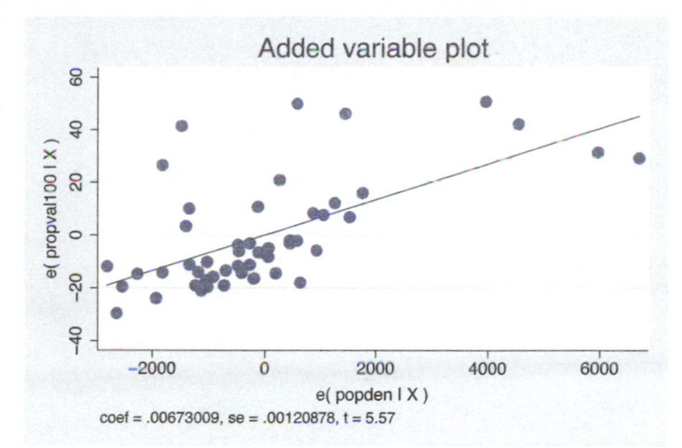

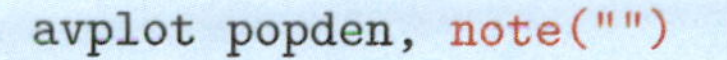

```
avplot popden, note("")
```

Here, we add the `note("")` option, which suppresses the display of the note at the bottom showing the coefficients for the regression model.

Uses allstates.dta & scheme vg_s2c

Before running the graph command, type

`reg propval100 popden pcturban`

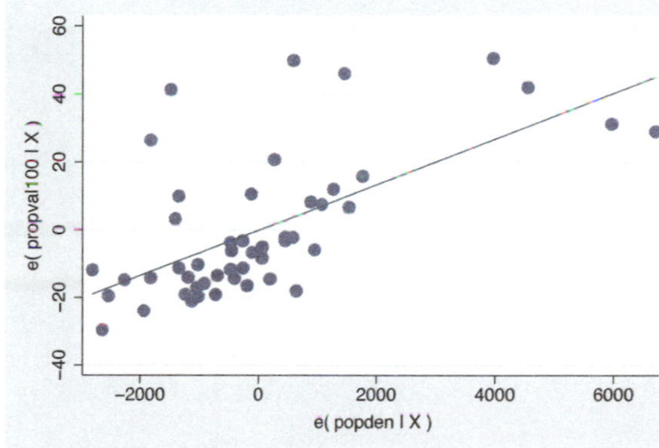


```
avplot popden, scheme(economist)
```

We can change the look of the graph by selecting a different scheme. Here, we use `scheme(economist)` to display the graph using the `economist` scheme. See Standard options : Schemes (318) for more details.

Uses allstates.dta & scheme vg_s2c

Before running the graph command, type

`reg propval100 popden pcturban`

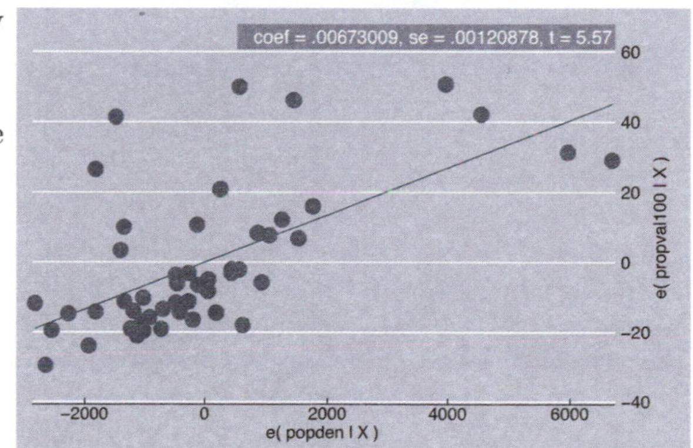

```
avplot popden, xsize(3) ysize(1) scale(1.3)
```

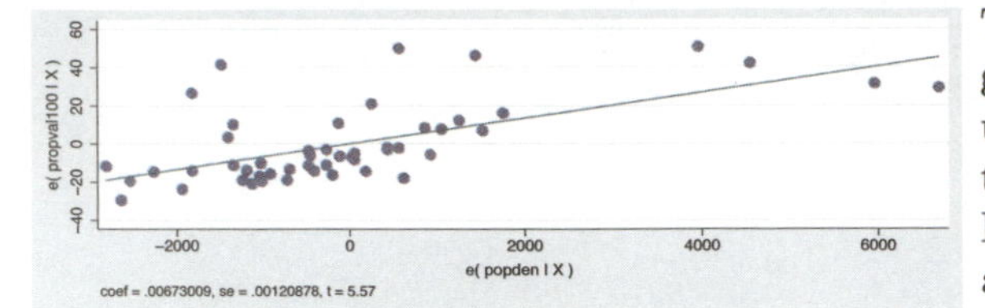

The section Standard options : Sizing graphs (322) describes options we can use to control the size of the graph and the scale of the contents of the graph. Here, we show the `xsize()`, `ysize()`, and `scale()` options.
Uses allstates.dta & scheme vg_s2c
Before running the graph command, type
`reg propval100 popden pcturban`

11.3 Saving and combining graphs, save/redisplay/combine

This section shows how to save, redisplay, and combine Stata graphs. We begin by showing how to save graphs either to disk or in memory. We also show how to redisplay the graph and, when we redisplay the graph, control the look of the graph.

```
twoway histogram urban, saving(hist1)
```

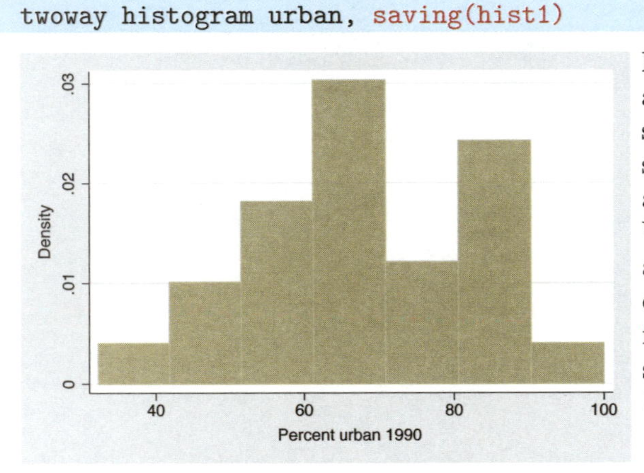

Most, if not all, Stata graph commands allow us to use the `saving()` option to save the graph as a Stata `.gph` file. We save this graph, naming it `hist1.gph`, and store it in the current directory. We will assume in these examples that all graphs are stored in the current directory, but we can precede the filename with a directory name and store it wherever we wish.
Uses allstates.dta & scheme vg_s2c

```
graph use hist1.gph
```

At a later time (including after quitting and restarting Stata), we can view the saved graph with the `graph use` command. If `hist1.gph` had been stored in a different directory, we would have to precede it with the directory where it was saved or use the `cd` command to change to that directory.
Uses allstates.dta & scheme vg_s2c

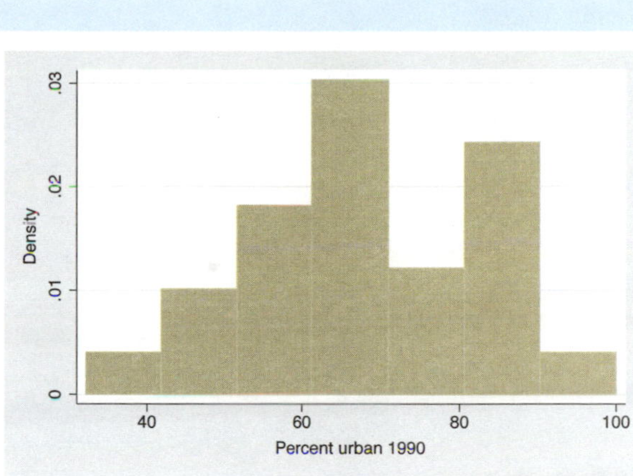


```
graph use hist1.gph, scheme(s1mono)
```

When we view the graph, we can add the `scheme()` option to view the same graph using a different scheme. Here, we view the last graph but use the `s1mono` scheme.
Uses allstates.dta & scheme s1mono

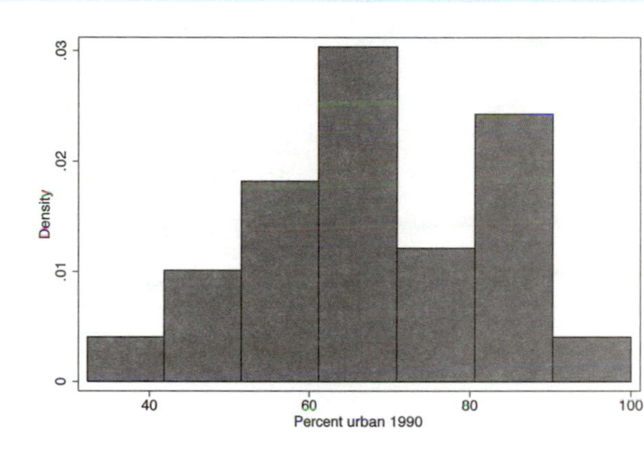


```
twoway histogram propval100, name(hist2)
```

The `name()` option is much like the `saving()` option, except that the graph is saved in memory instead of on disk. We can then view the graph later within the same Stata session, but once we quit Stata, the graph in memory will be gone.
Uses allstates.dta & scheme vg_s1c

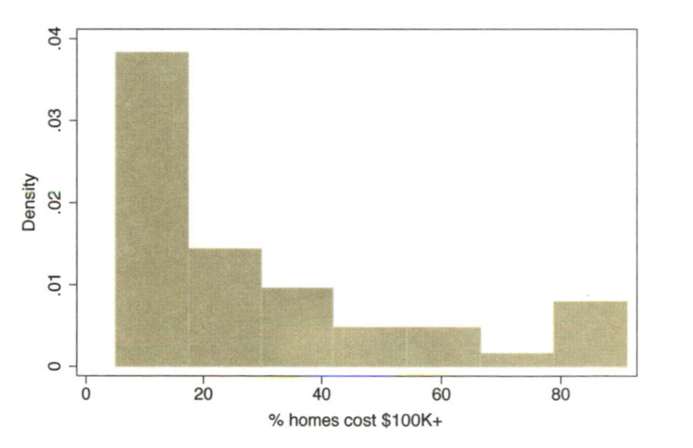

```
graph display hist2
```

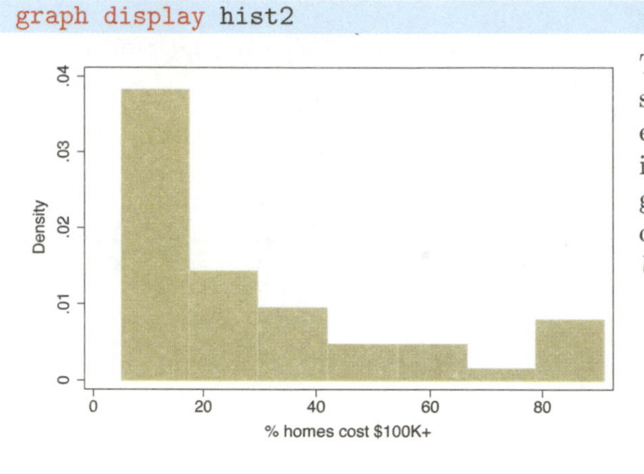

The graph display command is similar to the graph use command, except that it redisplays graphs saved in memory. Here, we redisplay the graph we created with the name(hist2) option.
Uses allstates.dta & scheme vg_s1c

```
graph display hist2, xsize(2) ysize(2)
```

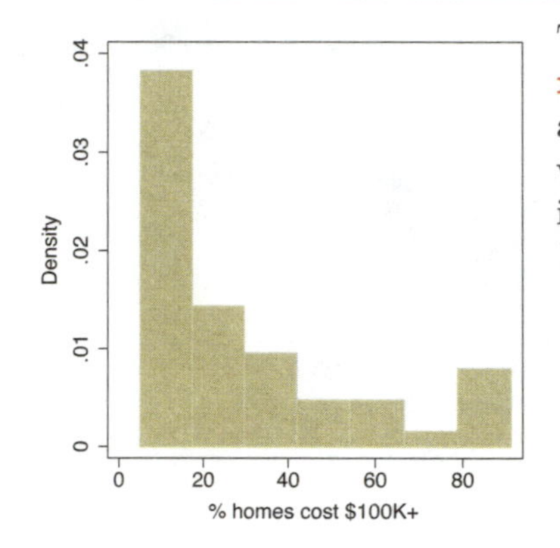

The graph display command allows us to use the xsize() and ysize() options to change the size and aspect ratio of the graph. Here, we redisplay the graph we named hist2 and make the graph 2 inches tall by 2 inches wide.
Uses allstates.dta & scheme vg_s1c

```
graph display hist2, scheme(s1mono)
```

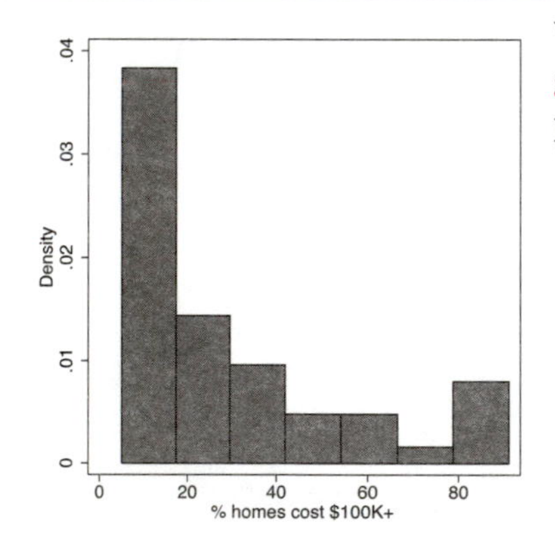

We can also use the scheme() option to view the same graph using a different scheme. Here, we view the previous graph but use the s1mono scheme.
Uses allstates.dta & scheme s1mono

Let's look at some examples to show how to combine graphs once they have been created and saved. First, we will see how to show two scatterplots side by side rather than overlaying them.

```
twoway scatter propval100 urban, name(scat1)
```

Using the `name(scat1)` option saves this scatterplot in memory with the name `scat1`.
Uses allstates.dta & scheme vg_s2c

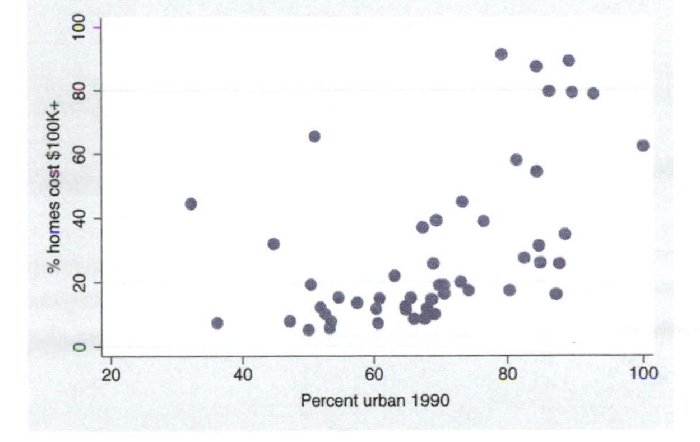


```
twoway scatter rent700 urban, name(scat2)
```

We save this second scatterplot with the name `scat2`.
Uses allstates.dta & scheme vg_s2c

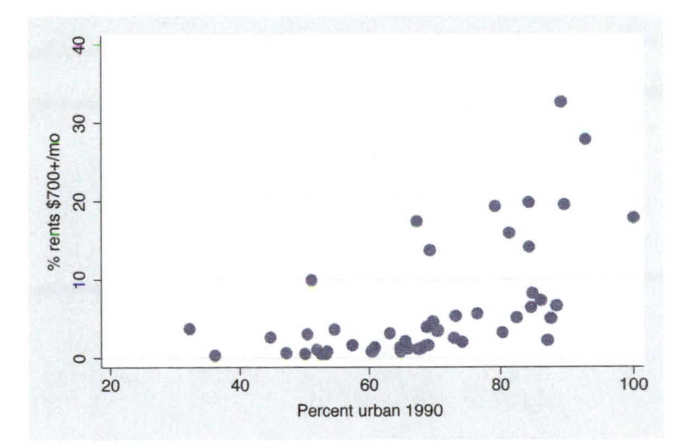

```
graph combine scat1 scat2
```

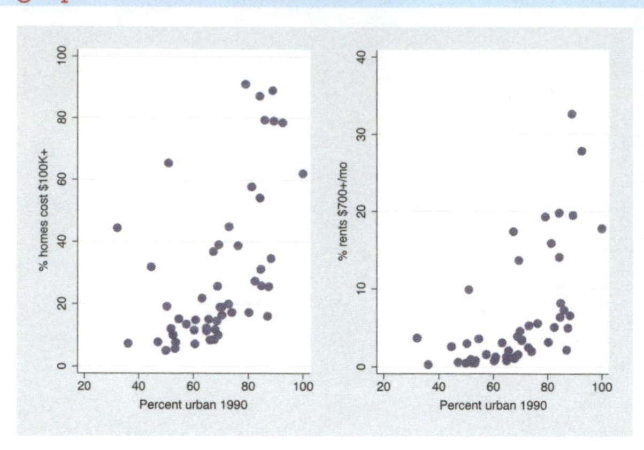

Using the `graph combine` command, we can see these two scatterplots side by side. In a sense, the y-axis is on a different scale for these two graphs since they are different variables. However, in another sense, the scale for the two y-axes is the same since they are both measured in percents.
Uses allstates.dta & scheme vg_s2c

```
graph combine scat1 scat2, ycommon
```

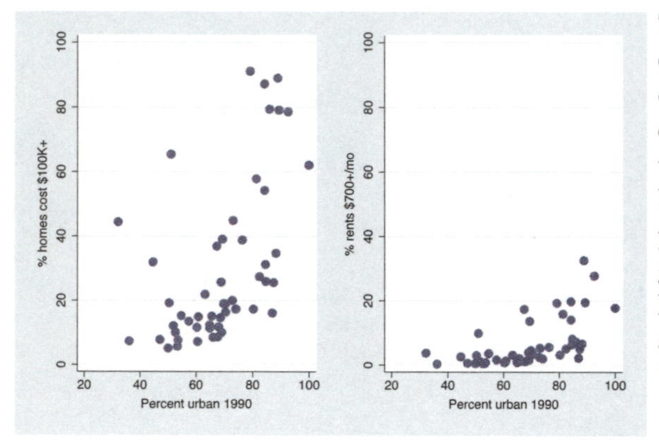

This graph is the same as the last one, except that the y-axes are placed on a common scale by using the `ycommon` option. This makes it easy to compare the two y-variables by forcing them to be on the same metric. Note that the `ycommon` option does not work when the graphs have been made using different kinds of commands, e.g., `graph bar` and `graph box`.
Uses allstates.dta & scheme vg_s2c

Let's look at more detailed examples showing how we can combine graphs and at options we can use in creating the graphs. The next set of examples uses the `sp2001ts` data file.

```
twoway rarea high low date, name(hilo)
```

We make a graph showing the high and low closing price of the S&P 500 for 2001 and save this graph in memory, naming it `hilo`.
Uses sp2001ts.dta & scheme vg_s2c

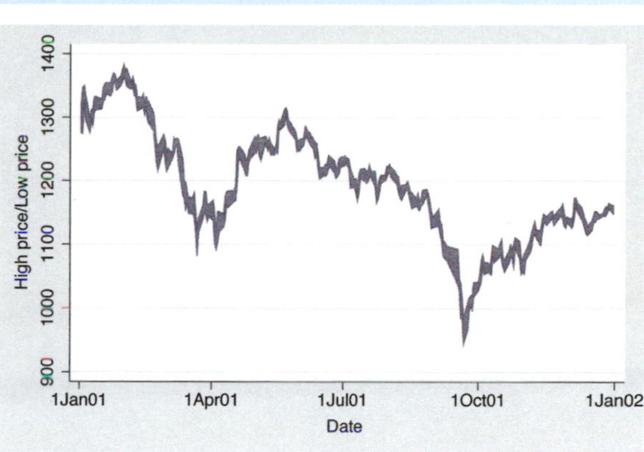


```
twoway spike volmil date, name(vol)
```

We can make another graph that shows the volume (millions of shares sold per day) for 2001 and save this graph in memory, naming it `vol`.
Uses sp2001ts.dta & scheme vg_s2c

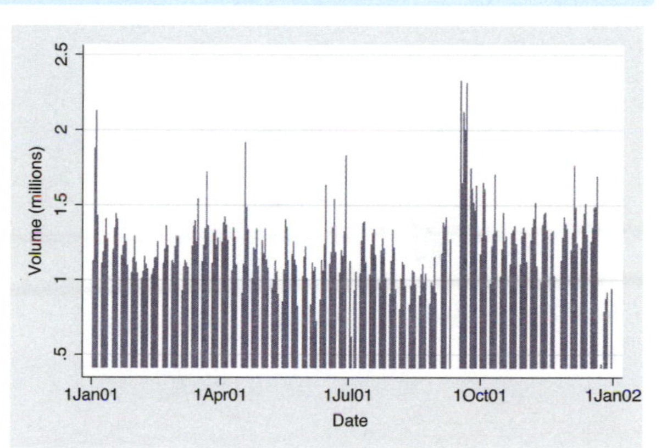


```
graph combine hilo vol
```

We can now use the `graph combine` command to combine these two graphs into a single graph. The graphs are displayed as a single row, but say that we would like to display them in a single column.
Uses sp2001ts.dta & scheme vg_s2c

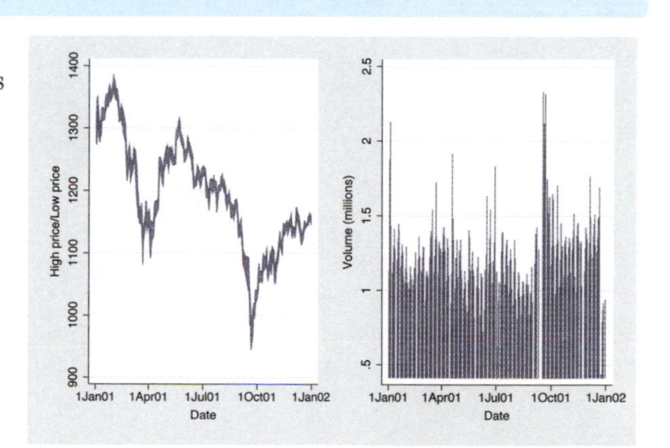

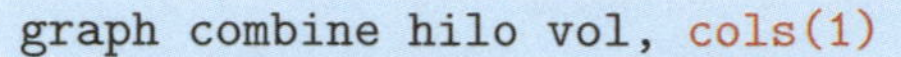

```
graph combine hilo vol, cols(1)
```

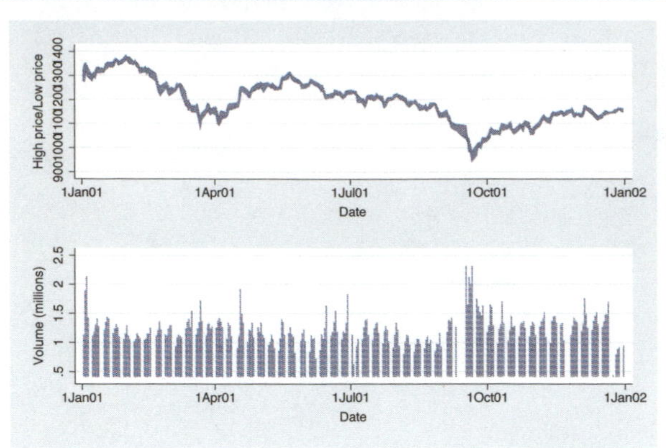

Using the `cols(1)` option, we can display the price above the volume. However, because the x-axes of these two graphs are scaled the same, we could save space and remove the x-axis scale from the top graph.
Uses sp2001ts.dta & scheme vg_s2c

```
twoway rarea high low date, xscale(off) name(hilo, replace)
```

Here, we use the `xscale(off)` option to suppress the display of the x-axis, including the space that would be allocated for the labels. We name this graph `hilo` again but need to use the `replace` option to replace the existing graph named `hilo`.
Uses sp2001ts.dta & scheme vg_s2c

```
graph combine hilo vol, cols(1)
```

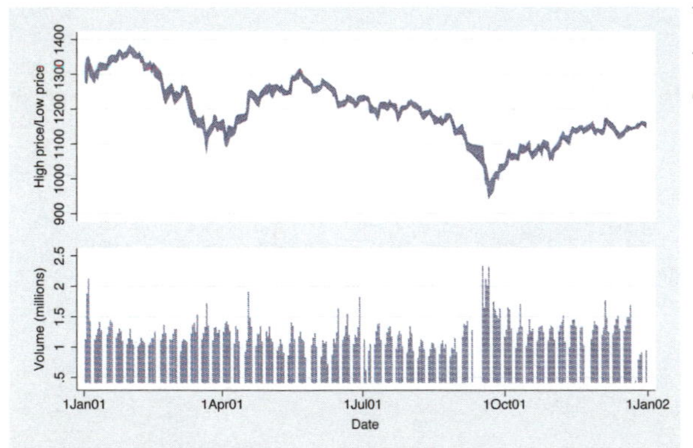

We combine these two graphs; however, we might want to push the graphs a bit closer together.
Uses sp2001ts.dta & scheme vg_s2c

```
graph combine hilo vol, cols(1) imargin(b=1 t=1)
```

Here, we use the `imargin(b=1 t=1)` option to make the margin at the top and bottom of the graphs to be very small before combining them. However, we might want the lower graph of volume to be smaller.
Uses sp2001ts.dta & scheme vg_s2c

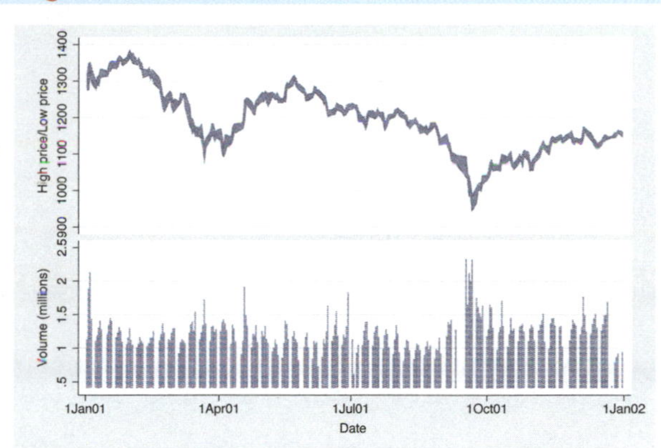


```
twoway spike volmil date, ylabel(1 2) fysize(25) name(vol, replace)
```

Using the `fysize()` (force y size) option makes the graph 25% of its normal size. We use this instead of `ysize()` because the `graph combine` command does not respect the `ysize()` or `xsize()` options. For aesthetics, we also reduce the number of labels. We save this graph in memory, replacing the existing graph named `vol`.
Uses sp2001ts.dta & scheme vg_s2c

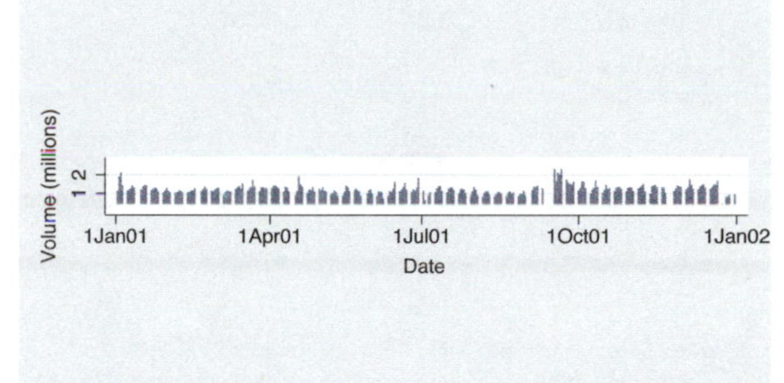


```
graph combine hilo vol, cols(1) imargin(b=1 t=1)
```

We combine these graphs again, and the combined graph looks pretty good. We might further tinker with the graph, changing the `xtitle()` for the volume graph to be shorter or modifying the `xlabel()` for the volume graph.
Uses sp2001ts.dta & scheme vg_s2c

11.4 Putting it all together, more examples

Most of the examples in this book have focused on the impact of a single option or a small number of options, using datasets that required no manipulation prior to making the graph. In reality, many graphs use multiple options together, and some require prior data management. This section addresses this issue by showing some examples that combine numerous options and require some data manipulation before making the graph.

```
twoway (scatter urban pcturban80) (function y=x, range(30 100)),
   xtitle(Percent Urban 1980) ytitle(Percent Urban 1990)
   legend(order(2 "Line where % Urban 1980 = % Urban 1990") pos(6) ring(0))
```

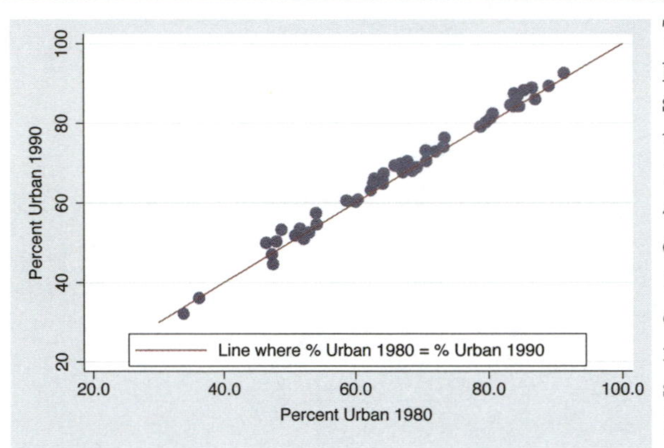

This graph shows the percentage of population living in an urban area of a state in 1990 against that of 1980. If there had been no changes from 1980 to 1990, the values would fall along a 45-degree line, where the value of y equals the value of x. Overlaying (`function y=x`), we can see any discrepancies from 1980 to 1990. The `range(30 100)` option makes the line span from 30 to 100 on the x-axis. *Uses allstates.dta & scheme vg_s2c*

```
twoway (lfitci ownhome borninstate) (lfitci ownhome borninstate,
   ciplot(rline) blcolor(blue) blwidth(thick) blpattern(dash))
   (scatter ownhome borninstate), legend(off) ytitle("% Own Home")
```

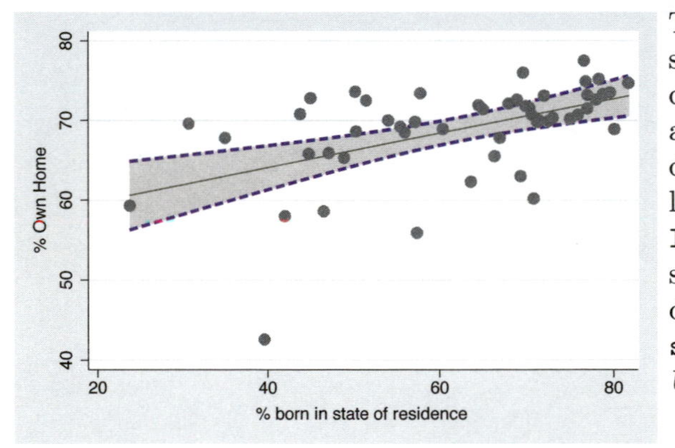

This example shows how we can make a scatterplot, a regression line, and a confidence interval for the fit shown as an area. We also add a thick, blue, dashed line showing the upper and lower confidence limits. The first `lfitci` makes the fit line and area; the second `lfitci` makes a thick, blue, dashed outline for the area; and `scatter` overlays the scatterplot. *Uses allstates.dta & scheme vg_s2c*

```
twoway scatter ownhome borninstate,
   by(nsw, hole(1) title("%Own home by" "%born in St." "by region",
   pos(11) ring(0) width(65) height(35) justification(center)
   alignment(middle)) note(""))
```

The hole(1) option leaves the first position empty when creating the graphs, and the title is placed there using pos(11) and ring(0). We use width() and height() to adjust the size of the textbox and justification() and alignment() to center the textbox horizontally and vertically. The note("") option suppresses the note in the bottom corner of the graph.
Uses allstates.dta & scheme vg_s2c

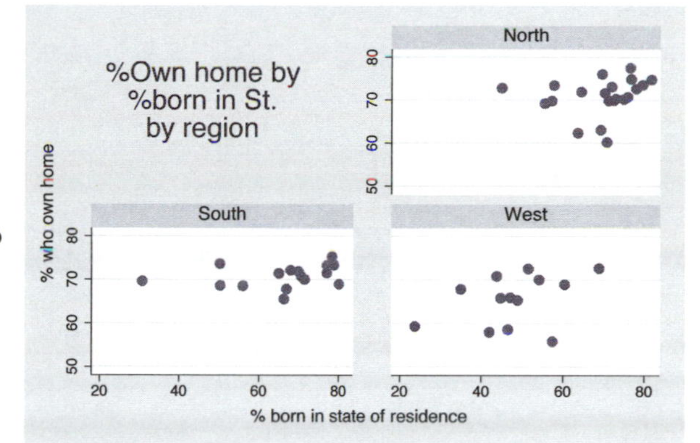


```
twoway (rspike hi low date) (rcap close close date, msize(medsmall)),
   tlabel(08jan2001 01feb2001 21feb2001) legend(off)
```

Before making this high/low/close graph, we first type tsset date, daily to tell Stata that date should be treated as a date in the tlabel() option. The rcap command uses close for both the high and the low values, making the tick line for the closing price, and the legend(off) option suppresses the legend. Using the vg_samec scheme makes the spikes and closes the same color.
Uses spjanfeb2001.dta & scheme vg_samec

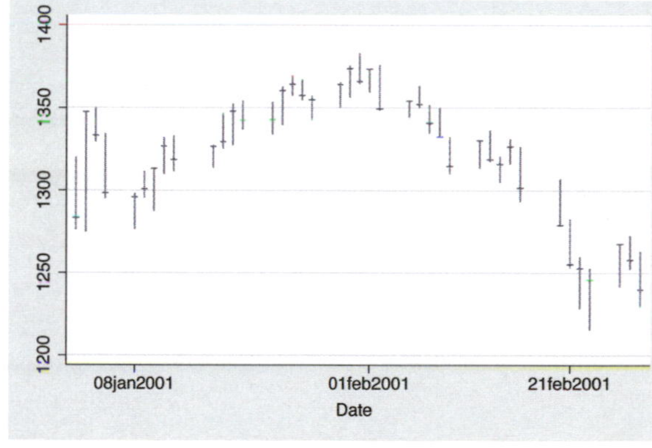

```
twoway (rspike hi low date) (rcap close close date, msize(medsmall))
   (scatteri 1220 15027 1220 15034, recast(line) clwid(vthick) clcol(red)),
   tlabel(08jan2001 01feb2001 21feb2001) legend(off)
```

This example is the same as above, except that this one uses `scatteri()` to draw a support-level line. Two y x pairs are given after the `scatteri`, and the `recast(line)` option draws them as a line instead of two points. The x-values were calculated beforehand using `display d(21feb2001)` and `display d(28feb2001)` to compute the elapsed date values.
Uses spjanfeb2001.dta & scheme vg_samec

The rest of the examples in this section involve some data management before we create the graph. For the next few examples, we use the `allstates` data file and run a regression command,

```
. vguse allstates
. regress ownhome propval100 workers2 popden
```

and then issue the

```
. dfbeta
```

command, creating DFBETAs for each predictor: `DFpropval100`, `DFworkers2`, and `DFurban`, which are used in the following graph.

```
twoway dropline DFpropval100 DFworkers2 DFurban statefips,
   mlabel(stateab stateab stateab)
```

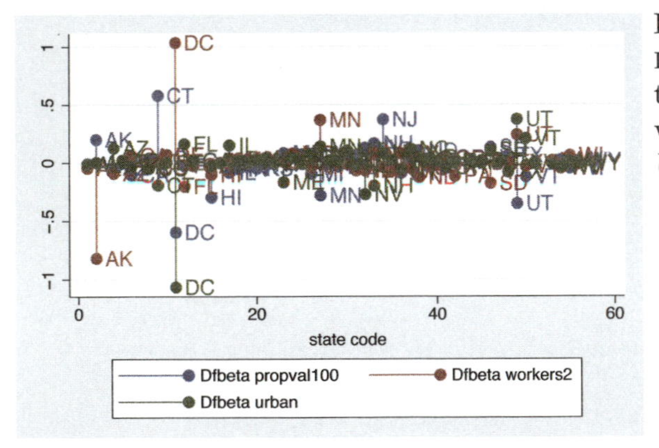

In this example, we show each of the DFBETAs as a `dropline` plot. We add the `mlabel()` option to label each point with the state abbreviation.
Uses allstates.dta & scheme vg_s2c

```
twoway (dropline DFpropval100 id if abs(DFpropval100)>.25, mlabel(stateab))
   (dropline DFworkers2 id if abs(DFworkers2)>.25, mlabel(stateab))
   (dropline DFurban id if abs(DFurban)>.25, mlabel(stateab))
```

This example is similar to the one above but simplifies the graph by showing only the points where the DFBETA exceeds .25. Note that we have taken the example from above and converted it into three overlaid `dropline` plots, each of which has an `if` condition.
Uses allstates.dta & scheme vg_s2c

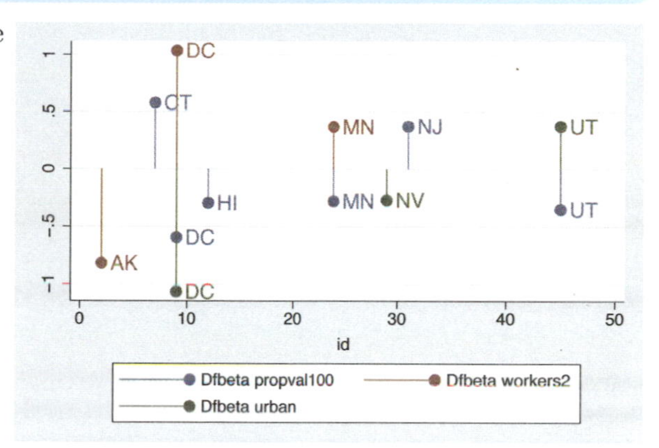

Before making the next graph, we need to issue three `predict` commands to generate variables that contain the Cook's distance, the studentized residual, and the leverage based on the previous regression command:

```
. predict cd, cook
. predict rs, rstudent
. predict l, leverage
```

We are then ready to run the next graph.

```
twoway (scatter rs id) (scatter rs id if abs(rs) > 2, mlabel(stateab)),
   legend(off)
```

This graph uses `scatter rs id` to make an index plot of the studentized residuals. It also overlays a second `scatter` command with an `if` condition showing only studentized residuals that have an absolute value exceeding 2 and showing the labels for those observations. Using the `vg_samec` scheme makes the markers the same for both `scatter` commands.
Uses allstates.dta & scheme vg_samec

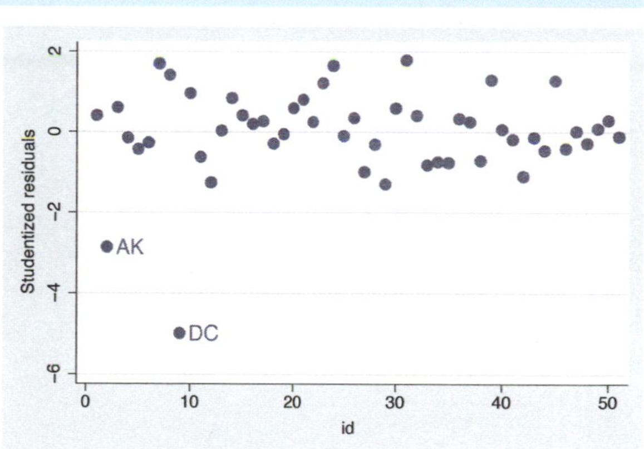

Introduction Twoway Matrix Bar Box Dot Pie Options Standard options Styles Appendix
Stat graphs Stat graph options Save/Redisplay/Combine More examples Common mistakes Customizing schemes Online supplements

```
twoway (scatter rs id, text( -3 27 "Possible Outliers", size(vlarge)))
   (scatteri -3 18 -4.8 10, recast(line))
   (scatteri -3 18 -3 3, recast(line)), legend(off)
```

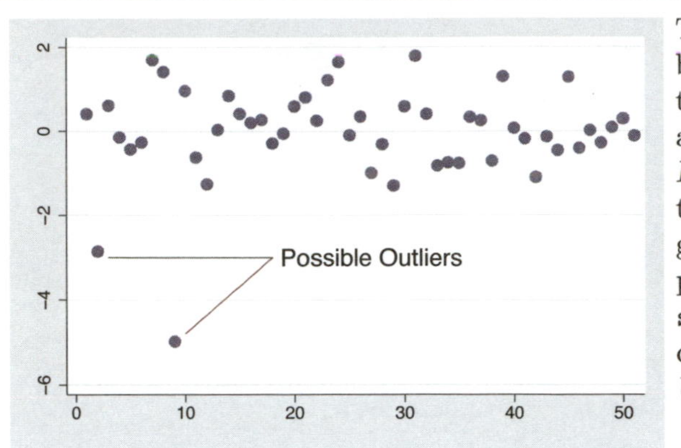

This graph is similar to the one above but uses the `text()` option to add text to the graph. Two `scatteri` commands are used to draw a line from the text *Possible Outliers* to the markers for those points. The y x coordinates are given for the starting and ending positions, and `recast(line)` makes `scatteri` behave like a line plot, connecting the points to the text.
Uses allstates.dta & scheme vg_s2c

```
twoway (scatter rs l [aw=cd], msymbol(Oh))
   (scatter rs l if cd > .1, msymbol(i) mlabel(stateab) mlabpos(0))
   (scatter rs l if cd > .1, msymbol(i) mlabel(cd) mlabpos(6)), legend(off)
```

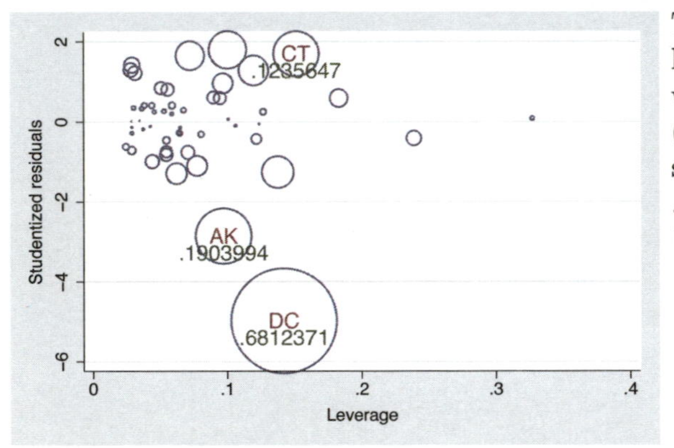

This graph shows the leverage-versus-studentized residuals, weighting the symbols by Cook's D (`cd`). We overlay it with a scatterplot showing the marker labels if `cd` exceeds .1, with the `cd` value placed underneath.
Uses allstates.dta & scheme vg_s2c

Imagine that we have a data file called `comp2001ts` that contains variables representing the stock prices of four hypothetical companies: `pricealpha`, `pricebeta`, `pricemu`, and `pricesigma`, as well as a variable `date`. To compare the performance of these companies, let's make a line plot for each company and stack them. We can do this using `twoway tsline` with the `by(company)` option, but we first need to reshape the data into a `long` format. We do so with the following commands:

```
. vguse comp2001ts
. reshape long price, i(date) j(compname) string
```

We now have variables `price` and `company` and can graph the prices by company.

```
twoway tsline price, by(compname, cols(1) yrescale note("") compact)
   ylabel(#2, nogrid) subtitle(, pos(5) ring(0) nobexpand nobox color(red))
   title(" ", box width(130) height(.001) bcolor(ebblue))
```

We graph `price` for the different companies with the `by()` option. Further, `cols(1)` puts the graphs in one column. `yrescale` and `ylabel(#2)` allow the y-axes to be scaled independently and labeled with about 2 values. The `subtitle()` option puts the name of the company in the bottom, right corner of each graph. The `title()` option creates an empty title that is thin, wide, and blue. Combined with the `compact` option, the title creates a border between the graphs.
Uses comp2001ts.dta & scheme vg_s2c

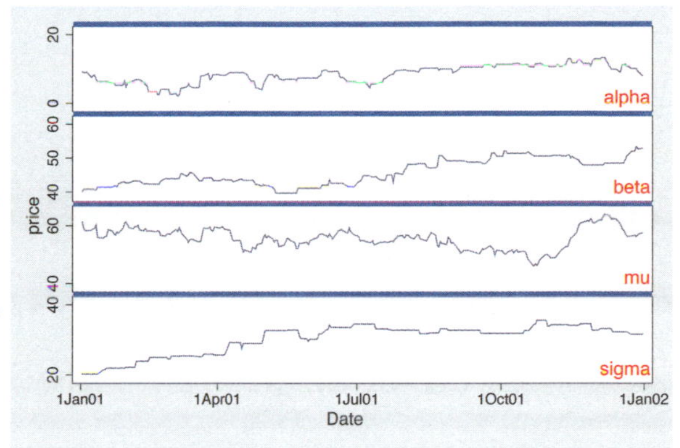

For the next graph, we want to create a bar chart that shows the mean of wages by occupation with error bars showing a 95% confidence interval for each mean. To do this, we first collapse the data across the levels of occupation, creating the mean, standard deviation, and count. Next, we create the variables `wageucl` and `wagelcl`, which are the upper and lower confidence limits, as shown below.

```
. vguse allstates
. collapse (mean) mwage=wage (sd) sdwage=wage (count) nwage=wage, by(occ7)
. generate wageucl = mwage + invttail(nwage,0.025)*sdwage/sqrt(nwage)
. generate wagelcl = mwage - invttail(nwage,0.025)*sdwage/sqrt(nwage)
```

After this, we are ready to execute the following command:

```
twoway (bar mwage occ7, barwidth(.5))
   (rcap wageucl wagelcl occ7, blwid(medthick) blcolor(navy) msize(large)),
   xlabel(1(1)7, valuelabel noticks) xscale(range(.5 7.5))
```

This bar chart is overlaid with a range plot showing the upper and lower confidence limits. The `xlabel()` option labels the values from 1 to 7, incrementing by 1. The `valuelabel` option indicates that the value labels for `occ7` will be used to label the x-axis. The `xscale()` option adds a margin to the outer bars, and the `barwidth()` option creates the gap between the bars.
Uses allstates.dta & scheme vg_s2c

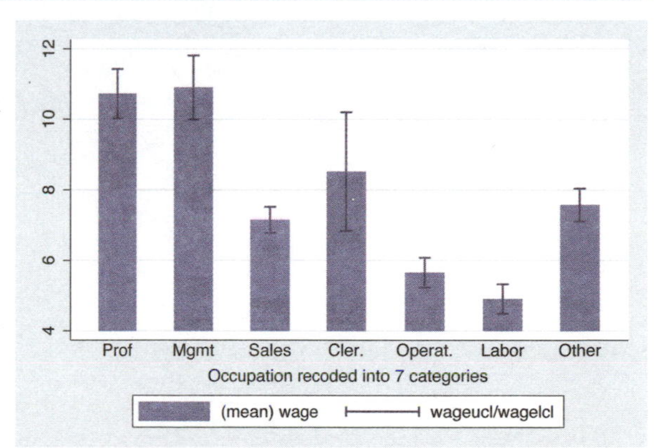

```
twoway (rcap wageucl wagelcl occ7, blwidth(medthick) msize(large))
   (bar mwage occ7, barwidth(.5) bcolor(navy)),
   xlabel(1(1)7, valuelabel noticks) xscale(range(.5 7.5))
```

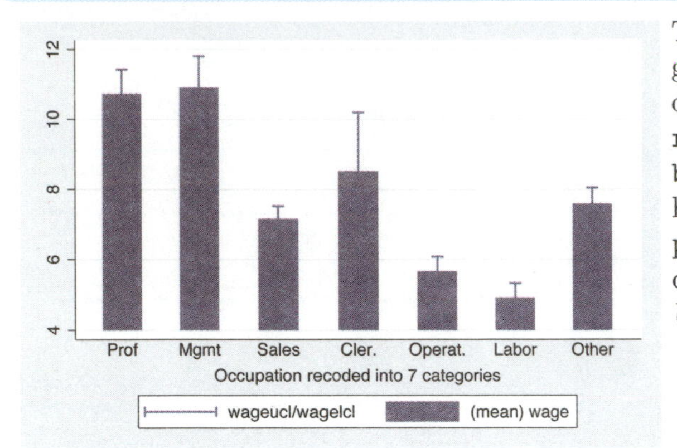

This graph is similar to the previous graph, except that we have reversed the order of the commands, placing the `rcap` command first, followed by the `bar` command. As a result, only the top half of the error bar is shown. As in the previous example, the `xlabel()` option determines the labels on the x-axis. *Uses allstates.dta & scheme vg_s2c*

Suppose that we wanted to show the mean wages with confidence intervals broken down by occupation and whether one graduated college. We use the `collapse` command to create the mean, standard deviation, and count by the levels of `occ7` and `collgrad`, and then we create the upper and lower confidence limits. Finally, the `separate` command makes separate variables for `mwage` based on whether one graduated college, creating `mwage0` (wages for noncollege grad) and `mwage1` (wages for college grad). These commands are shown below, followed by the command to create the graph.

```
. vguse nlsw
. collapse (mean) mwage=wage (sd) sdwage=wage
     (count) nwage=wage, by(occ7 collgrad)
. generate wageucl = mwage + invttail(nwage,0.025)*sdwage/sqrt(nwage)
. generate wagelcl = mwage - invttail(nwage,0.025)*sdwage/sqrt(nwage)
. separate mwage, by(collgrad)
```

```
twoway (line mwage0 mwage1 occ7) (rcap wageucl wagelcl occ7),
   xlabel( 1(1)7, valuelabel) xtitle(Occupation) ytitle(Wages)
   legend(order(1 "Not College Grad" 2 "College Grad"))
```

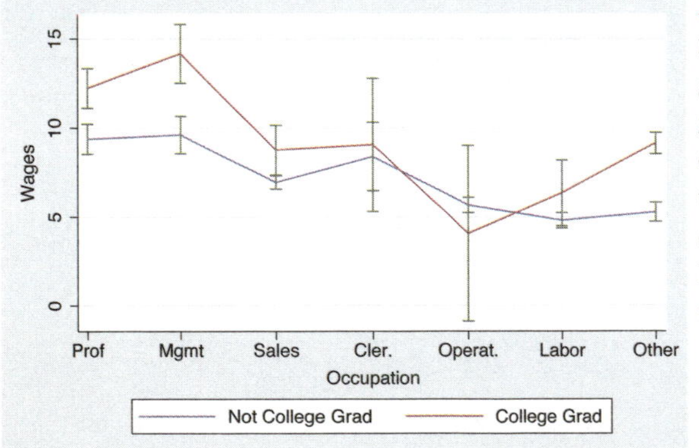

Here, we make a line graph showing the mean wages for the noncollege graduates, `mwage0`, and the college graduates, `mwage1`, by occupation. We overlay that with a range plot showing the confidence interval. The `xlabel()` option labels the x-axis with value labels, and the `legend()` option labels the legend. *Uses nlsw.dta & scheme vg_s2c*

This next graph shows a kind of scatterplot of the mean and confidence interval for **union** and **collgrad** for each level of occ7. To do this, we collapse the data file by occ7 and use those summary statistics to compute the confidence intervals below, followed by the command to create the graph.

```
. vguse nlsw
. collapse (mean) pct_un=un pct_coll=collgrad
     (sd) sd_un=union sd_coll=collgrad
     (count) ct_un=union ct_coll=collgrad, by(occ7)
. gen lci_un = pct_un - sd_un/sqrt(ct_un)
. gen uci_un = pct_un + sd_un/sqrt(ct_un)
. gen lci_coll = pct_coll - sd_coll/sqrt(ct_coll)
. gen uci_coll = pct_coll + sd_coll/sqrt(ct_coll)
```

```
twoway (rcap lci_coll uci_coll pct_un) (rcap lci_un uci_un pct_coll, hor)
  (sc pct_coll pct_un, msymbol(i) mlabel(occ7) mlabpos(10) mlabgap(5)),
  ylabel(0(.2).7) xtitle(% Union) ytitle(% Coll Grads) legend(off)
  titl e("% Union and % college graduates" "(with CIs) by occupation")
```

The overlaid **rcap** commands show the confidence intervals for both **union** and **collgrad** for each occupation. The **scatter** command uses an invisible marker and labels each occupation at the 10 o'clock position with a larger gap than normal.
Uses .dta & scheme vg_s2c

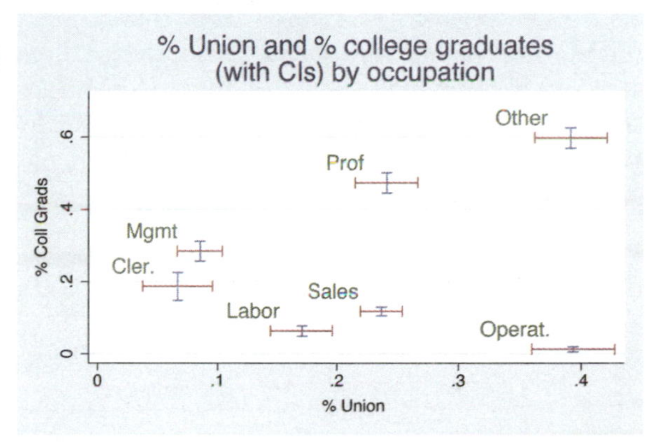

This section concludes with a graph adapted from an example on the Stata web site. The graph combines numerous tricks, so rather than show it all at once, let's build it up a piece at a time. Below is the ultimate graph we would like to create. It shows the population (in millions) for males and females in 17 different age groups, ranging from "Under 5" up to "80 to 84". The blue bar represents the males, and the red bar represents the females.

Introduction Stat graphs Twoway Stat graph options Matrix Save/Redisplay/Combine Bar Box Dot Pie More examples Options Standard options Common mistakes Styles Customizing schemes Appendix Online supplements

```
graph display
```

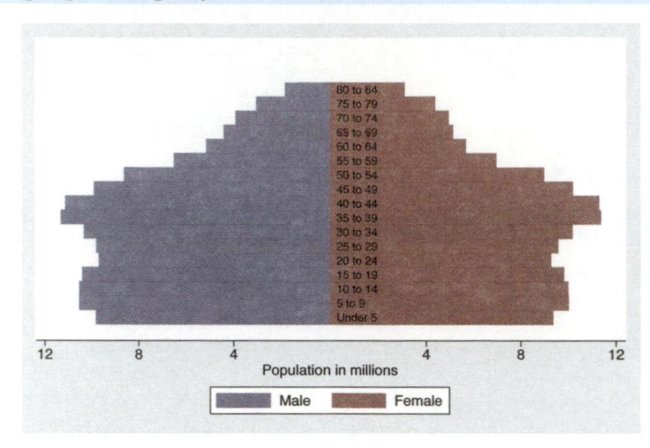

This is the graph that we wish to create. For now, we simply use the **graph display** command to display the graph. Because this is displayed using the **s2color** scheme, the size of the text is not enlarged as in the other vg_ schemes, so the text may be hard to read.
Uses pop2000mf.dta & scheme s2color

To build this graph, we first use the data file **pop2000mf**, which contains 17 observations corresponding to 17 age groups (for example, "Under 5", "5 to 9", "10 to 14", and so forth). The variables **femtotal** and **maletotal** contain the number of females and males in each age group. After using the file, we create **femmil**, which is the number of females per million, and **malmil**, which is the number of males per million, but this is made negative so that the male (blue) bar will be scaled in the negative direction. We also make a variable **zero**, which contains 0 for all observations.

```
. vguse pop2000mf
. gen femmil = femtotal/1000000
. gen malmil = -maletotal/1000000
. gen zero = 0
```

We now take the first step in making this graph.

```
twoway (bar malmil agegrp) (bar femmil agegrp)
```

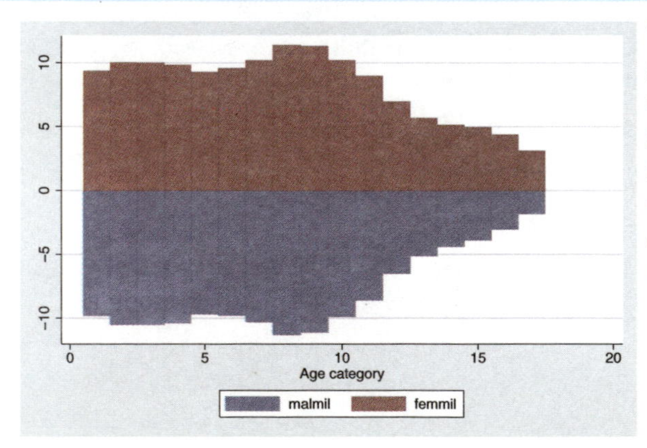

This is our first attempt to make this graph by overlaying the bar chart for the males with the bar chart for the females. The **agegrp** variable ranges from 1 to 17 and forms the x-axis, but we can rotate this as shown in the next example.
Uses pop2000mf.dta & scheme s2color

```
twoway (bar malmil agegrp, horizontal) (bar femmil agegrp, horizontal)
```

Adding the horizontal option to each bar chart, we can see the graph taking shape. However, we would like the age categories to appear inside of the red (female) bars.
Uses pop2000mf.dta & scheme s2color

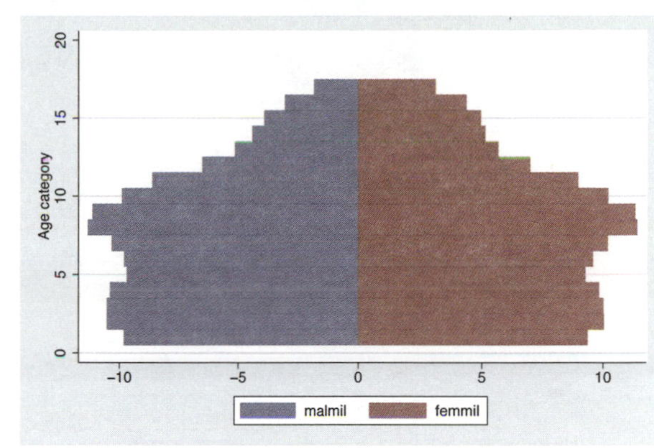


```
twoway (bar malmil agegrp, horizontal) (bar femmil agegrp, horizontal)
    (scatter agegrp zero, msymbol(i) mlabel(agegrp) mlabcolor(black))
```

This scatter command uses **agegrp** (ranging from 1–17) as the y-value and **zero** (0) for the x-value, leading to the stack of 17 observations. Using the msymbol(i) and mlabel() options suppresses the symbol but displays the name of the age group from the labeled value of **agegrp**. Next, we will fix the label and title for the x-axis.
Uses pop2000mf.dta & scheme s2color

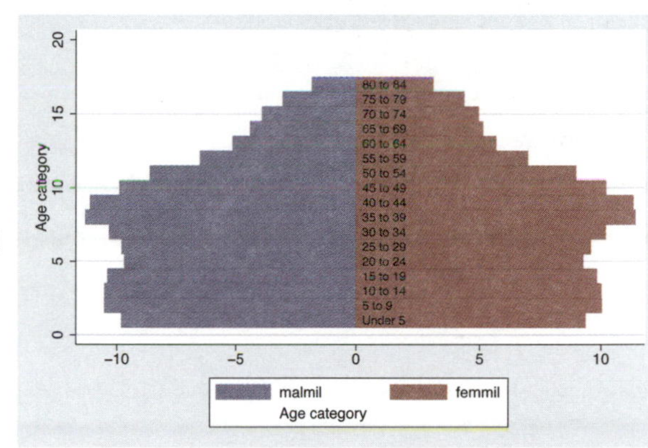


```
twoway (bar malmil agegrp, horizontal) (bar femmil agegrp, horizontal)
    (scatter agegrp zero, msymbol(i) mlabel(agegrp) mlabcolor(black)),
    xlabel(-12 "12" -8 "8" -4 "4" 4 8 12) xtitle("Population in millions")
```

We use the xlabel() to change −12 to 12, −8 to 8, −4 to 4, and to label the positive side of the x-axis as 4, 8, and 12. We also add a title for the x-axis. Next, let's fix the y-axis and the legend.

Uses pop2000mf.dta & scheme s2color

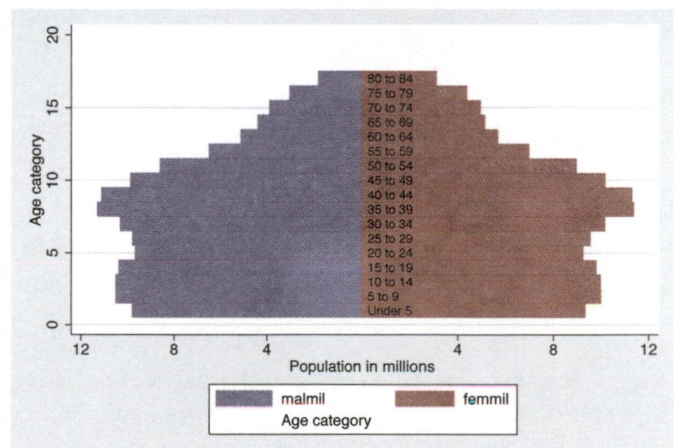

```
twoway (bar malmil agegrp, horizontal) (bar femmil agegrp, horizontal)
   (scatter agegrp zero, msymbol(i) mlabel(agegrp) mlabcolor(black)),
   xlabel(-12 "12" -8 "8" -4 "4" 4 8 12) xtitle("Population in millions")
   y scale(off) ylabel(, nogrid) legend(order(1 "Male" 2 "Female"))
```

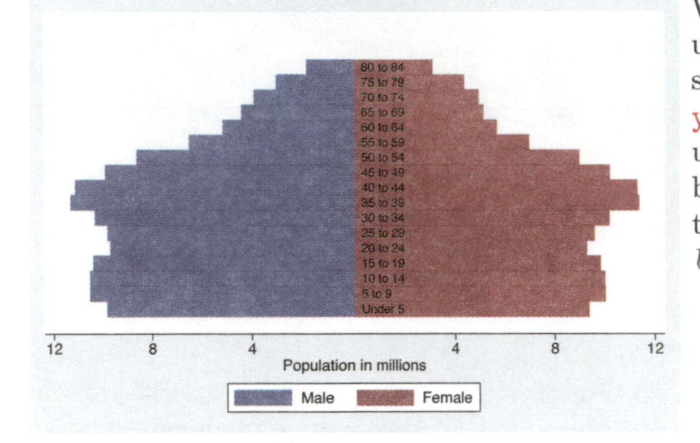

We suppress the display of the y-axis using the `yscale(off)` option and suppress the grid lines with the `ylabel(, nogrid)` option. Finally, we use the `legend()` option to label the bars and suppress the display of the third symbol in the legend.
Uses pop2000mf.dta & scheme s2color

11.5 Common mistakes

This section discusses mistakes that are frequently made when creating Stata graphs.

Using Stata 7 syntax

No matter how long we have been using Stata 8 (or later), we might revert back to old habits and type a graph command in Stata 7 style. Consider this example:

```
. graph propval100 rent700
```

Stata replies with this error message:

```
propval100graph_g.new rent700: class member function not found
r(4023);
```

Clearly, the easiest solution is to convert the command to the proper Stata 8 syntax.

Commas with graph options

With Stata 8, graph options can accept their own options (sometimes referred to as *suboptions*); for example,

```
. twoway scatter propval100 popden rent700, xtitle("My Title", box)
```

Note that the `xtitle()` option allows us to specify the x-title followed by a comma and a further suboption that places a box around the x-title. If we had been content with the existing x-title, we could have issued this command:

```
. twoway scatter propval100 popden rent700, xtitle( , box)
```

The box option places a box around the title. Note that we place a comma before the box option. Now, suppose that we are content with the existing legend but wish to make the legend display in a single column.

```
. twoway scatter propval100 popden rent700, xtitle( , box) legend(cols(1))
```

Based on the syntax from the title() option, we might have been tempted to have typed legend(, cols(1)), but that would have led to an error. Some options, like the legend() option, simply take a list of options with no comma permitted.

Using options in the wrong context

Consider the example below. Our goal is to move the labels for the x-axis from their default position at the bottom of the graph to the alternate position at the top of the graph.

```
. twoway scatter propval100 rent700, xlabel( , alternate)
```

This command executes, but it does not have the desired effect. Instead, it staggers the labels of the x-axis, alternating between the upper and lower positions. In this context, the alternate option means something different than we had intended. What we really wanted to specify was xscale(alternate):

```
. twoway scatter propval100 rent700, xscale(alternate)
```

This command moves the entire scale of the x-axis to the alternate position and has the desired effect. Another mistake we might have made was to put the alternate option as an overall option. This command is shown below with the result:

```
. twoway scatter propval100 rent700, alternate
option alternate not allowed
invalid syntax
r(198);
```

In this case, we are half-right. There is an option alternate, but we have used it in the wrong context, yielding the syntax error. In such cases, remember that the option we are specifying may be right, but we just need to put it into the right context.

Options appear to have no effect

When we add an option to a graph, we generally expect to see the effect of adding the option. However, sometimes adding an option has no effect. Consider this example:

```
. twoway scatter propval100 rent700, mlabpos(12)
```

This command executes, but nothing changes as a result of including the mlabpos(12) option, which would change the position of the marker labels to the 12 o'clock position. There are no marker labels in the graph, so adding this option has no effect. We would have to use the mlabel() option to add marker labels before we saw the effect of this option.

Consider another example, which is a bit more subtle. We would like to make the line (periphery) of the marker thick. When we run the following command, we do not see any effect from adding the mlwidth(thick) option:

```
. twoway scatter propval100 rent700, mlwidth(thick)
```

The reason for this is that the marker has a `line` color and a `fill` color, and by default, they are the same color, so it is impossible to see the effect of changing the thickness of the line around the marker. However, if we make the `line` and `fill` colors different, as in the following example, we can see the effect of the `mlwidth()` option:

```
. twoway scatter propval100 rent700, mlwidth(thick)
        mlcolor(black) mfcolor(gs13)
```

Options when using by()

Using the `by()` option changes the meaning of some options. Consider the following example:

```
. twoway scatter propval100 rent700, by(north) title(My title)
```

We might think that the `title()` option will provide an overall title for the graph, as it would when the `by()` option is not included. However, actually, each graph will have "My title" as the title; the graph as a whole will not. Instead, to provide an overall title for the graph, we would specify the command this way:

```
. twoway scatter propval100 rent700, by(north, title(My title))
```

When using the `legend()` option combined with the `by()` option, we should place options that affect the position of the legend within the `by()` option. Consider this example:

```
. twoway scatter propval100 popden rent700,
        by(north, legend(pos(12))) legend(cols(1))
```

Here, the `legend(pos(12))` option controls the position of the legend, placing it at the 12 o'clock position, so we place it within the `by()` option. On the other hand, the `legend(cols(1))` option does not affect the position of the legend, so we place it outside of the `by()` option. For more details on this, see Options : By (272).

Altering the wrong axis

When we use multiple x- or y-axes, it is easy to modify the wrong axis. Consider this example:

```
. twoway (scatter  propval100 ownhome)
        (scatter rent700 ownhome, yaxis(2) ytitle(Rents over 700))
```

We might think that the `ytitle()` option will change the title for the second y-axis, but it will actually change the first axis. Because `ytitle()` is an option that concerns the overall graph, we should place it at the very end of the graph command, as shown below.

```
. twoway (scatter  propval100 ownhome)
        (scatter rent700 ownhome, yaxis(2)), ytitle(Rents over 700, axis(2))
```

Note that we use the `axis(2)` option to indicate that `ytitle()` should be modified for the second y-axis.

When all else fails

I hope that, by describing these errors, I can help you avoid some common errors. Here are some additional ideas and resources to help you when you are struggling:

- Build graphs slowly. Rather than trying to make a final graph all at once, try to build the graph slowly adding, one option at a time. This is illustrated in Intro : Building graphs (29), where we took a complex graph and built it one piece at a time. Building slowly helps us isolate problems to a particular option, which we can then further investigate.
- When possible, model graphs from existing examples. This book strives to provide examples to model from. For additional online examples, see Appendix : Online supplements (382) for the companion web site for the book, which links to additional examples.
- For more detailed information about the syntax of Stata graphics, see [G] **graph**. Please remember that some of the graph commands available in Stata were added after the printing of [G] **graph** but are documented via the `help graph` command. See also Appendix : Online supplements (382), which has links to the online help that are organized according to the table of contents of this book.
- Reach out to fellow Stata users, either local friends, friends at Statalist, or friends at Stata tech support. See *http://www.stata.com/support/* for more details.

11.6 Customizing schemes

This section shows how to customize your own schemes. Although schemes can look complicated, it is possible to easily create some simple schemes on our own. Let's look at the `vg_lgndc` scheme as an example. This scheme is based on the `s2color` scheme but changes the legend to display at the 9 o'clock position, in a single column, with the keys stacked on top of the symbols. Here are the contents of that scheme:

```
#include s2color // start with the s2color scheme

clockdir legend_position  9   // put the legend in the 9 o'clock position
numstyle legend_cols      1   // make the legend display in 1 column
yesno legend_stacked      yes // stack the keys & symbols on top of each other

gsize legend_key_gap    half_tiny // very, very small gap between key and label
gsize legend_row_gap    small     // somewhat larger gap between key/label pairs
```

Rather than creating the `vg_lgndc` scheme from scratch, which would be very laborious, we use the `#include s2color` statement to base this new scheme on the `s2color` scheme. The subsequent statements change the position of the legend and the number of columns in the legend and stack the legend keys and symbols upon each other.

Say that we liked the `vg_lgndc` scheme but wanted to make our own version in which the legend is in the 3 o'clock position instead of the 9 o'clock position, naming our version `legend3`. To do this, we would start the Stata do-file editor, for example, by typing `doedit` and then type the following into it: (Of course, the scheme will work fine if we omit comments after the double slashes.)

```
#include s2color // start with the s2color scheme

clockdir legend_position  3   // put the legend in the 3 o'clock position
numstyle legend_cols      1   // make the legend display in 1 column
yesno legend_stacked      yes // stack the keys & symbols on top of each other

gsize legend_key_gap    half_tiny // very, very small gap between key and label
gsize legend_row_gap    small     // somewhat larger gap between key/label pairs
```

We can then save the file as `scheme-legend3.scheme`, and we are ready to use it. We can then use the `scheme(legend3)` option at the end of a graph command or type `set scheme legend3`, and Stata will use that scheme for displaying our graph. Below, we show an example using this scheme. (Note that the `legend3` scheme is not included among the downloadable schemes.)

```
twoway (scatter propval100 rent700) (lfit propval100 rent700),
  scheme(legend3)
```

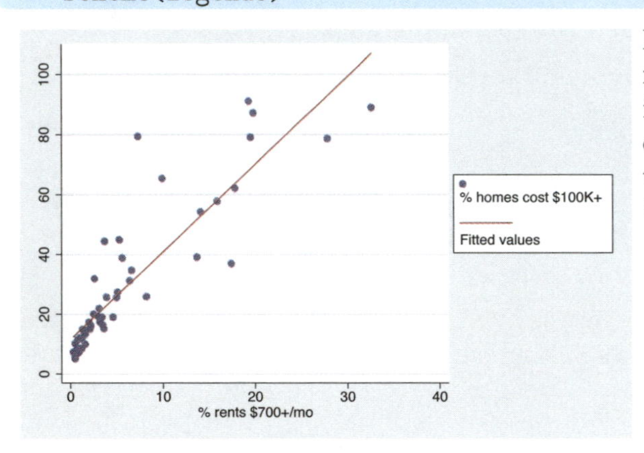

Here, we see an example using our newly created `legend3` scheme, and indeed, we see the legend in the 3 o'clock position, in a single column, with the legend stacked.
Uses allstates.dta & scheme legend3

So far, things are going great. However, note that Stata will only know how to find the newly created `scheme-legend3.scheme` while we are working in the directory where we saved that scheme. If we change to a different directory, Stata will not know where to find `scheme-legend3.scheme`. If, however, we save the scheme into our `PERSONAL` directory, Stata would know where to find it regardless of the directory we were in. For example, on my computer, I used the `sysdir` command, and it showed me the following information:

```
. sysdir
   STATA:  C:\Stata8\
 UPDATES:  C:\Stata8\ado\updates\
    BASE:  C:\Stata8\ado\base\
    SITE:  C:\Stata8\ado\site\
    PLUS:  c:\ado\plus\
PERSONAL:  c:\ado\personal\
OLDPLACE:  c:\ado\
```

From this, I know that my `PERSONAL` directory is located in `c:\ado\personal\`, so if I store either `.ado` files or `.scheme` files there, Stata will be able to find them. So, if instead of saving scheme-legend3.scheme into the current directory, we save it into our PERSONAL directory, Stata will be able to find it. (If we have already saved scheme-legend3.scheme to the current directory and also save it to the PERSONAL directory, we may want to remove the copy from the current directory.)

So far, this section has really focused on the mechanics of creating a scheme but has not said much about the possible content that could be placed inside a scheme. This is beyond the scope of this little introduction, but here are three other places where you can find this kind of information:

First, the help for schemes via help schemes will tell us about schemes in general. Also, help scheme_files contains documentation about scheme files and what we can change using schemes.

Second, looking at other schemes can help us find ideas, for example, the downloaded schemes for this book (see Appendix : Online supplements (382)). Say that we wanted to look at the vg_rose scheme. We could type which scheme-vg_rose.scheme, and that would tell us where that scheme is located. Then, we could use any editor (including the do-file editor) to view that scheme for ideas.

Third, we can look at the built-in Stata schemes, such as s1color, s2color, or economist. Looking at these schemes shows us the menu of items that we can fiddle with in our own schemes, but these schemes should never be modified directly. We can use the strategy outlined above where we make our own scheme and use #include to read in a scheme, and then we can add our own statements to modify the scheme as desired.

Schemes that other people have created and the schemes built into Stata will contain statements that control some aspect of a graph, but we may not know which aspect they control. For example, in the vg_rose scheme there is the statement

```
color      background eggshell
```

which obviously controls the color of some kind of background element, but we might not be sure which element it controls. We can find out by making a copy of the scheme and then changing eggshell to some other nonsubtle value, such as red, and then make a graph using this new scheme (using scheme(*schemename*), not set scheme schemename). The part of the graph that becomes red will indicate the part that is controlled by the color background statement.

Of course, we have just scratched the surface of how to create and customize schemes. However, this should provide the basic tools needed for making a basic scheme, storing it in the personal directory, and then playing with the scheme. Because schemes are so powerful, they can appear complicated, but if built slowly and methodically, the process can be straightforward, logical, and, actually, quite a bit of fun.

11.7 Online supplements

This book has a number of online resources associated with it. I encourage all readers to take advantage of these online extras by visiting the web site for the book at

http://www.stata-press.com/books/vgsg.html

Resources on the web site include

- Programs and help files. You can easily download and install the programs and help files associated with this book. To install these programs and help files, just type

  ```
  . net from http://www.stata-press.com/data/vgsg
  . net install vgsg
  ```

  After installing the programs and help, type `whelp vgsg` for an overview of what has been installed.

- Data files. All the data files used in the book are available at the web site for downloading. I encourage you to download the data files used in this book, play with these examples, and try variations on your own to solidify and extend your understanding. If you visit the website, you can download and save all the data files at once. You can quickly download all the datasets into your current working directory from within Stata by typing

  ```
  . net from http://www.stata-press.com/data/vgsg
  . net get vgsg
  ```

  If you prefer, you can obtain any of the data files over the Internet with the `vguse` command. Each example concludes by indicating the data file and scheme that was used to make the graph. For example, a graph may conclude by saying

  Uses allstates.dta & scheme vg_s2c

  This indicates that you can type `vguse allstates` and Stata will download and use the data file over the Internet for you (assuming that you have installed the programs).

- Schemes. This book uses a variety of schemes, and when you download the programs and help files (see above), the schemes used in this book are downloaded as well, allowing you to use them to reproduce the look of the graphs in this book.

- Hopefully, a very short or empty *Errata* will be found at the web site. Although I have tried very hard to make this book true and accurate, I know that some errors will be found, and they will be listed there.

- Links to the online *Stata Graphics Reference Manual*, which are organized according to the structure of the table of contents of this book.

- Other resources that may be placed on the site after this book goes to press, so visit the site to see what else may appear there.

Subject index

N

O

P

Q

R

U

W

80025 75540